TOWARDS A UNIFIED PICTURE OF NUCLEAR DYNAMICS

AIP CONFERENCE PROCEEDINGS 250

TOWARDS A UNIFIED PICTURE OF NUCLEAR DYNAMICS

NIKKO, JAPAN 1991

EDITORS:

Y. ABE
KYOTO UNIVERSITY

S. M. LEE
TSUKUBA UNIVERSITY

F. SAKATA
UNIVERSITY OF TOKYO

AIP

American Institute of Physics **New York**

L.C. Catalog Card No. 92-70143
ISBN 0-88318-930-5
DOE CONF-9106158

Printed in the United States of America.

Contents

FISSION DECAY AND COLLECTIVE DYNAMICS OF HOT NUCLEI

MICROCLUSTER IMPACT FUSION

MAXIMUM EXCITATION ENERGY AND ANGULAR MOMENTUM OF HOT NUCLEI

FROM BINARY DECAY TO MULTIFRAGMENTATION

SUPER HEAVY ELEMENTS

TOWARDS A UNIFIED PICTURE

Preface

The International Symposium, "Towards a Unified Picture of Nuclear Dynamics," was held at the Nikko Prince Hotel overlooking the beautiful lake Chuzenji, Nikko, from June 6 to 8 in 1991. A total of 102 scientists from 13 countries participated in the symposium.

The purpose of the symposium was to discuss the heavy-ion reaction dynamics intensively from low to intermediate energy region covering a broad spectrum of subjects, ranging from sub-barrier fusion to multifragmentations. In addition, there was a special session devoted to the microcluster impact fusion which was one of the most fascinating current topics.

The symposium aimed to reveal the underlying common physics among a variety of different phenomena and even to clarify novel concepts.

A major success, we heard from many participants, of the symposium was indeed to find critical points of discussion towards a unified picture of dynamics of the nuclear many-body system. We thank all the speakers and the participants for the valuable talks and the lively discussions during the meeting. We hope that the participants were satisfied with some deeper understanding and stimulating ideas. We believe that proceedings will certainly remind all the participants of such a nice atmosphere and will hopefully convey the important information to those who could not attend the symposium.

We would like to thank the members of the organizing committee for their fruitful collaboration, and particularly the young students and secretaries for their tremendous contributions before and during the meeting.

Finally, we express our deepest appreciation for the funding of the symposium to University of Tsukuba Fund for International Exchange, The Commemorative Association for the Japan World Exposition (1970), The Kajima Foundation, and many industrial companies.

Y. Abe
S. M. Lee
F. Sakata

ORGANIZING COMMITTEE

Y. Abe (Yukawa Institute, Kyoto University) Co-chairman
S. M. Lee (University of Tsukuba), Co-chairman
K. Imamura (FAIS)
J. Kasagi (Tokyo Institute of Technology)
T. Marumori (University of Tsukuba)
T. Motobayashi (Rikkyo University)
F. Sakata (Institute for Nuclear Study, University of Tokyo)
N. Takigawa (Tohoku University)
H. Tanaka (Sapporo Gakuin University)

HOST INSTITUTES

Institute of Physics, University of Tsukuba
Department of Physics, Tokyo Institute of Technology
Foundation for Advancement of International Science (FAIS)

SUPPORTED BY

University of Tsukuba, Fund for International Exchange,
The Commemorative Association for the Japan World Exposition (1970),
The Kajima Foundation,
and many industrial companies

PROCEEDINGS SECRETARIES

Y. Abe
S. M. Lee
F. Sakata

International Symposium NIKKO '91
Towards a Unified Picture of Nuclear Dynamics
June 6–8, 1991, Nikko, Japan

Schedule

Thursday, 6 June

9:00–9:10	Opening Address: Y. Abe
9:10–10:40	Sub-Barrier Fusion: Chairman—H. J. Kim
10:40–11:00	Coffee Break
11:00–1:00	Orbiting and Fusion–Fission Phenomena Chairman—S. M. Lee
1:00–2:00	Lunch
2:00–3:30	Direct Reaction Approaches and Beyond Chairman—B. Giraud
3:30–3:50	Coffee Break
3:50–5:50	Fission Decay and Collective Dynamics of Hot Nuclei Chairman—M. Di-Toro
5:50–8:00	Dinner
8:00–10:30	Special Session: Microcluster Impact Fusion Chairman—B. Tamain

Friday, 7 June

9:00–10:40	Maximum Excitation Energy and Angular Momentum of Hot Nuclei Chairman—J. B. Natowitz
10:40–11:00	Coffee Break
11:00–1:20	From Binary Decay to Multifragmentation Chairman—J. Peter

Saturday, 8 June

9:00–10:10	Super Heavy Elements Chairman—U. Schroeder
10:10–10:30	Coffee Break
10:30–12:00	Towards a Unified Picture Chairman—R. H. Siemssen

List of Participants

Y. Abe
Yukawa Institute for
Theoretical Physics
Kyoto University
Kitashirakawa, Sakyo-ku
Kyoto 606
JAPAN

Y. Achiba
Faculty of Science
Tokyo Metropolitan University
Fukazawa 2-1-1, Setagaya-ku
Tokyo 158
JAPAN

D. Ackermann
INFN-laboratori Nazionali
01 Legnano-VIA Romea 4-35020
Legnaro (Padova)
ITALY

Y. Akeboshi
Faculty of Science
Tokyo Institute of Technorogy
Ohokayama 2-12-1, Meguro-ku
Tokyo 152
JAPAN

T. Arai
Institute of Applied Physics
Tsukuba University
Ibaraki 305
JAPAN

Y. Bae
Molecular Physics Laboratory
SRI International
Menlo Park, CA 94025
U.S.A.

M. Baldo
INFN, Sez. Catanina
Corso Italia 57
95129 Catnia
ITARY

W. Bauer
Nat'l Superconducting
Cyclotron Laboratory
Michigan State University
East Lansing, MI 48824-1321
U.S.A.

R.R. Betts
Physics Division
Argonne National Laboratory
Argonne,Illinois, IL 60439
U.S.A.

A. Bonaccorso
Dipartimento di Fisica
Univ. di Pisa
Pizza Torricelli
56100 Pisa
ITALY

N. Carjan
Centre d'Etudes Nuclearies
de Bordeaux
Le Haut Vigneau
33175 Gradignan Cedex
FRANCE

L. Corradi
INFN-Laboratori Nazionali
01 Legnaro-VIA Pomea 4-35020
Legnaro(Padova)
ITALY

G.X. Dai
Institute of Modern Physics
No.57 Nanchung Road
Lanzhou
The People's Republic of China

J.N. De
VEC Center
Calcutta 700064
1/4F Salt Lake
INDIA

M. Di-Toro
INFN, Departoment di Fisica
57 Corso Italia
95129 Catania
ITALY

L. Ferreira
Universiyta di Catania and INFN,
Sez Catania,Corso Italia 57,
95129-Catania
ITALY

G. Friedlander
Chemistry Department
Brookhaven Natioinal Laboratory
Upton NY 11973
U.S.A.

W.A. Friedman
Physics Department
University of Wisconsin
Madison, Wisconsin 53706
U.S.A.

H. Funaki
Research Laboratory
for Nuclear Reactors
Tokyo Institute of Technorogy
Ohokayama 2-12-1, Meguro-ku
Tokyo 152
JAPAN

K. Furuno
Institute of Physics
Tsukuba University
Ibaraki 305
JAPAN

K. Furutaka
Faculty of Science
Tokyo Institute of Technorogy
Ohokayama 2-12-1, Meguro-ku
Tokyo 152
JAPAN

Y. Futami
Institute of Physics
Tsukuba University
Ibaraki 305
JAPAN

B. Giraud
SPT CEN-SACLAY
BP2 F91191
Gif/Yvette
FRANCE

D. Guerreau
GANIL
BP 5027
14021 Caen Cedex
FRANCE

K. Hagel
Cyclotron Institute
Texas A&M University
College Station, Texas 77843
U.S.A.

W. Henning
G.S.I.
Postfach 110552
D-6100 Darmstadt
GERMANY

B. Heusch
Centre de Recherchens Nucleaires
BP 20
67037 Strasbourg Cedex
FRANCE

D.J. Hinde
Department of Nuclear Physics
Research School of
Physical Sciences
The Australian Nat'l Universty
GPO BOx4,Canberra ACT 2061
AUSTRALIA

Y. Honjo
Institute of Physics
Tsukuba University
Ibaraki 305
JAPAN

A. Iwamoto
Tokai Research Establishment
JAERI
Thokai, Naka-gun
Ibaraki 319-11
JAPAN

C.L. Jiang
China Institute of Atomic Energy
P.O. Box 275(80), Beijing
The People's Republic of China
CHINA

Y. Kanada
Faculty of Science
Kyoto University
Kitashirakawa, Sakyo-ku
Kyoto 606
JAPAN

S.S. Kapoor
Nuclear Physics Divition
Bhaba Atomic Research Center
Bombay 400085
INDIA

J. Kasagi
Faculty of Science
Tokyo Institute of Physics
Ohokayama 2-12-1, Meguro-ku
Tokyo 152
JAPAN

B.T. Kim
Sung Kyun kwan University
Department of Physics
Suwon 440-746
KOREA

H.J. Kim
Tokai Research Establishment
JAERI
Tokai, Naka-gun
Ibaraki 319-11
JAPAN

Y.D. Kim
NSCL/CYCLOTRON LABORATORY
Michigan State University
East Lausing, MI 48824-1321
U.S.A.

Y.E. Kim
Department of Physics
Purdue University
West Lafayette,IN 47907
U.S.A.

M. Kuratani
Faculty of Science
Tohoku University
Aramaki, Sendai
Miyagi 980
JAPAN

K. Kwiatkowski
HUCF/Accelerator Boilding
Indiana University
Bloomington, IN 47405
U.S.A.

S.M. Lee
Institute of Physics
Tsukuba University
Ibaraki 305
JAPAN

G.R. Lee
Department of Mathmatics
Koshin University
Pusan
KOREA

Z.X. Li
Institute of Atomic Energy
Academia Sinica
P.O. Box 275(18)
Beijing 102413
CHINA

G.A. Lobov
Institute of Theoretical and
Experimental Physics
MOSCOW
U.S.S.R.

M. Louvel
L.P.C.
Universite de Caen
14050 Caen Cedex
FRANCE

J. Lu
Institute of Physics
Tsukuba University
Ibaraki 305
JAPAN

K. Maruyama
Faculty of Science
Kyoto University
Kitashirakawa, Sakyo-ku
Kyoto 606
JAPAN

T. Matsuse
Facilty of Textile Science
and Technology
Shinshu University
Tsuneda 3-15-1, Ueda-shi
Nagano 386
JAPAN

T. Mikumo
Chuoucho 4-10-5,
Higashikurume
Tokyo 180-03
JAPAN

T. Mizota
Institute of Physics
Tsukuba University
Ibaraki 305
JAPAN

T. Murakami
Faculty of Science
Tokyo Institute of Technorogy
Ohokayama 2-12-1, Meguro-ku
Tokyo 152
JAPAN

Y. Nagashima
Institute of Physics
Tsukuba University
Ibaraki 305
JAPAN

K. Yuasa-Nakagawa
Institute of Physics
Tsukuba University
Ibaraki 305
JAPAN

T. Nakagawa
Cyclotron Laboratory
RIKEN
Hirosawa 2-1, Wako-shi
Saitama 351-01
JAPAN

J.B. Natowitz
Cyclotron Insutitute
Texas A\& M Universty
College Station,TX 77843-3366
U.S.A.

K. Niita
Tokai Research Establishment
JAERI
Thokai, Naka-gun
Ibaraki 319-11
JAPAN

Yu.Ts. Oganessian
Laboratory of Nuclear
Reactions
JINR,141980 Dubna
U.S.S.R.

K. Ohkushi
Institute of Physics
Tsukuba University
Ibaraki 305
JAPAN

A. Ohnishi
Faculty of Science
Kyoto University
Kitashirakawa, Sakyo-ku
Kyoto 606
JAPAN

M. Ohta
Faculty of Science
Konan University
Okamoto 8-9-1,
Higashinada-ku
Hyogo 658
JAPAN

M. Okamoto
Research Laboratory
for Nuclear Reactors
Tokyo Institute of Technorogy
Ohokayama 2-12-1, Meguro-ku
Tokyo 152
JAPAN

Yu.E. Penionzhkevich
Laboratory of Nuclear
Reactions
JINR,141980 Dubna
U.S.S.R.

J. Peter
L.P.C.
Universite de Caen
F-14032 Caen Cedex
FRANCE

M. Petrovici
Institute for Physics and
Nuclear Engineeing
Haevy Ions Physics Department
POB-MG6 Bucharest-Magurele
ROMANIA

M. Ploszajczak
GANIL
PB 5027, F-14021 Cedex
FRANCE

Y.H. Pu
Institute of Physics
Tsukuba University
Ibaraki 305
JAPAN

B.J. Qi
China Institute of Atomic Energy
P.O. Box 275(80),
Beijing
CHINA

A. Ray
Oak Ridge National Laboratory
Nuclear Division
P.O. Box X
Oak Ridge, TN 37835
U.S.A.

N. Rowley
Daresbury Laboratory
Daresbury, Warrington WA4 4AD
England

B. Remaud
Laboratoire de Physique Nucleaire
Universite de Nantes
F-44072,Nantes Cedex
FRANCE

J. Richert
C.R.N.
BP 20
67037 Strasbourg Cedex
FRANCE

C. Samanta
Saha Institute of Nuclear Physics
Af,Sector 1, Bidhannagar
Calcutta-700064
INDIA

S. Sawada
Faculty of Science
Kyoto Univasity
Kitashirakawa, Sakyo-ku
Kyoto 606
JAPAN

W.U. Schroder
Department of Chemistry and
Nuclear Structure Research
Laboratory
University of Rochester
Rochester, NY 14627
U.S.A.

V.R. Shaginyan
Leningrad Nuclear
Physics Institute
Catchina Leningrad
District 188350
U.S.S.R.

K. Shima
Institute of Applied Physics
Tsukuba University
Ibaraki 305
JAPAN

T. Shinozuka
Cyclotron and Radioisotope
Center
Tohoku University
Aramaki, Sendai
Miyagi 980
JAPAN

R.H. Siemssen
K.V.I.
Univ. of Groningen
Groningen
The Netherlands

K.S, Sim
Department of Physics
Korea University
Seou 136-701
KOREA

K. Sueki
Faculty of Science
Tokyo Metropolitan University
Fukazawa 2-1-1, Setagaya-ku
Tokyo 158
JAPAN

I. Sugai
Institute for Nuclear Study
University of Tokyo
Midori-cho 3-2-1, Tanashi
Tokyo 188
JAPAN

Y. Sugiyama
Tokai Research Establishment
JAERI
Thokai, Naka-gun
Ibaraki 319-11
JAPAN

T. Suzuki
Aichi Prefectural University
Takada-cho 3-28, Nagoya
Aichi 467
JAPAN

A. Szanto De Toledo
Department of Physics
University of Sao Paulo
Sao Paulo
Brasil

Y. Tagishi
Institute of Physics
Tsukuba University
Ibaraki 305
JAPAN

G. Takeda
Faculty of Technology
Tohoku Gakuin University
Cyuou 1-13-1, Tagajyo
Miyagi 985
JAPAN

N. Takigawa
Faculty of Science
Tohoku University
Aramaki, Sendai
Miyagi 980
JAPAN

B. Tamain
L.P.C.
Universite de Caen
F-14032 Caen Cedex
FRANCE

H. Tanaka
Sapporo Gakuin University
Bunkyodai 11, Esashi-shi
Hokkaido 069
JAPAN

G.M. Ter-Akopian
Laboratory of Nuclear Reactions
JINR, 141980 Dubna
U.S.S.R.

J.P. Thomas
UCBL-43 Bd.du 11 November 1918
69622 Villevrbanng Cedex
FRANCE

S. Tomita
Institute of Physics
Tsukuba University
Ibaraki 305
JAPAN

H. Toyokawa
Institute of Physics
Tsukuba University
Ibaraki 305
JAPAN

W. Trautmann
G.S.I.
Postfach 110552
D-61 Darmstabt
GERMANY

M. Ueda
Faculty of Science
Tohoku University
Aramaki, Sendai
Miyagi 980
JAPAN

E. Uegaki
College of Education
Akita University
Tegatagakuen-cho 1-1
Akita 010
JAPAN

V. Viola
Indiana University
Cyclotron Facility
Bloomington,IN 47405
U.S.A.

R. Wada
Cyclotron Institute
Texas A\&M University
College Station,
TX 77843-3366
U.S.A.

T. Wada
RIKEN
Hirosawa 2-1, Wako-shi
Saitama 351-01
JAPAN

H. Wolter
Sektion Physik
Munchen Universtat
Am Coulombwall 1
D-8046 Garching
GERMANY

X.Z. Wu
China Institute of
Atomic Energy
Academia Sinica
P.O. Box 275(18)
Beijing 102413
CHINA

K. Yabana
Faculty of Science
Niigata University
Igarashininocho 8050
Niigata 950-21
JAPAN

A. Yajima
Faculty of Science
Tokyo Institute of Technology
Ohokayama 2-12-1, Meguro-ku
Tokyo 152
JAPAN

A. Yamagoshi
Faculty of Science
Kyoto University
Kitashirakawa, Sakyo-ku
Kyoto 606
JAPAN

K. Yoshida
Faculty of Science
Kyoto University
Kitashirakawa, Sakyo-ku
Kyoto 606
JAPAN

M. Yoshimura
Faculty of Science
Kyoto University
Kitashirakawa, Sakyo-ku
Kyoto 606
JAPAN

H.Q. Zhang
China Institute of Atomic Energy
P.O. Box 275(80), Beijing
Beijing
CHINA

SUB-BARRIER FUSION

SUB-BARRIER FUSION - AN EXPERIMENTAL REVIEW

R. R. Betts
Physics Division, Argonne National Laboratory, Argonne, IL 60439*

ABSTRACT

This paper contains a review of the current status of the experimental study of heavy-ion fusion at sub-barrier energies. Emphasis is placed on the comparison of the experimentally observed quantities with theoretical expectations. Results of measurements of the spin distributions of the composite systems formed following fusion are critically examined with a view to understanding the large discrepancies between theory and experiment which exist for some systems.

INTRODUCTION

Fusion of heavy ions at energies in the vicinity of the classical barrier has been actively studied for the last decade. The observation which sparked this tremendous activity having been the initial observation[1,2] of cross-sections at sub-barrier energies far in excess of those predicted by simple models. The physics underlying these observations has been known in a general way for quite some time, namely, quantum tunneling of the two ions in the presence of couplings to other degrees of freedom. Nevertheless, there continue to exist significant discrepancies between experiment and theory which act as a stimulus for further investigation. The basic physics to be studied via sub-barrier fusion is one of general application which appears in many other areas of physics. For example, tunneling phenomena are abundant in condensed matter studies and the role of coupling to other degrees of freedom in enhancing barrier penetration is strikingly obvious in the case of catalysis of chemical reactions.

The nuclear case, studied via heavy ion fusion is a rich field. The variety of nuclear shapes and other collective degrees of freedom which can be varied at will by suitable choice of target and projectile allows systematic study of the influence of changing couplings vice-versa, it is possible that we may gain new insight into the properties of the dinuclear system itself through the application of these studies.

CROSS SECTIONS

The first observation of enhanced cross-sections for sub-barrier fusion were reported by Stokstad et al.[1] for ^{16}O + Sm. These data, augmented by data for isotopes not measured in the original work[3] are shown in Fig. 1. The results are well-known; a progression towards higher cross-section in the sub-barrier region is seen to occur for increasing target mass, correlated with the transition from spherical ^{144}Sm to well deformed ^{154}Sm. The curves shown superimposed on the data are the results of a calculation

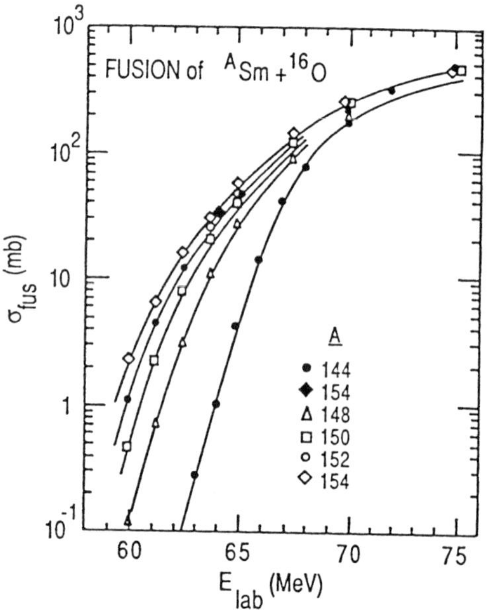

Fig. 1. Fusion cross-sections for ^{16}O + Sm. The solid curves are fits to the data using the model of Wong.

using a model due to Wong[4] in which the barrier between the two ions is modified by the static deformation of the target nucleus, increased deformation resulting in an overall lowering of the barrier when averaging of the (random) orientations of the target nucleus is taken into account. More sophisticated models (coupled-channels) in which the coupling between the entrance channel and inelastic channels is handled quantum-mechanically are also able to account for the observed cross-sections for these systems. This approach has the further merit that couplings to inelastic channels not represented by static ground-state deformation can also be included. Thus, even for spherical target and projectile nuclei, the sub-barrier fusion is enhanced due to barrier fluctuations introduced by the coupling to strong quadrupole and octupole vibrational excitations. This is demonstrated by data for Ni + Ni due to Beckerman et al.[2] shown together with a variety of calculations in Fig. 2. The solid lines are the results of coupled channels calculations by Esbensen and Landowne[5] in which all inelastic couplings of the one- and two-phonon excitations of target and projectile were included. For the symmetric $^{58}Ni + ^{58}Ni$ and $^{64}Ni + ^{64}Ni$ systems, the agreement is excellent. Large discrepancies were found for the $^{58}Ni + ^{64}Ni$ system however. It was suspected for some time that this discrepancy might arise from a neglect of neutron transfer channels which, for the asymmetric system, have well-matched Q-values. Inclusion of these channels in the calculation[6] leads to an improvement in the agreement, although still worse than the symmetric cases. This same calculation also shows good agreement with the measured[7] neutron transfer cross-sections as shown in Fig. 3. It should not be forgotten that

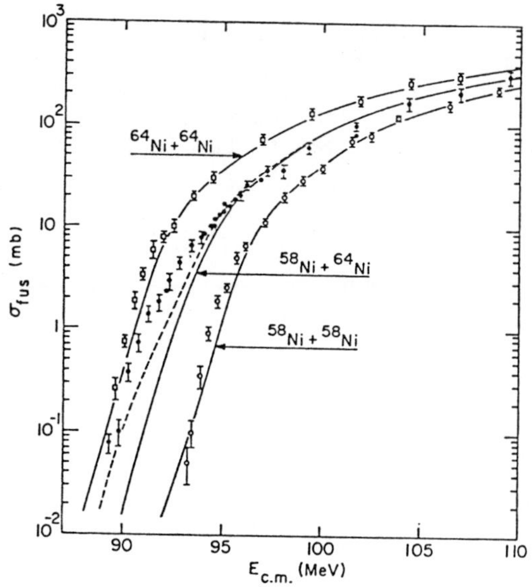

Fig. 2. Fusion cross-sections for Ni + Ni isotopes. The solid curves are coupled channels calculations including one- and two-phonon vibrational excitations. The dashed curve shows the effect of including one- and two-neutron transfer channels.

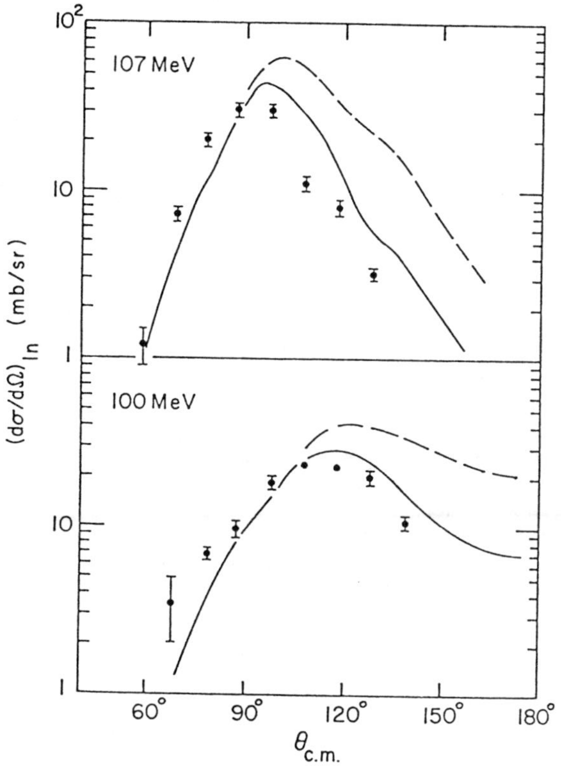

Fig. 3. Angular distributions for one-neutron transfer in the $^{58}Ni + ^{64}Ni$ system. The curves show the results of coupled channels calculations with (solid) and without (dashed) the inclusion of second-step neutron transfer.

coupled-channels theories of fusion must also give a good description of the reaction channels themselves and care should be taken to ensure that a good description of the fusion channel is not obtained at the expense of some other important channel.

An interesting aspect of the data is demonstrated in Fig. 4

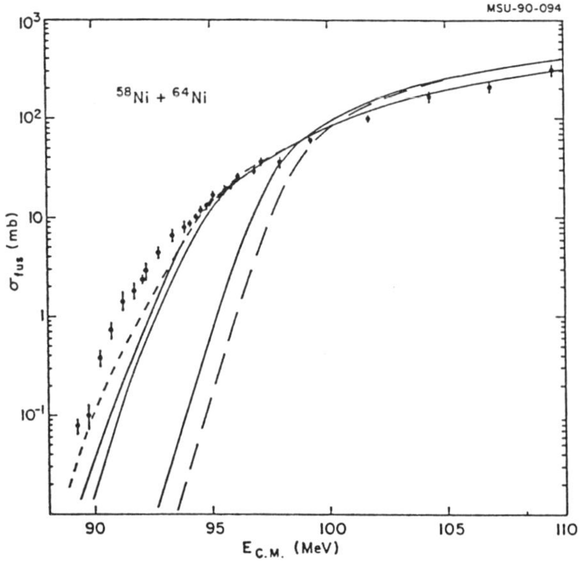

Fig. 4. Data for $^{58}Ni + ^{64}Ni$ fusion. The lower dashed curve is the no-coupling limit. Note the above barrier suppression in comparison with this limit and the correct description of the data by the coupled-channels calculations both above and below the barrier.

where the effects of including different couplings on the calculated cross-sections is shown in comparison with the data for $^{58}Ni + ^{64}Ni$. In addition to the couplings producing an enhancement at sub-barrier energies, we see that these same couplings produce a suppression of the fusion in the energy region immediately above the barrier. Thus, reaction channels which are strong above the barrier and there deplete the fusion channel are responsible for enhancements in the sub-barrier region. A more graphic demonstration of this feature of fusion cross-sections if found in data[8] for much heavier systems shown in Fig. 5. In this case, the measured evaporation residue cross-sections have been translated into a "Fusion Probability" by correcting the data for fission decay of the compound nucleus using statistical model calculations and by assuming an ℓ-independent probability for fusion. It is striking to see again the suppression of fusion relative to the expectations of either the Bass Model (dotted curves) or calculations with barriers shifted to account for the lowest energy data (dashed curves). The strong dependence of the suppression on the exact nuclear system is suggestive of the influence of nuclear structure, perhaps due to shell effects in the di-nuclear system[9] or to dissipation in the approach phase leading

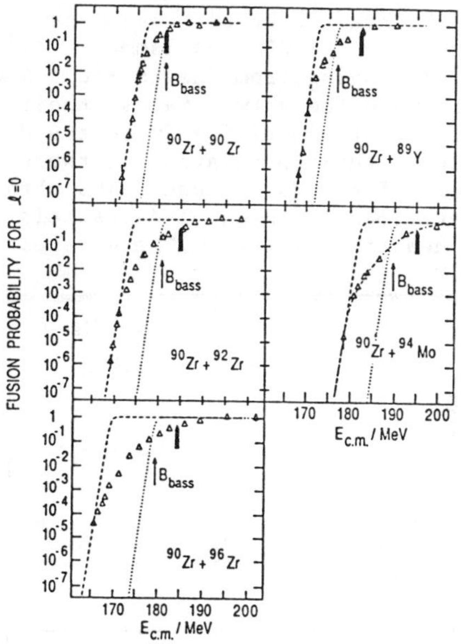

Fig. 5. Fusion probabilities for heavy symmetric systems showing, when compared with no-coupling calculations using the Bass model (dotted curve), the above barrier suppression and below barrier enhancement.

to strongly damped or deep-inelastic processes which compete with fusion. For these very heavy systems, the coupled channels approach fails due to the large number of complex reaction channels which are open. It is nevertheless a challenge to find a theoretical link which will allow the understanding of all the data on a common footing.

SPIN DISTRIBUTIONS

Although the cross-sections for fusion and other reaction channels are the first quantities which must be accounted for by any theory, it is clear that these may not distinguish between different models. For example, it has been argued[10] that an energy dependent potential derived from the elastic scattering channel may give a good account of sub-barrier fusion cross-sections. It was subsequently pointed out[11] that such an approach leads to a different prediction for the spin distribution of the fused system than does a full coupled-channels calculation. The motivation for attempting such measurements is therefore clear. Three different techniques have so far been employed to gain information on the spin distributions; isomer ratios, gamma-ray multiplicities and fission fragment angular distributions. The first and last methods have been used to provide information on the mean or r.m.s. value of the fused system whereas the measurement of gamma-ray multiplicities can, in principle, give much more detailed information on the distribution of angular momenta.

Isomer Ratios: DiGregorio et al. have measured the relative population of the ground ($J^{\pi}=3/2^{+}$) and isomeric ($J^{\pi}=11/2^{-}$) states in ^{137}Ce with projectiles ranging from ^{3}He to ^{12}C. Due to the large difference in spin between these two states, their population originates from different regions of spin in the decaying compound nucleus and the ratio of their population is therefore sensitive to the spin distribution. The deduced mean spins obtained by this method are in good agreement with coupled-channels predictions over a range of energies near the barrier. These results are summarized in Fig. 6.

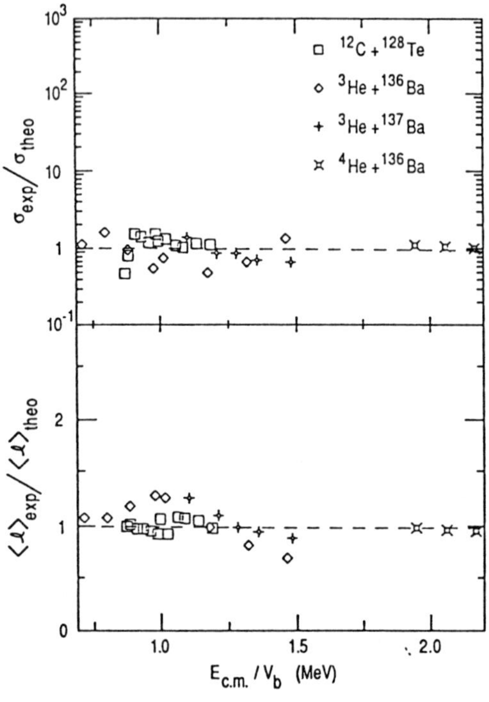

Fig. 6. Comparison of experimental and theoretical cross-sections and average angular momenta for systems measured using the isomer ratio technique.

Gamma Multiplicities: Measurements of this type can be divided into two categories. Those using a relatively small number of gamma-ray detectors and those using highly-segmented 4π arrays. A number of measurements using the former technique have been carried out by Vandenbosch, Gil and collaborators. Evaporation residues were identified either by detection of characteristic gamma-rays with a high-resolution detector or using a charged particle detector in conjunction with an electrostatic deflector to detect evaporation residues directly. Data for a variety of projectiles; α, ^{12}C, ^{16}O and ^{28}Si incident on ^{154}Sm have been used to deduce the mean angular momentum of the decaying compound nucleus[13,14] as a function of bombarding energy. These data are shown in Figs. 7 and 8 in comparison with calculated values from the model of Wong[4]. The overall agreement is impressively good.

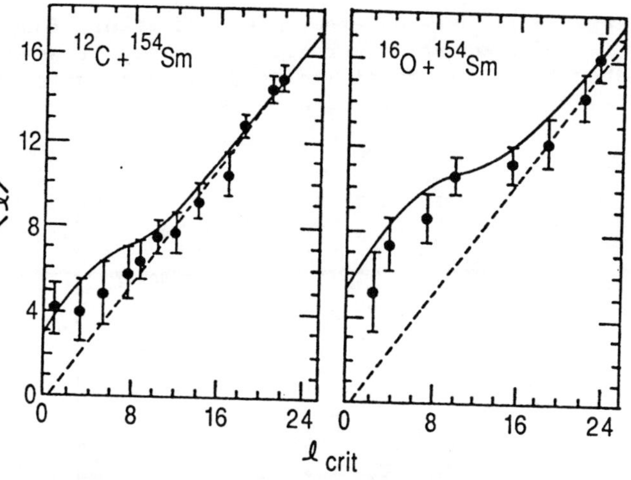

Fig. 7. Average angular momenta deduced from gamma-ray multiplicities for the $^{12}C + ^{154}Sm$ and $^{16}O + ^{154}Sm$ systems shown in comparison with no-coupling (dashed) and Wong model (solid) calculations.

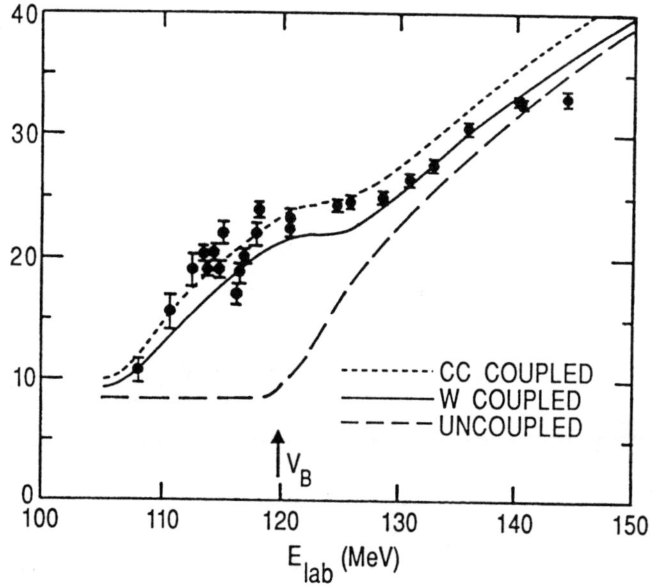

Fig. 8. Average angular momenta deduced from gamma-ray multiplicities for $^{28}Si + ^{154}Sm$ shown in comparison with no-coupling (dashed), Wong (solid) and coupled-channels (short dashed) model calculations.

The good agreement between experiment and theory found for the above studies is not however universal. A summary of essentially all the published gamma-ray multiplicity data as compiled in Ref. 12 is shown in Fig. 9. Two notable exceptions to the overall good

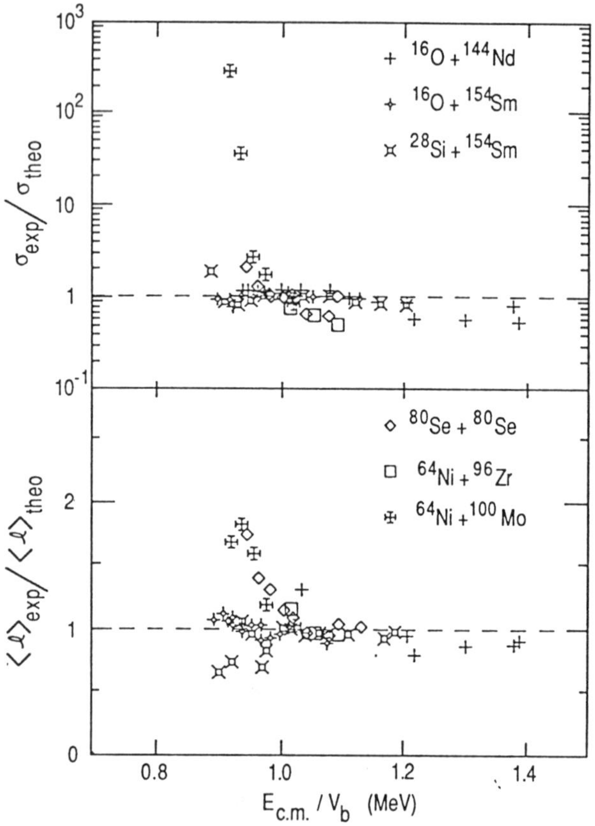

Fig. 9. Comparison of experimental and theoretical cross-sections and angular momenta for systems measured using gamma-ray multiplicities. Note the large values for $^{80}Se + ^{80}Se$ and $^{64}Ni + ^{100}Mo$.

agreement are seen - $^{80}Se + ^{80}Se$[15] and $^{64}Ni + ^{100}Mo$[16]. Concentrating on the $^{64}Ni + ^{100}Mo$ case - which is the more complete experiment and analysis of the two - we see that the coupled-channels calculation fails to predict both the experimental cross-sections and mean spin in the sub-barrier region. Even when the inelastic coupling strengths are arbitrarily increased by 50% to account for the sub-barrier cross-section, the mean and r.m.s. spins are still underpredicted in the sub-barrier region as shown in Fig. 10. This result therefore may represent a very significant discrepancy between theory and experiment.

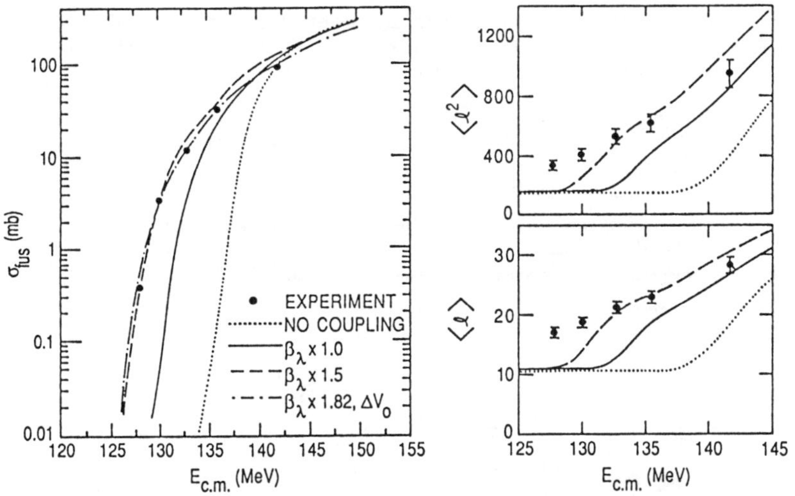

Fig. 10. Cross-sections and mean angular momenta for the ^{64}Ni + ^{100}Mo system. The curves show the results of various calculations identified in the figure.

It is therefore important to critically examine the experimental method and analysis in this and the ^{80}Se + ^{80}Se cases. Both experiments employ the same technique, namely, a large array of NaI (^{64}Ni + ^{100}Mo) or BGO (^{80}Se + ^{80}Se) detectors, used in conjunction with Ge detectors with which evaporation residues are identified by their characteristic discrete lines. Fold or hit distributions in the array for the various evaporation channels are thus obtained by gating on the discrete lines in the Ge detectors. In order that the weak evaporation channels can be seen in the presence of much more intense Coulomb excitation and transfer channels, the low folds must be suppressed electronically. Although this is corrected for in the analysis, it is possible that a bias towards higher folds is thus placed on the data. Another possible experimental difficulty comes from the relative weakness of the fusion gamma rays which may result in the missing of weak channels and also necessitates background subtraction to obtain clean fold distributions. Finally, in the analysis of these data, a unique transformation from gamma-ray multiplicity to spin was used when it is clear that the relation must depend on spin, especially at low values of the spin.

To overcome these difficulties, we have recently carried out measurements[17] of ^{16}O + ^{152}Sm fusion using a high efficiency system to directly detect evaporation residues in conjunction with a 42 element BGO array (70% of 4π) also surrounded by 8 Compton-suppressed Ge detectors. The evaporation residue detector consisted of a 25 element silicon array downstream of an electrostatic deflector. This system provides effective suppression of scattered beam particles while giving an overall detection efficiency of approximately 20% for evaporation residues. At an energy where the fusion cross-section is only 1 mb (5 MeV below the barrier), excellent statistics data were obtained in only two hours running.

Data were obtained at five energies between 60 and 80 MeV. The experimental fold distributions are shown in Fig. 11. These fold

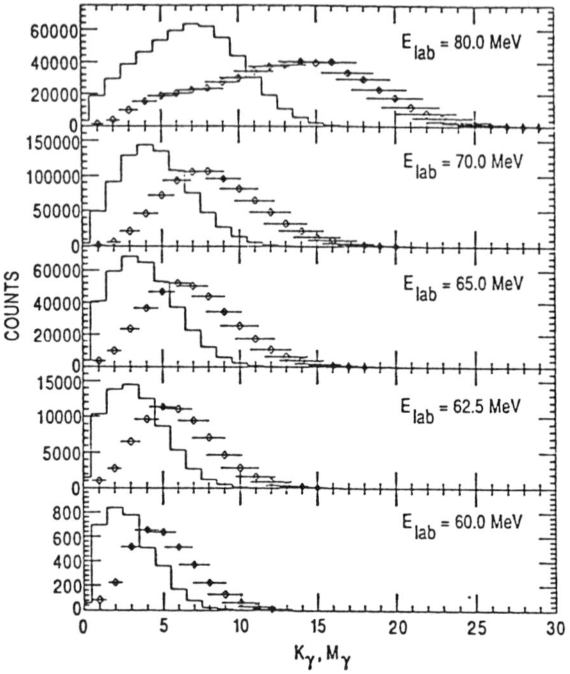

Fig. 11. Fold (histograms) and multiplicity (diamonds) distributions for ^{16}O + ^{152}Sm fusion as a function of bombarding energy.

distributions were converted to multiplicity distributions using an iterative Monte-Carlo procedure. The resulting multiplicity distributions are also shown in Fig. 1. It is clear that, in the present case, suppression of low folds would severely distort the measured fold distributions and consequently affect the deduced multiplicity distributions.

In order to compare these data with model predictions we have chosen to model the transformation of the theoretical spin distributions to gamma-ray multiplicities as this is conveniently done using existing statistical codes. The angular momentum and excitation energy dependence of the relation between multiplicity and spin is thereby automatically incorporated. The calculations were carried out using the code PACE2S with theoretical spin distributions from either coupled-channels or energy-dependent barrier calculations. Further details are given in Ref. 17. The energy dependence of the cross-section for fusion and moments of the multiplicity distributions are given in Fig. 12 in comparison with the results of the model calculations. The multiplicity data clearly favor the full coupled-channels results over those of the energy-dependent barrier although both give an equally good account of the cross-sections.

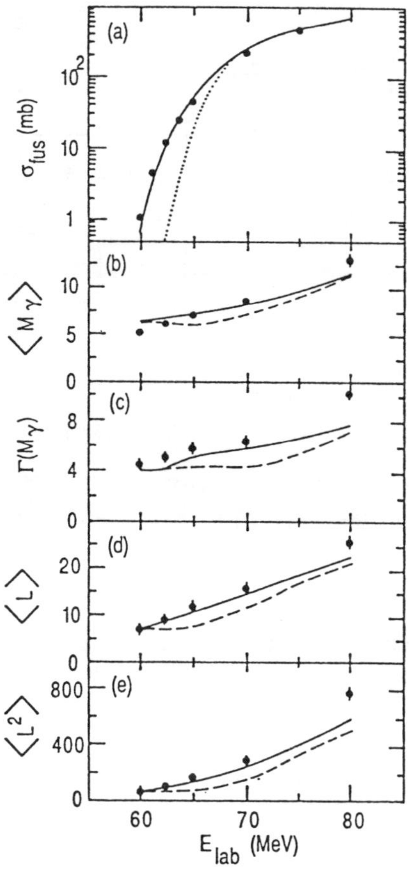

Fig. 12. Cross-sections and moments of multiplicity and spin distributions for ^{16}O + ^{152}Sm fusion. The results of no-coupling (dots) adjusted-barrier (dashes) and coupled-channels (solid) calculations are shown with the data.

To investigate possible deficiencies in the procedures followed elsewhere whereby multiplicity is converted to spin we have used the code PACE2S to estimate the coefficients in the linear transformation usually used to relate multiplicity and spin. Variation of parameters and input spin distributions produced little variation in the relative numbers of non-statistical and statistical gamma-rays and the numbers of neutrons and charged particles, the resulting mean spin obtained differing by less than 8% in the near barrier region. The moments of the spin distributions thus obtained are shown in Fig. 12 - again showing good agreement with the coupled channels results rather than with the energy-dependent barrier calculations. Finally, the deduced spin distributions are shown in Fig. 13 together with both sets of calculations. The level of agreement with the coupled-channels calculations, particularly in the barrier region, is impressive.

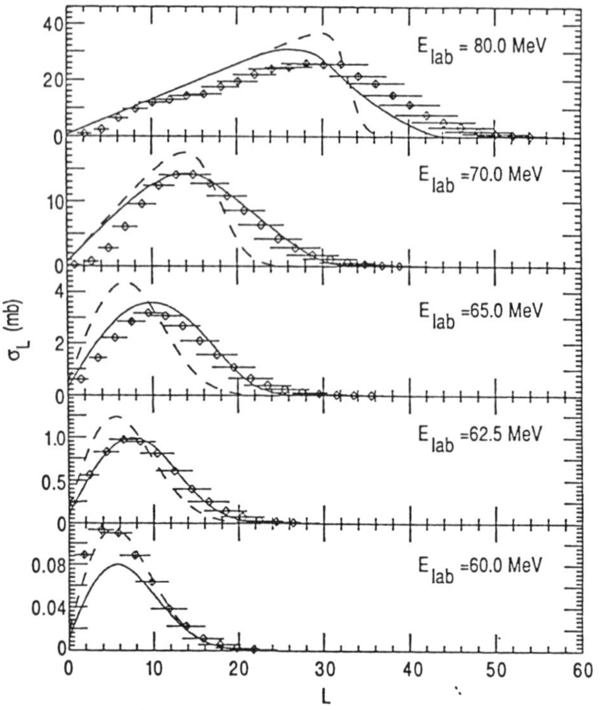

Fig. 13. Spin distributions for ^{16}O + ^{152}Sm at various bombarding energies. The curves are the results of adjusted-barrier (dashes) and coupled-channels (solid) calculations.

Returning to the critical questions raised regarding the ^{64}Ni + ^{100}Mo and ^{80}Se + ^{80}Se data. It seems unlikely that the multiplicity to spin transformation is a problem, especially as the compound system ^{164}Yb is quite close to that studied in detail above. The suppression of low folds may distort the spin distribution somewhat but, as the angular momenta involved are higher than in the ^{16}O + ^{152}Sm case, the effects should be smaller than would be the case for a lighter projectile. Last, the use of discrete lines to generate the complete multiplicity distribution has been checked with our data by forming multiplicity distributions gated on discrete lines. The resulting distributions were not significantly different. We therefore conclude that the observed extremely high values of the average spin from the ^{64}Ni + ^{100}Mo and ^{80}Se + ^{80}Se are indeed real and therefore represent a real challenge in our understanding of low energy fusion.

Fission Angular Distributions: According to the transition-state model of fission, the anisotropy of the fission-fragment angular distribution can, under some assumptions, be related to the mean-square angular momentum of the fissioning system. The only parameters which then enter into this determination are related to the moments of inertia of the saddle-point shape and the nuclear temperature. This model has been used to investigate[18] the spin distributions of a number of fissioning systems. The results of these analyses are given in Fig. 14. Uniformly, the experimental

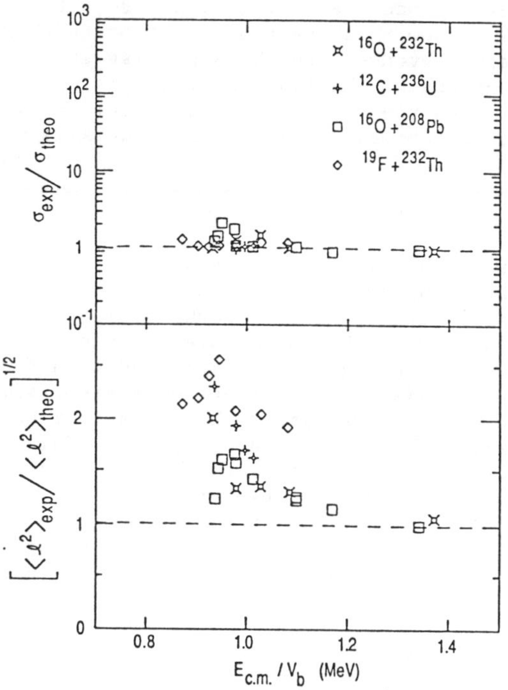

Fig. 14. Comparison of experimental and theoretical cross-sections and average angular momenta deduced from measured fission fragment angular distributions.

spin values are too high as compared to the coupled channels predictions. The question therefore arises again - is this a real discrepancy or some deficiency in the experiment and analysis? Experimentally, it has been suggested that the fission yields can be contaminated by fission occurring following transfer reactions. This has been studied experimentally[19] in the case of $^{16}O + ^{232}Th$ and, although present, the transfer fission was not found to be responsible for the observed effects at low energies.

Taking another point of view, the success of the coupled channels formalism in describing spin distributions in fusion of ^{16}O etc. on quite deformed targets must make it improbable that it is the theory that is at fault in the case of these fissioning systems - more likely is our model of the mechanism responsible for the production of the fission fragments.

Anomalously large fission fragment anisotropies are well known. In the case of projectiles incident on targets such that the fissioning system no longer has a fission barrier, large anisotropies were measured even though the standard fission models would predict isotropy[20]. Further study identified the responsible process as "fast-fission" in which dissipative processes outside the entrance channel barrier produced a damped rotating complex which subsequently reseparates into fission-like fragments. It has been usually thought that this process would not occur for projectiles as light as ^{16}O. Nevertheless, it is possible that, at sub-barrier energies, such processes may occur even for light projectiles. It is problematic to think of how this question might be further investigated. One possible avenue might be a precise study of the mass and energy distributions of the fission fragments which show the large anisotropies, with the hope that these might be different from fragments produced in "normal" fission. More work is needed to resolve this discrepancy.

SUMMARY

The current status of the experimental study of sub-barrier fusion of heavy ions has been reviewed. For not too heavy systems both the sub-barrier cross-sections and compound nuclear spin distributions are very well described by coupled channels calculations which include coupling to inelastic and transfer degrees of freedom. For heavier, more symmetric, systems similar calculations are unable to describe either the cross-sections or the large values of the average spin of the fused system. This discrepancy thus provides a very real challenge to our understanding of heavy ion reaction dynamics in the sub-barrier region. The uncovering of the degrees of freedom responsible for the additional enhancements and their incorporation in a consistent theory of heavy ion interactions will therefore be a focus of effort in the coming period.

REFERENCES

1. R. G. Stokstad et al. Phys. Rev. Lett. 41, 465 (1978).
2. M. Beckerman et al. Phys. Rev. Lett. 45, 1472 (1980).
3. D. E. DiGregorio et al. Phys. Lett. 176, 322 (1986).
4. C. Wong, Phys. Rev. Lett. 31, 766 (1973).
5. H. Esbensen and S. Landowne, Phys. Rev. C35, 2090 (1987).
6. H. Esbensen and S. Landowne, Nucl. Phys. A492, 473 (1989).
7. K. E. Rehm et al. Phys. Rev. Lett. 55, 280 (1985).
8. J. G. Keller et al. Nucl. Phys. A452, 173 (1986).
9. D. Berdichevsky et al. in Proceedings of Workshop on Nuclear Structure and Heavy Ion Reaction Dynamics, Notre Dame May 26-30, 1990 ed R. R. Betts and J. J. Kolata (IOP, Bristol, 1991) p. 105.
10. M. A. Nagarajan and G. R. Satchler, Phys. Lett. 173B, 29 (1986).
11. C. H. Dasso et al. Phys. Lett. 217B, 25 (1989).
12. D. E. DiGregorio and R. G. Stokstad, Phys. Rev. C43, 265 (1991).
13. S. Gil et al. Phys. Rev. C31, 1752 (1985).
14. S. Gil et al. Phys. Rev. Lett. 65, 3100 (1990).
15. P. J. Nolan et al. Phys. Rev. Lett. 54, 2211 (1985).
16. M. L. Halbert et al. Phys. Rev. C40, 2558 (1989).
17. A. H. Wuosmaa et al. Phys. Lett. (in press).
18. R. Vandenbosch et al. Phys. Rev. Lett. 57, 1499 (1987).
19. B. B. Back et al. in Proceedings of Sixth Winter Workshop on Nuclear Dynamics, Jackson Hole, February 17-24, 1990.
20. B. B. Back et al. Phys. Rev. Lett. 50, 818 (1983).

*Work supported by the U.S. Department of Energy, Nuclear Physics Division, under contract W-31-109-ENG-38.

A STUDY OF ELASTIC AND INELASTIC COLLISIONS OF ^{58}Ni + ^{90,94}Zr AT TWO NEAR-BARRIER ENERGIES

L.Corradi,D.R.Napoli,A.M.Stefanini
INFN-Laboratori Nazionali di Legnaro,I-35020 Legnaro (Padova),Italy

C.Signorini
Dip. di Fisica,Universita' di Padova and INFN-sezione di Padova,Italy

Y.Sugiyama, H.Ikezoe,Y.Tomita,K.Ideno
Dep. of Physics,Japan Atomic Energy Research Institute (JAERI),Japan

H.J.Kim
JAERI and Oak Ridge National Laboratory,Oak Ridge, Tennessee,USA

Y.Nagashima,T.Masuda,H.Fujiwara,Yue Hu Pu,Y.Tagishi
Tandem Accelerator Centre, University of Tsukuba, Japan

T.Sugimitsu
Faculty of Physics, University of Kyushyu, Japan

B.Qi
Low Energy Atomic Physics Institute, Beijing, China

ABSTRACT

Pure elastic scattering cross sections for the systems $^{58}Ni+^{90,94}Zr$ have been measured at energies of 247.5 and 232.5 MeV, making use of the ENMA spectrograph which is particularly suited for heavy-ion reactions. Also the differential cross sections for the excitation of the first 2^+ inelastic state of ^{58}Ni in the $^{58}Ni+^{90}Zr$ combination were extracted. Simple coupled channel calculations successfully reproduce the data at 247.5 MeV while some improvement is needed for the lower energy. Calculations with the code CC-FUS, based on the real optical model parameters extracted from the elastic scattering measurements, are compared with the previously measured fusion cross sections at energies near and below the Coulomb barrier.

INTRODUCTION

A lot of experimental and theoretical efforts have been made in recent years in studying the connections between heavy-ion reaction mechanisms and nuclear structure at energies close to the Coulomb barrier[1]. In particular it has been understood that fusion enhancement and threshold anomaly in the elastic scattering are intimately correlated, being both consequences of channel coupling effects[2]. The energy dependence of the real part of the optical potential one needs to reproduce fusion cross sections below the barrier, has the same dependence needed to fit elastic scattering cross sections at different energies[3]. Moreover, complete coupled channel (CC) calculations should automatically reproduce the energy dependence of the optical potential[4].

Anyway there are very few systems for which both a wide set of experimental cross sections of the reaction channels and a proper theoretical treatment are available. Therefore the demand for more investigations is clear, especially for heavy systems, where for example the role played by deep inelastic reactions on the fusion enhancement has not been clearly considered yet. Also, the threshold anomaly effect in the elastic scattering for heavy systems must be studied to verify whether the rather consistent picture we get for light and medium masses persists also in the heavy mass region.

Our attention has been focalized in recent years[5,6,7] on the Ni+Zr combinations; in particular for the systems here considered complete data sets already exist about evaporation residue and quasi-elastic transfer cross sections in the energy range near the Coulomb barrier[5]. These are typical "transitional" systems for which the role of complicated channels like e.g. deep inelastic are coming in, but also for which, at least at sub-barrier energies, the number of open channels to be considered is not exceedingly high so that one still hopes to be able to do a reliable CC calculation. Anyway an essential ingredient to do such a calculation is the measurement of pure elastic scattering and this was also the main motivation of our experiment.

EXPERIMENT

The measurement of pure elastic scattering for heavy systems is not an easy task: several factors enter into play like strong kinematic energy spreads, the necessity of using very thin and homogeneous targets, the ability to discriminate reaction channels other than the elastic one. Conventional time-of-flight systems have limited energy resolutions that can be compensated by long flight paths but at the price of reduced solid angles; kinematic coincidence set ups also do not always meet the necessary resolutions requested to separate the elastic events from the inelastic excitations of the lowest levels, which in our case are at 1.45 MeV (2^+ of ^{58}Ni) and 920 KeV (2^+ of ^{94}Zr). For these reasons we made use of the heavy-ion magnetic spectrograph ENMA[8]. This device, through the combined effect of its magnetic elements, effectively corrects for the kinematic energy spreads which typically amount to 2-3 MeV/deg for ^{58}Ni particles.

The ^{58}Ni beams were obtained from the JAERI tandem accelerator at energies of 232.5 and 247.5 MeV; $^{90,94}Zr$ target thicknesses amounted to $10 - 20\mu g/cm^2$, evaporated on $\simeq 10\mu g/cm^2$ carbon backing. The targets and two monitor detectors for absolute normalization were placed inside a sliding seal scattering chamber connected to the spectrograph. An ionization chamber filled with freon gas was used as focal plane detector, equipped with different anode sections for ΔE and E_{res}, and two positional wires, thus giving information also on the entrance angle of the ions.

Ni ions and/or recoil Zr ions were analyzed by ENMA and detected at the focal plane with an energy (from position) resolution of $\simeq$ 750 KeV. This was sufficient to get pure elastic scattering in the case of ^{90}Zr target, while for the ^{94}Zr case, the presence of the 2^+ excitation at 920 KeV makes it difficult to discriminate that level from the elastic peak, especially at backward angles where the cross sections are small and the statistics rather poor, due also to the use of very thin targets. The charge state distributions, made at different angles, agree with the semiempirical formula of Sayer[9] modified with the Shima[10] prediction of the most probable charge state for non carbon-targets.

In fig.1 it is shown a position spectrum obtained after setting proper windows in $\Delta E - E_{tot}$ and Position-E_{tot} matrices; the experimental FWHM is roughly in agreement with the theoretical one calculated taking into account path length differences, energy loss straggling, multiple scattering, ENMA intrinsic resolution and beam emittance.

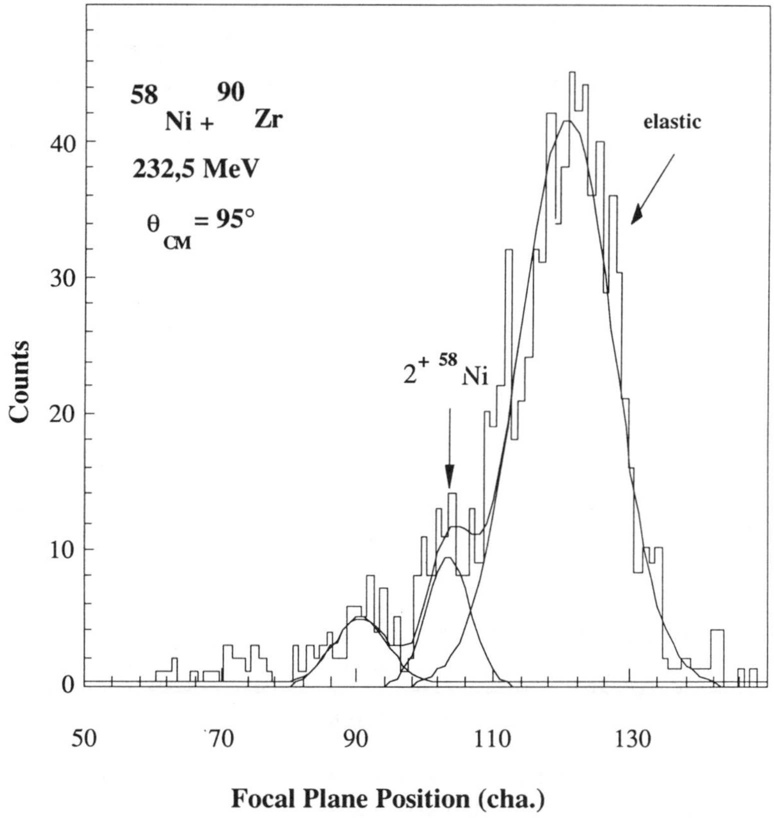

Fig. 1. Position spectrum for ^{58}Ni particles. The lines are multigaussian fits for the determination of the elastic and inelastic cross sections.

RESULTS AND DISCUSSION

In this section some preliminary results for the $^{58}Ni+^{90}Zr$ system will be presented. In fig.2 are shown the angular distributions for pure elastic scattering at both energies.

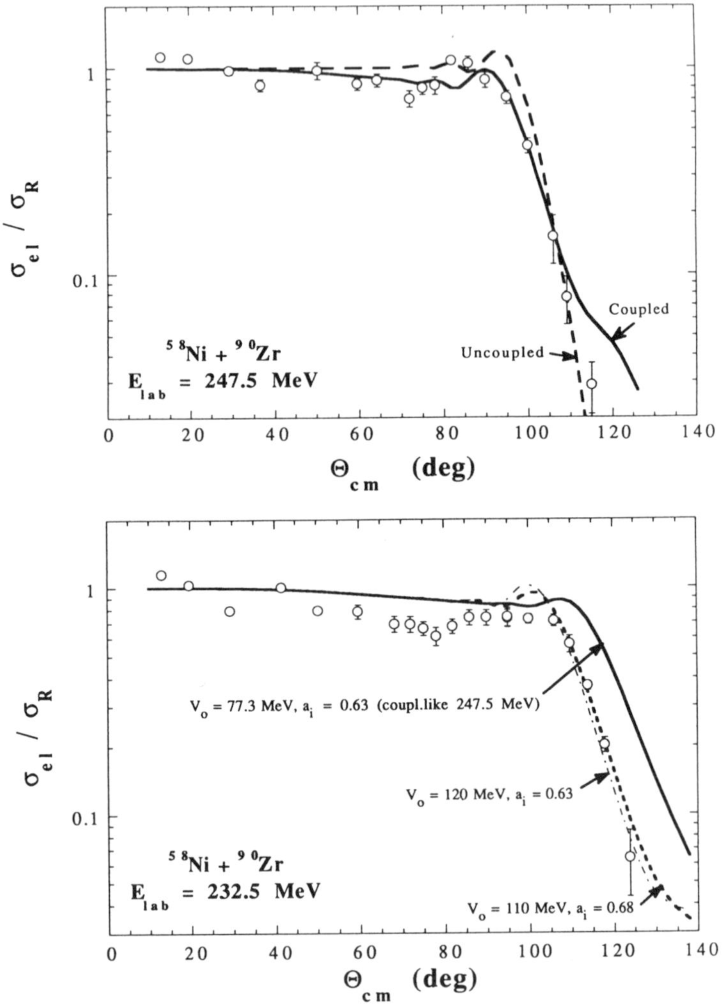

Fig. 2. Elastic scattering cross sections at the two measured energies. Upper panel: the optical potential parameters used for the calculations are V_O=77.3 MeV,V_I=10.0 MeV,R_O=R_{IO}=1.179, R_{CO}=1.30 fm,A=A_I=0.63 fm. Lower panel: the parameters are the same except those explicitly reported.

A drop of the cross sections below the Rutherford limit is evident at angles forward of the grazing peak; this is already an indication that an appropriate theoretical treatment of the data needs considering explicitly the coupling to at least the Coulomb excited states.

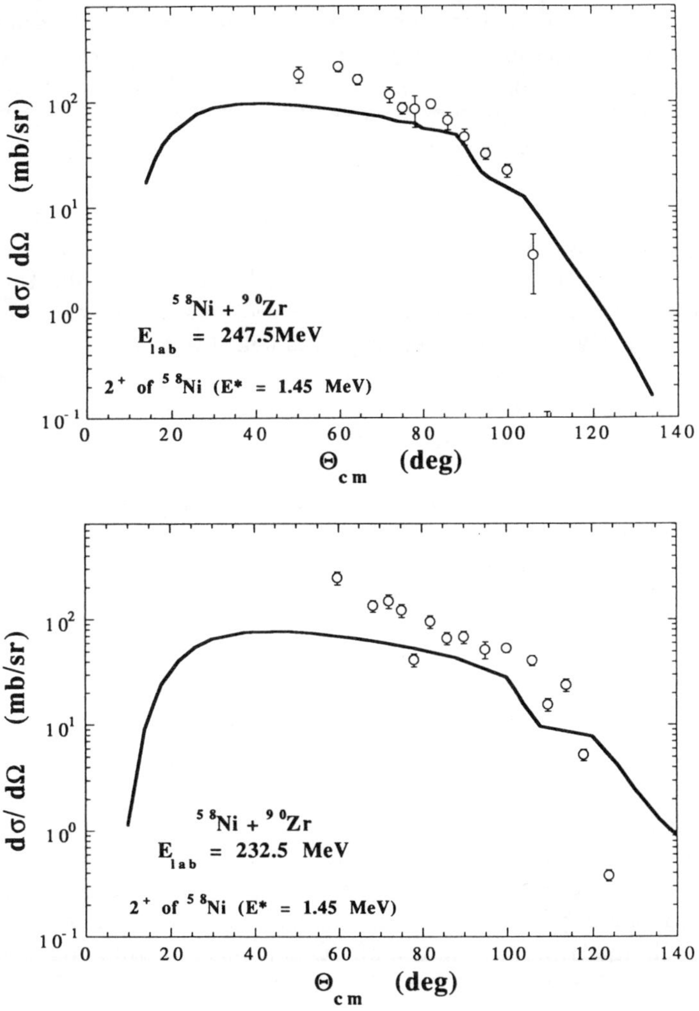

Fig. 3. Inelastic differential cross sections for the lowest 2+ level of ^{58}Ni. The lines are the results of CC calculations described in the text

In a first step a calculation was performed using the CC code PTOLEMY[11] and coupling the first $2^+, 3^-$ excited states of both projectile and target. The nuclear potential was a volume Woods-Saxon function with standard optical model parameters[12] and with the same geometry for the real and the imaginary parts. Fig.2 (upper) shows the comparison of the calculations with the data which are reproduced quite reasonably. The same set of parameters do not fit the data at the lower energy (fig.2-lower); to improve the things, without introducing additional pieces in the potential like e.g. surface parts and mantaining the same values for the radii, one needs to deepen the real depth. For comparison also a further calculation is shown with a different imaginary diffuseness.

As one can observe the calculations at this energy require further improvements, probably considering second order couplings and the inclusion of transfer channels, although these should not be important for the ^{90}Zr case since the experimental total transfer cross sections are only $\simeq$1/4 of the inelastic ones. Anyway the experimental points near the grazing and at backward angles correspond to an angular momentum range which overlaps with fusion and transfer channels; the attempt to describe simultaneously all these channels should give interesting informations on the interconnections between the various processes.

As said before, the energy resolution allows us to obtain cross sections for the excitation of the lowest 2^+ level of ^{58}Ni; in fig.3 the experimental points are plotted together with the CC calculations; the agreement is reasonable at 247.5 MeV while at 232.5 MeV a global underestimation of a factor $\simeq$1.5 is evident.

CONNECTION WITH FUSION

The main aim of this experiment was to see whether a simultaneous description of all reaction channels is possible within the CC procedure successfully adopted in other cases like the $^{58}Ni + ^{64}Ni$ system[13]. Especially interesting is the fusion enhancement below the barrier which for similar systems like $^{64}Ni + ^{100}Mo$ is not reproduced by the calculations[14]. With our rather heavy systems one may consider alternative descriptions in terms of macroscopic models involving neck formation[15] or friction[16], because the presence of many open channels makes the use of CC calculations difficult. Moreover the effect of the absorption under and beyond the Coulomb barrier[17] is still unclear and in CC calculations one has to take care of the size of the fusion radius.

Nonetheless we used the simple CC code CCFUS[18], slightly modified to give in input the real optical model parameters (depth, radius and diffuseness) as derived from the elastic scattering measurements at the higher energy.

We coupled a total of 6 channels: the first $2^+, 3-$ excited states of ^{58}Ni and ^{90}Zr plus 2 transfer channels, the pick up of 1 and 2 neutrons, with

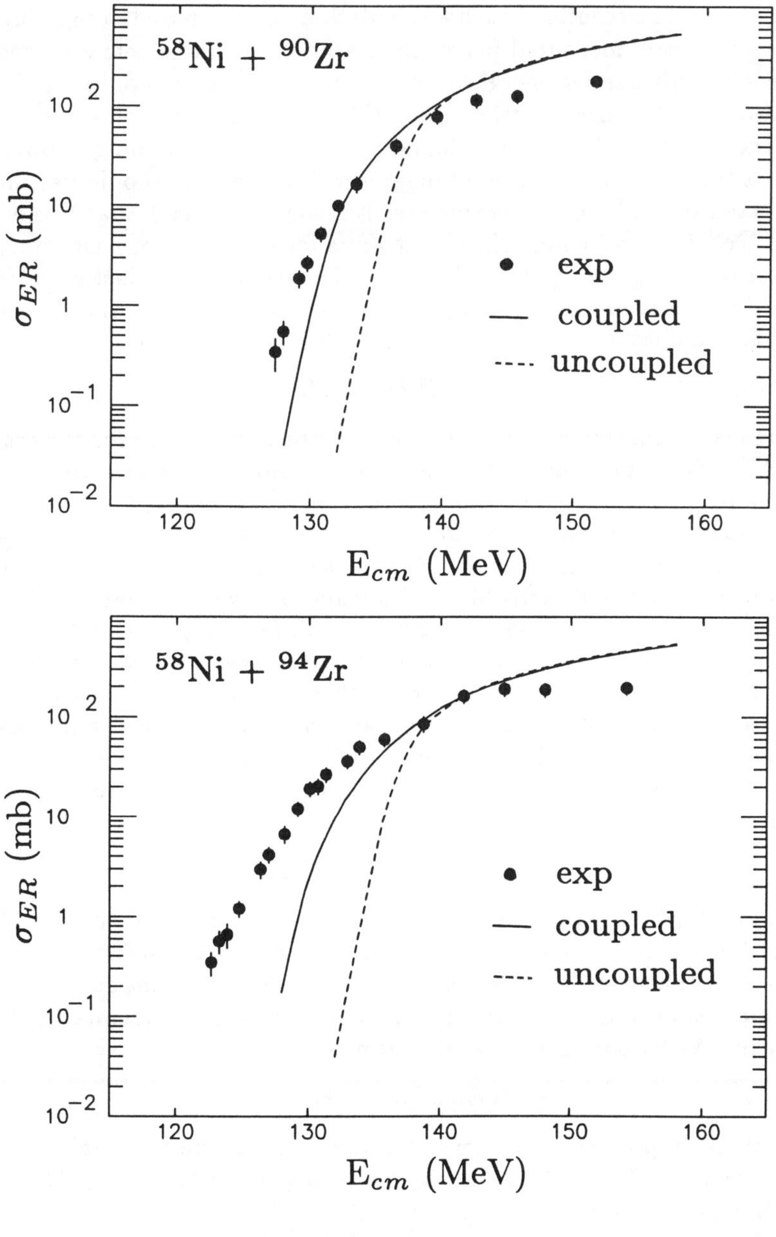

Fig. 4. Comparison of the previously measured evaporation residue cross sections[5] with CCFUS calculations. See text for details.

Q-values and form factors deduced from the experimental transfer angular distributions[5]. The results of such a calculation are compared in fig.4 (upper) with the previously measured fusion cross sections and one gets a reasonable agreement at sub-barrier energies; above the barrier one should add to the evaporation residue cross sections the still unmeasured fission cross section for a full comparison. The same calculation for ^{94}Zr, shown in fig.4-lower, underpredict the data by one order of magnitude but here we should use the not yet extracted optical model parameters. Anyway we remark that this is only a sort of "consistency" comparison that waits for further improvements, since the use of the imaginary potential in the calculation of the elastic scattering cross sections , already takes partly into account the effects of the absorption into the fusion channel.

CONCLUSIONS

Pure elastic scattering cross sections have been measured for the systems $^{58}Ni + ^{90,94}Zr$ at two energies close to the Coulomb barrier. For the ^{90}Zr target the resolution of the set-up was sufficient to separate the elastic peak from the lowest inelastic states, at variance with the ^{94}Zr case where the presence of the first 2+ level at 920 KeV makes it difficult a clear separation. The comparison of the PTOLEMY CC calculations with the data shows that, although there is fair agreement at 247.5 MeV, for the 232.5 MeV energy a variation of the optical model parameters is necessary. The calculation of the fusion cross sections below the barrier with the simple code CCFUS shows that while for the ^{90}Zr case the comparison is rather consistent, for the ^{94}Zr case other channels should be included. A consistent description of the different open channels will only be possible when more complete calculations are available.

ACKNOWLEDGMENTS

We thank Mr.A.Dal Bello for the precious technical support and Mr.G. Manente for the preparation of the targets. Three of us (L.C.,D.R.N.,C.S.) would like to thank the staff members of the JAERI tandem accelerator for their professional work; we appreciate as well the warm hospitality during our stay in JAERI and Tsukuba laboratories.

REFERENCES

1. Proc.Workshop on H.I.Collisions at Energies Near the Coulomb Barrier, Daresbury(GB),July 5-7 1990,ed.M.A.Nagarajan(IOP,Bristol 1991)
2. G.R.Satchler,Phys.Rep.199,148(1991)
3. C.Mahaux et al.,Nucl.Phys.A449,354(1986)
4. J.Diaz et al.,Nucl.Phys.A494,311(1989)
5. F.Scarlassara et al.,Zeit.Phys.A338,171(1991)
6. A.M.Stefanini et al.,Phys.Lett.B252,43(1990)

7. A.M.Stefanini,Invited talk at the 4th Int.Conf. on Nucleus-Nucleus Collisions,Kanazawa(Japan)June 10-14,1991, Proc. to be published
8. Y.Sugiyama et al.,Nucl.Instr.and Meth.A281,512(1989)
9. R.O.Sayer,Revue de Phys.Appl.12,1543(1977)
10. K.Shima et al.,Nucl.Instr.and Meth.200,605(1982)
11. M.H.Macfarlane and S.C.Pieper,Argonne National Laboratory Report ANL-76-11(Rev.1),1978;M.Rhoades-Brown,M.H.Macfarlane,and S.C.Pieper,Phys.Rev.C21,2417(1980)
12. H.Esbensen and S.Landowne,Phys.Rev.C40,126(1989)
13. A.M.Stefanini et al.,Phys.Lett.B240,306(1990)
14. M.L.Halbert et al.,Phys.Rev.C40,2558(1989)
15. C.E.Agujar et al.,Phys.Lett.B201,22(1988)
16. P.Fröbrich et al.,ref.1,p.223
17. T.Udagawa et al.,Phys.Rev.C39,1840(1989)
18. C.H.Dasso and S.Landowne,Phys.Lett.B183,141(1987)

PRESENT AND FUTURE
OF THE SUB-BARRIER FUSION REACTIONS

N. Takigawa
Department of Physics, Tohoku University, 980 Sendai, Japan

ABSTRACT

The large enhancement of the fusion cross section of heavy ion collisions at energies below the Coulomb barrier has stimulated many experimental as well as theoretical studies in the past decade. I overview the present status of the theoretical research of this subject and discuss some prospects for the future.

1. INTRODUCTION

It is by now well-established that heavy-ion fusion cross-sections below the Coulomb barrier are several orders of magnitude larger than one would expect from a one-dimensional barrier penetration picture[1]. After extensive theoretical investigations[2–7] it was concluded that this enhancement can be attributed to the coupling of the translational motion to additional degrees of freedom such as surface vibrations, deformation, rotation, nucleon transfer, or neck formation.

There remain, however, many unsettled problems. One of them is the spin distribution of the compound nucleus[2,7]. The principal intrinsic degree of freedom that governs the enhancement of the fusion cross section is also not yet fully identified. Nevertheless, the present understanding is advanced enough to predict several interesting phenomena which should be explored in the future.

This paper is meant to illustrate the present theoretical status of the sub-barrier fusion reactions (see also ref.2, which describes the interconnection between different phenomena) and to discuss problems which are expected to develop in the near future. There are many theoretical approaches which fall into three broad categories: potential approaches, coupled-channels approaches and microscopic approaches. The microscopic approaches are, however, so far applied to the description of light heavy-ion fusion reactions, such as the $^{12}C+^{12}C$ and $^{16}O+^{16}O$ systems in terms of the adiabatic Hartree Fock approximation[8]. I do not touch upon them in this paper. In sect.2 I give a few introductory remarks referring to potential approaches. The coupled-channels theories are the most standard approaches to understand the large enhancement of the sub-barrier fusion cross section. Roughly speaking, they can be put into two different points of view depending on which one is stressed to be more important: the effects of nuclear structure in the entrance channel or the properties of the unified system. In sect.3 I point out the virtue and the limitation of each point of view. In sect.4 a coupled-channels problem is converted into a set of one dimensional Schrödinger equations under the no-Coriolis and the adiabatic approximations and the effects of channel-coupling are interpreted in terms of a barrier distribusion. It offers a simple explanation for why the channel-coupling enhances the fusion cross section at energies below the Coulomb barrier and hinders at above the barrier for a certain range of energies. I then touch on two topics from recent studies in sect.5. One of them is the determination of the distributed potential barriers directly from experimental fusion excitation functions. The other is the description of the nuclear intrinsic degrees of freedom in terms of the interacting boson model. Among the future problems, I discuss three subjects. They are the fusion of a halo nucleus, the synthesis of superheavy elements by neutron rich beams and the effects of nuclear intrinsic degrees of freedom on the spectral distribution of partcles emitted from a hot compound nucleus. Sect.6 is devoted to discussing these topics.

2. ENHANCEMENT OF THE FUSION CROSS SECTION AND POTENTIAL APPROACHES

If $P_F(\ell)$ denotes the probability of fusion in the collision with a partial wave ℓ, then the fusion cross section is given by,

$$\sigma_{CF} = \frac{\pi}{k^2} \sum_{\ell} \ (2\ell + 1) \ \ P_F(\ell), \tag{1}$$

A simple estimate of the fusion cross section can be obtained by a potential model where $P_F(\ell)$ is determined by numerically solving the Schrödinger equation for the radial motion with an appropriately chosen potential under, e.g., the incoming wave boundary condition. Alternatively, one can use the WKB formula in the uniform approximation,

$$P_F^{(0)}(\ell) = \frac{1}{1 + e^{2\int_{r_2}^{r_3} \kappa(r)dr}}, \tag{2}$$

where $\kappa(r)$ is the "local wave number", and r_2 and r_3 are the inner and the outer classical turning points at the potential barrier. If one approximates the fusion barrier by a parabolic function, then

$$P_F^{(0)}(\ell) = \frac{1}{1 + \exp[\frac{2\pi}{\hbar\Omega_\ell} \cdot (V_B(\ell) - E)]}, \tag{3}$$

where $V_B(\ell)$ and Ω_ℓ are the height and the curvature parameter of the fusion barrier for the partial wave ℓ, respectively. Ignoring the ℓ dependence of Ω and of the barrier position R_B, one can obtain the following Wong formula[9],

$$\sigma_F^{(0)}(E) = R_B^2 \cdot \frac{\hbar\Omega}{2E} \cdot \ln[1 + \exp(\frac{2\pi}{\hbar\Omega} \cdot (E - V_{0B}))], \tag{4}$$

where V_{0B} is the height of the fusion barrier for the s-wave scattering. In eqs.(2) through (4) the upper suffix (0) designates the potential model. In the classical limit, where $\Omega \rightarrow 0$, or if $E \gg V_B$,

$$\sigma_F^{(0)}(E) \approx \pi R_B^2 \cdot (1 - \frac{V_{0B}}{E}), \tag{5}$$

while if $E \ll V_B$, then

$$\sigma_F^{(0)}(E) \approx R_B^2 \cdot \frac{\hbar\Omega}{2\pi} exp(\frac{2\pi}{\hbar\Omega} \cdot (E - V_{0B})) \quad . \tag{6}$$

The observed fusion cross section at energies above the Coulomb barrier can be well fitted by eq. (5) by estimating R_B and V_{0B} based on a folding potential. However, the fusion cross section at energies near and below the Coulomb barrier is much larger than the prediction of the potential model given by eqs.(4) or (6)[1]. This enhancement phenomenon prevails over many medium weight systems. As an example, fig.1 shows the fusion excitation function of the ^{58}Ni $+^{58}$Ni system at energies near and below the Coulomb barrier. The dots are experimental data taken from ref.10. The solid line is the prediction of a potential model, which was obtained based on Wong's formula by determining the potential curvature parameter $\hbar\Omega$ from the slope of the fusion excitation function at low energies and the height and the position of the s wave fusion barrier, V_{0B} and R_B, by fitting data at high energies[11]. The arrow along the abscissa shows the position of the Coulomb barrier V_{0B}.

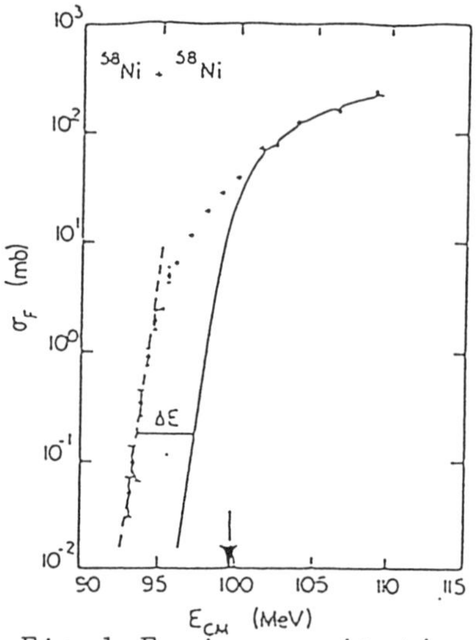

Fig.1 Fusion excitation function and the asymptotic energy shift (taken from ref.11)

Many theoretical studies were undertaken to explain the large enhancement of the fusion cross section at sub-barrier energies. Though a standard approach is to use one of the various versions of the coupled-channels approaches, potential approaches were also developed. As mentioned above, Wong's formula gives a simple estimate of the fusion cross section based on a potential model. In addition, it offers a starting point for many advanced theories. In early analyses $\hbar\Omega$ was varied to fit the observed fusion excitation function. The effective curvature thus obtained was about 10 MeV, which is much larger than 3 to 4 MeV predicted by the standard folding potential. The problem is then to explain why the effective fusion barrier appears so thin.

The dispersion relation between the real and the imaginary parts of the optical potential was extensively used in the past several years to explain the energy dependence of the real part of the optical potential, i.e. the deviation of the real part of the optical potential from the folding potential at low energies. This phenomenon, which is called the threshold anomaly, can be associated with the effects of channel-coupling. The relation of the threshold anomaly to barrier penetration for fusion was examined, and it was shown that the simple barrier penetration model can exhibit the strong enhancement of the subbarrier fusion cross section arising from channel-coupling if the threshold anomaly is taken into account[12]. Tamura and Udagawa raised a question concerning the place where the fusion takes place, more precisely the place which influences the fusion probability. They determined the imaginary part of the optical potential that is associated with fusion by simultaneously fitting the experimental fusion and elastic cross sections for the $^{32}S + ^{58,64}Ni$ at energies near the Coulomb barrier[13]. And they found that it extends outside the potential barrier. This suggests the important role played by the channel-couplng in fusion.

One intriguing approach is the inversion method, with which the effective potential barrier for fusion can be determined directly from the experimental fusion cross sections. Assuming the fusion at sub-barrier enrgies is described by a one-dimensional energy-independent local potential barrier, we can derive the following formula that determines the thickness of the potential at height V using the experimental fusion cross sections for the interval $V \leq E \leq B$, B being V_{0B}

$$t(V) = -\frac{2}{\pi}[\frac{\hbar^2}{2m}]^{1/2} \int_V^B \frac{dS_0/dE}{(E-V)^{1/2}} dE, \tag{7}$$

with

$$S_0(E) = \frac{1}{2} log[[\frac{d}{dE}[\frac{E\sigma}{\pi R^2}]]^{-1} - 1] \quad . \tag{8}$$

Applying this technique, Balantekin et al.[14] found that the potential obtained by the inversion method does not become thicker as one goes away from the top of the potential barrier downwards as one expects. Actually, the potential barrier so obtained is not a single valued function of the separation distance. This can be taken as a strong suggestion that the channel-coupling or other mechanisms play a decisive role in the fusion reactions involving medium mass nuclei.

Stelson[15] and Rowley et al.[16] have recently applied the inversion method in a different sense. Following a conclusion based on the coupled-channels approaches, they assumed that the fusion cross section is given by a sum of the fusion cross section from a distributed potential barriers. They have then determined the distribution of the potential barriers directly from the experimental data. I will describe their works in more detail in sect.5.1.

Finally, I wish to mention that Tsuda and myself applied the same inversion procedure as ref.14 to extracting the screening potential by bound electrons in the low energy ^{3}He+ d reaction[17].

3. COUPLED-CHANNELS APPROACHES: TWO "DIFFERENT" POINTS OF VIEW

Many theoretical studies have been developed based on coupled-channels approaches. They differ either in the principal intrinsic degrees of freedom, to which they wish to attribute the large enhancement of the fusion cross section, and/or in their theoretical framework, and in various levels of approximation. I listed them in Table 1, so that one can have a quick outline of the theoretical studies so far made.

Table 1. Various Coupled-Channels Approaches

1. Inelastic scattering
   - 1.1 Direct Numerical Solution (ECIS, PTOLEMY,....)
     - (A) Bohr Model
   - 1.2 Simplified C.C. (no Coriolis + Adiabatic approximations)
     - (A) Bohr Model
     - (B) IBM
   - 1.3 Simplified C.C. (constant coupling model)
2. Particle transfer (C.R.C.)
   - 2.1 direct numerical solution (FRESCO)
   - 2.2 pair transfer in a pairing vibration model
3. Eigenchannel method
4. Neck Formation
5. Path integral
   - 5.1 influence functional method
   - 5.2 perturbative path integral
6. Extended WKB

In the sense of the principal intrinsic degree of freedom that enhances the fusion cross section at sub-barrier energies, various coupled-channels approaches can be considered roughly from two different points of view. The one places much importance on the effects of nuclear structure in the entrance channel. The other stresses the importance of the properties of the unified system. In other words, the former emphasizes the important role played by the inelastic excitations of the target or the projectile or the transfer of a few nucleons between them.

The latter attributes the large enhancement of the sub-barrier fusion cross section to the effects of neck formation, which can probably be regarded as the macroscopic view of the effects of multiple-transfer of nucleons between the colliding partners and of many inelastic couplings.

As one of the representatives of the former, fig.2 shows the results of the coupled-channels calculations for the fusion excitation function for the ^{58}Ni + ^{64}Ni system taken from ref.18. The lower dashed curve shows the no coupling limit. The three solid lines show, in increasing order, the result of including the one-neutron transfer couplings

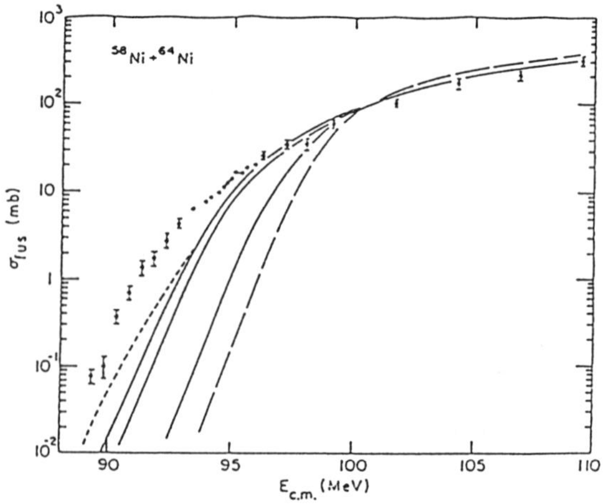

Fig.2 Comparison between the experimental fusion excitation function and the results of coupled-channels calculations (taken from ref. 18)

alone, the vibrational excitations alone and the combined effect of both processes. The upper dashed curve includes a direct two-neutron transfer coupling based on a macroscopic model for two particle transfer[19]. The figure shows that the effects of inelastic excitations dominate the enhancement. Adding the effects of particle transfer, one can reproduce the enhancement of the sub-barrier fusion cross section to a large extent. Some other mechanism is still needed to fully reproduce the large enhancement shown by the experimental data. In general, the coupled-channels approach based on the entrance channel configurations has difficulty to reproduce the data for heavy systems, say heavier than the present example.

Now I wish to turn to a systematics, which indicates the validity of the second point of view. In fig.1 we notice that the data points are shifted by a constant energy ΔE at extreme sub-barrier energies. Aguiar et al.[11] plotted this constant energy shift for various systems as a function of the system size parameter measuresd in terms of an effective fissility parameter,

$$\zeta = (Z^2/A)_{eff} = \frac{4Z_1Z_2}{A_1^{1/3}A_2^{1/3}(A_1^{1/3}+A_2^{1/3})}, \tag{9}$$

where the indices 1 and 2 refer to the target and the projectile, respectively.

The result is shown in fig.3. The solid line is their estimate of the decrease of the fusion barrier due to neck formation. The neck formation seems to reproduce fairly well the overall trend and the magnitude of the asymptotic energy shift for many systems in a wide range of the fissility parameter.

Two systems significantly deviate from the general trend. They are the $^{40}Ar+^{154}Sm$ and the $^{74}Ge+^{74}Ge$ systems. The large static deformation of the target nucleus ^{154}Sm could be the cause of the large deviation for the former, while, as Iwamoto and myself[20] have pointed out, the dynamic deformation of ^{74}Ge induced during the fusion reaction and the shell correction energy cooperatively working to induce an extra lowering of the fusion barrier could be the cause for

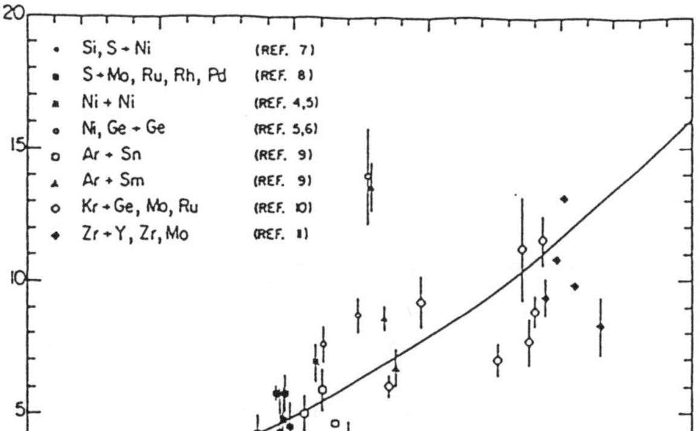

Fig.3 Comparison between the asymptotic energy shift and the lowering of the fusion barrier due to a neck formation (taken from ref. 11)

the latter. Though the idea of the neck formation is very attractive, the associated theory is still at a very primitive stage compared to the coupled-channels approaches. It is highly desirable to develop a reaction theory that can reliably calculate not only the decrease of the fusion barrier, but also the fusion cross section itself based on the idea of the neck formation.

4. SIMPLIFIED COUPLED-CHANNELS APPROACH: THE DISTRIBUTED POTENTIAL BARRIERS

A simplified treatment of the coupled-channels equations is one of the important theoretical developments in the past several years in studying heavy ion collisions at energies near and below the Coulomb barrier. Though there exist advanced computer codes, such as the ECIS-, the PTOLEMY- and the FRESCO-codes, that can perform very extended calculations, the simplified treatment makes it easier to deal with the coupled-channels equations. Also, it offers an insight into the effects of nuclear intrinsic degrees of freedom in terms of the potential renormalization. Here I illustrate how it works by taking the rotational coupling as an example.

In the case of exciting the members of $K^{\pi} = 0^+$ ground state rotational band of an axially-symmetric deformed nucleus, the coupled- channels equations read in the standard notations:

$$\begin{aligned}
&\{-\frac{\hbar^2}{2\mu}\frac{d^2}{dr^2} + \frac{\hbar^2}{2\mu r^2}\ell_1(\ell_1+1) + U(r) - E_{I_1}\} f_{I_1,\ell_1}(r) \\
&+ F_\lambda(r) \sum_{I_2,\ell_2} (-)^{J+\ell_2} i^{I_2+\ell_2-I_1-\ell_1} \left[\frac{(2\ell_1+1)(2I_1+1)(2I_2+1)}{2\lambda+1}\right]^{\frac{1}{2}} , \qquad (10) \\
&\times < I_1 I_2 00|\lambda 0 >< \ell_1 \lambda 00|\ell_2 0 > \begin{Bmatrix} \ell_1 I_1 J \\ I_2 \ell_2 \lambda \end{Bmatrix} f_{I_2,\ell_2}(r) = 0,
\end{aligned}$$

where I denotes the intrisic angular momentum of the rotational state and ℓ the angular momentum for the relative motion. The optical potential $U(r)$ contains an imaginary part in order to take into account the effects of the other intrinsic degrees of freedom on the scattering. The multipolarity λ of the intrinsic excitation is two if we restrict to quadrupole deformation. The coupling form factor $F(r) = F_2(r)$ consists of Coulomb and nuclear parts,

$$F(R) = F_c(R) + F_n(R), \qquad (11)$$

where the Coulomb part is[21–24]

$$F_c(R) = \frac{3}{\sqrt{20\pi}}\beta\, Z_1 Z_2 e^2 \frac{R_c^2}{R^3} \qquad (R \geq R_c) \qquad (12a)$$

$$= \frac{3}{\sqrt{20\pi}}\beta\, Z_1 Z_2 e^2 \frac{R^2}{R_c^3} \qquad (R < R_c) \qquad (12b)$$

and the nuclear part is

$$F_n(R) = -\sqrt{\frac{5}{4\pi}}\beta R_V \frac{dV_N}{dR} \quad . \tag{13}$$

In the standard coupled-channels approach, one solves eq.(10) by imposing an appropriate boundary condition, such as the incoming wave boundary condition at some point inside of the potential barrier, to obtain S matrices. The fusion cross section is then obtained based on the unitarity relation.

The basic idea of the simplified coupled-channels approach is to decouple eq.(10). It can be achieved simply by introducing two approximations[21,23,24]. The first is the no-Coriolis approximation, where the change in the centrifugal potential due to the finite multipolarity of the intrinsic excitation is ignored. This amounts to replacing $\ell_1(\ell_1+1)$ in eq. (10) by $J(J+1)$, J being the total angular momentum in the entrance channel. The second is the adiabatic approximation, where the finite excitation energy of the rotational motion is neglected, i.e. the E_{I_1} in eq.(10) is replaced by the c.m. energy E in the entrance channel. Subjecting eq.(10) to the unitary transformation which diagonalizes the coupling matrix, one then obtains a set of decoupled equations,

$$[-\frac{\hbar^2}{2\mu}\frac{d^2}{dr^2} + \frac{\hbar^2}{2\mu r^2}J(J+1) + U(r) + \lambda_\alpha F_2(r) - E]\phi_\alpha(r) = 0, \tag{14}$$

In eq.(14), $\{\lambda_\alpha\}$ denote the eigenvalues of the coupling matrix, and the wave functions $\phi_\alpha(r)$ in the "pseudo-channels" are related to the physical wave functions through the unitary transformation that diagonalizes the coupling matrix. If the description of the rotational band is truncated at $J = J_{max}$, the number of decoupled Schrödinger equations (14) to be solved is given by $\alpha_{max} = J_{max}/2+1$[25]. For example, if we restrict the possible excitation to the $J^\pi = 2^+$ member of the rotational band, the number of equations to be solved is $\alpha_{max} = 2$ (whereas there were 4 coupled equations in the physical channels). The eigenvalues λ_α which determine the strength of the correction to the diagonal potential in each pseudo-channel are given by[25]:

$$\lambda_\alpha = P_2(cos\ \theta_\alpha), \tag{15}$$

where $cos\ \theta_\alpha$ are the zeros of $P_{J_{max}+2}$ with $\theta_\alpha < \frac{\pi}{2}$. The boundary condition of the scattering problem, the properties of the unitary matrix which connects the physical and the pseudo-channel wave functions and the unitarity relation then lead to the following formula that gives the fusion cross section as a sum of the fusion cross section for each pseudo-channel, $\sigma_F^{(\alpha)}$,

$$\sigma_F = \sum_{\alpha=1}^{\alpha_{MAX}} w_\alpha \sigma_F^{(\alpha)} = \sum_{\alpha=1}^{\alpha_{MAX}} w_\alpha \sigma_F^{(0)}(E; U(r) + \lambda_\alpha F_2(r)), \tag{16}$$

where $\sigma_F^{(0)}(E; V(r))$ is the fusion cross section in the potential model for energy E when the potential is given by $V(r)$. The weight factors w_α are those in the $(J_{max}+2)$-points Gauss-Legendre quadrature formula (in fact only half of these is needed by symmetry). Based on eqs.(15) and (16), ref.25 gave a geometric interpretation to the fusion process: the fusion cross section is given by an appropriate sum of the fusion cross section for each fixed orientation of the deformed target with respect to the beam direction. The relevant orientations are determined by the maximum J-value at which the truncation of the rotational band is made, notice, however, that these orientations only determine the strength of the potential to be used in each pseudo-channel and do not correspond to anisotropic potentials.

We are thus treating the scattering from a distributed potential barriers for each partial wave. The important fact is that some of the effective potential barriers have lower heights than the original potential. As an example, fig.4 shows how the original potential barrier for the s wave scattering (dashed line) splits into two effective potential barriers (solid lines) when one truncates at the 2^+ member of the ground state rotational band of ^{238}U in the coupled-channels calculations in the scattering of ^{41}K from ^{238}U, which is discussed in sect.6.2. Assume that the incident energy is slightly lower than the original potential barrier, but higher than the barrier top of the lower effective potential. The fusion is then classically forbidden if there is no coupling. The coupling allows fusion to take place for the portion that corresponds to the lower potential barrier. This is the reason why the fusion cross section at sub-barrier energies is enhanced by the channel coupling.

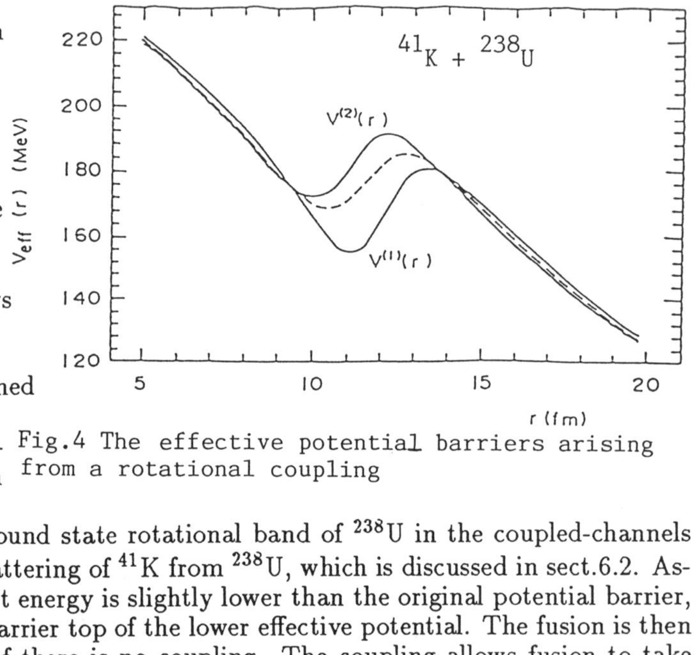

Fig.4 The effective potential barriers arising from a rotational coupling

The distribution of the potential barriers, i.e. the position and the weight of each potential barrier, reflects the properties of the involved nuclear intrinsic degrees of freedom, those of the coupling Hamiltonian and the highest states excited during the collision. It can be predicted by the zero point motion formula, which takes the quantal fluctuation of the nuclear intrinsic degrees of freedom such as the nuclear surface vibration into account. In this connection, table 2 collects the zero point motion formulae of the fusion cross section for several intrinsic degrees of freedom that couple to the translational motion. The form of the coupling Hamiltonian is also indicated. The derivation can be found in refs.6 and 24. In ref.6 the influence functional method was used to derive these formulae without paying attention to the angular momentum algebla. Ref.24 uses Green's function method and clarifies how the angular momentum algebra is handled. It also discusses the extension of the zero point motion formulae to the case where the intrinsic degrees of freedom are not in their ground states, but are in thermal equilibrium with a heat bath[26]. Ref.27 gives the zero point motion formula for the IBM coupling. Similar to eq.(16), $\sigma_F^{(0)}(E;U(R))$ in table 2 means the fusion cross section in a potential model when the potential is given by $U(R)$ and the incident energy is E. Note that the Gauss finite points quadrature formula eq.(16) is the exact formula for the rotational coupling if the rotational excitation is truncated, while eq.(T2-4) presumes that there is no truncation. They coincide, of course, when $J_{max}=\infty$. A discrete set of barriers then change into continuous barriers. Likewise, eq.(T2-2) is replaced by the corresponding finite points Hermite quadrature formula if the vibrational mode of excitation is truncated at somewhere[25].

Table 2: Zero Point Motion Formulae

(1) Linear and Quadratic Coupling to a Vibration

$$H_{int} = \alpha_0 f(R)(a^\dagger+a)+\alpha_0^2 h(R)(a^\dagger+a)^2 \tag{T2.1}$$

$$\sigma_F(E) = \frac{1}{\sqrt{2\pi}}\int_{-\infty}^{\infty} dx e^{-\frac{1}{2}x^2}\sigma_F^{(0)}(E;U(R)+x\alpha_0 f(R)+x^2\alpha_0^2 h(R)), \tag{T2.2}$$

(2) Coupling to a Rotor

$$H_{int} = \beta P_2(cos\theta)f(R) \tag{T2.3}$$

$$\sigma_F(E) = \int_0^1 dcos\theta\, \sigma_F^{(0)}(E;U(R)+\beta_2 P_2(cos\theta)f(R)), \tag{T2.4}$$

(3) Coupling to a Spin System

$$H_{int} = f(R)(J_+ + J_-) \tag{T2.5}$$

$$\sigma_F(E) = \frac{1}{4j}\sum_{k=0}^{2j}\sigma_F^{(0)}(E;U(R)+2(j-k)f(R)), \tag{T2.6}$$

(4) Coupling to IBM States : SU(3) Limit

$$H_{int} = \alpha(R)[s^\dagger d_0+d_0^\dagger s]+\sum_{m=-2}^{2}\phi_m(R)d_m^\dagger d_m\,, \tag{T2.7}$$

$$\sigma_F(E) = \sum_{k=0}^{N}\frac{(-1)^k}{(2k+1)}\frac{N!}{(N-k)!}\sum_{m=0}^{k}\frac{(-1)^m}{(k-m)!m!}\sigma_F^{(0)}(E;U(R)+\frac{N-3m}{2\sqrt{N(N+3/2)}}f(R)) \tag{T2.8}$$

5. RECENT DEVELOPMENTS

5.1. DETERMINATION OF THE BARRIER DISTRIBUTION

As already mentioned, Stelson[15] and Rowley et al.[16] have recently attempted to determine the barrier distribution directly from the experimental data. They assumed that the fusion cross section σ at energy E is given by

$$\sigma(E) = \int_0^\infty \sigma^{(0)}(E,B)D(B)dB \quad , \tag{17}$$

where $\sigma^{(0)}(E,B)$ is the fusion cross section (summed over all partial waves ℓ) for a single barrier B. The function $D(B)$ represents the barrier distribution to be determined from the data by inversion. It obeys the unitarity condition $\int_0^\infty D(B)dB = 1$.

One can show that

$$\frac{1}{\pi R^2}\frac{d^2(E\sigma)}{dE^2} = D(E) \tag{18}$$

if one uses the classical formula eq.(5) for $\sigma^{(0)}(E,B)$. If one uses Wong's formula eq.(4) instead, then

$$\frac{1}{\pi R^2}\frac{d^2(E\sigma)}{dE^2} = \bar{D}(E) \equiv \int_0^\infty D(B)G(X)dB = \sum_\alpha w_\alpha G(X_\alpha) \tag{19}$$

$$G(X) = \frac{2\pi}{\hbar\Omega}\frac{e^X}{(1+e^X)^2}, \tag{20}$$

with $X = \frac{2\pi}{\hbar\Omega}\cdot(E-B)$. Inserting the experimental data of σ in eqs.(18) or (19), one can directly determine the barrier distribution. The dots with bars in fig.5 show the results in ref.16 for $^{16}O+^{154}Sm$ system. These experimental barrier distributions were then compared with those predicted by the coupled-channels calculations that include the coupling to the 2^+ state of ^{154}Sm (top figure), or to the 4^+ as well as 2^+ states of ^{154}Sm (bottom figure). In fig.5, these theoretical predictions are indicated by the thick vertical lines in the case where the quantum tunneling is ignored, i.e. when eq.(18) is used, and by a smoothed curve

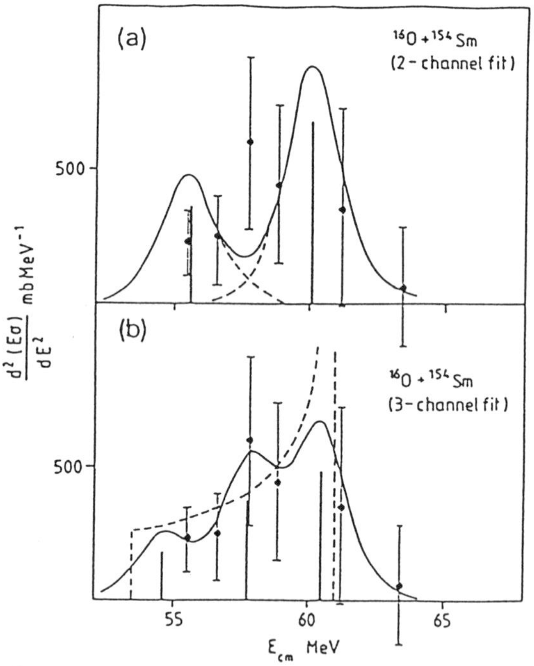

Fig.5 Distribution of the potential barriers (taken from ref. 16)

when the quantum tunneling is taken into account, i.e. when eq.(19) is used. The dashed line in the bottom figure represents the case when there is no truncation of the rotational band. It is remarkable that the coupled-channels calculations using only 3-channels can reproduce not only the rather broad width well but also the general trend of the experimental barrier distribution. Recently, much refined data were taken[28]. The analysis[29] of those refined data shows that the barrier distribution very well resembles the classical barrier distribution given by the dashed line in the bottom figure. In this connection, note that the barrier distribution of the 3-channels formalism already resembles very well the barrier distribution in the classical limit except for a few local structures.

Preceding ref.16, Stelson[15] has attempted to determine the barrier distribution for the ^{58}Ni+^{58}Ni, ^{64}Ni+^{64}Ni and ^{58}Ni+^{64}Ni fusion reactions based on eq.(18). He found a very broad barrier distribution for each system, which is characterized also by a sharp cutoff at the low energy end. Also, the barrier distribusion was found to be nearly flat for the first two systems. He related the lowest potential barrier to the separation energies of the valence neutrons and to the distance at which the potential barrier separating the single particle potentials of the projectile and of the target becomes low enough to allow for neutrons to flow between the colliding nuclei as they approach each other. He thus claimed that the neutron-flow between the colliding partners is the principal mechanism to enhance the fusion cross section and that inelastic excitations play only a secondary role. This contradicts the results of the coupled-channels calculations shown in Fig.2 , which indicates that the inelastic excitations play the major role in enhancing the sub-barrier fusuon cross section. Further studies will be required to disentagle these contradictory conclusions.

5.2. THE IBM APPROACH - THE SU(3) CHAIN

The geometrical model of Bohr and Mottelson has been used in almost all theoretical studies so far made of the effects of inelastic excitations on the fusion cross section. As an alternative, here I wish to refer to one of our recent works, in which the effects of nuclear structure are incorporated in the description of fusion reactions near the Coulomb barrier using the Interactiong Boson Model[27].

We consider the SU(3) limit. In the adiabatic limit, the total fusion cross-section is given by eq.(T2-8), where N is the boson number and the coupling form factor $f(R)$ is identified with $F_2(r)$ given in eqs.(11) through (13) in the geometric model.

First, I wish to remark on the similarity between the IBM and the geometric approaches. As an example, let us consider the case when the boson number N is one. The allowed angular momenta are then $L = 0$ and 2, and the total fusion cross section is given by

$$\sigma = \frac{2}{3}\sigma(E; U(R) - \frac{1}{\sqrt{10}}f(R)) + \frac{1}{3}\sigma(E; U(R) + \sqrt{\frac{2}{5}}f(R)) \qquad (21)$$

The eigenvalues of the coupling matrix and the corresponding weights in eq.(16) are therefore $\lambda_1 = -0.316$, $\omega_1 = 0.667$ and $\lambda_2 = 0.633$, $\omega_2 = 0.333$. The corresponding geometric model is the rotational excitation where $J_{max} = 2$. Similar to eq.(21), the total cross section is given by a sum of two terms, where the eigenvalues of the coupling matrix and the corresponding weights are easily found to be $\lambda_1 = -0.327$, $\omega_1 = 0.652$ and $\lambda_2 = 0.613$, $\omega_2 = 0.348$ from the zeros of $P_4(z)$ and the 4-points Gauss-Legendre quadruture formula. The fusion cross section predicted by the IBM model and by the geometric model are thus expected to resemble each other very much.

In Figure 6 we exhibit the schematic behavior of the fusion cross-section for different values of the boson number N for the $^{16}O + ^{154}Sm$ scattering. The dotted line is for $N = 0$ (bare potential), while the dashed and the solid lines are for $N = 1$ and for $N = 11$ (the actual boson number of ^{154}Sm), respectively. In calculating these cross-sections Woods-Saxon potential, whose depth, range and the diffuseness parameters are $V_0 = 68$ MeV, $R_o = 8.39$ fm and $a = 0.95$ fm, respectively, and $\beta_2 = 0.35$ were used for the nuclear part of the heavy-ion interaction potential and for the deformation parameter of ^{154}Sm. This value of the deformation parameter was deduced from the B(E2) values given by Raman et al.[30]. The experimental data taken from Stokstad et al.[31] are indicated by black dots. The arrow at 58.2 MeV indicates the barrier height. A salient feature of Figure 6 is the quick convergence of the calculated results as N (or the number of channels) increases. $N = 1$ and the large values of N yield the same cross section.

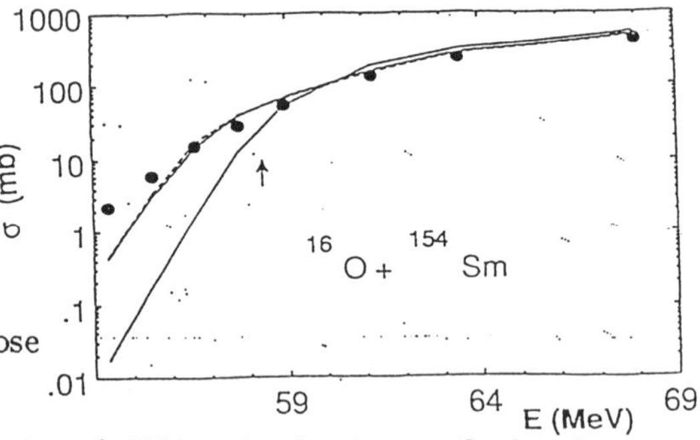

Fig.6 IBM calculations of the fusion excitation function

The coupling of the translational motion to intrinsic degrees of freedom affects not only the total fusion cross section but also the spin distribution of the coumpound system. Naturally, the insensitivity of the fusion excitation function to the details of intrinsic degrees of freedom made it popular to analyze the spin distribution of the compound nucleus in order to distinguish different models . Experimentally, the average angular momentum

$$< L >= \left[\sum_L L\sigma_L\right] / \sigma, \qquad (22)$$

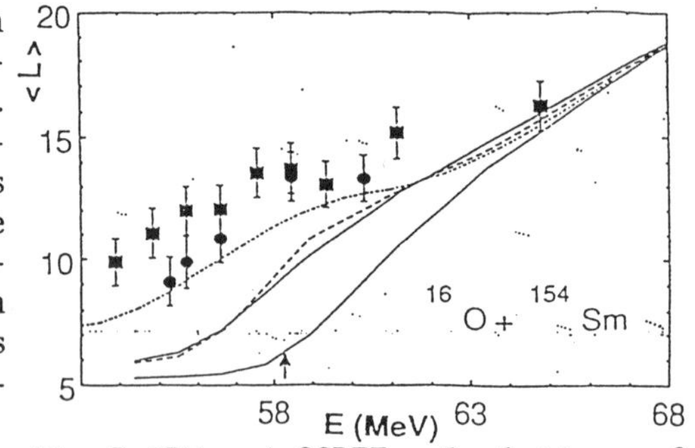

Fig.7 IBM and CCDEF calculations of the mean spin of the compound nucleus

can be determined reasonably accurately from the gamma ray multiplicities[32,33].

In Figure 7, we present the average angular momentum for the system ^{16}O + ^{154}Sm calculated with $N = 0$ (the dotted line), with $N = 1$ (the dashed line), and with $N = 11$ (the solid line). The same potential parameters as in Figure 6 were used. Once again one observes that the average angular momentum converges rather quickly. Our analysis confirms the results of Ref. 16 and 21, where it was shown that it is sufficient to include a small number of channels in the coupled-channel calculations to obtain the correct distribution of potential barriers. One generally expects, in a rather model independent way, the coupled-channel calculations to converge rapidly as the basis size is increased. Vandenbosch and his collaborators recently determined mean angular momentum for this reaction by two different techniques. They either observed low-lying gamma transitions in the evaporation residues or detected the residues directly by using an electrostatic deflector[33]. The first method has been applied with higher sensitivity than before and the second method enabled measurement at a lower energy than is possible with gamm-ray tagging techniques. The mean L-values they deduced from these recent measurements are substantially higher than their previous data[32] and, as shown in Fig.7, a bump is prominant in the new data. The circles represent data[33] determined from Ge-tagged multiplicities and the squares those determined from the deflector-tagged multiplicities. Also shown is a full coupled-channels result (the dash-dotted line), reported in Ref. 33, calculated using the coupled-channels code CC DEF[34], including quadrupole and hexadecapole deformations and octupole vibrations. One observes that the full coupled-channels calculation is in better agreement with experiment, but, like our calculation, it still underestimates the $< L >$ values near the maximum barrier energy, indicated by an arrow. The difference between the CC DEF result and our calculation is easy to understand. Assuming the exact SU(3) limit we neglected hexadecapole deformations. Similarly we excluded the octupole vibrations in our model. To describe octupole collectivity in IBM necessitates the introduction of f ($L = 3$) bosons.

6. FUTURE PROSPECTS

6.1. FUSION OF A HALO NUCLEUS

Many recent experimental, as well as theoretical, studies have shown that nuclei near the neutron drip line have very extended neutron distribution[35] and very low lying vibrational modes of excitation[36]. The neutron halo and the soft dipole mode of excitation of ^{11}Li are the typical examples of the novel structure. These novelties are expected to introduce many interesting new phenomena to heavy ion collisions, which will certainly be explored in the near future both experimentally and theoretically.

As an example, here I discuss the fusion reaction of ^{11}Li with ^{208}Pb, which we have recently studied[37]. Several problems have to be settled in order to discuss the features of the fusion cross section induced by a halo nucleus. First, it is essential to construct a realistic interaction Hamiltonian. We constructed it in such a way that it reflects the separation energy of the valence neutrons. For this purpose, we used the double folding procedure. We assumed that the ^{11}Li consists of a ^{9}Li core and two neutrons around it. The interaction potential then splits into two parts. The one describes the interaction between the core part of the projectile and the target, while the other represents the interaction of the valence neutrons with the target. We replaced the former by the standard Akyüz-Winther potential[38], which is used with success in describing the scattering of normal beams. To determine the latter, we assumed the Michigan Three-Range-Yukawa (M3Y) force[39] as the nucleon-nucleon interaction. The crucial point is how to represent the density of the valence neutrons. We expressed it in an exponential form,

$$\rho_{\Delta N}(s) = \rho^{(0)}_{\Delta N}\frac{e^{-2\kappa s}}{s^2}; \quad \rho^{(0)}_{\Delta N} = \frac{\kappa}{2\pi}\cdot \Delta N, \tag{23}$$

where s is the distance between the center of the valence neutrons and that of the core part of the projectile. The diffuseness parameter κ is related to the separation energy of the valence neutrons $\epsilon_{\Delta N}$, the number of valence neutrons ΔN, the mass number of the core part of the projectile A_C and the nucleon mass M_N. In eq.(23) and in what follows, the lower suffix ΔN is used to designate quantities refering to the valence neutrons.

Fig.8 shows the total s-wave potential obtained as a function of the separation distance between ^{11}Li and ^{208}Pb in the barrier region. Each line or symbol corresponds to different values of $\epsilon_{\Delta N}$ as indicated in the figure. The density of ^{208}Pb and the Coulomb radius parameter were chosen in a standard way(see ref.37 for details). The figure shows that the potential barrier drastically decreases when the separation energy is 0.2 MeV, which is close to the empirical two neutron separation energy of ^{11}Li[40].

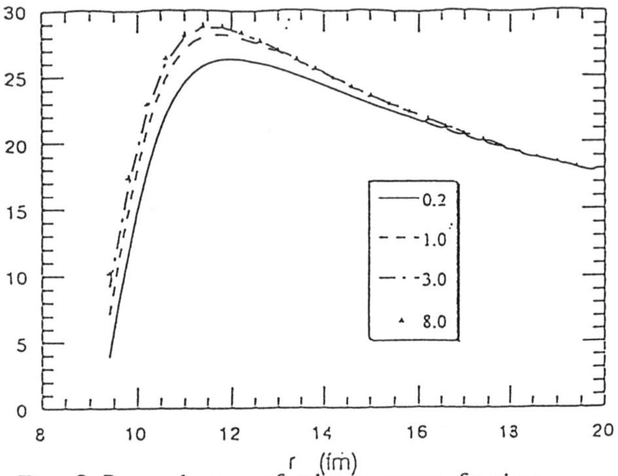

Fig.8 Dependence of the s-wave fusion fusion barrier for a halo nucleus collision on the separation energy of valence neutrons

The next step is to calculate the fusion cross section by taking the effects of coupling of the translational motion to soft vibrational modes of excitation into account. We took the effects of a soft dipole mode of excitation, i.e. the dipole excitation of the ^{9}Li core relative to the valence two neutrons, into account. We calculated the fusion probability for each partial wave based on the constant coupling model by fixing the strength of the coupling Hamiltonian to its value at a particular value of r, $r = r_c$. Then, following the same procedure to derive the Wong formula[9], the fusion cross section becomes

$$\sigma_F(E) = R_B^2 \cdot \frac{\hbar\Omega}{2E} \cdot \sum_i |A_{i\alpha}|^2 ln[1 + exp(\frac{2\pi}{\hbar\Omega} \cdot (E - \epsilon_i - V_B))], \tag{24}$$

where V_B, R_B and $\hbar\Omega$ are the barrier height, the barrier position and the potential curvature for the s-wave, respectively. Values of these parameters can be easily determined once the potential is obtained in the way described above. Two crucial parameters are the excitation energy of the soft dipole oscillation and the strength of the coupling Hamiltonian. They determine the energy shift ϵ_i and the transition matrix elements $A_{i\alpha}$ from the initial state α to intermediate states i.

We estimated the excitation energy of the soft dipole mode based on the sum rule method[41] for the vibration between the ^{9}Li core and the valence di-neutrons. Assuming that the energy weighted E1 strength is exhausted by a single state, we obtain the following simple relationship between the excitation energy of the dipole oscillation and the separation energy of the valence neutrons,

$$E_{SDM} = 6\epsilon_{\Delta N} \tag{25}$$

where the lower suffix SDM stands for soft dipole mode. It is interesting to recall that the peak energy of the photo-disintegration of deuterons is also proportional to the separation energy, though the coefficient is 2 instead of 6[42]. We determined the coupling Hamiltonian based on the linear coupling model. The coupling form factor then consists of three terms; i.e. the nuclear and the Coulomb couplings between the core of the projectile and the target, and the nuclear coupling between the valence neutrons in the projectile and the target. This contrasts to the case of usual stable beams, where the coupling form factor generally consists of two terms only, i.e. the nuclear and the Coulomb excitation terms. Consequently, the charge product $Z_C Z_T$, Z_C and Z_T being the atomic numbers of the core part of the projectiel and the target, respectively, is not the only parameter that determines the strength of the coupling Hamiltonian at the barrier position. Another important feature is that the nuclear coupling between the valence neutrons in the projectile and the target extends far outside when the valence neutrons form an extended halo. Note also that the strength of the coupling form factor is inversely proportional to the square root of E_{SDM}. This indicates that the excitation energy of the dipole oscillation has to be small in order to induce a sizable effect on fusion.

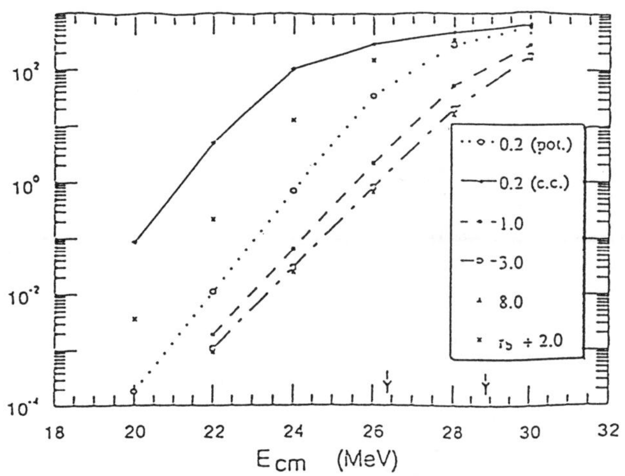

Fig.9 Enhancement of the sub-barrier fusion cross section in a halo nucleus collision (cluster model)

Fig.9 shows our numerical results of the fusion excitation function. A two-channel model was adopted by truncating the excitation of the soft dipole mode by its one-phonon state. A huge enhancement (up to 4 orders of magnitude) of the fusion cross section is observed for the case of $\epsilon_{\Delta N}$=0.2 MeV (the solid line) over that for $\epsilon_{\Delta N}$=3 and 8 MeV. Similarly to the case of the potential barrier, the enhancement due to the valence neutrons is small in the latter cases. The right and the left arrows along the abscissa indicate the position of the s-wave potential barrier for the cases of $\epsilon_{\Delta N}$=8.0 and 0.2 MeV, respectively. The enhancement of the fusion cross section originates partly from this lowering of the fusion barrier and partly from the coupling of the translational motion to the soft dipole mode. In order to see each effect separately, the dotted line with open circles show results calculated by the potential model, i.e. by ignoring the coupling to the dipole mode of excitation. All of these calculations were performed by assuming the barrier position $r_b = r_c$. It is clear that the results depend on the choice of r_c in the constant coupling model. In order to have an idea on how the numerical fusion cross section depends on the choice of r_c, we have included in fig.9 the fusion cross section calculated by assuming $r_b + 2fm = r_c$ (the stars). The separation energy was taken to be 0.2 MeV. Though the difference between the solid line and the stars indicate the necessity of more complete calculations to obtain more quantitative prediction of the fusion cross section, it is clear that the halo neutrons significantly enhance the fusion cross section at low energies.

Experimentally, the separation energy of the di-neutrons in ^{11}Li is found to be a small number around 200 to 300 keV. One can thus expect a large enhancement of the fusion cross section at low energies for the actual system involving ^{11}Li. It would be very interesting to experimentally confirm this large enhancement. In general, our calculations imply that a large enhancement of the fusion cross section can be expected only when then separation energy of the valence neutrons is extremely small.

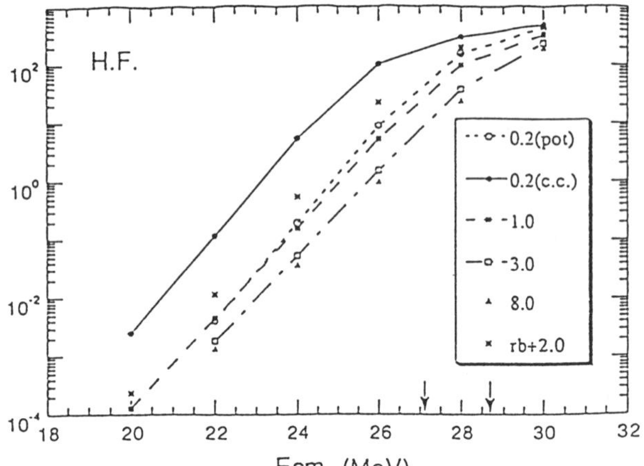

Fig.10 Same as fig.9 but independent particle model

Before I move to the next subject, I wish to comment on the dependence of our results on the intrinsic properties of the valence neutrons. So far an extreme cluster model was assumed for the two valence neutrons in ^{11}Li. The actual situation lies somewhere between the cluster model and the independent particle model. We therefore calculated the fusion cross section by assuming the other limit, i.e. the limit of the independent particle model where the density of the valence neutrons are determined based on the Hartree-Fock calculations with the SGII force[43].

We adjusted the central part of the Hartree-Fock potential to realize various values for the binding energy of the valence neutrons. The results are shown in Fig. 10. We find that the enhancement factor of the fusion cross section in this model for $\epsilon_{\Delta N}$=0.2 MeV is one to two order of magnitude smaller than that in the extreme cluster model. This shows that the effects of the valence neutrons on the fusion cross section very much depend on their intrinsic structure.

6.2. SYNTHESIS OF SUPERHEAVY ELEMENTS BY NEUTRON RICH BEAMS

I now turn to the problem of synthesizing superheavy elements by heavy ion collisions. The interest is to see whether some advantages can be gained by using a neutron rich radioactive beam as the projectile. This study was motivated by the fact that the cross section for producing heavy elements by stable beams gets very small with increasing atomic number. The cross section for making $Z = 109$, the heaviest element so far made, by the ^{58}Fe + ^{209}Bi reaction is only $10pb$[44]. It is therefore highly desirable to find a new way that has a larger cross section in order to synthesize heavier elements.

Superheavy elements are purely shell stabilized systems. Hence, it is essential to create them as cold as possible. But, near barrier fusion cross section gets larger with increasing bombarding energy. One therefore has to seek for an optimum condition. The system which has the lowest excitation energy for the same amount of fusion cross section is the most favorable.

At first sight, the large enhancement of the sub-barrier fusion cross section discussed in the previous section tempts us to use a halo nucleus as the projectile. However, we will soon encounter two difficulties. The first is the problem of a target. Since all the halo nuclei so far produced are light nuclei, an appropriate target that leads to a large enough atomic number of the compound system would not be easily found. Otherwise, heavier halo nuclei have to be made first, which may not be easy either. The second problem is the Q-value obstacle. Fig.11 shows the Q-value for the ^{A}Li + ^{238}U $\rightarrow$ $^{238+A}$Am reactions. As we see, the Q-value becomes a large positive number if one uses an isotope close to the neutron drip line. The large positive Q-value might destroy the shell-stabilized fission barrier needed to make superheavy elements. The extra Q-value of about 15 MeV shown in Fig.11 by using ^{11}Li as the projectile is much larger than the gain of about 5 MeV due to the enhancement of the sub-barrier fusion cross section shown in Fig.9. This implies that it is not promising to use a nucleus which is too close to the neutron drip line for the

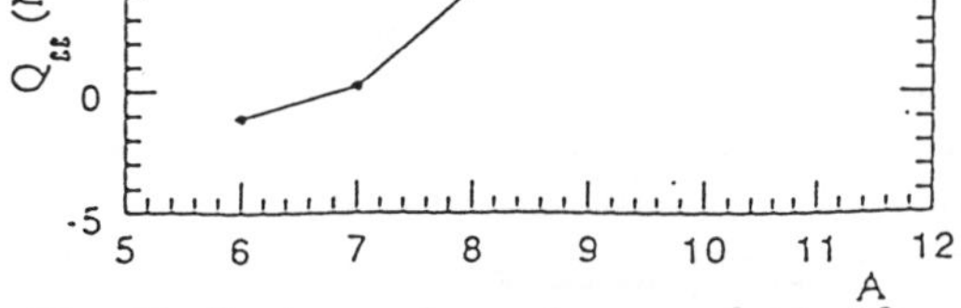

Fig.11 Isotope dependence of the Q-value

purpose of synthesizing a superheavy element, though as discussed below, other factors including the survival probability might alter the situation.

Here I discuss more in detail the features of reactions involving neutron rich beams by taking the synthesis of Z=111 element as an example(see refs.45 and 46 for more details). I compare the fusion of ^{46}K +^{238}U with that of ^{41}K +^{238}U. These belong to the hot fusion family. I then compare them to a corresponding cold fusion, ^{64}Ni +^{209}Bi. We choose K isotopes as the projectile of our study because alkali beams are the most promising for the RI beam project of ISOLDE type, which is planned in the Japan Hadron Project. The expected beam intensities relevant to the present study are 10^{12} pps of ^{41}K, 10^7 pps of ^{46}K for the proton beam of 1 GeV and 1 μA intensity and 10^{12} pps of ^{64}Ni.

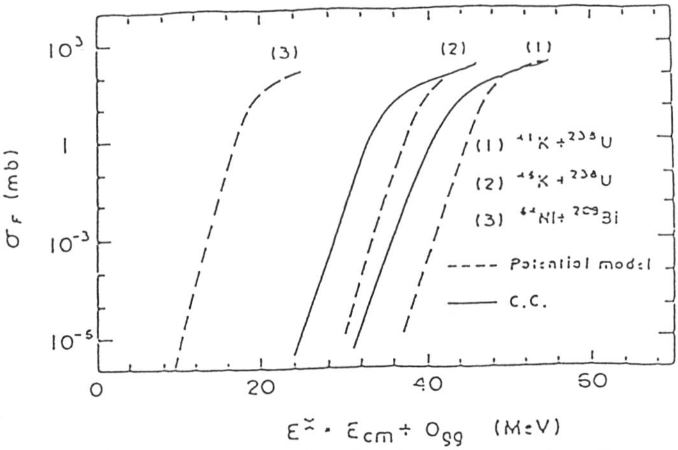

Fig.12 Comparison of the fusion excitation function for a neutron rich unstable beam and a normal beam. A corresponding cold fusion system is also shown.

Fig.12 shows our theoretical prediction of the fusion excitation function for the three systems mentioned above. The abscissa is the excitation energy of the compound nucleus E^*, instead of E_{cm}. The dashed lines are the fusion excitation functions predicted by the potential model, where the Akyüz-Winther potential was assumed for the real part of the optical potential. The solid lines are the results of the coupled-channels calculations which take the rotational excitation of ^{238}U into account. The truncation was made at $J_{max} = 6$. The results do not depend so much on the choice of J_{max} as long as $J_{max} \geq 2$. The Möller-Nix mass formula[47] was used to estimate the Q-value Q_{gg}. As we expect, the coupling to the rotational excitation significantly enhances the fusion cross section at low energies.

The figure shows that one can obtain the same amount of fusion cross section with less excitation energy of the compound system by replacing the normal beam ^{41}K with a neutron rich beam ^{46}K. The decrease of the excitation energy is caused by the combined effects of the lowering of the s-wave fusion barrier V_B and the increase of the absolute value of the ground state Q value, which is negative. Actually, ^{46}K was chosen as the projectile, because ^{46}K + ^{238}U is the optimum system that has the lowest $Q_{eff} = V_B + Q_{gg}$ value among all ^{A}K + ^{238}U reactions. Note that the fusion barrier height V_B monotonically decreases with the neutron number, while Q_{gg} takes a minimum value at a certain isotope, which is not close to the neutron drip line.

It is remarkable that the fusion cross section for the ^{46}K $+^{238}$U collision is larger than that for the $^{41}K+^{238}U$ by about 5 orders of magnitude at low energies. This is sufficient to compensate for the low intensity of the radioactive beam. As we show later, neutron rich beams generally have, in addition, an extra advantage that the survival probability against fission becomes larger. Fig.12 shows, however, that the cold fusion with a normal stable nucleus as the projectile, ^{64}Ni $+^{209}$Bi, has lower excitation energy than the ^{46}K $+^{238}$U for the same fusion cross section.

The extra push energy, which is expected to be required for such heavy systems as we are considering now to fuse, becomes a problem. We estimate the fusion cross section in the presence of the extra push energy by simply shifting the incident energy by the amount of extra push energy E_{XX}: $\sigma_F(E_{cm}) = \sigma_F^{(cc)}(E_{cm} - E_{XX})$. We estimated E_{XX} based on a phenomenological systematics for asymmetric systems[48](see refs. 45 and 46 for details). It predicts that the extra push energy is larger for cold fusion systems than that for hot fusion systems. Consequently, the excitation energy of the ^{64}Ni $+^{209}$Bi system and that of the ^{46}K $+^{238}$U system for the same fusion cross sections become closer than shown in fig.12 if one takes the extra push energy into account, though the former is still smaller than the latter. Note, however, that this conclusion totally relies on our phenomenological estimate of the extra push energy.

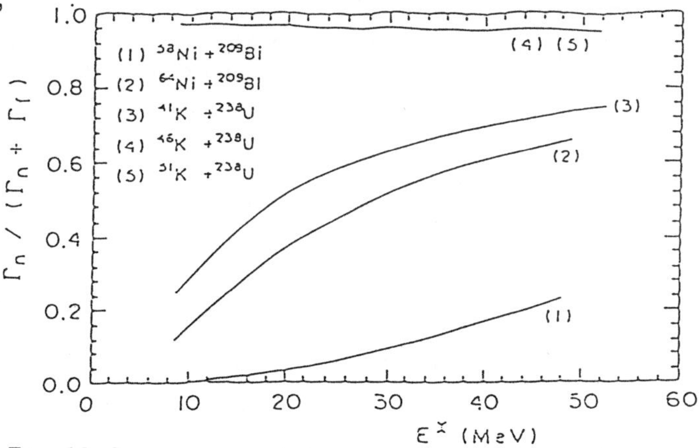

Fig.13 Comparison of the survival factor for various systems

Besides the fusion cross section, the survival probability of the (fused) compound system against fission is a crucial factor in choosing a favourable system to synthesize superheavy elements. We calculate the survival factor $\frac{\Gamma_n}{\Gamma_n+\Gamma_f}$ based on the standard formula[49]

$$\Gamma_n/\Gamma_f \approx 2TA^{2/3} \cdot exp[(V_f - S_n)/T]/K_0 \tag{26}$$

where $K_0 = \frac{\hbar^2}{2mr_0^2} \approx 14.2\ MeV$. The neutron separation energy S_n and the fission barrier height V_f can be estimated from the mass formula of Möller and Nix[47] and their calculations of V_f. The result is shown in fig.13 for five systems, ^{58}Ni $+^{209}$Bi, ^{64}Ni $+^{209}$Bi, ^{41}K $+^{238}$U, ^{46}K $+^{238}$U and ^{51}K $+^{238}$U, as a function of the excitation energy. The survival factor for the ^{46}K $+^{238}$U and the ^{51}K $+^{238}$U systems cannot be distinguished on the figure. The figure clearly shows that the survival probability becomes larger for systems with larger number of neutrons. This suggests that the (HI,xn) hot fusion processes, x being typically 4, do not suffer much from the reduction of the cross section due to the survival factor $(\frac{\Gamma_n}{\Gamma_n+\Gamma_f})^x$ if one uses neutron rich beams because of their large $\frac{\Gamma_n}{\Gamma_n+\Gamma_f}$ values. One can thus use higher bombarding energy for such systems in order to compete with the (HI,1n) cold fusion process.

6.3. ADDENDUM TO SECTS. 6.1. AND 6.2 (FUTURE DEVELOPMENTS)

Many problems remain to be studied concerning the fusion of a halo nucleus. I mention some of them here. The enhancement of the fusion cross section due to coupling to nuclear intrincic degrees of freedom is very sensitive to the detailed properties of the coupling form factor. The microscopic calculations of the transition densities for various multipole modes of excitation in a halo nucleus show that they look very different from those in the standard Tassie model[50]. It would thus be very interesting to see what happens if one uses the transition density based on microscopic calculations. So far we considered the dynamical effects of only the dipole oscillation. Microscopic calculations of the response functions show that there exist also soft monopole and soft quadrupole modes of excitations in halo nuclei. They will introduce additional enhancements of the fusion cross section at low energies. In all the calculations of the fusion cross sections shown above, we have totally left out the effects of neutron transfer or break-up process during the colllision. These effects will play very important role in the fusion of a halo nucleus. If the valence neutrons are transfered between the halo nucleus and the target, it will strongly enhance the fusion cross section. On the other hand, if they are lost by a break-up process, they might even hinder the fusion cross section. It would be extremely interesting to study which is more likely. I pointed out the sensitivity of the fusion cross scetion to the intrinsic properties of the valence neutrons in the halo nucleus. This suggests that detailed studies of the fusion cross section would provide useful information about the intrinsic structure of the valence neutrons.

Table 3. Reaction Q-values

$^{46}K+^{238}U$	$^{44}K+^{240}U$	-3.97 MeV
	$^{45}K+^{239}U$	-1.45 MeV
	$^{47}K+^{237}U$	1.56 MeV
	$^{48}K+^{236}U$	0.82 MeV
$^{41}K+^{238}U$	$^{39}K+^{240}U$	-7.57 MeV
	$^{40}K+^{239}U$	-5.61 MeV
	$^{42}K+^{237}U$	1.15 MeV
	$^{43}K+^{236}U$	5.94 MeV
$^{64}Ni+^{209}Bi$	$^{62}Ni+^{211}Bi$	-6.88 MeV
	$^{63}Ni+^{210}Bi$	-5.49 MeV
	$^{65}Ni+^{208}Bi$	-1.31 MeV
	$^{66}Ni+^{207}Bi$	0.77 MeV

Many problems have to be clarified also in order to identify an optimum system for synthesizing superheavy elements. One of the most crucial parameters is the maximum excitation energy, above which the shell stabilized fission barrier disappears. Our analyses using ^{A}K + ^{238}U reactions showed that one can choose a neutron rich beam that optimizes the effective Q-value, though it still leads to a higher excitation energy than a normal beam system that belongs to a cold fusion family. Also, the survival probability becomes large if one uses neutron rich beams. In order to conclude whether the optimum neutron rich beam (hot fusion) is more unfavorable than the optimum stable beam system (cold fusion), we have to study carefully the effects of the extra push energy and of nucleon transfer channels on fusion. In this connection, it should be noted that the Q-values of the one- and the two-neutrons pick-up reaction channels in all the three systems which we have discussed in sect. 6.2 are positive (see table 3). These channels will increase the fusion cross section at low energies. Though the discussions here were restricted to the use of radioactive beams, it could be interesting to examine stable beams as well, e.g. it would be interesting to compare the ^{48}Ca beam with the ^{50}Ca beam. The lesson from our analyses is that one might be able to opimize the fusion cross section, the effective Q-value, the extra push energy and the survival probability by a careful choice of a suitable isotope. Concerning the use of a halo nucleus, it would be premature to conclude that halo nuclei are not good choice because of the Q-value obstacle before we study the effects of the nucleon transfer on the fusion cross section and the survival probability.

6.4. DYNAMICAL EFFECTS ON DECAY: THE EVAPORATION SPECTRUM

The decay of a metastable state is a basic question in many fields of physics[51]. A hot subject in this connection is to clarify the effects of other degrees of freedom on the decay probability, which is intimately related to the dynamical effects on fusion. As my last subject, here I address the evaporation spectrum as an example of this problem.

The energy spectrum of particles emitted from a hot compound nucleus is usually analyzed in terms of the Weisskopf formula[52],

$$\sigma(E_\nu) \propto E_\nu\left(1 - \frac{B_\nu}{E_\nu} \right) \cdot e^{-(E\nu - B_\nu)/T_0}, \tag{27}$$

where the suffix ν refers to different species of the emitted particles, such as n, p, α ..... One of the underlying assumptions in such calculations is to estimate the decay probability, which is replaced by the inverse formation probability in actual calculations, by a potential model. This approach is successful in fitting experimental evaporation spectrum at high energy side by appropriately choosing a value for the temperature parameter T_0. In some cases, however, it significantly underestimates the cross section at low energy side. The energy spectrum of α-particles emitted in the ^{14}N+^{154}Sm scattering at E_{lab} =490 MeV is one of such examples[53].

From the discussions in the previous sections, it is natural to expect that the evaporation spectrum near and/or below the potential barrier is strongly influenced by the properties of the residual nucleus through dynamical effects on the decay probability. As an example, we studied the effects of a surface vibration of the residual nucleus on the energy spectrum of the evaporation particles by generalizing the zero point motion formula so as to take also the thermal fluctuation into account[26,24]. The thermal fluctuation should be taken into account because the particles are emitted from a hot compound nucleus. We have assumed that the residual nucleus is in thermal equilibrium with a heat bath of temperatute T. Similar to eq.(T2-2), the decay probability is found to be given by an integration of the decay probability for a given amplitude of the vibration. An important change is that the amplitude of the oscillation $\alpha^{(0)}$, the amplitude of the quantal zero point motion, is replaced by

$$\alpha_{eff}(T) = \alpha^{(0)} \cdot \frac{1}{\sqrt{tanh\frac{\hbar\omega}{2T}}} \quad . \tag{28}$$

One can easily show that $\alpha_{eff}(T)$ reduces to the amplitude of the quantum zero point motion at $T = 0$, and it becomes the amplitude of the thermal fluctuation at high temperatures.

Fig.14 shows an example of our calculations (see ref.26 for details). It shows the energy spectrum of alpha particles, ^{16}O and ^{40}Ca, which are evaporated from a compound nucleus with temperature T. The residual nucleus was assumed to be ^{188}Pt for all cases. The lines (1), (2) and (3) represent the energy spectrum for the cases, where the decay probability was calculated in the potential model, by taking the quantum fluctuation of the quadrapole surface vibration of ^{188}Pt into account, i.e. by replacing the decay probability ($1-\frac{B_\nu}{E_\nu}$) in eq.(27) by the value given by the quantal zero point motion formula, and by taking the thermal fluctuation into account as well. In the latter, the temperature was assumed to be $T = T_0 = 1MeV$. The figure shows that the coupling of evaporated particles with a surface vibration of the residual nucleus indeed enhances the cross section at low energy side when heavy

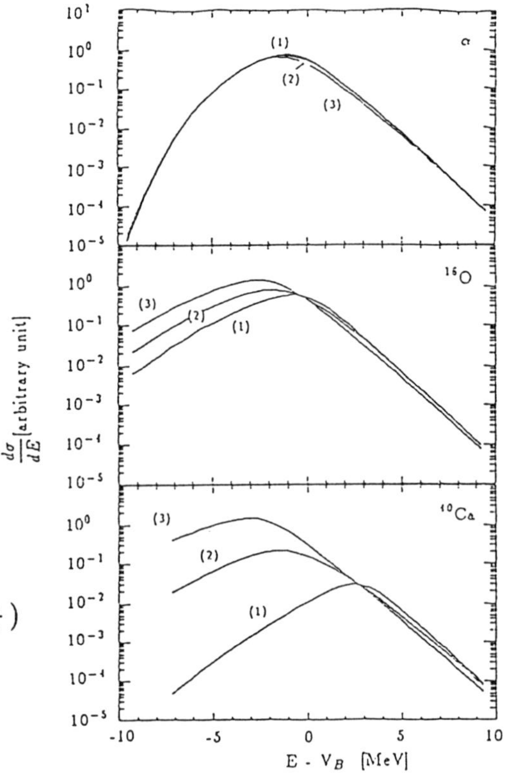

Fig.14 Effects of a surface vibration of the residual nucleus on the spectral distribuion of evaporation particles

particles are emitted. Also, it shows that the thermal and the quantal fluctuations of the vibration have comparable effects on the energy spectrum at $T = 1MeV$.

Concerning the energy spectrum of α-particles which motivated our study, however, one needs to find another explanation. For the α-particles emitted in the $^{14}N+^{154}Sm$ reaction, it could be more physical to study the effects of the deformation of the residual nucleus, though it is questionable that a static deformation of the nucleus in the ground state persists at such high temperatures (3 to 4 MeV) relevant to the present problem. Notice also that the effects of rotational coupling to inclusive decay probability is independent of the temperature if the deformation is rigid[24].

In general, it would be interesting to study the dynamical effects on the decay process. As a schematic case, Fig.15 shows the energy surface in a two-dimensional system, where the potential curvature in the direction of the y-coordinate depends on the value of the x-coordinate. The figure illustrates the possibility that this leads to a finite effective potential barrier(solid line) in the x-direction even if the bare potential (the dashed line) is flat. As an example, such dynamical effects might change the limiting angular momentum of a nucleus against fission[54].

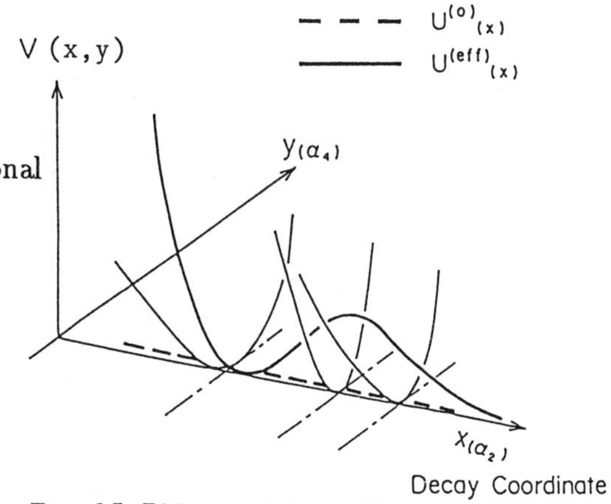

Fig.15 Illustration of an effective potential barrier arising from the coupling to other degrees of freedom

Acknowledgement: I should like to thank Y. Alhassid, A.B. Balantekin, A. Iwamoto, F. Michel, M.A. Nagarajan, G. Reidenmeister, N. Rowley and H. Sagawa for useful discussions. I am very grateful to H.J. Kim for many useful comments in preparing the manuscript.

1. M. Beckerman, Rep. Prog. Phys. 51 (1988) 1047.

2. G.R. Satchler, Proc. of the International Nuclear Physics Conference, São Paulo, Brazil, Aug. 20-26 (1989), World Scientific, p.541.

3. H. Esbensen, Nucl. Phys. A352 (1981) 147.

4. C.H. Dasso, S. Landowne, and A. Winther, Nucl. Phys. A405 (1983) 381; A432 (1985) 555.

5. H.J. Krappe, K. Mohring, M.C. Nemes, and H. Rossner, Z. Phys. A314 (1983) 23.

6. A.B. Balantekin and N. Takigawa, Ann. Phys. (NY) 160(1985) 441.

7. R. Betts, invited talk in this conference

8. P.-H. Heenen, Phys. Letts. B99(1981)298.

9. C.Y. Wong, Phys. Rev. Lett. 31(1973)766.
10. M. Beckerman et al., Phys. Rev. C23(1981)1581.
11. C.E. Aguir, V.C. Barbosa, L.F. Canto and R. Donangelo, Phys. Letts. B201(1988)22.
12. G.R. Satchler, M.A. Nagarajan, J.S. Lilley and I.J. Thompson, Ann. Phys. 178(1987)110.
13. T. Udagawa, T. Tamura and B.T. Kim, Phys. Rev. C39(1989)1840.
14. A.B. Balantekin, S.E. Koonin and J.W. Negele, Phys. Rev. C28(1983)1565.
15. P.H. Stelson, Phys. Letts. B205(1988)190.
16. N. Rowley, G.R. Satchler, and P.H. Stelson, Phys. Letts. B254(1991)25.
17. K.Tsuda and N. Takigawa, to be published
18. H. Esbensen and S. Landowne, Nucl. Phys. A492(1989)473
19. C.H. Dasso and G. Pollarolo, Phys. Letts. B155(1985)223; C.H. Dasso and A. Vitturi, Phys. Letts. B179(1986)337.
20. A. Iwamoto and N. Takigawa, Phys. Letts. B219(1989)176
21. N. Takigawa and K. Ikeda, Proc. Symp. on Many Facets of Heavy Ion Fusion Reactions, ed. W. Henning et al (Argonne: ANL-PHY-86-1) (1986), pp. 613-620.
22. N. Takigawa, F. Michel, A.B. Balantekin and G. Reidenmeister, Phys. Rev. C44 (1991) 477.
23. N. Takigawa, F. Michel, A.B. Balantekin and G. Reidenmeister, Phys. Letts. B262 (1991) 199.
24 N. Takigawa, Y. Alhassid and A.B. Balantekin, preprint, 1991
25. M.A. Nagarajan, A.B. Balantekin, and N. Takigawa, Phys. Rev. C34(1986)894.
26. M. Abe and N. Takigawa, Phys. Letts. B209 (1988)149
27. A.B. Balantekin, J.R. Bennett and N. Takigawa, Phys.Rev.C44 (1991) 145.
28. D.J. Hinde et al., private communication
29. N. Rowley, invited talk at the IV-th International Conference of Nucleus-Nucleus Collisions, Kanazawa, June 10-14, 1991.
30. S. Raman, et. al., At. Data Nucl. Data Tables 36(1987)1.
31. R.G. Stokstad, et. al., Phys. Rev. C21(1980)2427.
32. R. Vandenbosch, et. al., Phys. Rev. C28(1983)1161; S. Gil, et. al. Phys. Rev. C31(1985)1752.
33. S. Gil, et. al., Phys. Rev. C43(1991)701.
34. J. Fernandez Niello, C.H. Dasso, and S. Landowne, Comput. Phys. Commun. 54(1989)409.

35. I. Tanihata et al., Phys. Letts. B206(1988)592.

36. P.G. Hansen and B. Jonson, Europhys. Lett. 4(1987)409.

37. N. Takigawa and H. Sagawa, Phys. Letts. B in press; N. Takigawa, H. Sagawa and T. Shinozuka, talk at the IV-th International Conference of Nucleus-Nucleus Collisions, Kanazawa, June 10-14, 1991.

38. Ö. Akyüz and A. Winther, Enrico Fermi International School,1979.

39. G.F. Bertsch, J. Borysowicz, H. McManus and W.G. Love, Nucl. Phys. A284(1977)399; G.R. Satchler and W.G. Love, Phys. Rep. 55(1979)183; G.R. Satchler, Nucl. Phys. A329(1979)233.

40. C. Thibault et al., Phys. Rev. C12 (1975) 644; J.M. Wouters et al., Z. Phys. A331(1988) 229; T. Kobayashi et al., preprint.

41. Y. Suzuki, to appear in Nucl.Phys.; H. Sagawa and M. Honma, Phys. Letts. 251B(1990)17.

42. J.M. Blatt and V.F. Weisskopf, Theoretical Nuclear Physics (Wiley, New York,1952)p.609.

43. Nguyen van Giai and H. Sagawa, Phys. Letts. 106B(1981)379.

44. P. Armbruster, Ann. Rev. Nucl. Part. Sci. 35(1985)135; F.P. Heßberger et al., Proc. 5th Int. Conf. on Nuclei Far From Stability, Rosseau, Canada, 1987, ed. I.S. Towner, p.786; G. Münzenberg, Rep. Prog. Phys. 51(1988)57; Yu. Ts. Oganessian, Nucl. Phys. A488(1988) 65c.

45. N. Takigawa and T. Shinozuka, Proc. of the 18th INS Int. Syms. on Physics with High-Intensity Hadron Accelerators, Tokyo, March 14-16, 1990, (World Scientific)p.223;

46. N. Takigawa, F. Michel, A.B. Balantekin and G. Reidenmeister, Inst. Phys. Conf. Ser. No.110(1991)p.205.

47. P. Möller and J.R. Nix, At. Data and Nucl. Data Tables 39(1986)225

48. J.P. Blocki et al., Nucl. Phys. A459(1986)145

49. R. Vandenbosch and J.R. Huizenga, Nuclear Fission (Academic press, New York, 1973), p.229

50. H. Sagawa contribution to the IV-th International Conference of Nucleus-Nucleus Collisions, Kanazawa, June 10-14, 1991.

51. P. Hänggi, P. Talkner and M. Borkovec, Rev. Mod. Phys. 62(1990) 251; and references therein.

52 T. Ericson, Advances in Physics 9(1960)425.

53 G. Nebbia et al., Phys. Letts. B176(1986)20; J.B. Natowitz, invited talk at the IV-th International Conference of Nucleus-Nucleus Collisions, Kanazawa, June 10-14, 1991.

54. S. Cohen, F. Plasil and W.J. Swiatecki, Ann. Phys. 82(1974)557.

ORBITING AND FUSION–FUSION PHENOMENA

ORBITING: TOWARD A UNIFIED MODEL OF NUCLEUS-NUCLEUS COLLISION OR EXCITATION OF NUCLEAR MOLECULAR STATES IN HEAVY ION COLLISIONS?

A. Ray* and D. Shapira

Oak Ridge National Laboratory, Oak Ridge, TN 37831

*Partial support provided by Joint Institue for Heavy-Ion Research, Oak Ridge, TN 37831

Abstract It has been suggested that damped reaction reaction products at backward angles from the $^{28}Si+^{12}C$, $^{24}Mg+^{12}C$ and $^{20}Ne+^{12}C$ reactions result from the formation of long-lived orbiting complexes. This suggestion was originally made on the basis of the observation that the total integrated binary yields are much larger than the predictions of compound nucleus calculations although they follow a $1/\sin\theta_{c.m.}$ angular distribution near 180°. Subsequent studies have confirmed the noncompound origin of these yields through the observations of entrance channel dependence of back-angle yields and strong selectivity of natural parity states. However, it has also been found that the formation of an orbiting complex and the associated back-angle "anomalies" are not universal. It appears that the formation of orbiting complex dominates the back-angle yields for systems having a small number of open channels. This is similar to the systematic behavior that was observed for the anomalous back-angle elastic scattering and suggests that the same physical process might be responsible for back-angle anomalies in elastic and inelastic channels. Recently new insight has been obtained about orbiting reaction mechanism from the measurement of the complete gamma-ray angular correlations for the $^{28}Si+^{12}C$ orbiting system. It has been found that the measured density matrices for the $^{12}C(2^+)$ and ^{28}Si states are almost diagonal with respect to the direction of motion of the fragments. The measured density matrices and spin alignments are consistent with the picture of formation of a long-lived dinuclear complex undergoing orbiting, bending and wriggling motions, but not with those obtained from statistical compound nucleus or sticking model calculations.

INTRODUCTION

Elastic scattering of light heavy ions at energies above the Coulomb barrier typically exhibits a diffraction pattern and an exponential drop of yield[1] for angles larger than the grazing angle, characteristic of collisions between strongly absorbed particles. However anomalously large back-angle yields, strongly oscillatory angular distribution and highly regular gross structure modulated by much narrower structures in the excitation function have been observed[2] at back-angles for certain systems such as $^{28}Si+^{12}C$, $^{28}Si+^{16}O$ etc. Such back-angle "anomalies" are not universal features of light heavy ion systems in this energy regime and have been seen only in systems having a small number of open channels. Measurements[3,4,5] of the full spectrum of binary products at backward angles have shown that even larger yields exist for excited products and the spectra show the characteristics of a fully damped process. Fully-damped binary yields from $^{28}Si+^{12}C$ [3], $^{20}Ne+^{12}C$ [4], $^{24}Mg+^{12}C$ [5] at backward angles showed large cross-sections and the total integrated binary yields follow $1/\sin\theta_{c.m.}$ angular distribution near 180°. The integrated yield (20-30 mb) is a factor of 5 to 6 larger than the predictions of a compound nucleus model. These observations led to the suggestion[3] that a dinuclear orbiting reaction mechanism is responsible for damped back-angle yields, although a compound nucleus reaction mechanism can also lead to a $1/\sin\theta_{c.m.}$ angular distribution. According to the dinuclear orbiting model, instead of fusing, the two nuclei form a dinuclear intermediate state and then break apart producing a large yield in the entrance channel compared to statistical model calculations. A constrained phase-space model with a dinuclear sticking-model configuration has been developed by Shivakumar et.al[6]. This model, usually referred to as the equilibrated orbiting model, successfully explains the observed large, damped binary yields in several cases such as $^{28}Si+^{12}C$ [6], $^{28}Si+^{14}N$ [7]. Measurements of the $^{35}Cl+^{12}C$ [8] systems show smaller yields of ^{12}C and similar products. The analysis[8] suggests that asymmetric fission is adequate to account for the yields of ^{12}C and similar products. Asymmetric fission has also been found to be the dominant process in the emission of damped binary products from heavier systems such as $^{32}S+^{24}Mg$ [9]. There has been a controversy[8,9,10] in the literature regarding the equilibrated orbiting model and fusion-fission model. Both models are statistical. The controversy is mainly focussed on the calculation of absolute yields of damped ^{12}C and similar products. The statistical model calculation for the yield of these heavy fragments is sensitive to the level density parameter and to the optical model parameters used to calculate transmission coefficients, critical angular momentum and spin distribution of the compound nucleus. So it is difficult to reach a conclusion about the reaction mechanism on the basis of absolute yields only.

Entrance Channel Dependence

One of the criteria to distinguish between compound nuclear and orbiting reactions should be the entrance channel dependence of back-angle yields. If a compound nucleus has been formed, then its decay should be independent of how it was formed. On the other hand, the decay of an orbiting complex should depend strongly on how it was formed. We rely on this criterion to distinguish between compound nucleus and orbiting reactions, because this criterion is free from the uncertainties of calculation of absolute cross section for the light ion compound nucleus reaction. The entrance channel dependence of back-angle yields of $^{28}Si+^{12}C$ and $^{24}Mg+^{16}O$ reactions has been studied[11]. In these experiments, the ^{40}Ca nucleus was formed at the same excitation energy (47.9 MeV) by using $^{28}Si+^{12}C$ and $^{24}Mg+^{16}O$ reactions at $E_{c.m.}$=34.5 MeV and 31.8 MeV respectively. The spin distribution of the compound nucleus ^{40}Ca formed by these two reactions should be very similar, because the critical angular momenta for the two systems $^{24}Mg+^{16}O$ and $^{28}Si+^{12}C$ at $E_{c.m.}$=31.8 MeV and 34.5 MeV as obtained from trajectory model calculations[12] are the same (ℓ_{crit} =20$\hbar$) within 5%. The measured fusion cross sections for the $^{24}Mg+^{16}O$ system[13] and $^{28}Si+^{12}C$ system[14] at the corresponding energies are equal within 10%. So the critical angular momentum and spin distributions of ^{40}Ca formed by two different reactions are very nearly equal. It is found that the back-angle yields of oxygen and carbon particles from ^{40}Ca depend strongly on whether ^{40}Ca is formed by $^{24}Mg+^{16}O$ or $^{28}Si+^{12}C$. The absolute oxygen and carbon cross sections predicted by the statistical model[15] are sensitive to the values of ℓ_{crit} and diffuseness parameters of the compound nucleus spin distribution, because the contributions to the cross section for these heavy particle emissions come only from the tail of the compound nucleus spin distribution. However, the ratio of the two cross sections is comparatively insensitive to variation of different parameters. The ratio of the oxygen to carbon cross-sections decreases by (20-25)% for 10% decrease of ℓ_{crit} and it decreases by (30-35)% if the diffuseness parameter of the compound nucleus spin distribution is decreased by a factor of 4. In Fig. 1(a), we plot the ratio of back-angle oxygen to carbon yield as a function of exit channel excitation energy for both $^{28}Si+^{12}C$ and $^{24}Mg+^{16}O$ systems at $E_{c.m.}$=31.8 MeV and 34.5 MeV respectively and find large entrance channel dependence of back-angle yields. The observation of large entrance channel dependence implies that a noncompound process is responsible for the back-angle carbon and oxygen yields from $^{28}Si+^{12}C$ and $^{24}Mg+^{16}O$

reactions. We also know that a long-lived intermediate state is responsible for the back-angle yields because of the observation of $1/\sin\theta_{c.m.}$ angular distribution of the back-angle yield.

Beck et al.[8] measured yields of ^{12}C and similar products from the $^{35}Cl+^{12}C$ system and found small yields of the order of a few mb. Although such small yields are in agreement with the fusion-fission model, it is important to find out whether there is any significant entrance channel dependence of these reactions. The entrance channel dependence of back-angle yields of oxygen and carbon particles from ^{47}V nucleus has been studied in the present work. The ^{47}V nucleus has been formed by $^{35}Cl+^{12}C$ and $^{31}P+^{16}O$ reactions at E_{lab}=180 MeV and 135.6 MeV respectively. So the excitation energy of ^{47}V nucleus is 59.1 MeV for both systems and the ℓ_{crit} =(27-28)$\hbar$ for the two systems as obtained from a trajectory model calculation[12] is the same within 4%. In Fig. 1(b), we plot the ratio of oxygen to carbon cross-section for several MeV wide bins in excitation energy versus the corresponding exit channel excitation energy. We do not find any significant entrance channel dependence in our comparison of $^{31}P+^{16}O$ with $^{35}Cl+^{12}C$ systems. The total integrated yields for oxygen and carbon particles for both $^{31}P+^{16}O$ and $^{35}Cl+^{12}C$ reactions follow a $1/\sin\theta_{c.m.}$ angular distribution in the center-of-mass frame near 180°. So the observations of small yields and lack of any significant entrance channel dependence give strong evidence that the back-angle yields of ^{12}C and similar products from $^{35}Cl+^{12}C$ and $^{31}P+^{16}O$ reactions come predominantly from a fusion-fission process.

It is very interesting to contrast the dominance of entrance-channel dependence and orbiting yield in the $^{28}Si+^{12}C$ and $^{24}Mg+^{16}O$ systems and the lack of these effects in the $^{31}P+^{16}O$ and $^{35}Cl+^{12}C$ systems. The equilibrium orbiting model[6] can explain the large damped orbiting yield of binary fragments from the $^{28}Si+^{12}C$ system, but fails to explain the observed entrance-channel dependence of the orbiting yield and the lack of large damped orbiting yield for the $^{31}P+^{16}O$ and $^{35}Cl+^{12}C$ systems. One of the main assumptions of the equilibrium orbiting model[6], the formation of an equilibrated sticking configuration of orbiting nuclei, is questionable in view of these observations. However orbiting characteristics[3,11] as seen in certain systems are clearly manifestations of a long-lived non-compound process taking place in those systems. Although there is no overall quantitative understanding of these effects, we can still make a few interesting observations. It is known that a few partial waves near the grazing angular momentum are responsible for the back-angle binary fragment yields. An estimate of the number of open channels available to carry away the grazing angular momentum has been made[16,17] following the method suggested by Abe and Haas[18]. The number of open channels for each system was obtained by a triple summation over all possible two-body mass partitions in the exit channels, over all possible angular-momentum couplings, and over all possible energy distributions between fragments and normalized to 1 mb of the grazing partial wave flux. The calculation shows that for the $^{35}Cl+^{12}C$ and $^{31}P+^{16}O$ systems, the number of open channels available to carry away the grazing angular momenta is about 300 times larger than the number available in the $^{28}Si+^{12}C$ and $^{24}Mg+^{16}O$ systems where the orbiting process dominates the back-angle binary yields. So it appears that a long-lived noncompound orbiting process is associated with the back-angle yields of light heavy ion systems having a small number of open channels. This is similar to the systematic behavior that was observed for the anomalous elastic scattering at backward angles studied for a very large number of systems.

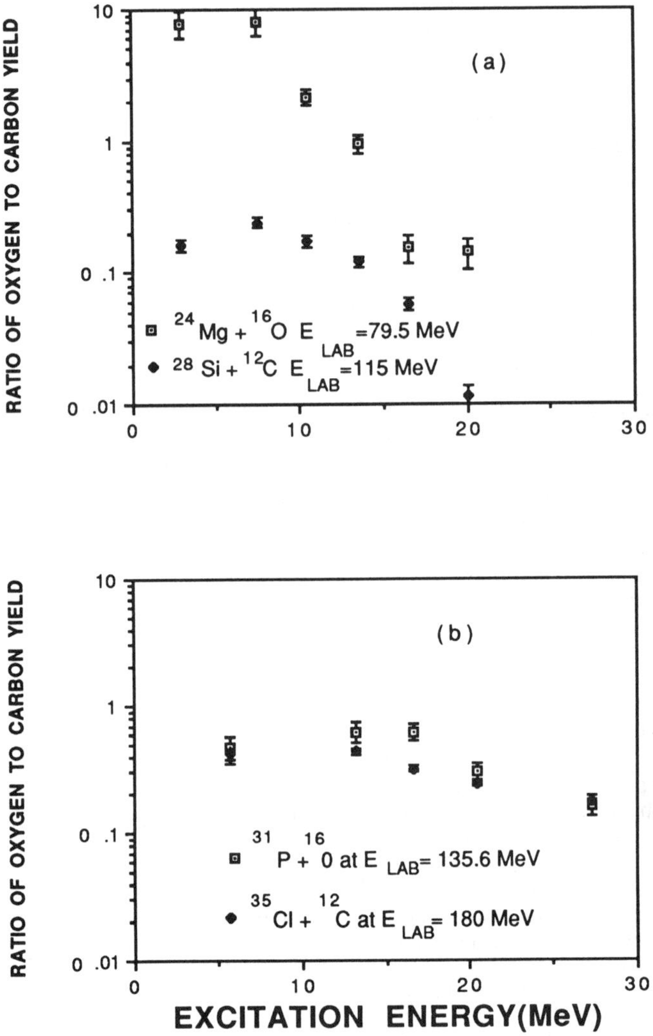

Fig. 1 Ratio of oxygen to carbon cross-sections versus excitation energy for the (a) $^{24}Mg+^{16}O$ and $^{28}Si+^{12}C$ systems, (b) $^{31}P+^{16}O$ and $^{35}Cl+^{12}C$ systems.

Selective Population of Natural Parity States

Let us now turn to other distinguishing features of orbiting reactions. Dunnweber et al[5] observed selective population of natural parity states by measuring back-angle oxygen,nitrogen and carbon spectra from $^{24}Mg+^{12}C$ orbiting reaction. Such selective population of natural parity states supports a dinuclear reaction mechanism for the back-angle yield of binary products from $^{24}Mg+^{12}C$ reaction. In compound nucleus reactions, both natural and unnatural parity transitions are allowed. Austern and Blair[19] have shown that in multi-step direct-reaction processes involving projectiles with zero spin, final states having only natural parities are selected if the following conditions are satisfied.

A) The collective coordinates of the system vary very slowly in the intermediate states so that the transition amplitude from the initial nuclear state to the final nuclear state does not depend explicitly on the collective coordinates of the system. As a result the relevant Green's function is independent of the nuclear Hamiltonian involving collective coordinates of the system.

B) The displacement operator of collective deformations of the nuclear surface commutes with the Green's function so that the interactions can be combined to give only one spherical harmonic of a single argument. Austern and Blair[19] have shown that this is a reasonable assumption for strongly absorbed projectiles which have short wavelength.

If we consider the formation of a long-lived dinuclear intermediate state where the two nuclei are connected by a narrow neck, then the collective coordinates should vary slowly and the above conditions are satisfied approximately. The observation of selective population of natural parity states can therefore be viewed as further evidence for orbiting.

Naive Sticking Model

We have learnt so far that the back-angle yields of certain light heavy ion systems having a small number of open channels are dominated by an orbiting process which enhances back-angle binary yields, creates strong entrance channel dependence of back-angle binary yields and preferentially populates natural parity states. A naive orbiting picture is shown in Fig.2. In this picture, the two orbiting nuclei form a sticking configuration, rotate like a rigid body and then break apart preferentially into the entrance channel. A strong alignment of the spin angular momentum vectors of the two nuclei along the normal to the reaction plane is expected from such a picture. In this model, if ℓ_i is the initial angular momentum, I_1, I_2 are the moments of inertia of the two nuclei, R is the distance of separation of the two nuclei and μ is the reduced mass of the system, then the amount of spin transferred to the nuclei from orbital angular momentum $\Delta\ell$, is

$$\Delta\ell = \left(\frac{I_1+I_2}{I_1+I_2+\mu R^2}\right)\ell_i \tag{1}$$

and the ratio of the fragments' spins is

$$\frac{J_1}{J_2} = \left(\frac{A_1}{A_2}\right)^{5/3} \tag{2}$$

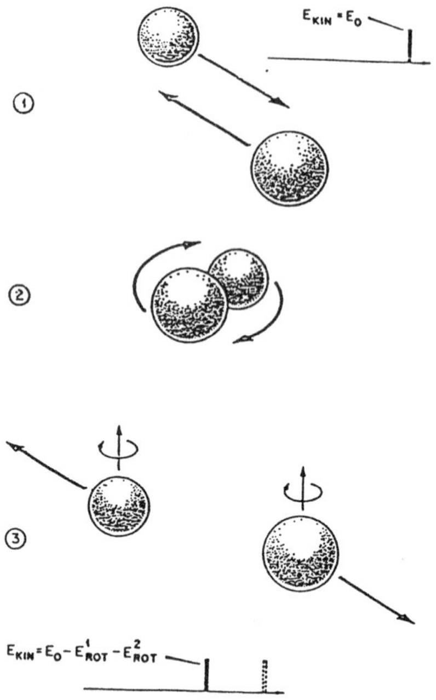

Fig.2. Classical Orbiting Picture.

where $J_1+J_2=\Delta\ell$. Here J_1 and J_2 are the spins of the outgoing nuclei and A_1 and A_2 are their masses. So in this simple model, both outgoing nuclei gain a fraction of the incident orbital angular momentum and hence should come out excited. γ-ray angular correlation experiments can measure spin alignment and find out whether such an orbiting sticking model really works.

γ-ray Angular Correlation Study

A gamma-ray angular correlation experiment was performed[20] initially at $\theta_{c.m.}=180°$ for the $^{28}Si+^{12}C$ reaction at $E_{c.m.}=43.5$ MeV to learn about the spin alignment and division of excitation energy between the fragments. Fig. 3 shows the γ-ray angular correlation at $\theta_{c.m.}=180°$. We find that m=0 magnetic substate with respect to the beam axis is predominantly excited in this reaction. The contributions[20] of m=± 1, ± 2 magnetic substates are less than 5% of the contribution of m=0 magnetic substate. A nonequilibrium division of excitation energy[20] was also observed. It was found that ^{12}C comes out in its ground state most of the time (70%) resulting in the entire excitation energy (15-20) MeV appearing in ^{28}Si. Some qualitative aspects of the reaction mechanism are fairly clear. During the reaction process, the kinetic energy of relative motion and the initial orbital angular momentum is converted into internal excitation energy of the reaction intermediate, by either collective excitation or particle exchange. Part of the initial orbital angular momentum are transferred into the internal rotational degrees of freedom. If m_C, m_{Si} be magnetic substates of ^{12}C and ^{28}Si with respect to the beam axis, then from the conservation of angular momentum $m_C+m_{Si} = 0$. However m_C and m_{Si} individually need not be zero. The observed large enhancement of m=0 magnetic substate for both ^{12}C and ^{28}Si with respect to the beam axis is very striking.

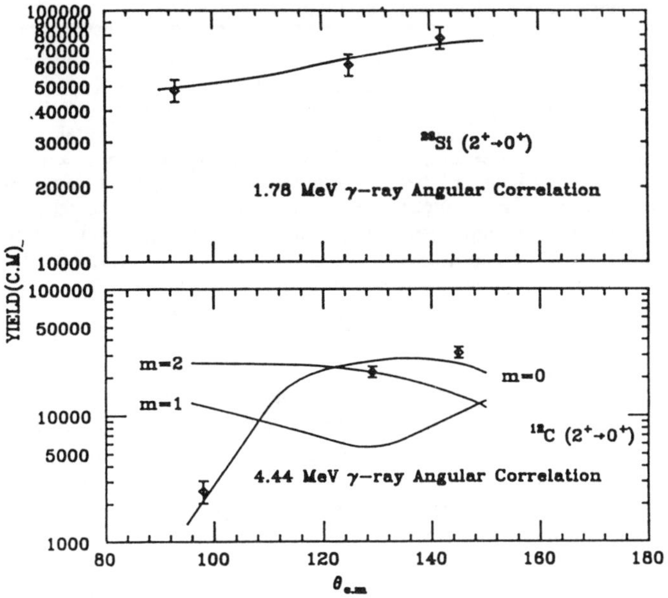

Fig. 3 The angular correlation for 1.78 MeV γ rays in $(-20<Q<0)$ MeV and that for 4.44 MeV γ-rays in $(-20 \leq Q \leq -11.6)$ MeV at E(si)=142.7 MeV.

If we consider that the rotational degrees of freedom are strongly coupled and the reaction intermediate rotates as a rigid body, i.e the sticking limit has been reached, then if only radial forces are present in the breakup, the binary reaction products will come out with their spin angular momentum vectors pointing normal to the reaction plane. This is not in disagreement with our angular correlation results at $\theta_{c.m.}=180^{\circ}$. According to the sticking model, the partition of the initial orbital angular momentum into intrinsic and final orbital components for a rigid rotation is simply proportional to the relevant moments of inertia for the dinuclear system. An average orbiting angular momentum of about $23\hbar$ is expected at $E_{c.m.}=42.7$ MeV from the measured fusion cross section. For a rigid dinuclear system approximated by spheres in contact, one expects the ^{12}C spin to be $1.6\hbar$ and the ^{28}Si to be $6.5\hbar$ from eq(1) and eq(2). The latter value is in reasonable agreement with the average value of $(7\text{-}9)\hbar$ that we have deduced in a separate experiment[20]. The interpretation of the expected spin of $1.6\hbar$ (or perhaps $1.2\hbar$ if a more extended dinuclear complex is assumed) is more difficult because our system is quantized. However, our qualitative expectation from this simple picture would be that ^{12}C should preferentially be emitted in 2^+ state. This is in contrast to the observation that ^{12}C is preferentially emitted in its ground state. So although the sticking model agrees with the observed γ-ray angular correlation measurement at $\theta_{c.m.}=180^\circ$, it disagrees with the observed division[20] of excitation energy between carbon and silicon nuclei. Our observations are also in disagreement with a statistical model calculation which predicts that the ratio of the yields of the m=± 1 magnetic substates to that of the m=0 magnetic substate is 1.4 and the ratio of the yields of m=± 2 to that of the m=0 magnetic substate is 0.4. So our observation that the contributions of m=± 1, m=± 2 magnetic substates of ^{12}C are less than 5% of the m=0 contribution is completely in disagreement with the statistical model calculation.

Therefore it was important to perform γ-ray angular correlation measurements at several angles away from the beam axis to learn about the nature of the dinuclear system formed in the orbiting reaction. One expects complete alignment of fragment spins with respect to the normal to the scattering plane on the basis of the simple sticking model. From statistical model calculations, one also expects large alignment of fragment spins with respect to the normal to the scattering plane and the spin orientation of the emitted fragments should be approximately cylindrically symmetric with respect to the normal to the reaction plane. Recently, a complete γ-ray angular correlation study was performed[21] with the $^{28}Si+^{12}C$ system at $E_{c.m.}=43.5$ MeV and 48 MeV, using the Oak Ridge National Laboratory Spin Spectrometer. Three ΔE-E silicon detector telescopes were placed at $\theta_{lab}= 10^\circ$, 15° and 25° to identify carbon particles using energy and energy loss signals. Fig. 4, 5 and 6 show examples of NaI spectra in coincidence with carbon particles detected at $\theta_{lab}=15^\circ$ at $E_{c.m}$ = 43.5 MeV for -10< Q< 0 MeV, -22< Q <-10 MeV and -40<Q<-22 MeV respectively. At all angles we observe Doppler-shifted 1.78 MeV and 2.84 MeV γ ray lines from the (2^+-0^+) and (4^+-2^+) transitions in ^{28}Si, 1.34 MeV γ rays from (2^+-0^+) transitions in ^{24}Mg and 4.44 MeV γ rays from ^{12}C (2^+-0^+) transitions. The 1.34-MeV peak becomes visible at high excitation energies, and presumably arises from the cooling down of ^{24}Mg formed from unbound ^{28}Si. The large Doppler shift of the 4.44-MeV γ rays indicates that they are mostly coming from fast moving ^{12}C. The observations of comparable yields at $\theta=174^\circ$, 88°, 102.3° for both 4.44-MeV and 1.78-MeV γ rays imply rather poor spin alignments for both $^{12}C(2^+)$ and $^{28}Si(2^+)$ states, in disagreement with the simple dinuclear sticking model[3,20] for which the expectation would be that the yield at $\theta = 88^\circ$ should be a factor of 7 larger than the yield at $\theta = 174^\circ$. We observe similar poor spin alignment for $^{12}C(2^+)$ and 28Si(2^+) states at $E_{lab}=160$ MeV also.

A complete study of the γ ray angular correlation has been done. The γ-ray angular correlation function can be written as[22]

$$W(\theta,\phi)=\sum_k\sum_q[4\pi/(2k+1)]^{1/2}\,t_{kq}(J)\,R_kY_{kq}(\theta,\phi) \tag{3}$$

where θ, ϕ are polar and azimuthal angles of the direction of emission of the γ rays, Y_{kq} is the spherical harmonic function, R_k is the radiation parameter, t_{kq} is the polarization tensor and J is the spin of the emitting nucleus. We obtain values of the radiation parameter R_k from the tabulations in Ref. 23 . We now choose a coordinate frame whose Z-axis is along the direction of motion of the nucleus emitting γ rays and whose Y-axis is normal to the reaction plane and consider γ-ray emission from the (2^+-0^+) transition. Then considering[22] symmetry properties of the polarization tensor and the conservation of parity

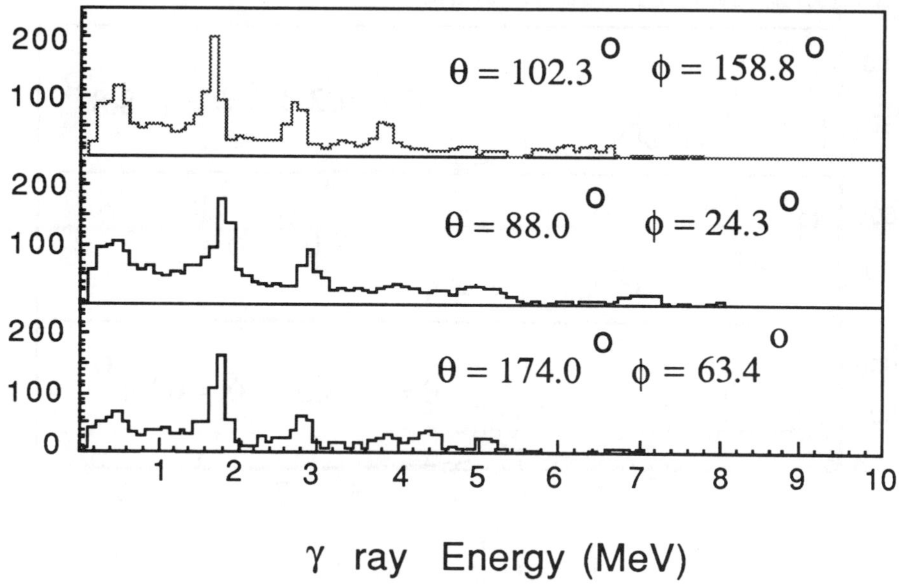

Fig. 4 Coincidence NaI spectra for the reaction $^{28}Si+^{12}C$ at $E_{c.m} = 43.5$ MeV and $-10<Q<0$ MeV at $\theta_{lab}=15^0$, $\phi_{lab}= 0^0$. Polar and azimuthal angles are measured in a coordinate frame whose Z-axis is normal to the reaction plane and X-axis is along the beam direction.

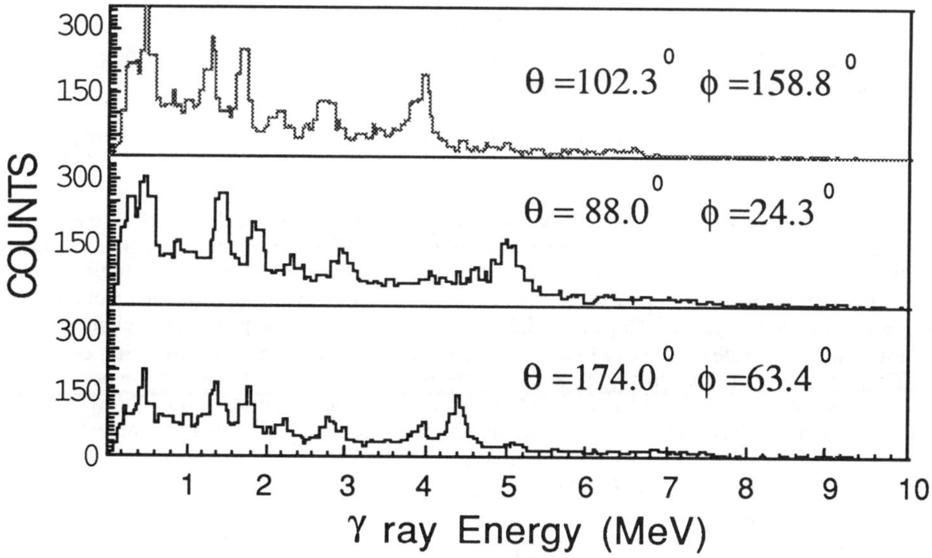

Fig. 5 Coincidence NaI spectra for the reaction $^{28}Si+^{12}C$ at $E_{c.m} = 43.5$ MeV and $-22<Q<-10$ MeV at

θ_{lab}=15°, ϕ_{lab}= 0°. Polar and azimuthal angles are measured in a coordinate frame whose Z-axis is normal to the reaction plane and X-axis is along the beam direction.

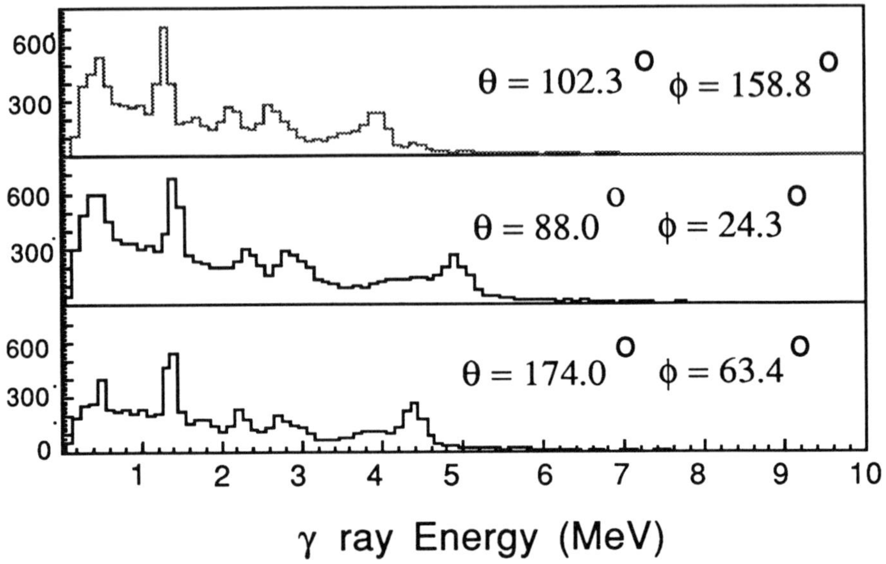

Fig. 6 Coincidence NaI spectra for the reaction $^{28}Si+^{12}C$ at $E_{c.m}$ = 43.5 MeV and $-40<Q<-22$ MeV at θ_{lab}=15°, ϕ_{lab}= 0°. Polar and azimuthal angles are measured in a coordinate frame whose Z-axis is normal to the reaction plane and X-axis is along the beam direction.

we find nine independent elements of the polarization tensor ($t_{00},t_{20},t_{21},t_{22},t_{40},t_{41},t_{42},t_{43},t_{44}$). These elements must be real numbers[22]. The elements of the density matrix ρ are given by the expression[22]

$$\rho_{mm'} = \sum_k \sum_q t_{kq}(J) < J,J,m',-m / k,q > (-1)^{J-m} \tag{4}$$

We used the 4.44 MeV and 1.78 MeV γ-ray yields from the NaI detectors of the Spin Spectrometer to determine the nine elements of each of the corresponding polarization tensors by fitting them with Eq. (3), after taking into account the finite solid angle of the NaI detectors and the Doppler shifts. The attenuation[24] due to the hyperfine interaction has been estimated to be (2-3)% and we did not correct for it. Subsequently, the elements of the density matrices of $^{12}C(2^+)$ and $^{28}Si(2^+)$ states were determined by using eq. (4). Then the matrices were rotated to a coordinate frame whose Z-axis is along the normal to the reaction plane defined by the beam axis and the direction of motion of the scattered carbon particle. The spin alignment parameter with respect to the Z-axis is defined[25] as

$$P_{zz} = \frac{3\sum_m m^2 \rho_{mm} - J(J+1)}{J(2J-1)}$$

The spin alignment parameter is conserved for stretched E2 transitions. Particle emission and non-stretched transitions tend to decrease P_{zz}. In $-10<Q<0$ MeV, ^{28}Si is predominantly (90%) excited to 6^+, 4^+ and 2^+ states. These states decay by stretched E2 cascades that preserve the alignment. So in $-10<Q<0$ MeV, the spin alignments of the $^{28}Si(2^+)$ state as obtained from the angular distribution of 1.78 MeV γ rays and those of the original parent states of ^{28}Si should be about the same. We measured the angular distribution of 4.44 MeV γ rays to determine the spin alignment and density matrices of the $^{12}C(2^+)$ state. All states of ^{12}C above 4.44 MeV are particle unbound and decay predominantly by particle emission. We find by use of the CASCADE statistical model[15] calculation that the ratio of cross-sections for populating the ^{12}C(7.564 MeV) and the ^{12}C(4.44 MeV) state is only 3% for the reactions being studied, and the probability of populating higher lying states in ^{12}C is even smaller. However the $^{12}C(2^+)$ state might be excited by the following sequential decay:
$^{28}Si+^{12}C \rightarrow ^{24}Mg^* + ^{16}O^* \rightarrow ^{24}Mg^* + ^{12}C(2^+) + ^4He$, if the ^{16}O is excited to at least (16-17) MeV. In the $(-22<Q<-10)$ MeV region, such a process is very unlikely to happen according to the CASCADE model calculations and present and previous[20] measurements. In Fig. 7, we show P_{zz} (Z-axis normal to the reaction plane) of the $^{12}C(2^+)$ state in $(-22<Q<-10)$ MeV and the $^{28}Si(2^+)$ state in $(-10<Q<0)$ MeV as a function of emission angle of the carbon particle at $E_{c.m}$= 43.5 MeV. In Fig. 8, we show P_{zz} for $^{12}C(2^+)$ and $^{28}Si(2^+)$ states in $(-16<Q<0)$ MeV and $(-26<Q<-16)$ MeV as a function of emission angle of the carbon particle at $E_{c.m}$ =48 MeV. The values and variations of P_{zz} as a function of emission angle are similar for the two bombarding energies. Error estimates were obtained by adding in quadrature the error in determining the area under the peak and the error in the relative gamma detector efficiencies. As a check, we also determined the magnetic substate population ρ_{11} of the $^{28}Si(2^+)$ state for $(-3<Q<-0.6)$ MeV using a coordinate system whose Z-axis is normal to the reaction plane. We find that ρ_{11} is consistent with zero within the statistical uncertainties, as required by the conservation of parity and rotational invariance. We also obtained small spin alignments (0.15) for $^{28}Si(2^+)$ and $^{24}Mg(2^+)$ states for $Q<-15$ MeV. However it is difficult to interpret the spin alignment of the original parent state in these Q-value bins because of nonstretched transitions and particle emissions. We find that the values of P_{zz} for $^{12}C(2^+)$ drop to about 0.10 ± 0.05 for $-40<Q<-22$ MeV and $-44<Q<-26$ MeV for both bombarding energies.

We now discuss the off-diagonal elements of the density matrices. For this purpose, we choose a coordinate frame whose Z-axis is along the direction of motion of the nucleus emitting γ rays and whose Y-axis is normal to the reaction plane. Let Z_{Si} and Z_C denote the directions of motion of ^{28}Si and ^{12}C nuclei respectively. In Table 1, we show the deduced density matrix of the $^{12}C(2^+)$ state for $\theta_{c.m} = 145^\circ$ in the range $(-22<Q<-10)$ MeV at $E_{c.m}$= 43.5 MeV. We find that the density matrix is almost diagonal with respect to the direction of motion of the particle. This means that the spin orientation of the $^{12}C(2^+)$ state is cylindrically symmetric with respect to its direction of motion Table 1 also shows the density matrix of the $^{28}Si(2^+)$ state for $\theta_{c.m} =148^\circ$ in the range $(-10<Q<0)$ MeV at $E_{c.m}$= 43.5 MeV. The orientation of the $^{28}Si(2^+)$ state is also found to be approximately cylindrically symmetric with respect to its direction of motion. Since the $^{28}Si(2^+)$ state is mostly populated by stretched E2 transitions in the $(-10<Q<0)$ MeV region, the orientations of the parent ^{28}Si states are also approximately cylindrically symmetric with respect to the direction of motion of the ^{28}Si ion.

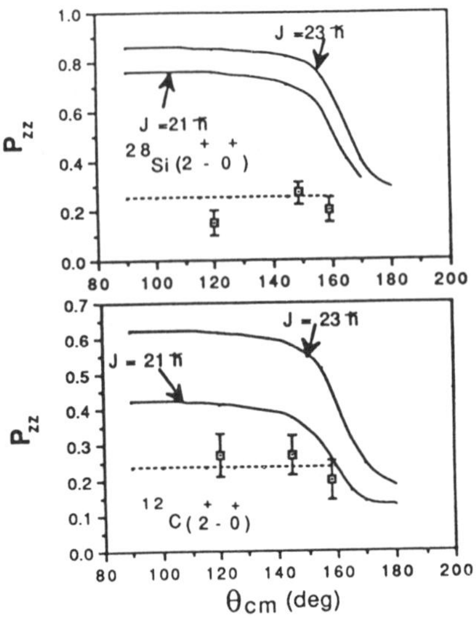

Fig. 7. The spin alignment parameter of 1.78 MeV γ rays for -10<Q<0 MeV and 4.44 MeV γ rays for -22 <Q<- 10 MeV at $E_{c.m.}$ = 43.5 MeV. The full curves and dashed lines are statistical and dinuclear orbiting model calculations as explained in the text.

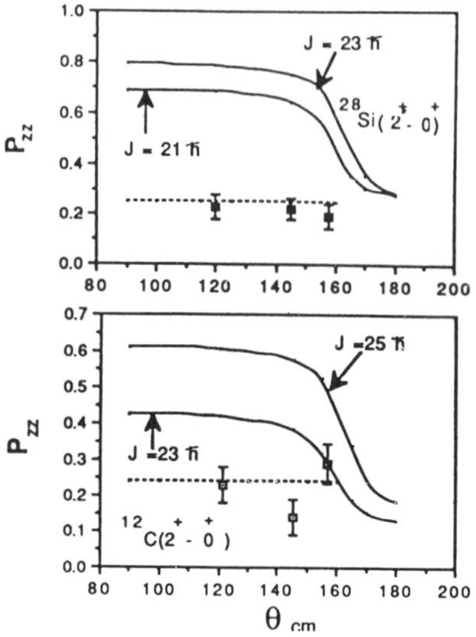

Fig. 8. The spin alignment of 1.78 MeV γ rays for -16<Q<0 MeV and of 4.44 MeV γ rays for -26<Q<-16 MeV at $E_{c.m.}$ = 48 MeV. The full curves and dashed lines are statistical and dinuclear orbiting model calculations respectively as explained in the text.

Similar behavior of the density matrix was found at other scattering angles and at the other energy. Table 2 shows the deduced density matrix of $^{12}C(2^+)$ state for $\theta_{c.m}$ =139.2° in the range -40<Q<-22 MeV at E_{cm} =43.5 MeV and for θ_{cm} =138.2° in the range -44<Q<-26 MeV at $E_{c.m}$ =48 MeV. We still find symmetry around the direction of motion of the carbon particle. However ρ_{22} along the direction of motion of ^{12}C has increased and so the value of the spin alignment parameter has decreased. We obtained the $^{12}C(2^+)$ yield by integrating over the 4.44-MeV γ rays and measuring the efficiency of detecting 4.44-MeV γ rays in the Spin Spectrometer. Table 3 shows the ratios of yields of $^{12}C(2^+)$ to ^{12}C(g.s) integrated over Q<-10 MeV at $E_{c.m}$= 43.5 MeV and Q<-15 MeV at $E_{c.m}$ = 48 MeV. Error estimates were obtained by adding in quadrature the error in determining the total number of 4.44-MeV γ rays and the error in determining the absolute efficiency of detecting 4.44-MeV γ rays in the Spin Spectrometer. The error in determining the absolute detection efficiency comes from the uncertainties in determining the target thickness, the total number of incident particles on the target and the uncertainties in the known γ-ray cross-section.

Sticking Model

The simple dinuclear sticking model[3,20] predicts that the spin alignment parameter P_{zz} (Z-axis normal to the reaction plane) is equal to unity, independent of angle. The sticking model also predicts[20] that the $^{12}C(2^+)$ state should almost always be excited. Both predictions disagree with our observations.

Table I

Density matrices for $^{12}C(2_1^+)$ and $^{28}Si(2_1^+)$ states

	Observation		Statistical Model[3]	Dinuclear Model[4]
	$^{12}C(2_1^+)$[1]	$^{28}Si(2_1^+)$[2]	$^{12}C(2_1^+)$	$^{12}C(2_1^+)$
ρ_{00}	0.27±0.02	0.29±0.019	0.31	0.29
ρ_{11}	0.28±0.013	0.23±0.011	0.235	0.24
ρ_{22}	0.08±0.02	0.13±0.014	0.11	0.11
ρ_{01}	0.014±0.012	0.005±0.017	0.012	0
ρ_{02}	-0 019±0.011	-0.028±0.01	-0.102	-0.013
ρ_{12}	0.018±.013	-0.027±.011	0.022	0
ρ_{1-1}	-0.017±.015	-0.05±.01	-0.132	-0.041
ρ_{1-2}	0.007±.01	-0.007±.009	0.0062	0
ρ_{2-2}	0.004±.012	0.012±.011	0.0095	0.0015

[1] Z_C as Z-axis, $-22<Q<-10$ MeV, $\theta_{cm}=145°$.

[2] Z_{Si} as Z-axis, $-10<Q<0$ MeV, $\theta_{cm}=148°$.

[3] Z_C as Z-axis, $-22<Q<10$ MeV.

[4] Orbiting, bending and wriggling modes, Z_C as Z-axis.

Table 2

Density Matrices for $^{12}C(2^+)$ state for large Q-value bins

	Observation		Dinuclear Model [3]
	$^{12}C(2^+)$ [1]	$^{12}C(2^+)$ [2]	$^{12}C(2^+)$
ρ_{00}	0.215 ± 0.024	0.23 ± 0.020	0.29
ρ_{11}	0.25 ± 0.014	0.22 ± 0.012	0.24
ρ_{22}	0.14 ± 0.02	0.16 ± 0.015	0.11
ρ_{01}	-0.005 ± 0.013	-0.007 ± 0.012	0
ρ_{02}	-0.0025 ± 0.011	-0.005 ± 0.010	-0.013
ρ_{12}	-0.0011 ± 0.013	-0.0095 ± 0.012	0
ρ_{1-1}	-0.037 ± 0.016	-0.033 ± 0.014	-0.041
ρ_{1-2}	0.0051 ± 0.0097	0.023 ± 0.013	0
ρ_{2-2}	0.027 ± 0.016	0.023 ± 0.0094	0.0015

[1] Z_c as Z-axis, $-40<Q<-22$ MeV, $\theta_{cm} = 139.2^\circ$, $E_{cm} = 43.5$ MeV

[2] Z_c as Z-axis, $-44<Q<-26$ MeV, $\theta_{cm} = 138.2^\circ$, $E_{cm} = 48$ MeV

[3] Z_c as Z-axis.

Table 3

Ratio of yields of $^{12}C(2^+)$ to ^{12}C(g.s.)

Lab Energy (MeV)	Lab Angle	$^{12}C(2^+)/^{12}C$(g.s.)	Statistical Model
145	10°	0.56±0.085	
145	15°	0.60 ±0.09	0.944
145	25°	0.56±0.0.85	
160	10°	0.94±0.15	
160	15°	0.87±0.13	1.24
160	25°	0.87±0.13	

Statistical Model

In Fig 7 and 8, the solid curves are the predictions of a compound-nuclear statistical model. We assume that ^{12}C and ^{28}Si form a compound nucleus and consider emission of ^{12}C from the compound nucleus. Let **J,** S_{Si}, S_C , **S** , ℓ be total angular momentum, silicon spin, carbon spin, channel spin and orbital angular momentum, respectively. Let m_J, m_{Si}, m_C, m_S and m_ℓ be their respective magnetic substates with respect to the normal to the scattering plane. Although the compound nucleus will be formed with a broad spin distribution, the contribution to the carbon yield comes from a few partial waves near the critical angular momentum ($J_{crit}=23\hbar$ at $E_{c.m}=43.5$ MeV, Jcrit $=25\hbar$ at Ec.m= 48 MeV). We obtain the following expression for the density matrix using the Hauser-Feshbach formalism[26] for a specific value of J and S_{Si}. When the carbon particle is detected at an angle $\theta_{c.m}$ with respect to the beam axis, then the elements of the density matrix for the $^{12}C(2^+)$ state in a coordinate frame whose Z-axis is normal to the scattering plane and X-axis along the beam direction is given by

$$\rho m_c m_{c'} = \sum_{S} \sum_{\ell} \sum_{m_{Si}} \sum_{m_J} \sum_{m_{J'}} < S_C, S_{Si}, m_C, m_{Si} | S, m_S >< S, \ell, m_s, m_\ell | J, m_J >$$
$$< S_{Si} S_C, m_{Si}, m_{c'} | S, m_{s'}, m_{\ell'} | J, m_{J'} > T_\ell(\varepsilon) \exp[2\{a(E_{cm} - V_c - E_{rot})\}^{1/2}] \qquad (5)$$
$$Y_{\ell,m_\ell}(\frac{\pi}{2}, \theta cm)\, Y^*_{\ell,m_{\ell'}}(\frac{\pi}{2}, \theta_{cm})\, Y_{J,m_J}(\frac{\pi}{2}, 0)\, Y_{J,m_{J'}}(\frac{\pi}{2}, 0)$$

Here a is the level density parameter, V_C is the Coulomb barrier, $E_{c.m}$ is the incident center-of-mass energy and Erot is the total rotational energy. T_ℓ is the transmission coefficient of ^{12}C for orbital angular momentum ℓ and exit channel kinetic energy ε. We obtained values of T_ℓ by using an optical model potential[27]. The normalization was obtained by multiplying the calculated density matrix by $(tr\rho)^{-1}$. A similar expression was used for ^{28}Si states also. The two curves in Fig. 7 were calculated for $J=21\hbar$, $23\hbar$ and $S_{Si}=8\hbar$ for ^{12}C and $S_{Si}=6\hbar$ for ^{28}Si. In Fig.8, we show calculations for $J=23\hbar,25\hbar$ and $S_{Si}=8\hbar$ for ^{12}C and $J=21\hbar,23\hbar$ and $S_{Si}=6\hbar$ for ^{28}Si at $E_{cm}=48$ MeV. The total rotational energy E_{rot} was calculated by assuming a prolate ellipsoidal shape[28] for the excited ^{28}Si and a spherical ^{12}C nucleus touching each other, so that the major axis of ^{28}Si is along the line joining the centers of two nuclei. The statistical model code CASCADE[15] predicts that most of the contributions to the carbon yield come from $J>23\hbar$ at $E_{c.m}=43.5$ MeV and $J>25\hbar$ for $E_{c.m.}=48$ MeV. So a weighted average over different values of J will significantly increase P_{zz} and worsen the already poor agreement with the experimental data. A weighted average over the distribution of S_{Si} obtained from CASCADE code calculations increases calculated P_{zz} by approximately 10%.

We conclude that the statistical model calculation predicts much higher alignment than observed. Variation of the level density parameter and the shape of the nuclei within reasonable limits do not change this conclusion. We also show statistical model calculations for $J=23\hbar,25\hbar$ and $S_{Si}=8\hbar$ at $E_{c.m.}=48$ MeV. Again we find large discrepancies between statistical model calculations and observations.

The discrepancies with the statistical model extend also to off-diagonal elements of the density matrix. Using Eq(5), we find that the statistical model predicts that the density matrix will be almost diagonal with respect to the normal to the reaction plane. To compare with the experimental results, we rotate the calculated density matrix to a coordinate frame where Z_C is the Z-axis and the Y-axis is normal to the reaction plane. The rotated density matrix has significant off-diagonal terms (ρ_{02}, ρ_{1-1}) as shown in the third column of Table 1 , in disagreement with our observations. The calculations shown in Table 1 have been performed for $J=23\hbar$. We also find large off-diagonal elements $\rho_{02} = -0.068$, $\rho_{1-1} = -0.085$ for the $^{12}C(2^+)$ state even for $J=21\hbar$. Table 3 shows the calculated ratios of $^{12}C(2^+)$ to ^{12}C(g.s) integrated over $Q<-10$ MeV and $Q<-16$ MeV. We find that the statistical model predicts considerably larger ratios than observed.

Dinuclear Orbiting Model

Let us consider a simple picture of two nuclei connected by a narrow neck living long compared to the damping time of their relative motion. The twisting motion[29] between them will be small compared to bending[29] and wriggling[29] modes in the case of a narrow neck[30]. If we assume that bending and wriggling modes dominate and sum incoherently over different orientations of the spin angular momentum vectors lying randomly in a plane perpendicular to the body-fixed symmetry axis, then the density matrix of $^{12}C(2^+)$ with respect to the body-fixed symmetry axis is given by

$$\begin{aligned} &\rho_{00} = 0.375,\ \rho_{11} = \rho_{-1-1} = 0.25,\ \rho_{22} = \rho_{-2-2} = 0.0625 \\ &\rho_{01} = \rho_{02} = \rho_{12} = \rho_{1-1} = \rho_{2-2} = \rho_{1-2} = 0 \end{aligned} \tag{6}$$

In the body-fixed frame,the dinuclear system breaks along the symmetry axis. Here the Coulomb energy V_C =18.1 MeV and orbital energy $V_\ell = \ell(\ell+1)/2I_R$ =6.2 MeV for $\ell=12\hbar$, where I_R is the moment of inertia of the relative motion. So in a space-fixed frame, the trajectory of emitted ^{12}C will be rotated by $\pm\tan^{-1}(V_\ell/V_C)^{0.5} = \pm 30^\circ$. with respect to the body-fixed symmetry axis. We rotate the density matrix (Eq.4) by $\pm 30^\circ$ and take an average. The results, shown in the last column of Table 1, are in better agreement with the experimental data than are the statistical model calculations as shown by the small off-diagonal (ρ_{02}, ρ_{1-1}) elements. A similar analysis for the ^{28}Si state also predicts approximate symmetry of ^{28}Si states with respect to the direction of motion of ^{28}Si. We also obtain alignment parameters of 0.24 and 0.25 (as shown by dashed lines in Fig 7,8), with respect to the normal to the reaction plane for $^{12}C(2^+)$ and ^{28}Si states, in agreement with our observation.

Our simple classical dinuclear model does not predict any variation of alignment parameter and density matrix with the angle of emission of carbon particles. This model is not quantitatively valid near $\theta_{c.m} = 180^\circ$, where we expect significant variation of alignment parameter and elements of density matrix with the angle of emission of carbon particles. However this model qualitatively agrees with the observed γ-ray angular correlation at $\theta_{c.m} = 180^\circ$ because, according to this model, the spin angular momentum vectors will lie in a plane perpendicular to the beam axis when ^{12}C is detected at $\theta_{c.m} = 180^\circ$, in agreement with observation[20]. Table 2 shows that in the more inelastic Q-value bin (-40<Q<-22) MeV, the measured value of ρ_{22} along the direction of motion is larger than expected from dinuclear orbiting model calculations. As a result, we see that in $-40<Q<-22$ MeV region, P_{zz} is smaller (0.10 ± 0.05) than expected from the dinuclear orbiting calculation. This can be understood because the angular-momentum misalignment along the body-fixed symmetry axis (twisting mode) increases as the Q-value decreases and therefore ρ_{22} along the direction of motion increases and the alignment parameter is reduced. Also, some of the excited carbon particles in this bin can come from ^{16}O and so will be less aligned.

Table 3 shows the disagreement between the calculated ratios from statistical model and measured ratios of yields of the $^{12}C(2^+)$ to ^{12}C(g.s). The lower values of $^{12}C(2^+)/^{12}C$(g.s) qualitatively support the dinuclear orbiting picture,where the non-excitation of the twisting mode will lower the ratio compared to statistical model calculations. Sanders et.al[31] measured the ratio of yields of the $^{12}C\ (2^+)$ to ^{12}C(g.s) for $^{32}S+^{24}Mg$ reaction and found agreement with the compound nucleus fusion-fission picture. The number of open channels for $^{32}S+^{24}Mg$ is much larger than that of $^{28}Si+^{12}C$ system. So we conclude that for systems such as $^{32}S+^{24}Mg$, $^{31}P+^{16}O$, $^{35}Cl+^{12}C$ having a large number of open channels, the back-angle damped binary yield is essentially dominated by the compound nucleus fusion-fission process. On the other hand, the back-angle damped yield of systems such as $^{28}Si+^{12}C$, $^{24}Mg+^{16}O$, $^{24}Mg+^{12}C$ having a small number of open channels is dominated by long-lived noncompound orbiting process.

What is orbiting?

We have learned that back-angle binary yield of systems having a small number of open channels is dominated by an orbiting process. We also find that the systems where anomalously large back-angle elastic yields have been seen, also have large damped back-angle binary yields which show orbiting characteristics. This empirical connection suggests that the same physical process might be responsible for both damped orbiting and anomalously large back-angle elastic yields as observed in certain light heavy ion systems. Many nonresonant and resonant processes have been proposed to explain the observed back-angle anomalies.

1) Nonresonant Process

Shivakumar et.al[6] considered orbiting as a nonresonant nucleon exchange process where the two nuclei form a dinuclear configuration and exchange nucleons. According to Randrup's diffusion model[32], the nucleon exchange rate between the two nuclei is $= (4/T_F)N\tau$, where $N=0.25\pi C^2\rho V$. Here ρ is the number of nucleons per unit volume, V is the Fermi velocity of nucleon, T_F is the Fermi energy and τ is the effective temperature. If we consider $^{28}Si+^{12}C$ dinuclear system connected by a neck of radius 1.25 fermi, then from Randrup's transport model[32], we find that the nucleon exchange rate $=0.17\times10^{+22}$ per sec. According to this model the system should reach the sticking limit in the time scale of t_{stick} given by the following equation

$$t_{stick} = \frac{I_A I_B I_R}{I_A + I_B + I_R} \cdot \frac{1}{mN\overline{\mu}R^2C^2} \qquad (7)$$

where I_A, I_B, I_R are moments of inertia of interacting nuclei A, B and the moment of inertia of their relative motion, m is mass of a nucleon, C is the radius of neck and R is the distance between centers of the two nuclei and $\overline{\mu}$ is defined as

$$\frac{1}{\bar{\mu}} = \frac{a^2}{J_A} + \frac{b^2}{J_B} + \frac{R^2}{J_R}$$

where a, b are the distances from the nuclear center to the perpendicular window plane separating two parts of the system. J_A, J_B, J_R are spin angular momenta of A, B and relative motion. We find that t_{stick} is of the order of 10^{-21} sec for $^{28}Si+^{12}C$ system and two or three nucleon exchanges are possible in this time scale. If we use a bigger neck radius then the nucleon exchange rate will increase and the system will reach the sticking limit faster. We know from the 1/sinθ angular distribution of the orbiting yield that the lifetime of the system should be greater than $(3\text{-}4)\times10^{-21}$ sec. So according to the transport model[32], the dinuclear system should reach the sticking limit and should be described by an equilibrated dinuclear model as developed by Shivakumar et al.[6] As we discussed earlier, such a model fails to explain important nonstatistical signatures of orbiting reactions such as entrance channel effects, nonequilibrium division of excitation energy, poor spin alignment etc. Such an equilibrated orbiting model also does not explain why orbiting has been seen only in certain systems having a small number of open channels. We find clear evidence that in a orbiting reaction, the dinuclear system does not reach the sticking configuration as expected from particle exchange models. However alternative nonresonant mechanisms are also possible. For example, the two nuclei might excite each other by collective excitations and might not reach the sticking configuration within the lifetime of the dinuclear system. Such models have not yet been quantitatively developed.

2) Resonant Process

Excitation of nuclear molecular states might be responsible for the large back-angle damped orbiting yields. Such resonant explanations have been forwarded to understand anomalously large elastic yields in the direct[33] and transfer channels[34]. Although the angle-integrated yield of individual low-lying states for the $^{28}Si+^{12}C$ reaction show the presence of gross structures[33] in their excitation functions, no significant structures[3] have been seen in the excitation function of angle-integrated total orbiting yield. Recently, Glaesner et.al[35] measured the angle-integrated excitation functions of damped back-angle yields from $^{24}Mg+^{12}C$ reaction integrated over 1 MeV Q-value bins. They find gross structures in the excitation function, but the amplitudes of oscillations decrease steadily as they go to higher excitation energy where a large number of states has been summed. However they still see weak structures in the excitation function of angle-integrated total damped back-angle yields. If it is a resonant process, then we are not dealing with a single isolated resonance, but several resonant amplitudes are most likely contributing to the process so that no significant structure is left in the excitation function when they are summed over. It may be worthwhile to carefully measure excitation functions of angle-integrated highly excited individual states contributing to back-angle damped yields and look for correlations among them. If we find that the amplitudes for different states are completely random, then that should imply that a nonresonant long-lived deep inelastic process is responsible for the orbiting yield. On the other hand, if we can identify several series of states strongly correlated among themselves, although no structure in the excitation function is left when all of them are summed over, then we may infer that several resonant amplitudes are contributing. We also know that several partial waves near ℓ_{crit} contribute to the orbiting yields. So if it is a resonant process, then clearly several resonant amplitudes should be present.

It would be interesting to undertake careful measurements of excitation functions of angle integrated yield of highly excited individual states to learn more about the orbiting reaction mechanism.

Oak Ridge National Laboratory is operated by Martin Marietta Energy Systems, Inc. for the U.S. Department of Energy under contract DE-AC05-84OR21400.

REFERENCES

1) J. G. Cramer, R. M. DeVries, D. A. Goldberg, M. S. Zisman and C. F. Maguire, Phys. Rev. **C14**, 2158 (1976)
2) P. Braun-Munzinger and J. Barrette, Phys. Rep. **87**, 209 (1982)
3) D. Shapira, R. Novotny, Y. D. Chan, K. A. Erb, J. L. C. Ford Jr, J. C. Peng and J. D. Moses, Phys. Lett. **B114**, 111 (1982)
4) D. Shapira, J. L. C. Ford Jr., and J. Gomez del Campo, Phys. Rev. **C26**, 2470 (1982)
5) W. Dunnweber, A. Glaesner, W. Hering, R. Ritzka, W. Trombik, J. Czakanski and W. Zipper, Phys. Rev. Lett. **61**, 927 (1988)
6) B. Shivakumar, S. Ayik, B. A. Harmon and D. Shapira, Phys. Rev. **C35**, 1730 (1987)
7) S. Ayik, D. Shapira and B. Shivakumar, Phys. Rev. **C38**, 2610 (1988)
8) C. Beck, B. Djerroud, B. Heusch, R. Dayras, R. M. Freeman, F. Haas, A. Hachem, J. P. Wieleczko and M. Youlal, Z. Phys. **A334**, 521 (1989)
9) S. Sanders, D. G. Kovar, B. B. Back, C. Beck, B. K. Ditcher, D. Henderson, R. V. F. Janssens, J. G. Keller, S. Kaufman, T. F. Wang, B. Wilkins and F. Videbaek, Phys. Rev. Lett. **59**, 2856 (1987)
10) D. Shapira Phys. Rev. Lett. **61**, 2153 (1988)
11) A. Ray, S. Gil, M. Khandaker, D. D. Leach, D. K. Lock and R. Vandenbosch, Phys. Rev **C31**, 1573 (1985)
12) R. Vandenbosch, Nucl. Phys. **A339**, 167 (1980)
13) S. L. Tabor, D. F. Geesaman, W. Henning, D. G. Kovar, Phys. Rev. **C17**, 2136 (1978)
14)K. T. Lesko, D. K. Lock, A. Lazzarini, R. Vandenbosch, V. Metag and H. Doubre, Phys. Rev. **C25**, 872 (1982)
15) F. Puhlhofer, Nucl. Phys. **A280**, 267 (1977)
16) C. Beck, B. Djerroud, R. M. Freeman, F. Haas, B. Heusch, A. Morsad, M. Youlal, A. Hachem, R. Dayras, J. P. Wieleczko, T. Matsuse and S. M. Lee, Proceedings of the Workshop on Nuclear Structure and Heavy Ion Reaction Dynamics, Notre Dame, USA, p.213 (1990)
17) Y. Abe Private Communication, 1990.
18) F. Haas and Y. Abe, Phys. Rev. Lett. **46**, 1667 (1981)
19) N. Austern and J. S. Blair, Ann of Physics, **33**, 15 (1965)
20) A. Ray, D. D. Leach, R. Vandenbosch, K. T. Lesko and D. Shapira, Phys. Rev. Lett. **57**, 815 (1986)
21) A. Ray, D. Shapira, M. Halbert, H. Kim, J. Gomez del Campo, J. P. Sullivan, J. T. Mitchell and B. Shivakumar, Phys. Rev. **C43**, 1789 (1991)
22) G. R. Satchler, Direct Nuclear Reactions, Oxford University Press (1983)
23) H. J. Rose and D. M. Brink, Revs. Mod. Phys. **39**, 306 (1967)
24) Z. Berant, M. B. Goldberg, G. Goldring, S. S. Hanna, H. M. Loebenstein, I. Plesser, M. Popp, J. S. Sokolowski, P. N. Tandon and Y. Wolfson, Nucl. Phys. **A178**, 155 (1971)
25) T. Dossing, Nucl. Phys. **A357**, 488 (1981)
26) H. Feshbach, Nuclear Spectroscopy Part B (F. Ajzenberg-Selove ed.), Academic Press, New York (1960)
27) C. M. Perey and F. G. Perey, Atomic and Nuclear Data Table **17**, 1 (1976)

28) F. Glatz, P. Betz, J. Siefert, F. Heidinger and H. Ropke, Phys. Rev. Lett. **46**, 1559 (1981)
29) R. P. Schmitt and A. J. Pacheco, Nucl. Phys. **A379**, 313 (1982)
30) R. Vandenbosch, Phys. Rev. **C20**, 171 (1979)
31) S. J. Sanders, B. B. Back, R. V. F. Janssens, D. G. Kovar, D. Habs, D. Henderson, T. L. Khoo, H. Korner, G. E. Rathke, T. F. Wang, F. L. H. Wolfs and K. B. Beard, Phys. Rev. **C41**, 1901 (1990)
32) J. Randrup , Nucl. Phys. **A383**, 468 (1982)
33) R. Ost, M. R. Clover, R. M. DeVries, B. R. Fulton, H. E. Gove and N. J. Rust, Phys. Rev. **C19**, 740 (1979)
34)S. J. Sanders, H. Ernst, W. Henning, C. Jachcinski, D. G. Kovar, J. P. Schiffer and J. Barrette, Phys. Rev. **C31**, 1775 (1985)
35) A. Glaesner , W. Dunnweber, M. Bantel, W. Hering, D. Konnerth, R. Ritzka, W. Trautmann, W. Trombik and W. Zipper, Nucl. Phys. **A509**, 331 (1990)

FUSION–FISSION LIKE STUDIES FROM MEDIUM HEAVY TO LIGHT COMPOUND SYSTEMS

B. HEUSCH*

Centre de Recherches Nucléaires et Université Louis Pasteur
B.P. 20, 67037 STRASBOURG Cedex FRANCE

INTRODUCTION

For many years we have focussed our measurements on the fission–like reaction mechanisms following the formation of the target plus projectile composite system A_{CN}. At the 1984 Tsukuba International Symposium I presented the study of the onset of symmetrical fission phenomena (1) observed for $A_{CN} > 90$ in agreement with the fact that we are above the Businaro–Gallone point. Since then we have been able to confirm the existence of fission–like deexcitation channels for energies well above the one where the fission barrier vanishes (2). To explain such behavior, processes related to the fusion dynamics have been used (3), in which the existence of conditionnal saddle points for the colliding system can play an important role. (See the extrapush model for example (4)).

More recently new approaches to fission studies has allowed different new behavior to be pinpointed. One of them is the contribution of incomplete fusion processes up to energies in the region of the Fermi energy where one no longer observes fusion. In this frame we studied extensively the system $A_{CN} = 108$ (5) ; similar conclusions have been obtained for $A_{CN} = 147$ (6) which show that there is a maximum excitation energy per nucleon, around 7 to 8 MeV, that a nucleus can sustain, in agreement with their binding energy. However, it should be point out that very recently new Ganil results (7) demonstrate that there is no such saturation of the maximum possible excitation energy in a nucleus.

Correlated to these new facets of fusion studies are the measurements of the emission of more and more massive fragments, called IMF (intermediate mass fragments) as the bombarding energy is increased. The evaporation of IMF, explained first by Moretto in 1975 (8), competes with the fission process even at low energy (around twice the Coulomb barrier) and could dominate the compound nucleus deexcitation modes at high energies. This latter situation was proposed to explain the outgoing fragment distributions measured for the $^{84}Kr + ^{27}Al$ leading to $A_{CN} = ^{111}In$ at 25, 35, and 45 MeV/nucleon (9). At low energy (5.9 MeV/nucleon) we observed (10) the symmetrical fission component as expected, together with an asymmetrical fission–like component which could originate from a fast–fission mechanism. The purpose of our new measurement for this $^{84}Kr + ^{27}Al$ reaction at RIKEN facility (11) at an intermediate energy of 10.6 MeV/nucleon should check the origin of this component knowing that the incomplete fusion contribution should be negligible. Later different calculations of sequential decay appeared, describing at once the whole energy relaxed fragment distributions (12, 13 and 14) one should measure as a function of excitation energy or temperature.

Perhaps the most interesting domain of fission study developments during these last years concerns its observation in the decay of light nuclei, typically from $A_{CN} = 90$ down to $A_{CN} = 40$. Here the competing mechanism is orbiting which can dominate this deexcitation mode (15).

EXPERIMENTAL PROCEDURE

One of the dominant difficulties of these measurements is related to the ambiguous signature if a given outgoing channel originates from a total fusion reaction mechanism or not, knowing that the main competing mechanisms are related to direct or semi–direct processes like fully relaxed deep inelastic collisions (DIC) or orbiting phenomena. To undertake such study the experimentalist has to measure and compare different typical behavior in the deexcitation of a given system looking for example at entrance channel effects, dependence on bombarding, i.e. excitation energy, angular mass Z–number and energy distributions, l dependence up to incident energies well above the vanishing of the fission barrier and so on. Moreover one has to measure at the same time the well identified outgoing direct channels which are quasi elastic and partially relaxed DIC as well as the compound nucleus evaporation and fission products.

In the experiments refered to here the compound system decay for $40 \leq A_{CN} \leq 111$ will be considered. The measurements are mainly focussed on the heavy fragment detection, in singles as well as in coïncidence. In the latter one it is mainly the binary nature which is checked. The experimental setups which are generally used are based on time–of–flight techniques associated with Z identification. These techniques are well known but the evolution during the last years is more focussed on the construction of larger and larger devices to assure the reasonable geometrical efficiency necessary to run coïncidence measurements. The associated detection of light particles becomes of great importance to determine the sharing of the total excitation energy. But probably the most straightforward results are obtained using neutron detection systems as it has been done by the Berlin group of Dr. Hilscher (See the review presented by Dr. Hinde (16)) to deduce the fission timescales. Such measurements are still not available for light A_{CN} systems ; certainly in the next two years new devices like the Texas A and M fractionated ball or DEMON (multi–neutron detector facility (17)) will bring new insight to this subject.

RESULTS AND DISCUSSION

A general review of all results obtained in the field is of course impossible. I will try to present and discuss the data we obtained together with similar conclusions drawn for other systems and for which a full presentation would be much better done by the authors themselves.

For the A_{CN} = 111 system studied using the $^{84}Kr + ^{27}Al$ reaction, we pointed out (10) that at 6 MeV/nucleon one observes the coexistence of a symmetric and an asymmetric fission component (see figure 1). Moreover, the emission of intermediate mass fragments (IMF) can also lead to an asymmetrical mass component. These reasons motivated us to study further this $^{84}Kr + ^{27}Al$ system at higher energy, to check the energy dependence of these components. At the chosen incident energy of 10.6 MeV/nucleon we were higher than the fission barrier vanishing energy. On the other hand total fusion should always dominate this process as it has been observed up to 25 MeV/nucleon (9) where the dominant contribution is related to the evaporation of IMF by the compound nucleus. The experiment has been done using a 10.6 MeV/nucleon ^{84}Kr beam delivered by the RIKEN facility

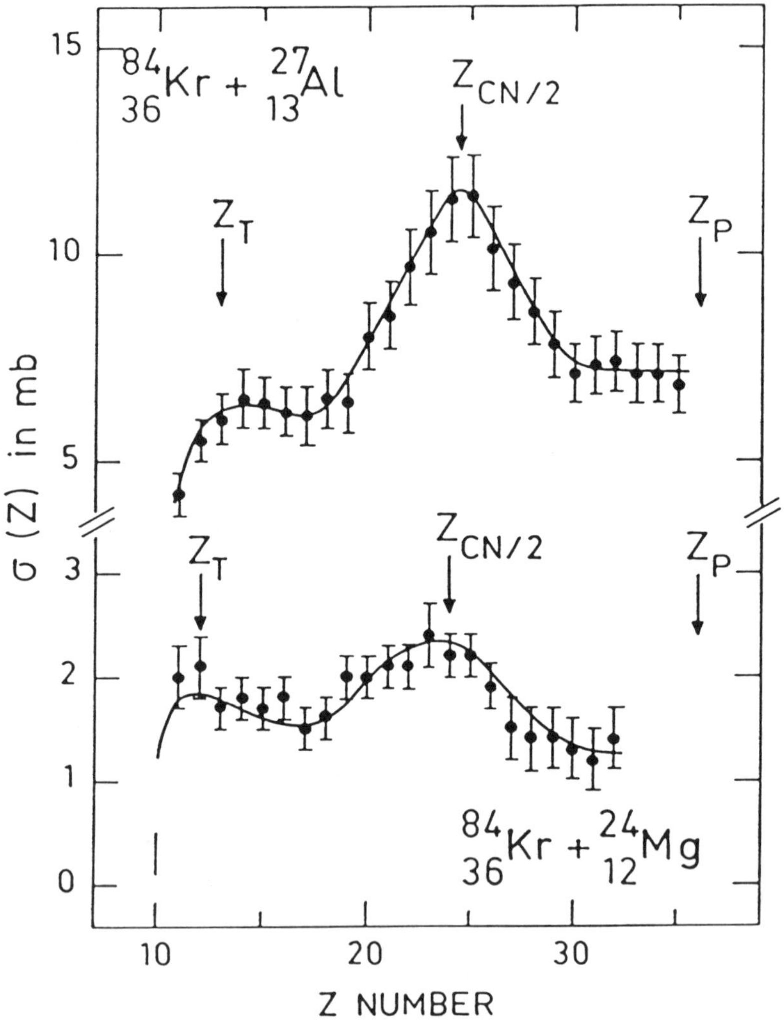

Fig. 1 : Fully relaxed fragment Z distributions measured at the same ^{84}Kr bombarding energy (490 MeV) for both reactions. Z_P, Z_T and $Z_{CN/2}$ refer to projectile, target and half compound nucleus atomic numbers, respectively.

and the results are presented elsewhere (18). Here I want only to pinpoint the arguments which confirm the dominance of the compound nucleus deexcitation modes. Figure 1 illustrates perfectly this assumption : symmetrical fission cross section, centered at the $Z_{CN/2}$ for both systems studied at exactly the same bombarding energy (490 MeV), is nearly 4 times larger for ^{84}Kr + ^{27}Al compared to ^{84}Kr + ^{24}Mg (19). This factor can be reproduced roughly using compound nucleus deexcitation calculations. However the asymmetrical component scaling factor is suprisingly the same in contradiction to a direct process origin which should have nearly the same amplitude for both systems. An overall view for the study of the system ^{84}Kr + ^{27}Al is presented in figure 2 where the total fragment cross section are plotted on the same linear scale between 5.9 and 45 MeV/nucleon. The most striking observations, when the energy increases, are related (i) to the large increase of intermediate mass fragment emission (up to Z ≃ 8) on one side and (ii) to the large increase of fragmentation like products (Z > 20) which "replace" the symmetrical fission peak. The fact that cross sections for fragments typically between Z = 10 and Z = 15 are nearly constant over such a large bombarding domain can be accidental, due to compensating contributions.

Three different calculations of compound nucleus sequential decay have been applied to these results. The first one called EDCATH has been presented by Auger et al. (9) ; in figure 3 the cross section predictions are compared to the data measured for fusion evaporation (20) as well as fully relaxed fragments (10). The same calculation reproduces quite well the data obtained at high energy for ^{84}Kr + ^{27}Al as well as for ^{84}Kr + ^{12}C (21) : see figure 4. The second calculation has been carried out using the code developped by Charity et al. (13), called GEMINI. Figure 5 shows that overall satisfactory agreement is found, again for evaporation as well as fission products at 5.8 and 10.6 MeV/nucleon. It should however be point out that both calculations fail for the details of fully relaxed fragment cross sections : the Gemini code underestimates by a factor 2 the measured data at both energies. Such disagreement has also bee observed by Han et al. (39) for the ^{63}Cu + ^{27}Al study done at a similar energy (12.6 MeV/nucleon). Using the code EDCATH for ^{84}Kr + ^{27}Al (fig.3) the asymmetrical component is mainly underestimated by a factor nearly 4. This latter result suggested to us (10) the existence of a fast fission contribution where only the mass asymmetrical degree of freedom is not relaxed. The third calculation result is presented in figure 6. The details of this Extended Hauser–

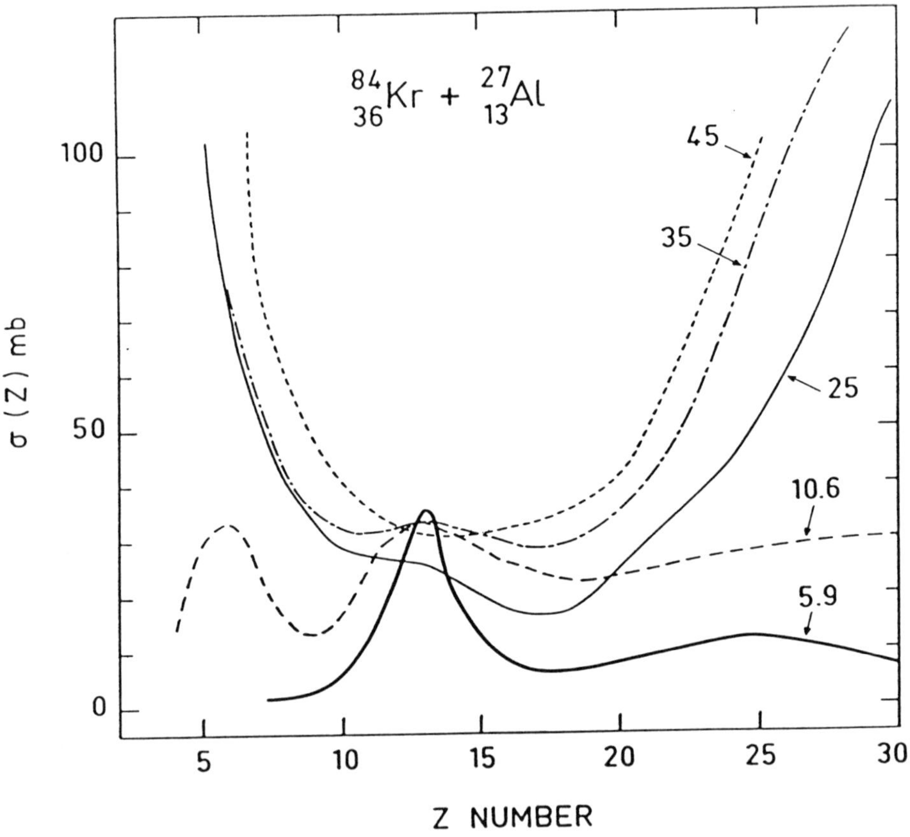

Fig. 2 : Total fragment cross section distributions measured at the 5 indicated ^{84}Kr incident energies quoted in MeV per nucleon.

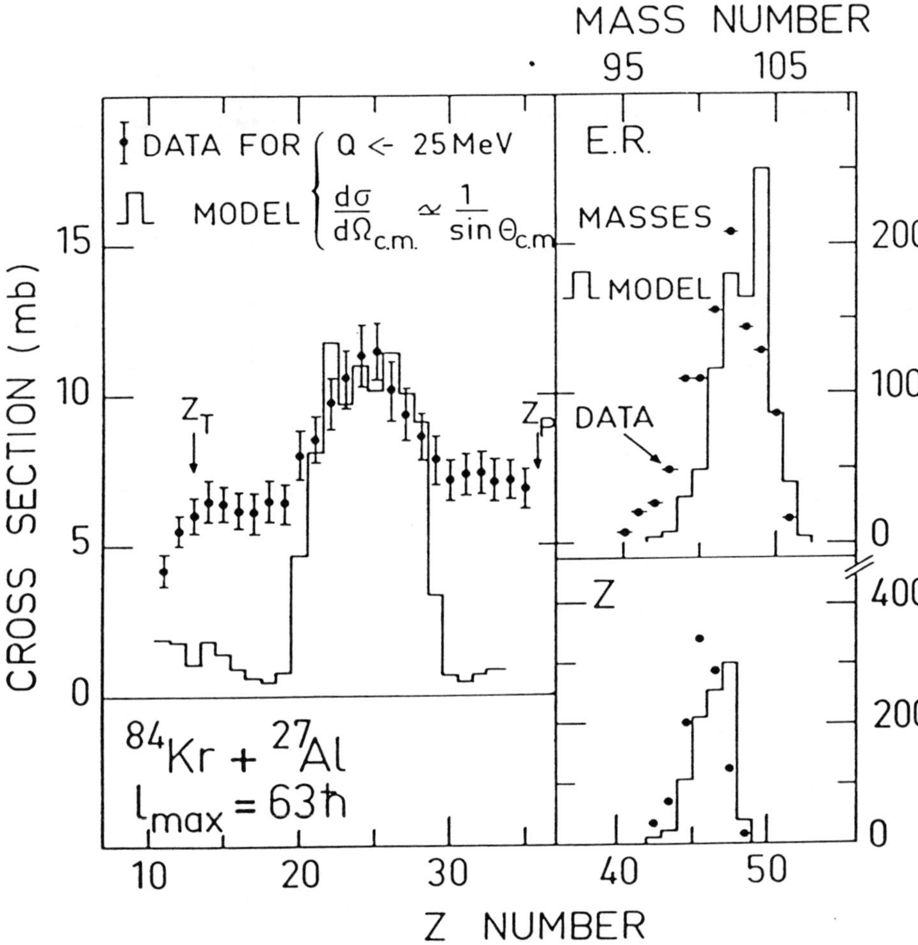

Fig. 3 : Comparison between the measured ^{84}Kr (490 MeV) + ^{27}Al fully relaxed fragments (Q < − 25 MeV) and evaporation residue E.R. (20) Z distributions to those calculated (histograms) using the code EDCATH by Auger et al. (9).

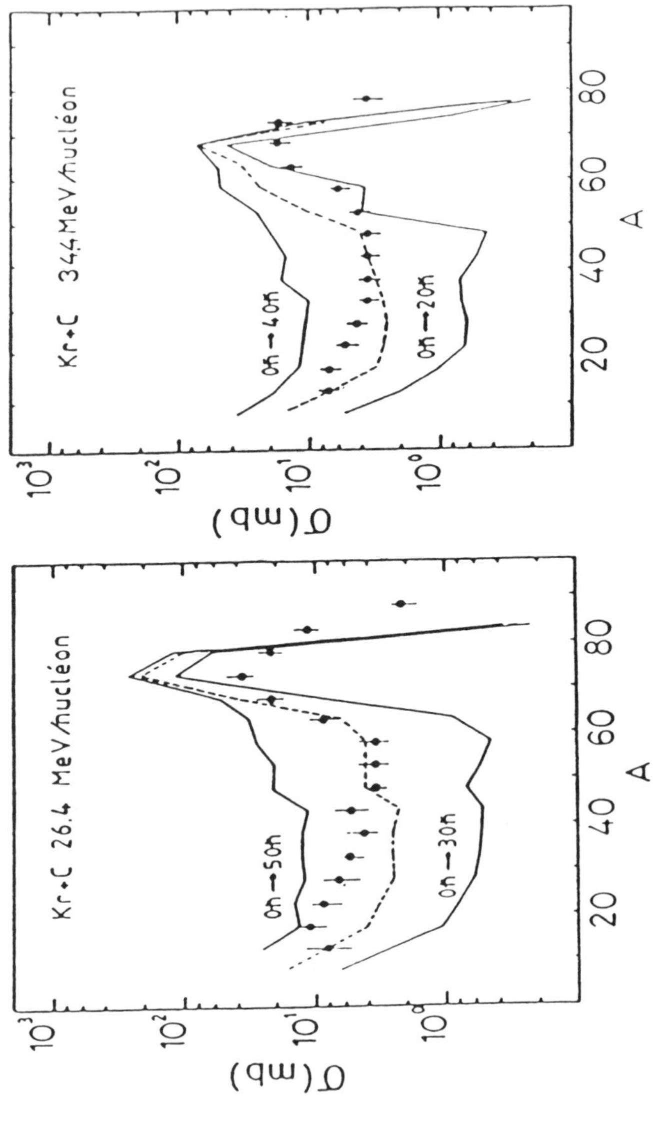

Fig. 4 : Comparison between experimental and calculated outgoing fragment mass distributions using EDCATH code for different range of angular momenta values.

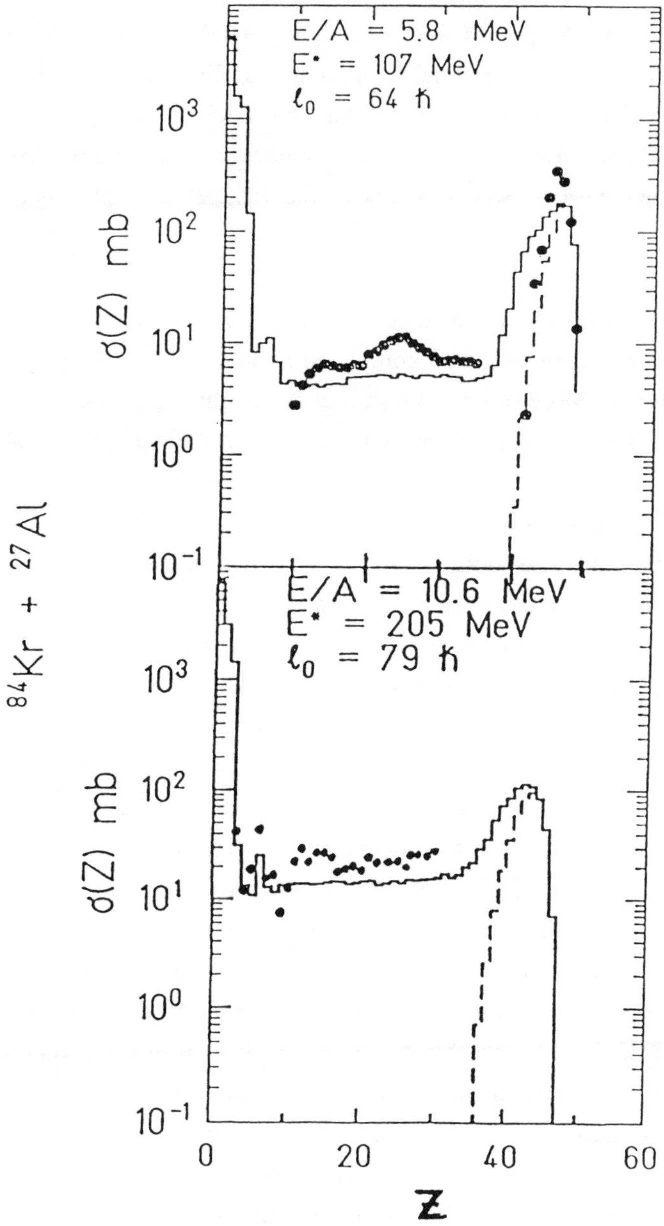

Fig. 5 : GEMINI code predictions (histograms) compared to data measured for $^{84}Kr + ^{27}Al$ at 5.8 and 10.6 MeV per nucleon.

Feshbach model (EHFM) are given by Matsuse (22) ; let me just point out how good is the agreement with experimental data (see figure 6) as well as for the very different entrance channel ^{3}He + ^{nat}Ag (11). This strongly suggests that there is no need for other mechanisms besides the sequential evaporation of light particles, IMF and symmetrical fission originating all from the compound nucleus. Surely the rotating liquid drop model (RLDM) fails deeply in explaining how one can form the CN ^{111}In with an angular momentum of 86ħ, well above the RLDM calculated maximum value of 68ħ.

Knowing how successful a CN sequential decay model can be it becomes very interesting to move to much lighter compound systems for which symmetrical fission should disappear. Moreover the suggestion that orbiting phenomena could dominate the fission–like outgoing channels motivated us to study a lighter system as ^{56}Ni for which fusion–fission has been observed (23), and which could be studied by different entrance channels. We choose A_{CN} = ^{47}V, formed in the ^{35}Cl + ^{12}C reaction (24) studied at the Strasbourg 18 MV–MP tandem accelerator, for which a general discussion and comparison to similar data has been presented by C. Beck one year ago (25). Since then we studied the symmetrical entrance channel ^{23}Na + ^{24}Mg : the integrated cross sections are plotted in figure 7 together with those obtained for ^{35}Cl + ^{12}C. The main result is that these cross sections scale by a factor around 8, but the relative values as well as the odd even structure are conserved from one system to the other. In figures 8 and 9 these values are compared to the EHFM predictions (22) and again one can consider the general agreement with data very satisfactory. The fact that the cross–sections are larger for ^{23}Na + ^{24}Mg is related (i) to the larger l values brought in and (ii) to the fact that fusion can better occur for this symmetrical system, compared to the asymmetrical one ^{35}Cl + ^{12}C, due to the larger radii product. The compound nucleus origin of these cross sections can be checked by looking at the existence or not of an entrance channel effect. In this respect, together with a check of a possible carbon contamination of the ^{24}Mg target, we analysed the carbon to oxygen ratio in the study of the ^{31}P + ^{16}O entrance channel, that we ran in Oak Ridge (26) using a ^{16}O jet target and a ^{31}P beam. The ^{47}V compound system was formed at a same excitation energy (E^{*} = 64 MeV) as that studied by the ^{35}Cl + ^{12}C system. In figure 10 are shown the results in the lower part : obviously there is nearly no entrance channel effect whereas for the A_{CN} = 40 system (27) studied by ^{24}Mg + ^{16}O and ^{28}Si + ^{12}C reactions a very large difference was observed. A Ray will (15)

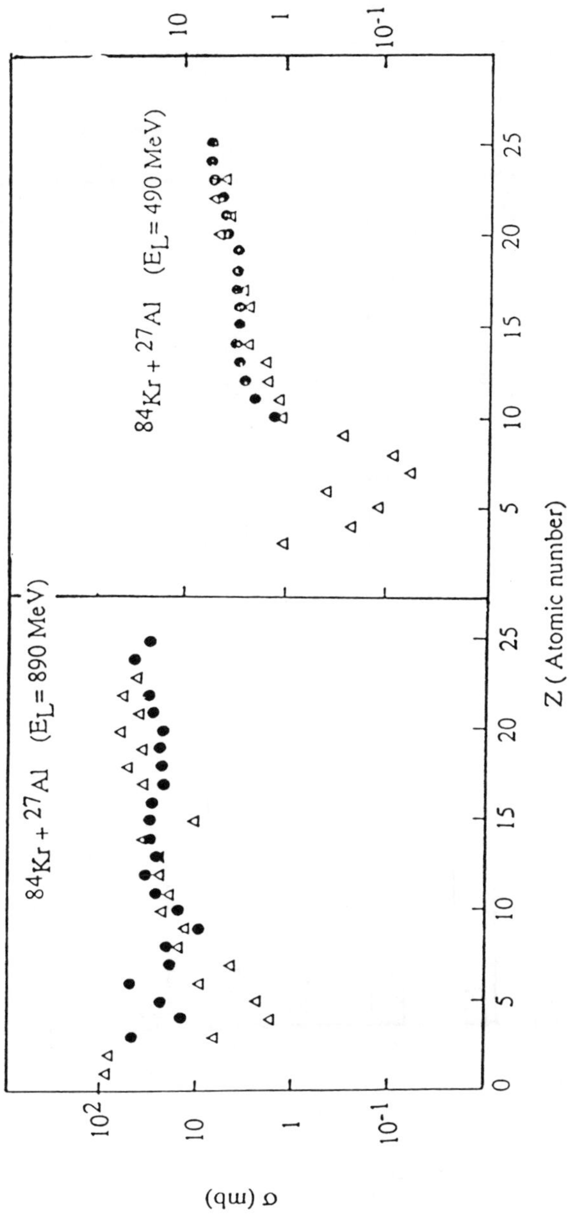

Fig. 6 : Comparison between experimental (●) and theoretical (Δ) outgoing fragment Z distributions using the EHFM calculation.

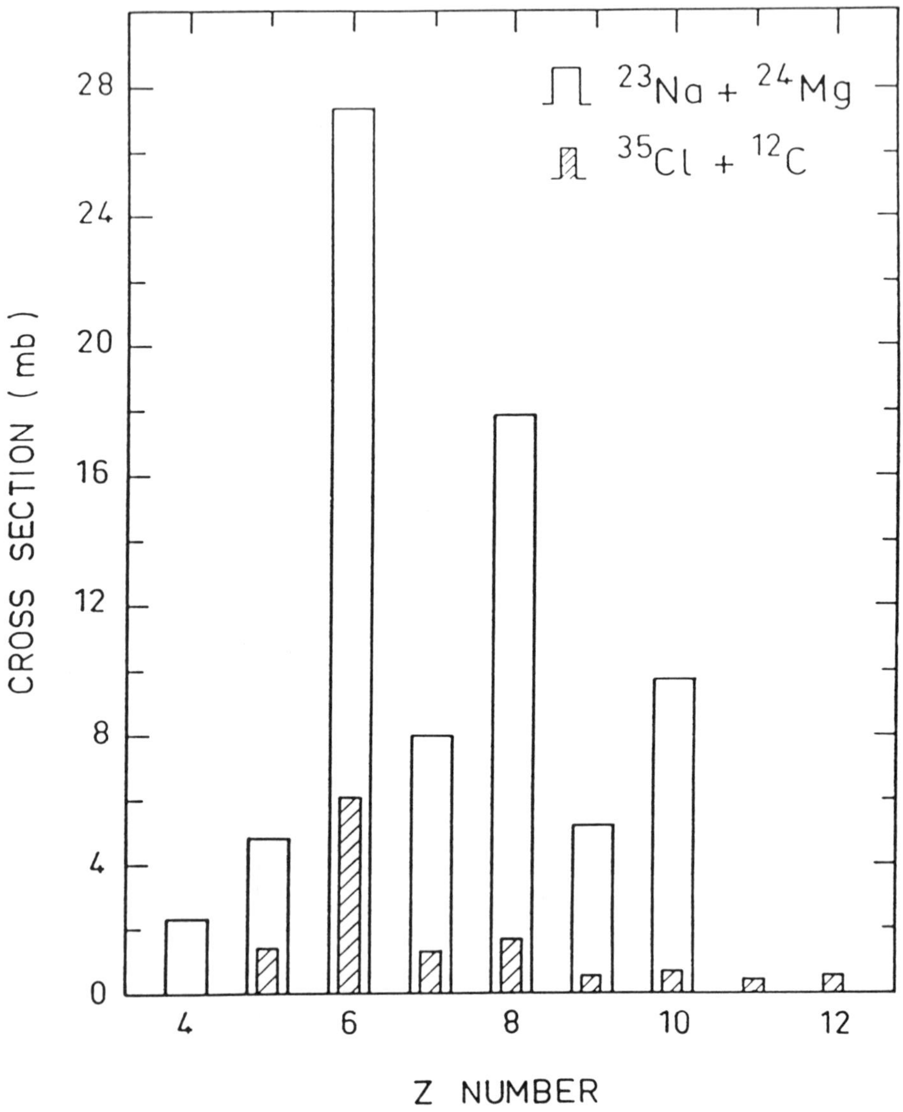

Fig. 7 : Total Z number fragment cross sections measured for both entrance channels leading to the same excitation of the ^{47}V compound nucleus of 64 MeV.

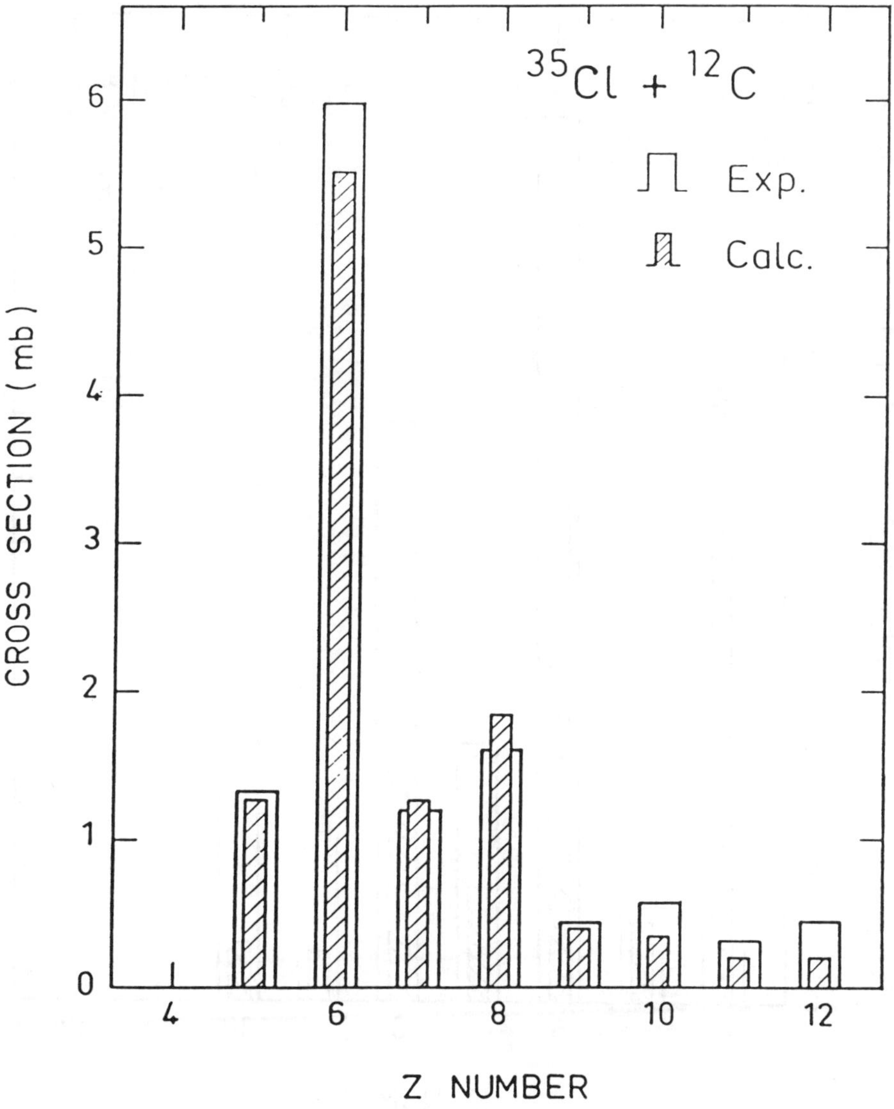

Fig. 8 : Comparison of the measured and the calculated fragment cross sections using the EHFM for the ^{35}Cl (200 MeV) + ^{12}C reaction.

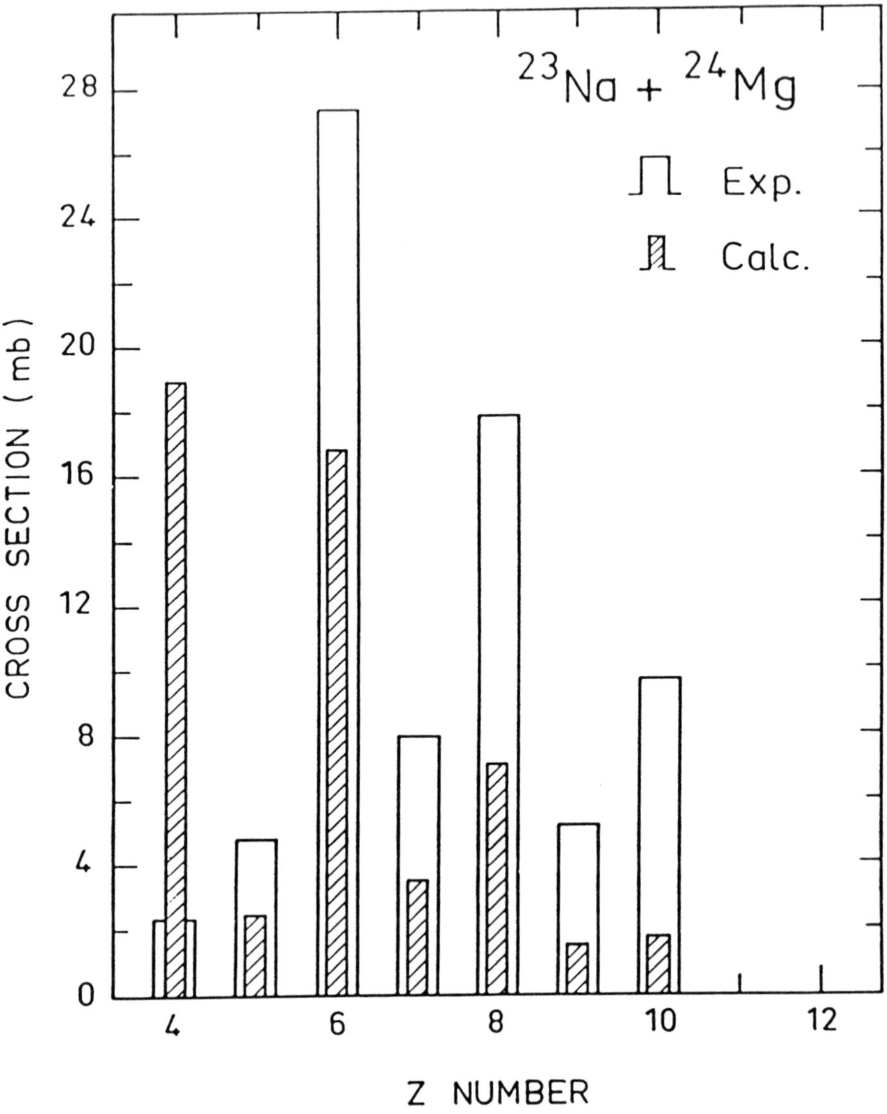

Fig. 9 : Comparison of the measured and the calculated fragment cross sections using the EHFM for the ^{23}Na (89 MeV) + ^{24}Mg reaction.

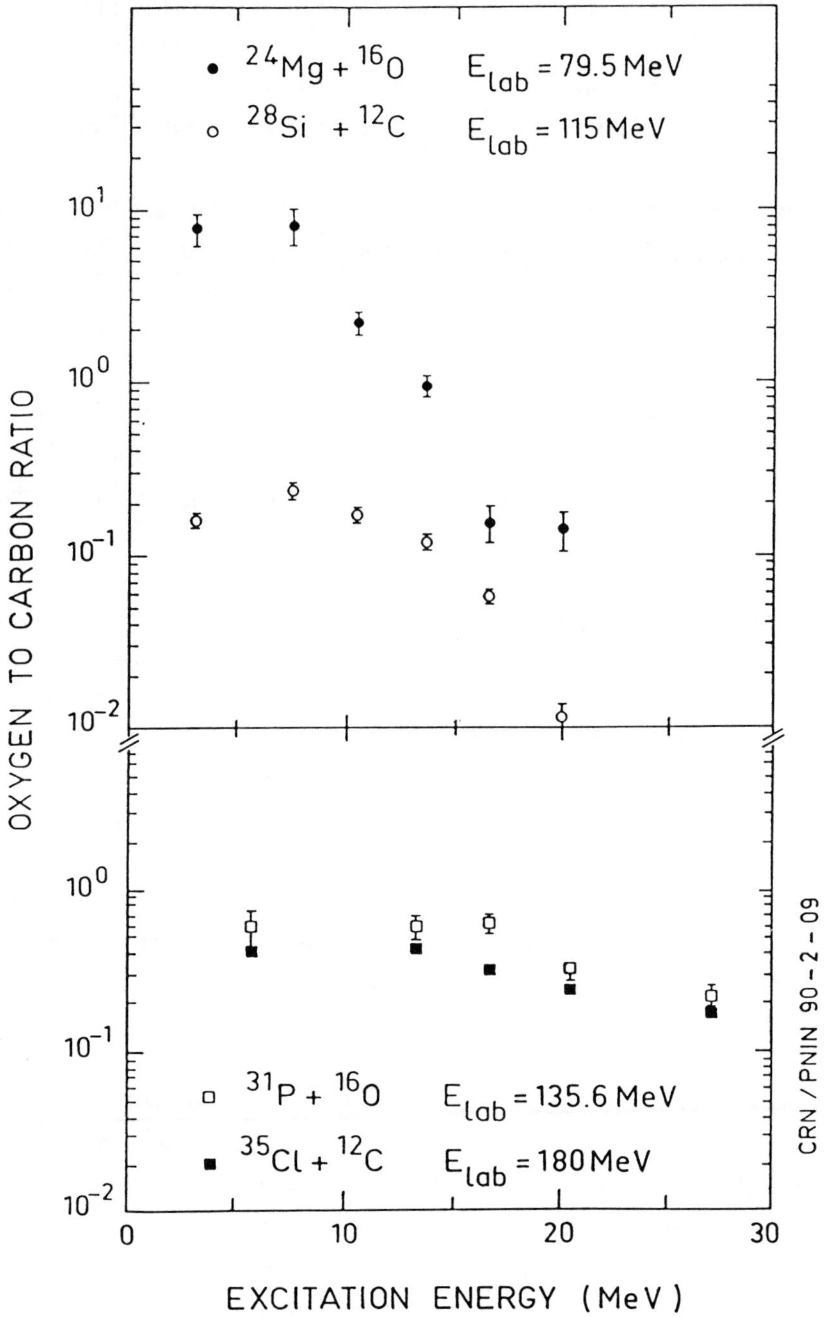

Fig. 10 : Oxygen to Carbon ratio measured for different entrance channels forming the A_{CN} = 40 (uper part) and A_{CN} = 47 (lower part) at the same excitation energy.

discuss at length this difference during this conference. I will just mention that this effect is correlated to the difference in the number of open channels available for these systems (figure 11) which leads to an enhancement of the orbiting channels for systems having a low number of available channels for the fusion decay–process.

After these two extreme cases of fission plus IMF emissions by compound nuclei of mass A_{CN} = 111 and A_{CN} = 47, I would like to focuss on "intermediate" A_{CN} system results. At high energies (above 20 MeV per nucleon) the sequential decay has been well accounted for by different calculations. Figure 4 present the EDCATH predictions for Kr induced reactions on ^{12}C and ^{27}Al. The general trend of sequential decay model calculations with increasing temperature of the compound nucleus appears very clearly in figure 12 and 13 : the nucleus ^{100}Mo and ^{40}Ar are considered in these two examples calculated by Richert and Wagner (27). The model is presented and discussed by Richert at this Conference (28).

An interesting comparison of entrance channel and/or neutron number effects has been recently studied at JAERI (29). Figure 14 present the data obtained for A_{CN} = 97 (^{35}Cl + ^{62}Ni reaction) and A_{CN} = 101 (^{37}Cl + ^{64}Ni and ^{12}C + ^{89}Y reactions) the number of protons for both systems being equal to 45. One can clearly conclude from this figure that there is no effect related to the 4 neutron mass difference. Symetric fission is however only visible in the case of the very asymmetric ^{12}C + ^{89}Y system ; for the two other reactions the asymmetric splitting hides partially the symmetric fission component. These all observations suggest a contribution from the fast fission processes ; the predictions of EHFM should be therefore very instructive. Moreover the study of the ^{32}S + ^{59}Co reaction lead us (32) and Iori et al. (33) to very similar results.

In this coherent ensemble of results and theoretical interpretations it is very unlikely that the analysis presented by Yokota et al. (34) is correct. These authors present an analysis of the A_{CN} = ^{64}Zn system formed by ^{37}Cl + ^{27}Al and ^{16}O + ^{48}Ti entrance channels and interpret their relaxed data as only originating from DIC processes. From Fig. 15, taken from their paper (figure 6 of reference 34), one cannot conclude that symmetric fission is not observed because the Z distribution is not measured far enough ($Z_{CN/2}$ = 15) although a maximum at Z = 14 is observed. The same applies also for ^{37}Cl + ^{12}C and ^{37}Cl + ^{48}Ti (A_{CN} = 47 and 85 respectively).

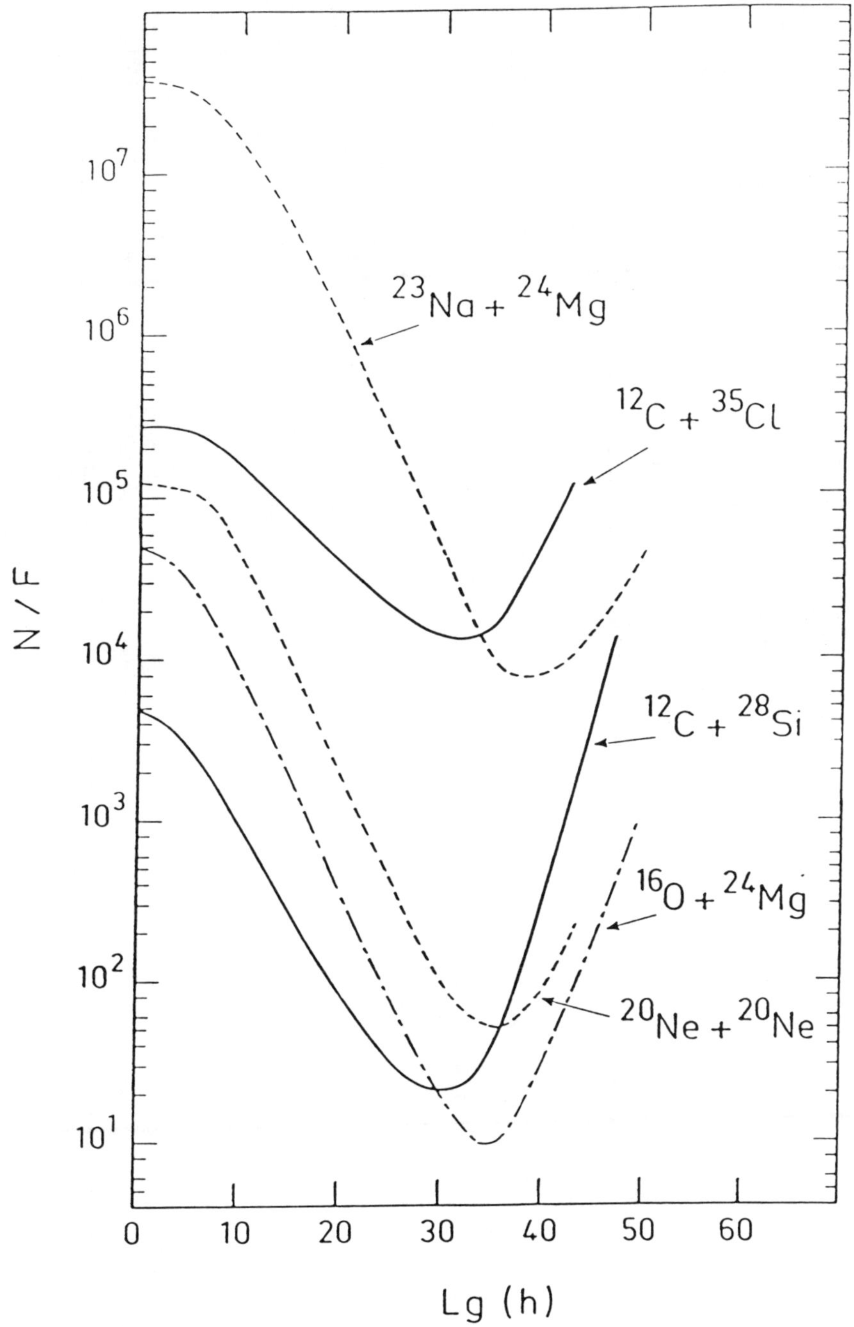

Fig. 11 : Calculated numbers of open channels available, for a given angular momentum value, to deexcite satistically the ^{47}V and ^{40}Ca compound nucleus.

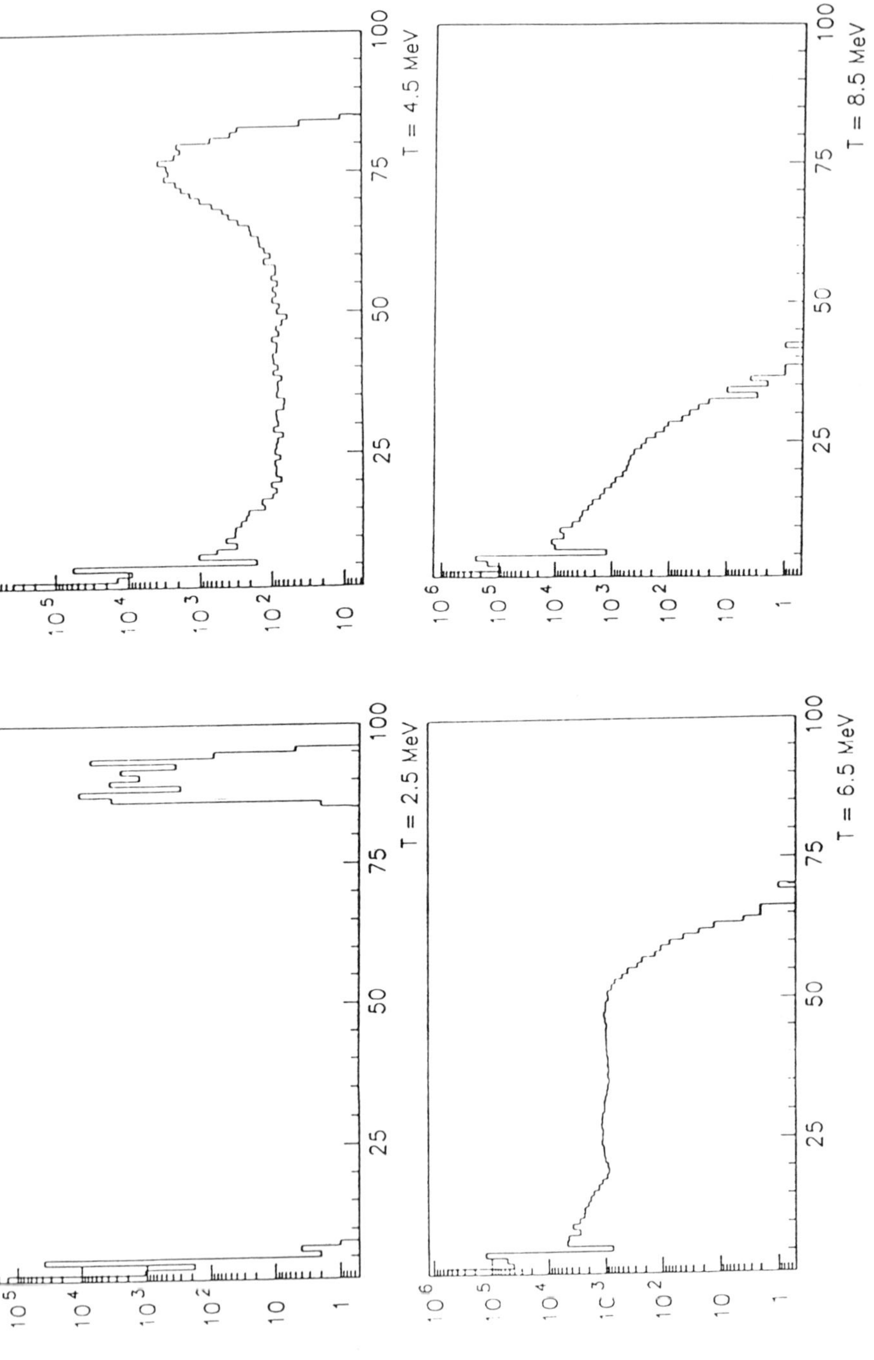

Fig. 12 : Sequential decay calculation by Richert and Wagner (27) for ^{100}Mo nucleus as a function of temperature.

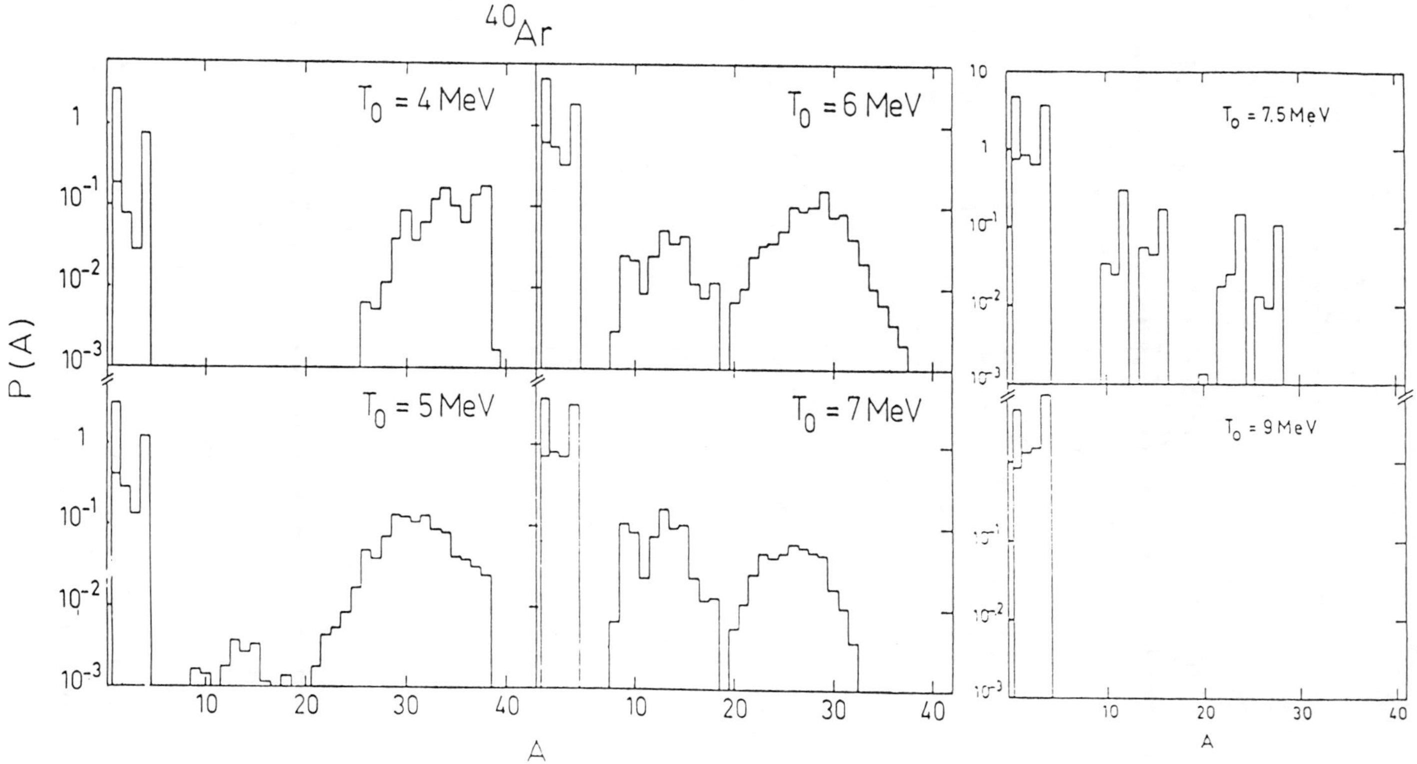

Fig. 13 : Fragment mass distributions calculated by Richert and Wagner (27) for the sequential decay of ^{40}Ar nucleus at the indicated T_0 temperatures.

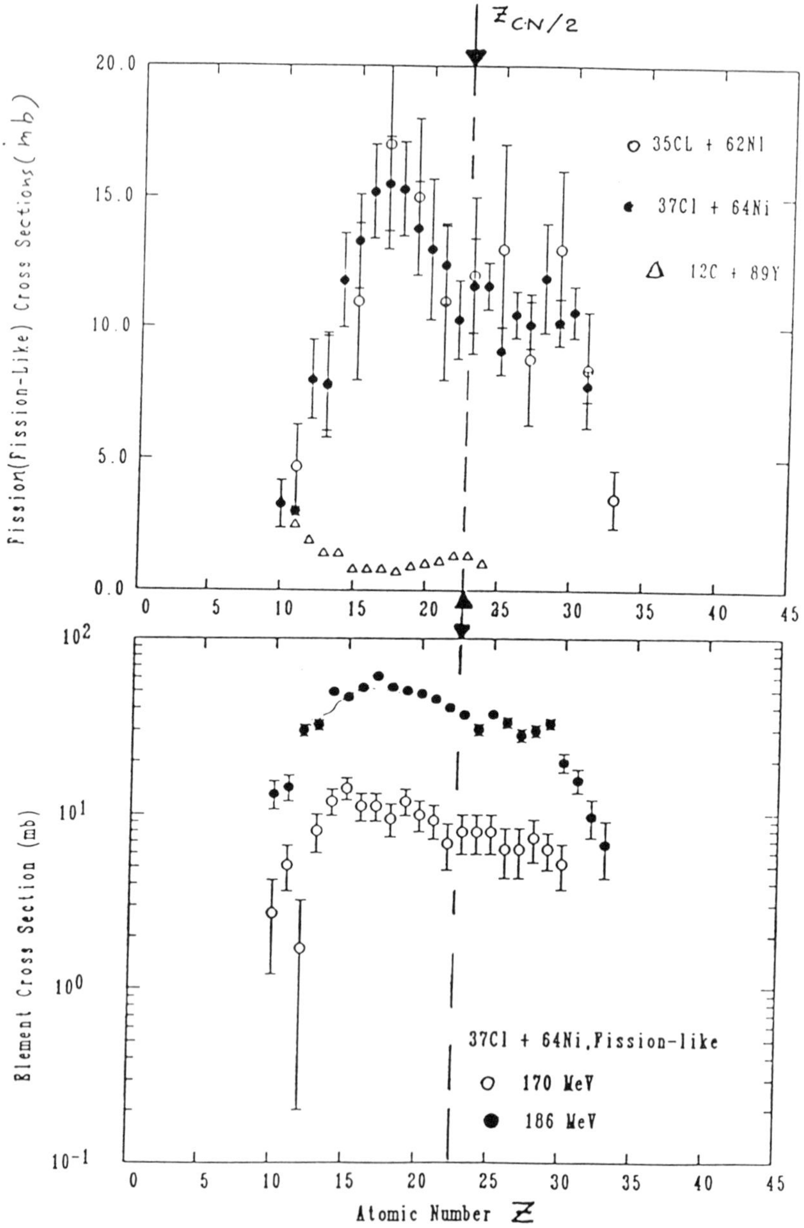

Fig. 14 : Fragment atomic Z distributions measured for $A_{CN} = 97$ and $A_{CN} = 101$ for which the total number of protons is the same ($Z_{CN} = 45$). The $^{35}Cl + ^{62}Ni$ and $^{12}C + ^{89}Y$ are taken in ref. (30) and (31) respectively.

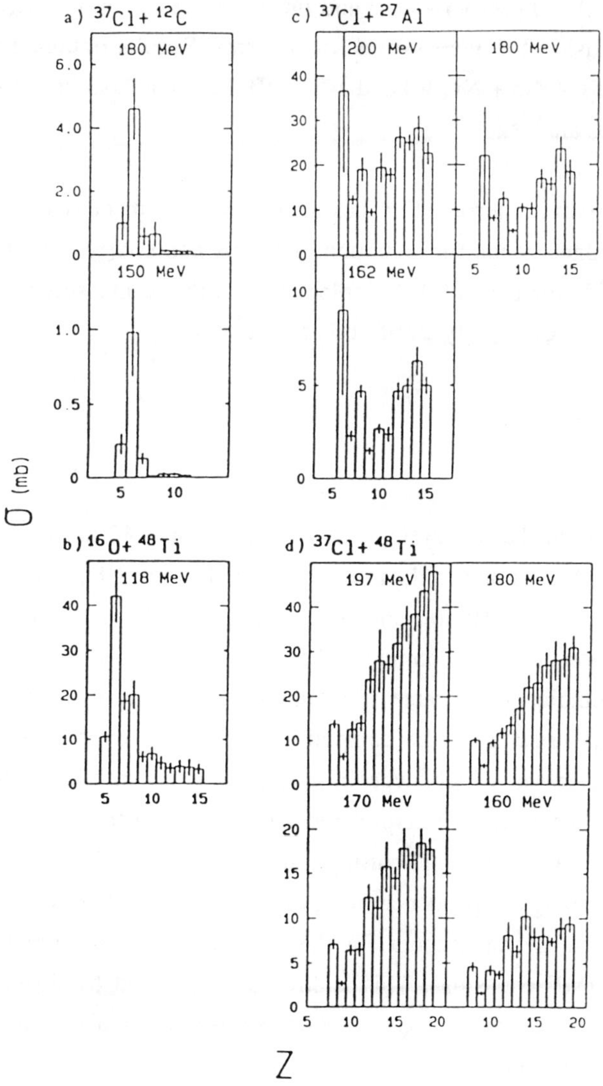

Fig. 15 : Fully relaxed fragment Z distributions measured by Yokota et al. (34) for the 4 indicated systems.

Moreover the larger Z = 6 cross section, interpreted as only due to partially damped DIC would show probably no entrance channel effect as for the present ^{47}V data discussed above. Finally the large fission–like scaling factor measured at E* (^{64}Zn) = 101 MeV between ^{37}Cl (200 MeV) + ^{27}Al and ^{16}O (118 MeV) + ^{48}Ti is again due to the enhancement of fusion for more symmetric systems. Results obtained elsewhere for ^{20}Ne + $^{50,54}Cr$ (35), ^{20}Ne + Ni (36) and ^{32}S + ^{50}Ti (37) can also be considered as an emphasizing of this analysis.

Such an alternative interpretation has also be widely discussed between the Argonne and Oak Ridge groups for compound systems nears A_{CN} = 50 to 60 (See reference (38) and (23) for example). Complementary precise measurements should bring a new insight and definitively clarify this controversy.

SUMMARY

It has been shown that for systems as light as A_{CN} = 47 up to systems just above the Businaro Gallone point in the mass region of 100 to 110 the probability for a system to deexcite by the fission channel, is not negligible. As predicted, the asymmetrical separation becomes dominant when the A_{CN} mass is decreasing but the symmetrical mode remains measurable. The ambiguities in the measured outgoing fragment distributions arise from the competition with IMF emissions as well as dynamical fission processes which depend strongly on the studied system. Fully relaxed DIC has also be used to interpret the results. I tried to show that precise checks on the behavior of two neighbouring systems as well as search for entrance channel effect and/or energy dependence bring evidence enough that the deexcitation of the compound nucleus can account for the symmetric and asymmetric fission channels as well as IMF emissions. This is strongly supported by different recent calculations all done in this frame. These all conclusions indicate also that the RLDM fails in the data interpretation.

The strength of the fission channel depends strongly on the possibilities a system has to deexcite. For very light systems especially the number of open channels available determines directly the flux repartition between direct or compound processes and therefore very large differences in the general behaviour of two neighbouring systems can be observed.

* Group report ; collaborators list name and Institutions :

C. BECK[1], D. BLUMENTHAL[2], R. DAYRAS[3], B. DJERROUD[1], R.M. FREEMAN[1], H. FUJIWARA[4], J. GOMEZ DEL CAMPO[5], F. HAAS[1], A. HACHEM[6], B. HEUSCH[1], K. IEKI[7], S.C. JEONG[4], H.J.KIM[5], S.M. LEE[4], T. MATSUSE[8], T. MIZOTA[4], A. MORSAD[1], T. NAKAGAWA[9], Y.H. PU[4], A. RAY[5], D. SHAPIRA[5], B. SHIVAKUMAR[2], T. SUGIMITSU[10], J.P. WIELECZKO[11], M. YOULAL[1], K. YUASA-NAKAGAWA[9]

1 CRN Strasbourg, France
2 Wright Nuclear Structure Laboratory, Yale University, USA
3 CEN Saclay, DPhN/BE, France
4 Tsukuba University, Japan
5 Oak Ridge National Laboratory, USA
6 Laboratoire de Chimie–Physique et de Radiochimie, Nice, France
7 Rikkyo University, Tokyo, Japan
8 Shinshu University, Ueda, Japan
9 Riken Laboratory, Tokyo, Japan
10 Kyushu University, Fukuoka, Japan
11 Ganil, Caen, France

REFERENCES

1) B. HEUSCH, Proc. of the Tsukuba Int. Sym. on Heavy Ion Fusion Reactions, World Scientific Publishing, (1985) 172

2) D. BANGERT et al., Nucl. Phys. **A456** (1986) 489

3) For a general review see Ch. NGO, Progress in Particles and Nuclei 16 (1985) 139

4) W.J. SWIATECKI . Phys. Scr. (1981) 113

5) A. FAHLI et al., Phys. Rev. **C34** (1986) 161

6) J. PETER, Proc. of Riken – IN2P3 Symposium on heavy–ion collisions, Shimoda (1987) 357

7) D. DURAND et al. ; to be published

8) L.G. MORETTO, Nucl. Phys. **A247** (1975) 211

9) F. AUGER et al., Phys. Rev. **C35** (1987) 190

10) B. HEUSCH et al., Z. für Physik **A322** (1985) 309

11) S.M. LEE et al., Nucl. Phys. **A519** (1990) 221c

12) S.M. LEE et al., Proc. of the Obernai Second IN2P3 RIKEN Symposium on heavy Ion–collisions, World Scientific Publishing (1990) 155

13) R.J. CHARITY at al., Nucl. Phys. A (1989)

14) J. RICHERT and P. WAGNER, Z. für Physik **A330** (1988) 283

15) A. RAY, Invited talk contribution to this Conference

16) D.J. HINDE et al. , Phys. Rev. **C39**(1989)2268

17) The DEMON projekt, B. HEUSCH, Second IN2P3 RIKEN Symposium ; World Scientific Publishing (1990) 239

18) K. YUASA–NAKAGAWA, contribution to this conference

19) B. HEUSCH et al, Z. für Phys. **A312** (1983) 109

20) W.F.N. SCHNEIDER et al., Nucl. Phys. **A371**(1981)493

21) B. FAURE–RAMSTEIN, Orsay, Thèse N° 463 (1988)

22) T. MATSUSE, contibution to this conference

23) S. SANDERS et al., Phys. Rev. Lett. **61** (1988) 2154 and Phys. Rev. **C40** (1989) 2091

24) C. BECK et al., Z. für Phys. **A334** (1989) 521

25) C. BECK et al., Proc. of the Workshop on the Interface between Nuclear Structure and H.I. Dynamics, Notre Dame, (1990)

26) A. RAY et al., Phys. Rev. july 1991, to be published

27) C. BARBAGALLO et al., Z. für Phys. **A324**(1986)97

28) J. RICHERT, Contibution to the 4th Int. Conf. on nucleus–nucleus collisions, Kanazawa (1991) and references therein

29) Y.H. PU et al., private communication

30) BISPLINGHOFF et al., Phys. Rev. **C17**(1978)177

31) J. NATOWITZ et al., Phys. Rev. **C13**(1976)171

32) J.P. COFFIN et al., Phys. Rev. **C30**(1984)537

33) I. IORI et al., Phys. Lett. **132B**(1983)304

34) W. YOKOTA et al., Z. für Phys. **A333**(1989)379

35) Y. NAGAME et al., Z. für Phys. **A317**(1984)31

36) F.E. OBENSHAIN et al., Phys. Rev. **C18**(1978)764

37) J. BARRETTE et al., Nucl. Phys. **A279**(1977)125

38) A. RAY et al., Phys. Rev. **C31**(1985) 1573

39) H.Y. HAN et al., Nucl. Phys. **A492**(1989)138

COMPLEX FRAGMENT MASS DISTRIBUTIONS IN $^{84}Kr + ^{27}Al$ AT E_{lab} = 10.6 MeV/u

K. Yuasa-Nakagawa, Y. H. Pu, S. C. Jeong, H. Fujiwara, T. Mizota, S. M. Lee
Institute of Physics, University of Tsukuba, Tsukuba, Ibaraki 305, Japan
T. Nakagawa
Cyclotron Laboratory, RIKEN, Wako, Saitama 351-01, Japan
B. Heusch
CRN-Strasbourg, 67037 Strasbourg Cedex, France
K. Ieki
Department of Physics, Rikkyo University, Toshima, Tokyo 171, Japan
T. Sugimitsu
Department of Physics, Kyushu University, Fukuoka 812, Japan
T. Matsuse
Faculty of Textile Science and Technology, Shinshu University, Nagano 386, Japan

ABSTRACT

To study the decay mechanisms of intermediate mass fragments and fission products from a hot compound nucleus, we made an experiment for $^{84}Kr + ^{27}Al$ at the energy of 10.6 MeV/u using Riken Ring Cyclotron Facility. The distributions of nuclear element Z and its isotope mass are extensively analyzed in terms of Extended Hauser-Feshbach Method (EHFM) and it is concluded that the sequential decay plays an important role in the decay process from the highly excited compound nucleus. The dependence of element distributions on angular momentum is intensively discussed by comparing the two systems $^{84}Kr + ^{27}Al$ and ^{3}He + Ag, which form the same compound nucleus with different angular momentum window at the same excitation energy.

INTRODUCTION

The study of formation, equilibration and decay mechanisms of composite systems that result from the interaction of energetic heavy ions has been the subject of considerable experimental and theoretical interest for many years[1-4]. However, the dynamics of these highly excited systems that possess large angular momenta are still not well understood.

Light particle evaporation and fission are recognized as the dominant decay modes for highly excited nuclear systems. The emission of low energy intermediate mass fragments (IMF), with typical nuclear charges of $Z \geq 3$ and kinetic energies close to the exit channel Coulomb barrier, is a process distinct from light particle evaporation or fission[5-9], and is generally associated with the most violent collisions in intermediate and high energy nuclear reactions (E/A > 200 MeV). Numerous mechanisms have been investigated to account for IMF emission, including coalescence[10], decay of a localized "hot spot"[11,12], a liquid-gass phase transition[13] and the shattering of cold nuclear matter[14]. However, intermediate-mass fragment emission has also been observed in the decay of statistically equilibrated sources at relatively low excitation energies[15]. For this reason IMF emission must be viewed as a broad general phenomenon which links the excitation energy extremes from cold to very hot nuclei.

From these points of view, we decided to make an experiment for the system ^{84}Kr + ^{27}Al and study intensively the correlation between the angular momentum of a compound nucleus and the mechanism of particle emission, furthermore discover the outlook of decay mechanism from a hot compound nucleus.

Before we performed this experiment, two experiments for the same system at different incident energies have been already done, one is at lower energy 5.8 MeV/u[16] and another one is at higher energy 35 MeV/u[17]. Figure 1 shows the Z-distribution of this system at these two energies. We see a large difference between the Z-distributions of fully damped binary reaction products. At the energy of 5.8 MeV/u we can see the bulk at atomic number of 25, half of the atomic number of the compound nucleus, but at 35 MeV/u there is no such peak and appears to be very smooth. Since the change of incident energy influences the element distribution so strong, it may be a good to choose ^{84} Kr + ^{27}Al system for studying the decay mechanism of hot nucleus.

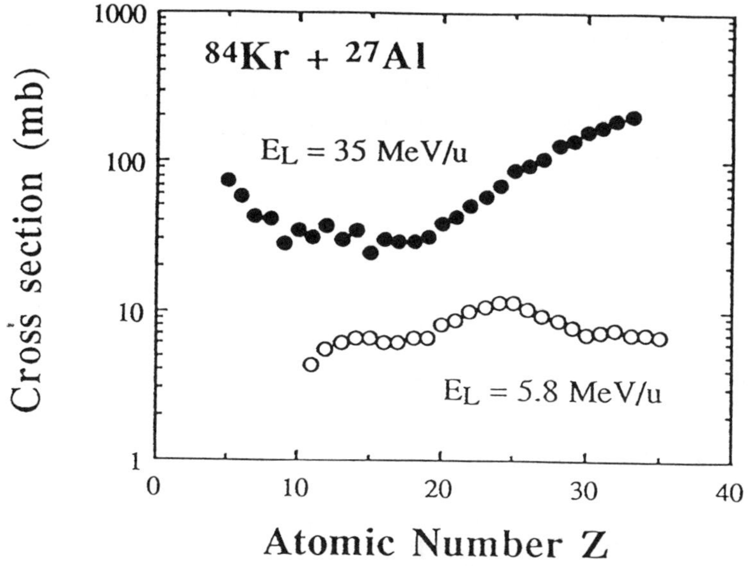

Fig.1 Z-distributions for the system ^{84}Kr + ^{27}Al at the incident energies of 5.8 and 35 MeV/u. Open circle shows the Z-distribution for E_L = 5.8 MeV/u and black circle for 35 MeV/u.

According to the empirical formula for complete and incomplete fusion[18], the incomplete fusion is already dominant at the incident energy of 35 MeV/u. Therefore it is very difficult to determine the initial mass and angular momentum distribution and excitation energy of the compound nucleus at this incident energy. This is the reason why we chose the lower energy, 10.6 MeV/u, where the complete fusion is still dominant. The preliminary results of this study have already been reported[19].

There is fortunately an experimental data for ^{3}He + Ag at the incident energy of 30 and 65 MeV/u[15,20]. This system forms the same compound nucleus of ^{111}In as ^{84}Kr + ^{27}Al does. Furthermore the excitation energies of ^{3}He + Ag at 30 and 65 MeV/u are almost the same as those of ^{84}Kr + ^{27}Al at 5.8 and 10.6 MeV/u, respectively (~100 and 200 MeV). Of course ^{3}He + Ag has the lower angular momentum than that of ^{84}Kr + ^{27}Al, if the excitation energy is the same. Figure 2 shows the Z-distribution for ^{3}He + Ag at 30 MeV/u and ^{84}Kr + ^{27}Al at 5.8 MeV/u. The excitation energy is about 100 MeV for both two systems and the maximum angular momenta are about 20ħ for ^{3}He + Ag and 63ħ for ^{84}Kr + ^{27}Al. We see how much the element distributions depend on the angular momentum comparing these two systems. It seems that the Z-distribution for the system ^{3}He + Ag is rather well described in terms of a simple power low[16], but for the system ^{84}Kr + ^{27}Al there are considerable element contributions originated from fission for the atomic number $Z \geq 10$. The difference of angular momentum influences the Z-distribution very much, as we see on these figures.

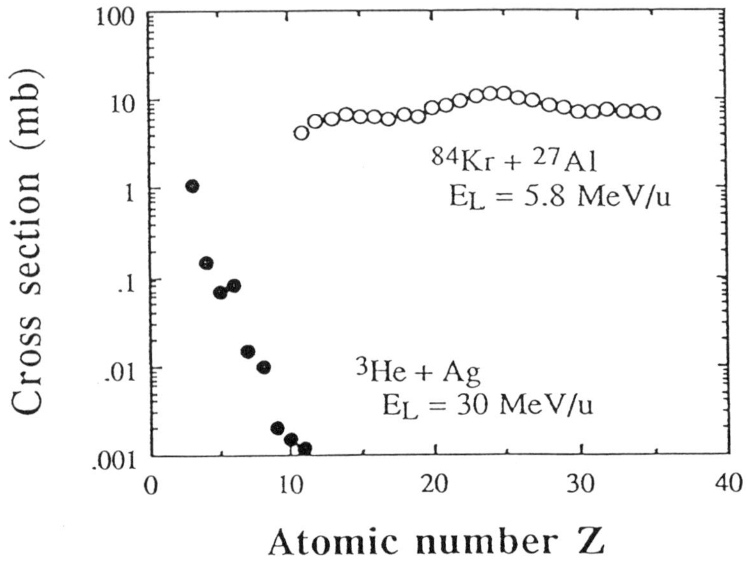

Fig.2 Z-distributions for the systems ^{84}Kr + ^{27}Al at the energy of 5.8 MeV/u (open circles) and ^{3}He + Ag at 30 MeV/u (black circles).

The aims of our experiments are as follows,
(1) How is the element distribution for ^{84}Kr + ^{27}Al at the incident energy of 10.6 MeV/u, where the complete fusion is still dominant?
(2) How does the high excitation energy and angular momentum of the compound nucleus influence element distributions?

EXPERIMENTAL SET-UP

This experiment has been performed in RIKEN Ring Cyclotron Facility. The self supporting ^{27}Al target of ~ 610 μg/cm^2 in thickness is bombarded with an 890 MeV ^{84}Kr beam (E_L/A = 10.6 MeV/u). Figure 3 shows the experimental set-up using a large cylindrical scattering chamber ASCHRA (3.0 m × 4.8 m)[21].

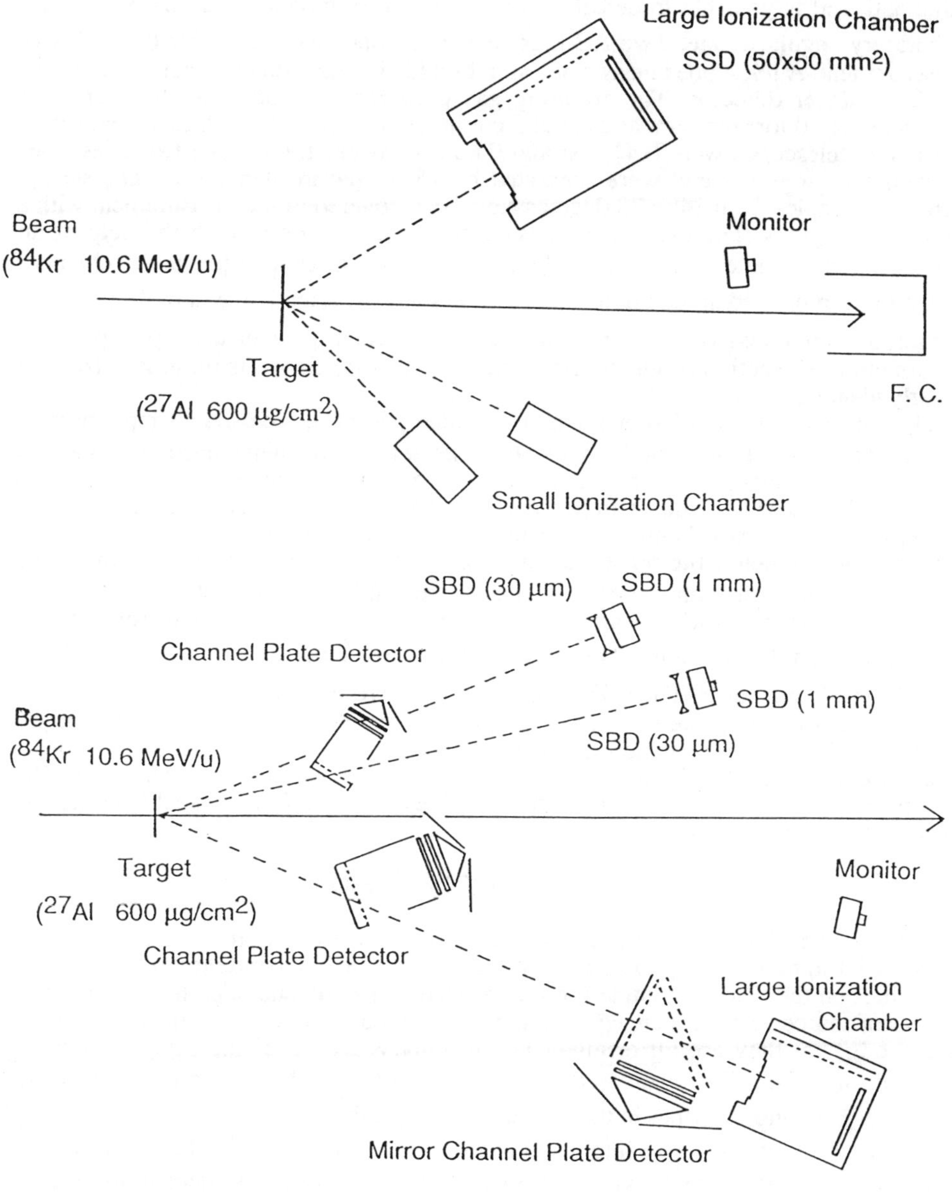

Fig.3 The experimental set-up used in a large cylindrical scattering chamber ASCHRA in RIKEN.

At first we concentrated to measure the element distribution of emitted particles and make a coincidence measurement between two fission fragments. By utilizing reverse-kinematics, it is possible to detect all of the source fragments in a relatively small laboratory angular range. Two ΔE-E counter telescopes were used for the inclusive measurement. A telescope consists of a small gridded ionization chamber and a silicon surface barrier detector. The counting gas used for the both gas chambers was isobutane of 30 torr pressure and about 5 cm in length. The solid angles subtended by these two telescopes were 0.41 msr and 0.83 msr, respectively. The two telescopes were moved together and were separated by 15 degree in plane. With this set-up, laboratory angles from 14 to 70 degrees have been covered in the measurement with a step of 2 degrees. The exclusive measurement was also done with the help of an additional gas counter telescope[22] which consists of a large area ionization chamber and an ion-implanted silicon detector which has a large effective area of 46 mm × 46 mm and a thickness of 300 μm. This large gas chamber can also give position information of reaction products. Using these detectors, elements up to Z = 16 were clearly identified.

For measurements of isotope mass of emitted particles, we used a simple time of flight method. The time of flight system consists of a channel plate detector[23] as a start detector and a surface barrier detector as a stop one. This channel plate detector is so compact that measurements at angles as forward as possible can be achieved. Another system consists of two channel plate detectors as a stop and start detector. These are a compact type which is the same type mentioned above and mirror type channel plate detector[24]. The flight path was 86 cm. The time resolution was almost 250 ps.

Furthermore, two surface barrier detector telescopes were used to measure atomic number Z of emitted particles. The thicknesses of surface barrier detectors were 30 μm for ΔE and 1 mm for E. The solid angles of these two telescopes were 0.13 msr and 0.2 msr, respectively. The telescopes sampled the angular range of laboratory angles from 6 to 19 degree, at which measurements were not done with the gas counter telescopes. We could identify Z up to 40 by using these silicon telescopes. The component corresponding to evaporation residues was clearly separated from fission products at the most forward angles.

RESULTS AND DISCUSSION

From the kinematical coincidence measurement, the source velocities of the observed fragments were extracted. We checked the independence of the source velocities on the detection angle for several elements and concluded that the average values of source velocities are equal to the recoil velocity of complete fusion products, indicating that they are all originated from binary decays of the same compound nucleus of ^{111}In. The measured angular distributions for fission products are essentially of $(\sin\theta)^{-1}$ shapes in the reference frame of the emitter.

Figure 4 shows the angle integrated cross section for ^{84}Kr + ^{27}Al at 10.6 MeV/u. The Z-distribution does not exhibit smooth $Z^{-\tau}$ behavior for this system at this energy but we can see the some Z-dependence of the absolute cross section for the lighter elements. For the heavier particles the cross section becomes smooth and the distribution has a heights resulting from fission. It is clear that the cross section for

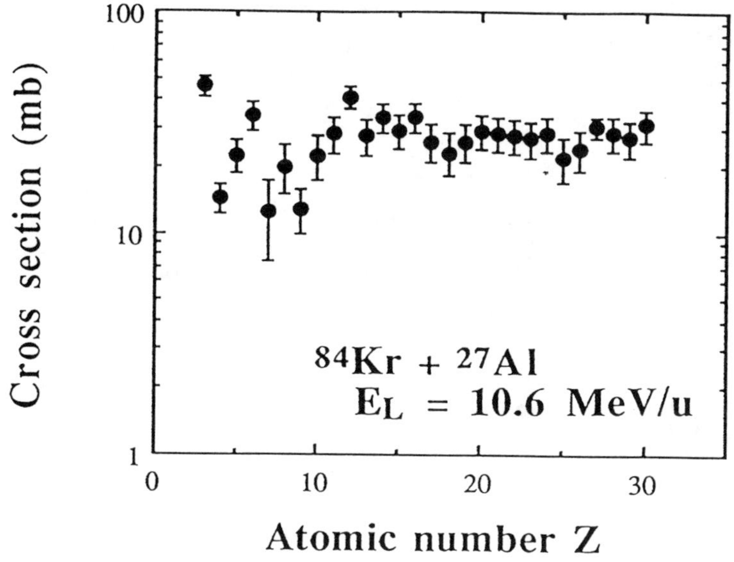

Fig.4 The obtained Z-distribution for ^{84}Kr + ^{27}Al at 10.6 MeV/u.

intermediate mass fragments obtained from ^{84}Kr + ^{27}Al is much larger than that from ^{3}He + Ag, even if the excitation energy is the same: in other words the difference of angular momentum plays an important role for the emission of intermediate mass fragments.

Separating evaporation residues from fission products at the forward angles, we obtained the fusion-evaporation cross section of 550 ± 100 mb. From the result of the absolute cross section mentioned above, we got the cross section for Z ≥ 5 of 500 ± 50 mb and total fusion cross section of 1050 ± 180 mb for ^{84}Kr + ^{27}Al at the incident energy of 10.6 MeV/u. In Figure 5 the observed reaction cross sections for ^{84}Kr + ^{27}Al at two energies vs $1/E_{CM}$ are indicated. The white circle means a total fusion cross section and black one for evaporation one. Solid lines show the estimation of the successive critical distance model[25]. The critical angular momentum corresponding to this total fusion cross secion is 83 ± 7 $\hbar$ and that of evaporation becomes 63 $\hbar$ using the sharp cut-off model.

The other purpose of this study is to try to interpret the apparent differences of the nuclear element distributions observed in the compound nucleus of ^{111}In formed through ^{84}Kr + ^{27}Al at 5.8 MeV/u and by ^{3}He + Ag at 30 and 65 MeV/u consistently in terms of statistical compound nucleus model. We used the Extended Hauser-Feshbach Method (EHFM)[26]. It is first of all essential to reproduce these two experimental data simultaneously and to give a reasonable interpretation why the ^{84}Kr + ^{27}Al system enhances the fission products part but the ^{3}He + Ag system does the intermediate mass fragment alone. Figure 7 and 8 show the calculated Z-distribution of fission fragments for ^{84}Kr + ^{27}Al at the incident energy 5.8 MeV/u and for ^{3}He + Ag

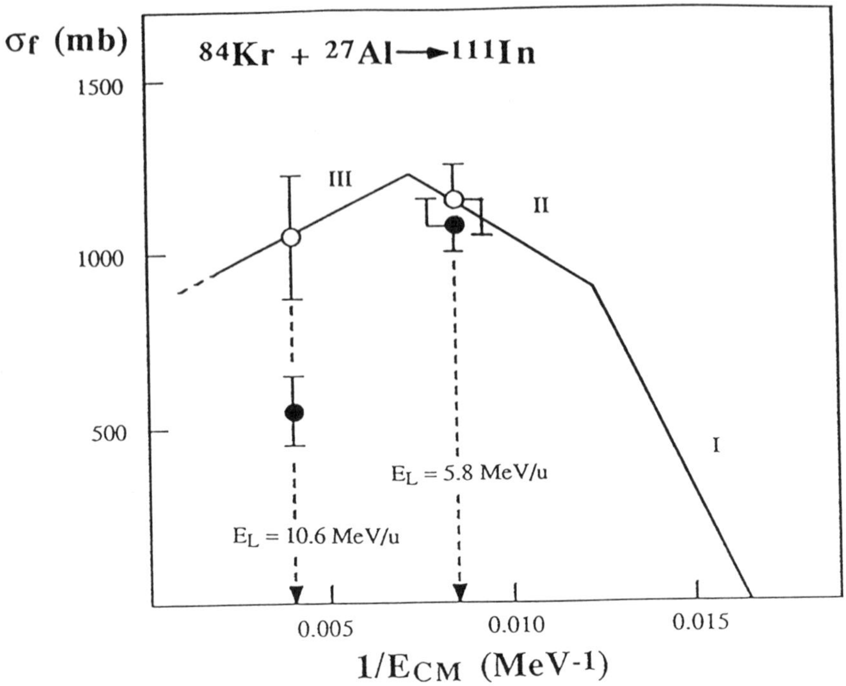

Fig.5 The fusion cross section vs. $1/E_{CM}$: the solid lines are calculated by the model (Ref.25), open circles show the experimental fusion cross sections and black circles the experimental fusion-evaporation cross sections (Ref.27 and the present work).

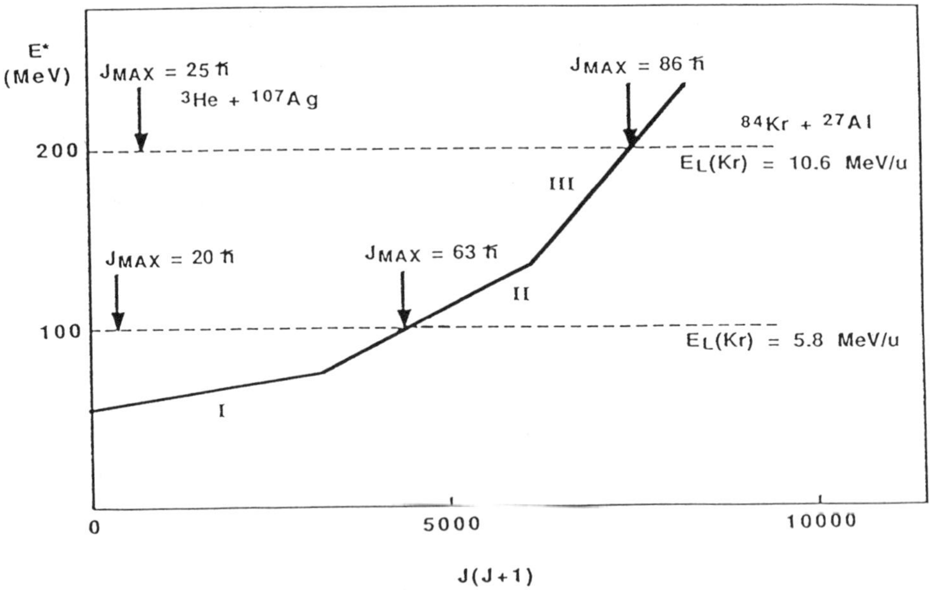

Fig.6 The representative plot of total angular momentum J vs. the excitation energy E^* in the compound nucleus In.

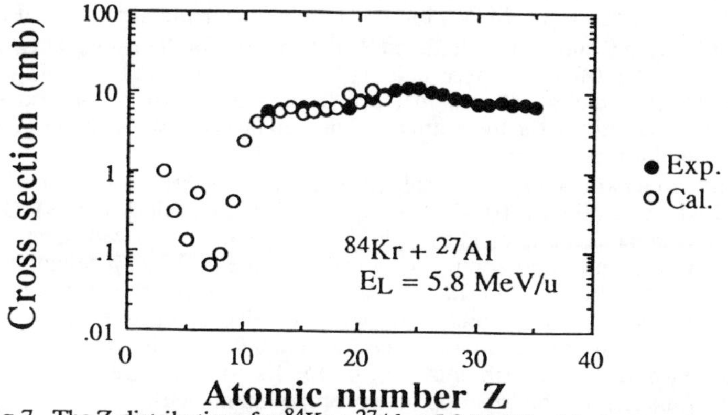

Fig.7 The Z-distributions for $^{84}Kr + ^{27}Al$ at 5.8 MeV/u : black circle are the experimental data (Ref.16) and open circles the calculated results using Extended Hauser-Feschbach Method.

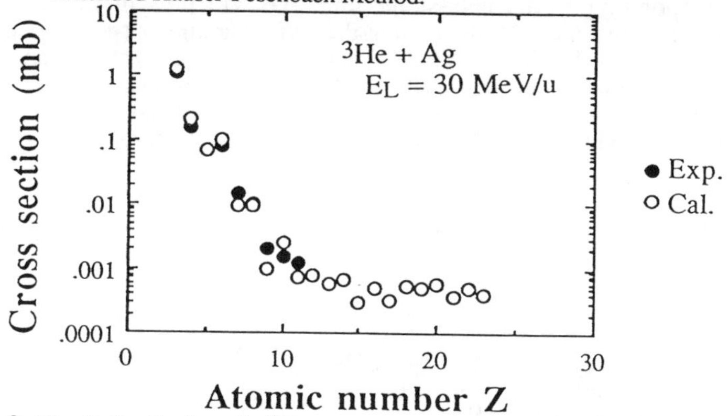

Fig.8 The Z-distributions for ^{3}He + Ag at 30 MeV/u : black circles are the experimental data (Ref.15) and open circles the calculated results.

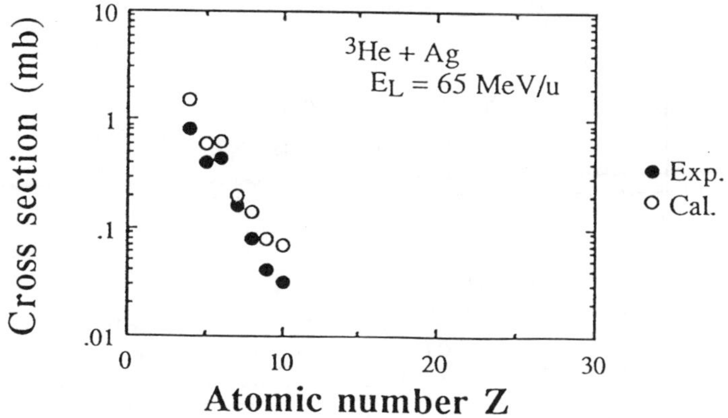

Fig.9 The Z-distributions for ^{3}He + Ag at 65 MeV/u : black circles are the experimental data (Ref.20) and open circles the calculated results.

at 30 MeV/u with the experimental data, respectively. These two systems have the same excitation energy (100 MeV) but the angular momenta are different (20 ℏ and 63 ℏ). Figure 8 and 9 show the calculated Z-distribution for ^{3}He + Ag at two different energies 30 and 65 MeV/u. These calculations are performed keeping the same parameter set. It is clearly shown in these figures that this calculation is able to reproduce Z-distributions for the compound nucleus having wide range of excitation energy and angular momentum.

Using the same parameter set, we calculated the case of ^{84}Kr + ^{27}Al at 10.6 MeV/u. The result is shown in Figure 10 with the experimental data (black points). Open circle shows the first-step calculation and triangle shows the calculation with sequential decay process, i.e., multi-step calculation. We can see that the multi-step calculation has the better agreement with the experimental data especially for the intermediate mass fragments. In this multi-step calculation we have also taken into account the IMF emission after the evaporation decay. It should be noted that the discrepancies between the calculation and the experimental results for fission part are mainly due to the uncertainty of momentum distribution in the compound nucleus.

Now we will discuss about the isotope distribution. Figure 11 (a), (b) and (c) show the isotope distributions for each atomic number. The peaks for each particle are almost corresponding to the beta-stable-line. We will present the calculation of Extended Hauser-Feshbach Method using the same parameter set as the case of the absolute cross section. In Figure 12 (a), (b) and (c) the isotope distributions of carbon, potassium and chlorine are presented, respectively . It is clear that the peak of the first-step calculation is much larger than that of experimental data, but the multi-step calculation can reproduce the peak of the experimental data very nicely especially in the cases of

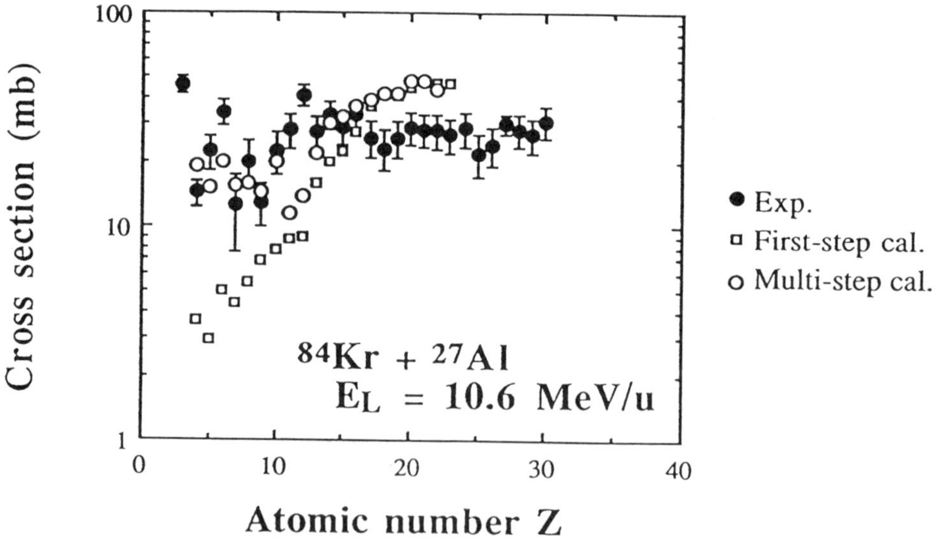

Fig.10 The calculated results of Z-distribution for ^{84}Kr + ^{27}Al at 10.6 MeV/u : black points are the experimental data obtained in this work and open squares show the results of first-step calculation and open circles the multi-step calculation.

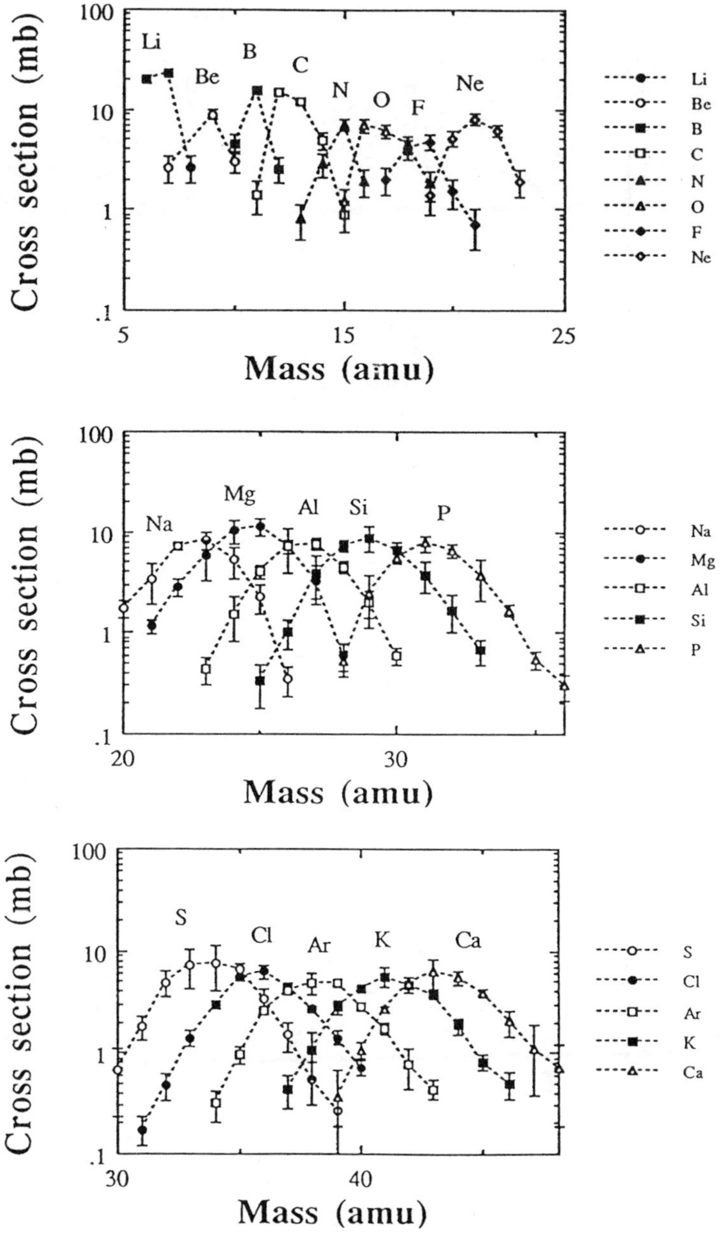

Fig.11 The isotope mass distributions for $^{84}Kr + ^{27}Al$ at 10.6 MeV/u : (a) shows from Li to Ne, (b) from Na to P and (c) from S to Ca.

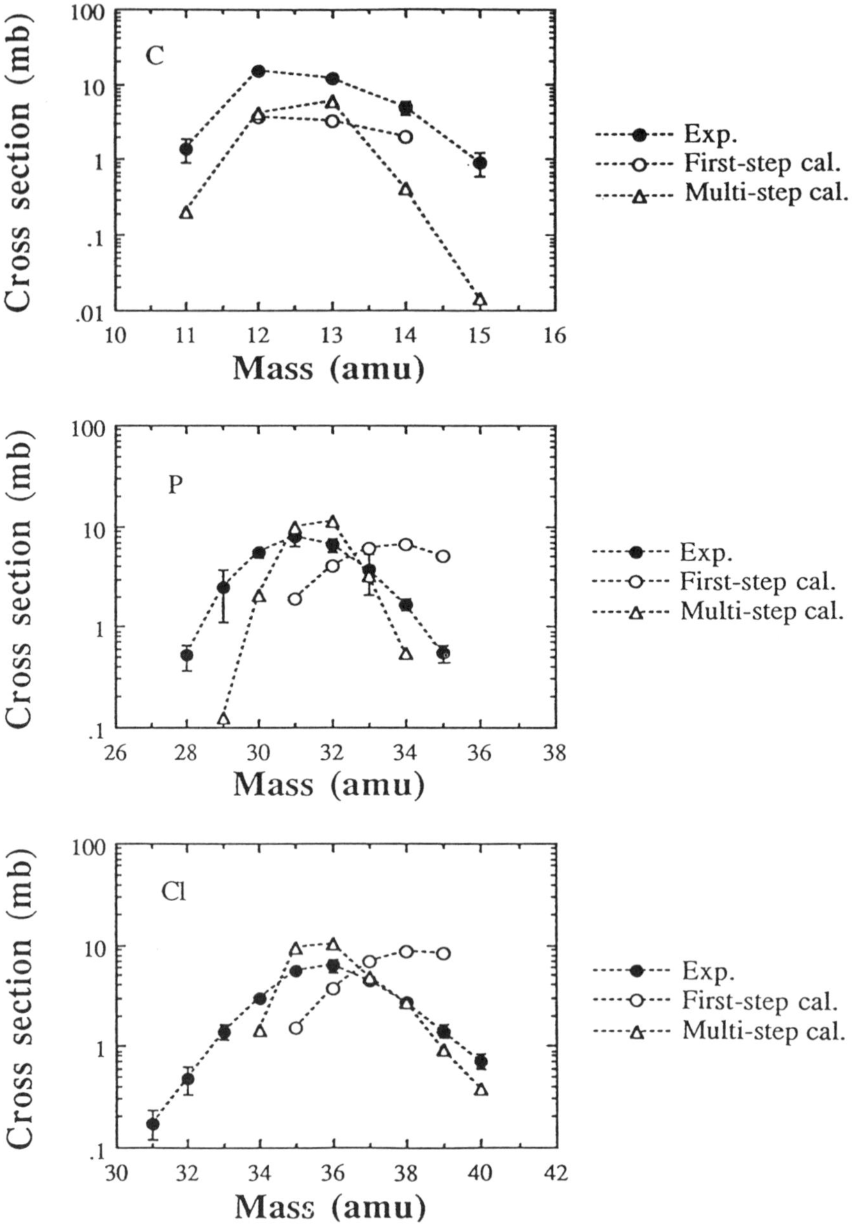

Fig.12 The calculated results of the isotope mass distributions : black points show the experimental data obtained in this work, open circles are the results of the first-step calculation, and open triangles the multi-step calculation. (a) shows the case of carbon, (b) potassium and (c) chlorine.

potassium and chlorine, which must be originated from the fission process. That means the sequential decay process plays an important role in the decay mechanism through fission at this energy, i.e., fission products also deexcite by emitting light particle (post-fission evaporation).

CONCLUSION

From the calculation of mass distribution, it is concluded that the sequential decay process plays an important role in the reaction of $^{84}Kr + ^{27}Al$ at higher incident energy.

The experimental element and mass distributions are satisfactory reproduced by the calculation of the extended Hauser-Feshbach Method over the wide range of the excitation energy and angular momentum.

REFERENCES

1. C. K. Gelbke et al., Nucl. Phys. A269, 460 (1976).
2. T. Matsuse et al., Proc. of Tsukuba Fusion Symp. p113 (1984).
3. R. J. Charity et al., Nucl. Phys. A483, 371 (1988).
4. J. Gomez del Campo et al., Phys. Rev. C43, 2689 (1991).
5. A. M. Poskanzer et al., Phys. Rev. C3, 882 (1971).
6. G. D. Westfall et al., Phys. Rev. C17, 1368 (1978).
7. R. E. L. Green et al., Phys. Rev. C22, 1594 (1980).
8. J. E. Finn et al., Phys. Rev. Lett. 48, 1321 (1982).
9. R. W. Minich et al., Phys. Lett. 118B, 58 (1982).
10. D. Boal, Phys. Rev. C28, 2568 (1983).
11. D. J. Fields et al., Phys. Rev. C30, 1912 (1984).
12. W. A. Friedman et al., Phys. Rev. C28, 16 (1983).
13. P. J. Siemens, Nucl. Phys. A428, 189c (1984).
14. J. Aichelin et al., Phys. Rev. C30, 107 (1984).
15. L. G. Sobotka et al., Phys. Rev. Lett. 51, 2187 (1983).
16. B. Heusch et al., Z. Phys. A322, 309 (1985).
17. F. Auger et al., Phys. Rev. C35, 190 (1987).
18. H. Morgenstern et al., Phys. Rev. Lett. 52, 1104 (1984).
19. S. M. Lee et al., Nucl. Phys. A519, 221c (1990)
20. K. Kwiatkowski et al., Phys. Lett. 171, 41 (1986).
21. T. Nakagawa et al., Riken Accel. Prog. Rep. 22, 147 (1988).
22. Y. H. Pu et al., UTTAC Annual Rep. p13 (1988).
23. T. Mizota et al., UTTAC Annual Rep. p18 (1988).
24. T. Mizota et al., Nucl. Inst. Meth. A305, 125 (1991)
25. S. M. Lee et al., Proc. of Tsukuba Fusion Symp. p63 (1984).
26. T. Matsuse et al., Contribution in this symposium.
27. W. Schneider et al., Nucl. Phys. A371, 493 (1981).

STATISTICAL EMISSION OF COMPLEX FRAGMENTS FROM HIGHLY EXCITED COMPOUND NUCLEUS

T. MATSUSE , S. M. LEE* , Y. H. PU*, K. Y. NAKAGAWA*
C. BECK** and T. NAKAGAWA***

Faculty of Textile Science and Technology, Shinshu Univ., NAGANO 386, Japan.
* Institute of Physics, Univ. of Tsukuba, IBARAKI 305, Japan.
** Centre de Recherches Nucléaires, et Université Louis Pasteur
B. P. 20, F-67037 Strasbourg, CEDEX, France.
*** Cyclotron Laboratory, RIKEN, Wako, SAITAMA 351-01, Japan

Abstract

In order to study the mechanism of complex fragments production from highly excited light and medium compound nucleus induced by relatively low energy heavy ion reactions, the full statistical analysis have been performed in terms of Extended Hauser Feshbach (EHF) method.

At first the charge- and kinetic energy- distributions of $^{35}Cl +^{12} C$ reaction at (E_{Lab} = 180,200 MeV) and $^{23}Na +^{24} Mg$ reaction at (E_{Lab} = 89 MeV) which form the ^{47}V compound nucleus of almost same excitation energy are extensively investigated as a prototype of the light mass system. The variations in observed cross section from fragment to fragment are understood by the variations of binding energy of the lighter fragments of binary decay from the compound nucleus. The difference of the yield in the measured cross sections between the reactions is interpreted as the entrance channel effect that $^{23}Na +^{24} Mg$ channel has the larger critical angular momentum for fusion cross section than $^{35}Cl +^{12} C$ channel. The measured kinetic energy distributions in the laboratory system of the complex fragments are shown to be well reproduced by the EHF-method. Therefore the observed complex fragment production are understood as the statistical binary decay from the compound nucleus induced by heavy-ion reaction.

Next, the EHF-method is extensively applied to the study of the complex production from the ^{111}In compound nucleus which formed by the $^{84}Kr +^{27} Al$ reaction at (E_{Lab} = 890 MeV) . Because the complex fragments (fissioning fragments) in the first step decay from this compound nucleus have large spins and highly excitation energies, the complex fragments decay sequentially by emitting the light particles. It is shown that the effect of multi-step cascade decay of fissioning fragments is very important for reproducing the general trend of the observed quantities such as the observed isotope-mass distributions.

1. Introduction

Although, the study on the mechanism of complex fragment production from hot nucleus may be one of the interesting subjects in the heavy ion physics in the intermediate energy region, it seems to be not straight forward to realize how to understand the reaction mechanism and what is important. Of course several models and physical pictures may exist for the studies, but it is even now very important to recognize that to what extend the full equilibrate model is valid for reproducing the observed quantities. Then we concentrate our investigation on the statistical emission of the complex fragments from highly excited compound nucleus formed in the relatively low energy heavy ion reactions.

Though it is better to study the phenomena comprehensively, in order to make clear our aim, but we take the following two compound nuclei as the prototype in this study; ^{47}V compound nucleus of the lighter mass system and ^{111}In compound nucleus of the medium mass one. Because the general introduction of the aim of this study and reviews have been already presented by Heusch [1] and by Yuasa-Nakagawa [2], here some complementary introductions are given.

In the heavier mass system than $A_{CN} > 70$, it is well recognized that the fission-like phenomena really exist as the binary decay from the fused system in the low energy heavy-ion reactions such as $^{32}S +^{78} Ge$ [3] and $^{32}S +^{59} Co$ [4] reactions. Then it is expected to be measured the fission-like phenomena from the fused system in the lighter mass system $A_{CN} < 70$.

In recent years, Sanders et al. [5] have measured fully energy-dumped complex fragment production from the $^{32}S +^{24} Mg$ reaction at (E_{Lab} = 121,142 MeV) and concluded that the fragments can be understood as the binary decay (fission-like decay) from the compound nucleus. They have also discussed the reaction mechanism in terms of the transition-state model [6] and the equilibrium orbiting-model [7] and showed that the transition-state model reproduces well the general trend of the mass distributions. But it seems to be noticed that they have needed the introduction of the Wigner term to the mass of the decaying fragments with $\alpha -$ like *nuclei* in order to get the enhancements in the mass distribution.

In the more lighter mass system, the complex fragment's detection has been performed by means of a kinematical coincidence technique in the $^{35}Cl +^{12} C$ reaction at a bombarding energies of 180MeV [8],and 200 MeV [9] and the charge distributions have been obtained with the large fluctuations from fragment to fragment. It is also discussed [9] that the origin of the heavy fragment production is the fusion-fission of the ^{47}V compound nucleus rather than an orbiting mechanism.

Before the performance of these essential experiments [5,8,9] on the measurements of fission-like fragments, we have already proposed to extend the Hauser-Feshbach method (EHF-method) to the calculation of the cross section of decaying complex fragments (fission-like fragments) from the fused system induced by the heavy ion reaction [10]. The EHF method should be considered to attempt for treating light-particle emission and complex fragment production (fission-like production) in the equivalent statistical way from a view point of scission point model. Especially if we adequately use the observed binding energy of the fragments in the phase space calculation, it has shown that EHF-method is very valid for reproducing the fission-like cross sections of the medium mass systems [11]. Then the EHF-method have been applied to the interpretation of the fission-like phenomena of the reactions $^{32}S +^{24} Mg$ at (E_{Lab} = 121,142 MeV) and $^{35}Cl +^{12} C$ at (E_{Lab} = 180,200 MeV) and we have showed that the observed complex fragment production are understood as the statistical binary decay from the compound nucleus induced by heavy-ion reaction [13]. A brief presentation is given on the EHF-method for these lighter mass system in section 2.

More recently Djerroud et al [13] have performed the experiment of $^{23}Na +^{24} Mg$ reaction at (E_{Lab} = 89 MeV) which forms the ^{47}V compound nucleus of almost same excitation energy with that of the $^{35}Cl +^{12} C$ reactions [8],[9]. The important point of these experiments is that these two channels with different mass asymmetry forms the ^{47}V nucleus with the same excitation energy (E^*_{CN} = 64 MeV) and the full equilibrated cross sections (fusion-fission-like and fusion-evaporation) have been completely measured in the both channels. It is obtained that the fission-like fragments of the $^{23}Na +^{24} Mg$ reaction have the larger cross sections than that of the $^{35}Cl +^{12} C$ reaction but the measured charge distributions in the both channels show the almost same fluctuation from fragment to fragment. The discussion on the difference in the observed quantities between the $^{35}Cl + ^{12}C$ reaction at $E_{Lab} = 200$MeV [9] and $^{23}Na + ^{24}Mg$ reaction at $E_{Lab} = 89$MeV [13] is presented by showing the calculated results of EHF-method in Section 3.

On the complex fragment production from the ^{111}In compound nucleus, we concentrate how the EHF-method can reproduce the observed quantities which measured in the reaction of $^{84}Kr + ^{27}Al$ at $E_{Lab} = 890$MeV [2]. The preliminary results and discussion on this reaction have already been presented [14]. As can be seen the measured charge distribution (see Fig.4 of Ref[2]), not only the fission-like fragments have the large cross section and but also the significant enhancements have been measured in the mass region of intermediate fragments $Z < 10$. Then it is interesting that the obtained critical angular momentum J_{cr} for the total fusion cross section formation exceeds the limit of the Rotating Liquid Model (see the discussion in Ref.[2]). Furthermore the essential point of the experiment is that the isotope mass distributions for all complex fragment production have been measured (see Fig.11 of Ref[2]). These observed isotope mass distributions will give a good criterion for identifying the full equlibriated nature of the ^{111}In compound nucleus.

Because the formed compound nucleus ^{111}In of this experiment has the high excitation energy $E^* = 216$MeV , it is easily expected that the emitted complex fragments and evaporation residues at the first step emission have the highly excited energy and spin distributions and the multi-step cascade decays occur sequentially. In this report the effect of the multi-step cascade calculation are discussed by showing the calculated results of the charge, isotope-mass and kinetic energy distributions in section 4.

The detailed interpretations on the interrelation of this experiment and the other ones such as $^{84}Kr + ^{27}Al$ at $E_{Lab} = 5.8$ and 35MeV/u and $^{3}He + Ag$ at $E_{Lab} = 30$ and 65MeV/u are shown by Yuasa-Nakagawa [2].

2. Extended Hauser-Feshbach Method

The outline of Extended Hauser-Feshbach method is shown to calculate the cross sections of the full statistical binary decay from the highly excited compound nucleus. Incident channel with projectile (Z_P, A_P) and target (Z_P, A_T) of CM-energy E_{CM} forms the compound system (Z_{CN}, A_{CN}) of excitation energy E^*. Here, $E^* = E_{CM} + Q_0$, and Q_0 is Q-value of incident channel. The total angular momentum is represented by J. The formed compound state decays through the usual particle evaporation of n, p, α and the binary decay of complex fragment set $((Z_L, A_L)$ and $(Z_H, A_H))$ with angular momentum L and energy E of the relative motion and channel spin I.

Because we consider only full statistical decay from the compound nucleus, it is assumed that the averaged S-matrix $\overline{|S^{(J)}_{L,I}(E)|^2}$ is available for calculating the cross section. And the cross section of an exit channel is represented by the usual formula in the case of initial channel spin zero.

$$\frac{d^2\sigma}{d\Omega dE} = \frac{\pi\hbar^2}{2\mu_c E_{CM}} \sum_J (2J+1) \sum_{L,I} \overline{|S^{(J)}_{L,I}(E)|^2} \sum_M (I - MLM|J0)^2 |Y_{LM}(\theta\phi)|^2 \qquad (1)$$

Here μ_c denotes the reduced mass of the incident channel.

Extended Hauser-Feshbach method

In order to calculate the averaged S-matrix, the Hauser-Feshbach method is applied. By the use of the phase space $N^{(J)}_{LI}(E)$ of a exit channel, the averaged S-matrix is defined by the following formula

$$\overline{|S_{L,I}^{(J)}(E)|^2} = T_{(J)}^{(fus)}(E_{CM})\frac{N_{LI}^{(J)}(E)}{D^{(J)}} \tag{2}$$

Here $T_{(J)}^{(fus)}(E_{CM})$ is the transmission coefficient of the initial channel which forms fusion cross section. In this study we use following simple distribution for the transmission coefficient,

$$T_{(J)}^{(fus)}(E_{CM}) = \frac{1}{1+\exp((J-J_{cr})/\Delta_J)} \tag{3}$$

The critical angular momentum J_{cr} is determined by our fusion model (see reference [15] for details). The defuseness parameter Δ_J is used to be $1\hbar$ in the present calculation. The total phase space $D^{(J)}$ is represented by the following formula as the summation of whole decay channels f

$$D^{(J)} = \sum_f \int \sum_{(L,I)J} N_{LI}^{(J)}(E)dE + N_\gamma^{(J)}(E^*) \tag{4}$$

Here $N_\gamma^{(J)}(E^*)$ denotes the phase space of γ-ray decay. In our calculation giant-E1-transition is included.

Phase space of light particle evaporation

The usual Hauser-Feshbach Method is used to calculate the phase space of the light particle evaporation and the formula is represented as follows

$$N_{LI}^{(J)}(E) = \int \rho_I(\epsilon)T_L(E)\delta(\epsilon + E + Q - E^*)d\epsilon \tag{5}$$

Here $T_L(E)$ is the transmission coefficient of the n, p, α channel and are calculated by the use of the standard optical potential model. Level density formula $\rho_I(\epsilon)$ which has been shown by Bohr and Mottelson [16] is used

$$\rho_I(\epsilon) = \frac{1}{24}\left(\frac{a\hbar^2}{2g}\right)^{3/2}(2I+1)a\frac{e^{2\sqrt{X}}}{X^2} \tag{6}$$

Here X is defined by the following formula

$$X = a(\epsilon - \frac{\hbar^2}{2g}I(I+1) - \Delta_{pair}) \tag{7}$$

ϵ is the whole excitation energy of the residual nucleus.

In this calculation the shell structure effect of the nucleus is introduced to the level density parameter a as the temperature dependent one (see the details Ref[12]). Moment of inertia g is evaluated by the following rigid body one. $g = \frac{2}{5}AM_n(r_0A^{1/3})^2$ where $r_0 = 1.2$fm is used here and M_n is the nucleon mass. The pairing energy $\Delta_{pair} = \Delta_n + \Delta_p$ is evaluated by the use of the formula which is given in Ref.[16], The value of Q is the usual separation energy which is evaluated by the following formula,

$$Q = B_{GS}(N_{CN}, Z_{CN}) - B_{GS}(N_L, Z_L) - B_{GS}(N_H, Z_H) \tag{8}$$

where the $B_{GS}(N, Z)$ is the observed binding energy of Ref.[17]. Then it should be noticed that the calculated phase space depends strongly on the binding energy of the decaying fragments.

Phase-space of the complex fragment and fission

To calculate the phase space of the complex fragment, we categorize into two parts. The first one is the discrete low excitation below the threshold of the light particle's separation energy. The second one is the usual statistical excitation which is represented by the level density formula of eq.(6). Then the calculation of the phase space of the complex fragment and fission is divided into four parts as follows.

$$N_{LI}^{(J)}(E) = N_{LI}^{(J)}(E)^{(1)} + N_{LI}^{(J)}(E)^{(2)} + N_{LI}^{(J)}(E)^{(3)} + N_{LI}^{(J)}(E)^{(4)} \tag{9}$$

$$N_{LI}^{(J)}(E)^{(1)} = \int\int \sum_{(I_L, I_H)I} \rho_{I_L}(\epsilon_L)\rho_{I_H}(\epsilon_H)T_L(E)\delta(\epsilon_L + \epsilon_H + E + Q - E^*)d\epsilon_L d\epsilon_H \tag{10}$$

$$N_{LI}^{(J)}(E)^{(2)} = \sum_i \int \sum_{(I_{L_i}, I_H)I} \rho_{I_H}(\epsilon_H)T_L(E)\delta(\epsilon_{L_i} + \epsilon_H + E + Q - E^*)d\epsilon_H \tag{11}$$

$$N_{LI}^{(J)}(E)^{(3)} = \sum_j \int \sum_{(I_L, I_{H_j})I} \rho_{I_L}(\epsilon_L)T_L(E)\delta(\epsilon_L + \epsilon_{H_j} + E + Q - E^*)d\epsilon_L \tag{12}$$

$$N_{LI}^{(J)}(E)^{(4)} = \sum_i \sum_j \sum_{(I_{L_i}, I_{H_j})I} T_L(E)\delta(\epsilon_{L_i} + \epsilon_{H_j} + E + Q - E^*) \tag{13}$$

Here the discrete levels are represented by the use of i and j for lighter part and heavier part of binary fragment, respectively. The i-th discrete excitation energy and spin of lighter part is represented by $(\epsilon_{L_i}, I_{L_i})$. The j-th discrete excitation energy and spin of heavier one is represented by $(\epsilon_{H_j}, I_{H_j})$.

Evaluation of the transmission coefficient $T_L(E)$

The transmission coefficient $T_L(E)$ at the scission point of the relative motion between two fragments is simply represented by the following formula for all type of the excitation of the fragments

$$T_L(E) = \frac{1}{1 + \exp((V(L) - E)/\Delta_L)} \tag{14}$$

where Δ_L is the defuseness parameter and the value of 1MeV is used for all channels. In this calculation we use the Krappe-Nix potential [18] for evaluating the barrier height of $V(0)$. This $V(0)$ is the Coulomb barrier V_c of the relative motion between binary fragments. Then the barrier height $V(L)$ in the relative motion for angular momentum L is parameterized by the use of the following simple formula ,

$$V(L) = V(0) + \frac{\hbar^2}{2\mu_f R_s^2} L(L+1) \tag{15}$$

where μ_f is the reduced mass between fragments. Here R_s is the scission point between fragments.

Here, in order to make clear the physical picture of our model, we introduce the barrier height of the scission point V_s for total spin $J = 0$ and angular momentum $L = 0$ which is measured from the ground state of the compound nucleus as follows,

$$V_s = V_c + Q. \tag{16}$$

The schematic representation is given in the Fig.2(b). As can be understood from the formulae of phase space calculation which are shown in the above equations, the values of the phase space are essentially limited by this barrier height of V_s for the case of total spin $J = 0$.

3. Results and discussions for ^{47}V compound nucleus

The calculated charge distributions of $^{35}Cl +^{12} C$ and $^{23}Na +^{24} Mg$ reactions are shown in Fig.1 by comparing with the experimental ones in the wide charge range from fusion-fission-like to fusion-evaporation distributions [8,9,13]. On the charge and kinetic energy distributions of $^{35}Cl +^{12} C$ at E_{Lab} = 180MeV shown in Fig1.(a), we have already discussed in Ref.[12]. One of the main aim of this report is to discuss the difference in the observed quantities between the $^{35}Cl +^{12} C$ reaction at E_{Lab} = 200MeV [9] and $^{23}Na +^{24} Mg$ reaction at E_{Lab} = 89MeV [13] by comparing with the calculated results of EHF-method.

The important point is that these two channels with different mass asymmetry forms the ^{47}V nucleus with the same excitation energy ($E^*_{CN} = 64MeV$) and the full equilibrated cross sections σ_{fus} (fusion-fission-like σ_{FF} and fusion-evaporation σ_{FE}) have been completely measured in the both channels. The values of the J_{cr} for the total fusion cross section formation (see eq.(3)) are obtained by the use of the measured fusion cross sections σ_{fus} as follows; the $J_{cr} \simeq 24\hbar$ for the $^{35}Cl +^{12} C$ reaction at E_{Lab} = 200MeV [9] and $J_{cr} \simeq 29\hbar$ for the $^{23}Na+^{24} Mg$ reaction at E_{Lab} = 89MeV [13]. Then it is expected that the larger fusion fission-like cross sections in the $^{23}Na+^{24} Mg$ reaction than in $^{35}Cl +^{12} C$ reaction is reasonably understood by the difference of the observed J_{cr} -values.

It is very interesting that these obtained J_{cr} -values are well reproduced by our fusion model [15]. In the case of entrance channel with mass symmetric $^{23}Na +^{24} Mg$ at E_{Lab} = 89MeV, the fusion cross section is limited by the statistical yrast line, on the other hand the fusion cross section with asymmetric mass entrance channel of $^{35}Cl +^{12} C$ at E_{Lab} = 200MeV is determined by the successive critical distance [15]. Then we can discuss what is the entrance channel effect.

As can be seen in Fig.1, the over all of the observed charge distributions are well reproduced by the EHF-method calculation. The charge distributions with $Z < 11$ which correspond to the fusion fission-like one are the first-step binary decay from the ^{47}V compound nucleus. On the other hand the charge distributions with $Z > 12$ which correspond to the fusion evaporation are the result of multi-step cascade calculation. The reason why the EHF-calculation for these reactions can well reproduce not only the yields of the measured cross sections but also the variations from fragment to fragment is that the observed binding energy for the binary fragments at scission point is used to evaluate the separation energy Q -value (see eq.(8)). In Fig.2(a) it is shown how the barrier height of the scission point depends on the Q -values of the decaying fragments. The schematic interpretation of relation between the barrier height of scission point V_s defined in eq.(16) and the Coulomb barrier V_c in the relative motion is shown in Fig2.(b). Because so-called α -like nuclei in the p-shell and sd-shell region have strong binding energy, barrier height V_s of the α -like fragments becomes low. Then the strong

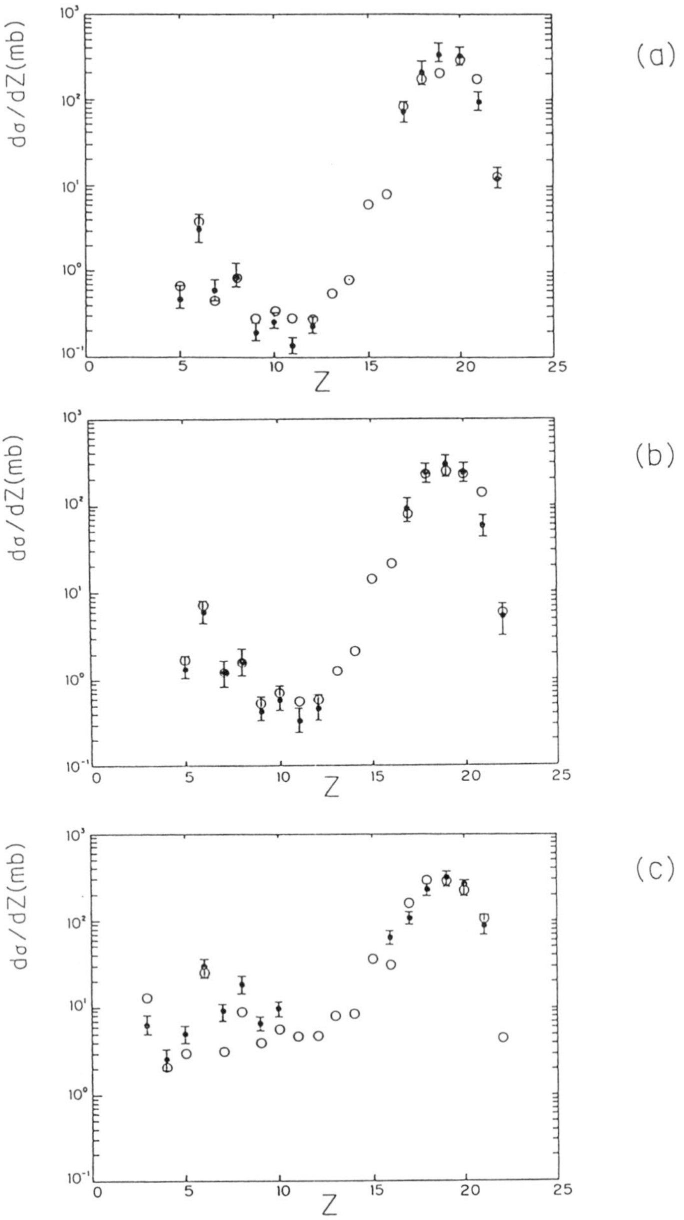

Fig.1 Comparisons of calculated charge distributions (open circles) by EHF-method with experimental ones (solid circles). (a) $^{35}Cl + ^{12}C$ reaction at E_{Lab} = 180MeV [8], (b) $^{35}Cl + ^{12}C$ reaction at E_{Lab} = 200MeV [9], (c) $^{23}Na + ^{24}Mg$ reaction at E_{Lab} = 89MeV [13].

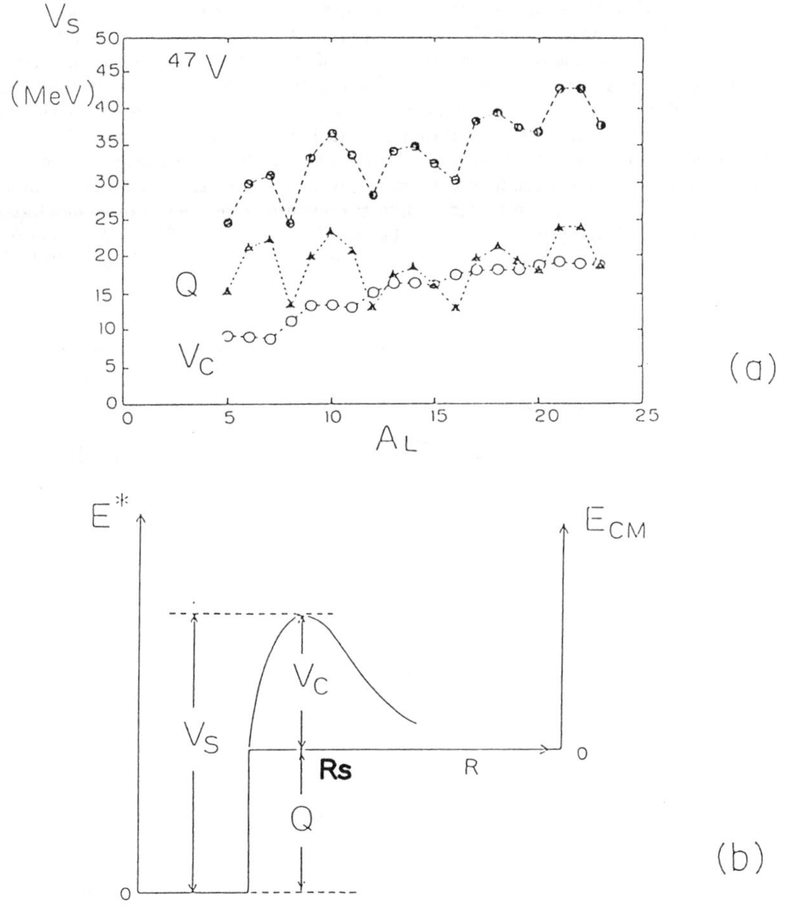

Fig.2 (a) Fission fragments dependence of barrier height of scission point for the ^{47}V-compound nucleus. Variable A_L shows the mass number of the lighter part of binary decay from compound nucleus. The plotted barrier height V_s (closed circles) of scission point are the lowest barrier in the available binary configurations for total spin $J = 0$ and anguler momentum $L = 0$. Open circles and triangles show the Coulomb barriers V_c and Q-values, respectively. (b) The schematic representation of relation between barrier height V_s of scission point and Coulomb barrier V_c. The V_s is measured from the ground state of the compound system and V_c is defined by the relative distance R between decaying binary fragments. The R_s is the distance of scission point. The Q denotes the separation energy of the decaying binary fragment from the compound nucleus.

enhancement with even Z in charge distributions is reasonably understood. Therefore the nascent fragment from the compound nucleus around the scission point are expected to have the almost same properties of the individual nucleus.

In order to show how the fusion fission-like cross section depends to the total spin J -values, the partial wave dependence of the competition between fission-like and evaporation cross section obtained by EHF-calculation are shown in Fig.3 and 4. In this calculation the cross sections with $4 < Z < 11$ are considered as the fission-like cross section. From these calculated results the onset partial wave of the fission-like fragments production is expected to be about $20\hbar$. In these energy region the ratio of the fission-like cross section is very small by comparing the total fusion cross section, but the competition between fusion-like and evaporation occurs in somewhat smoothly with wide width in spin space as can be seen in the inset of Fig.3. As concern with the reactions $^{35}Cl +^{12} C$ at $E_{Lab} = 200$MeV and $^{23}Na +^{24} Mg$ at $E_{Lab} = 89$MeV,

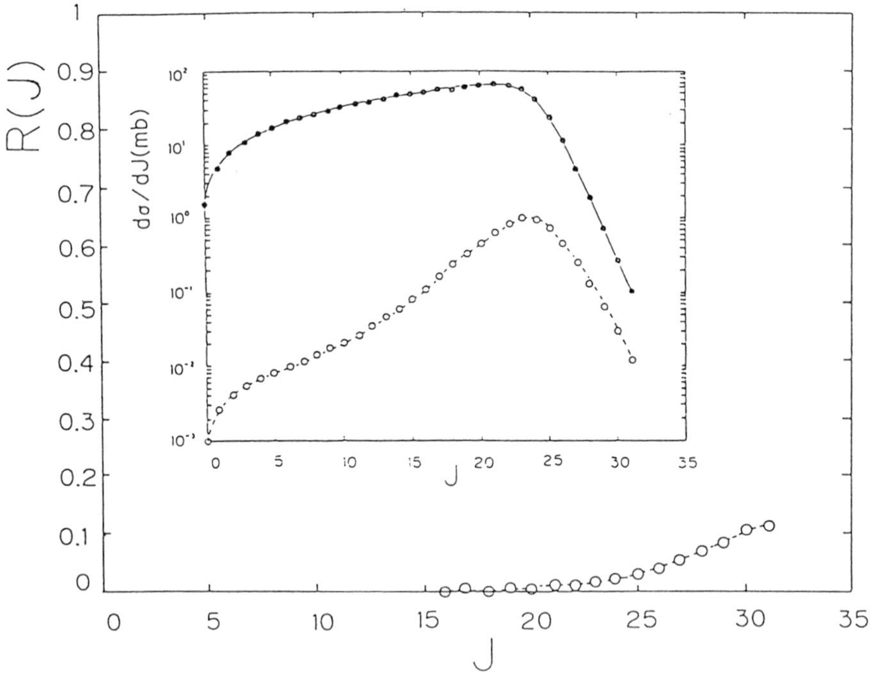

Fig.3 Partial wave dependence of competition between fission and evaporation cross sections obtained by EHF-method calculation for the $^{35}Cl +^{12} C$ reaction at $E_{Lab} = 180$MeV. The inset are the partial wave dependence of evaporation cross section (closed circles) and fission one (open circles). The solid line shows the total fusion cross section.

because the J_{cr} -values for the total fusion cross sections are different in the two

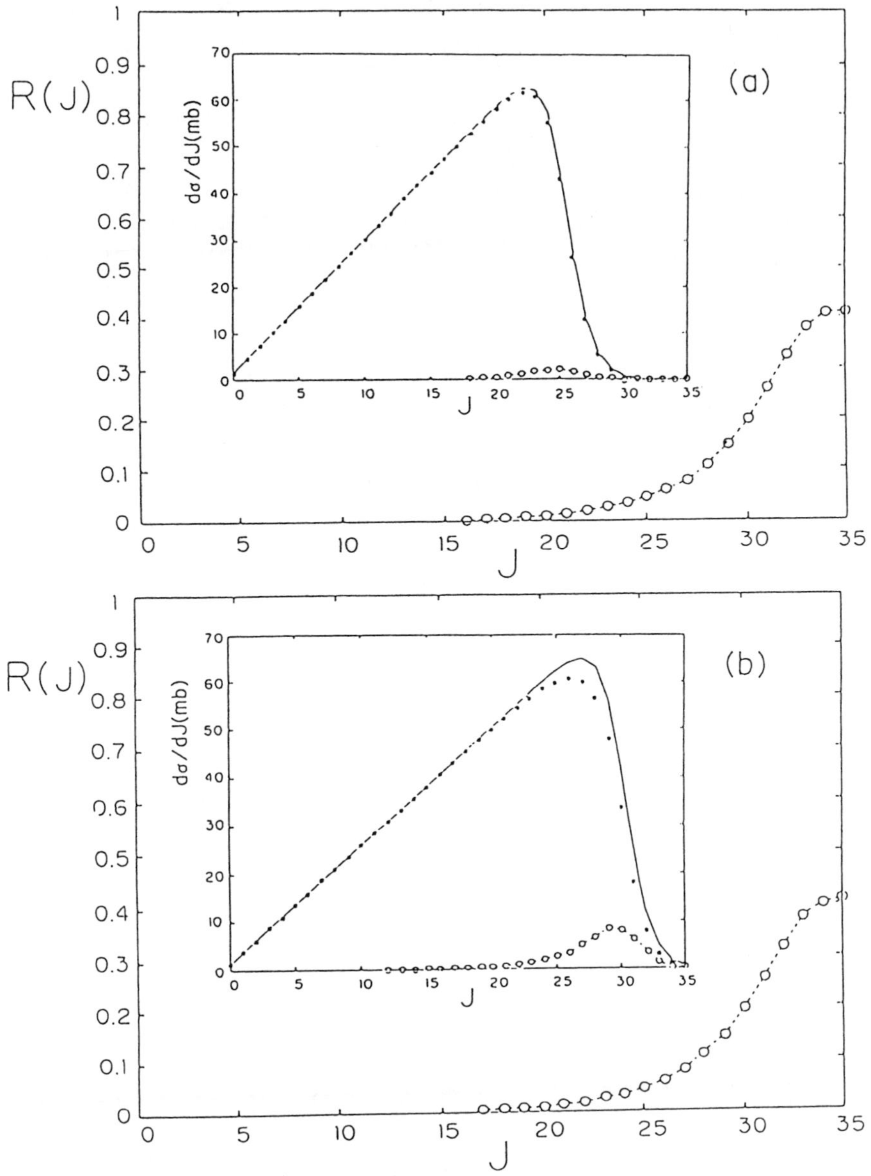

Fig.4 Partial wave dependence of competition between fission and evaporation cross sections obtained by EHF-method calculation. (a) $^{35}Cl+^{12}C$ reaction at E_{Lab} = 200MeV. (b) $^{23}Na+^{24}Mg$ reaction at E_{Lab} = 89MeV. The inset are the partial wave dependence of evaporation cross section (closed circles) and fission one (open circles). The solid line shows the total fusion cross section.

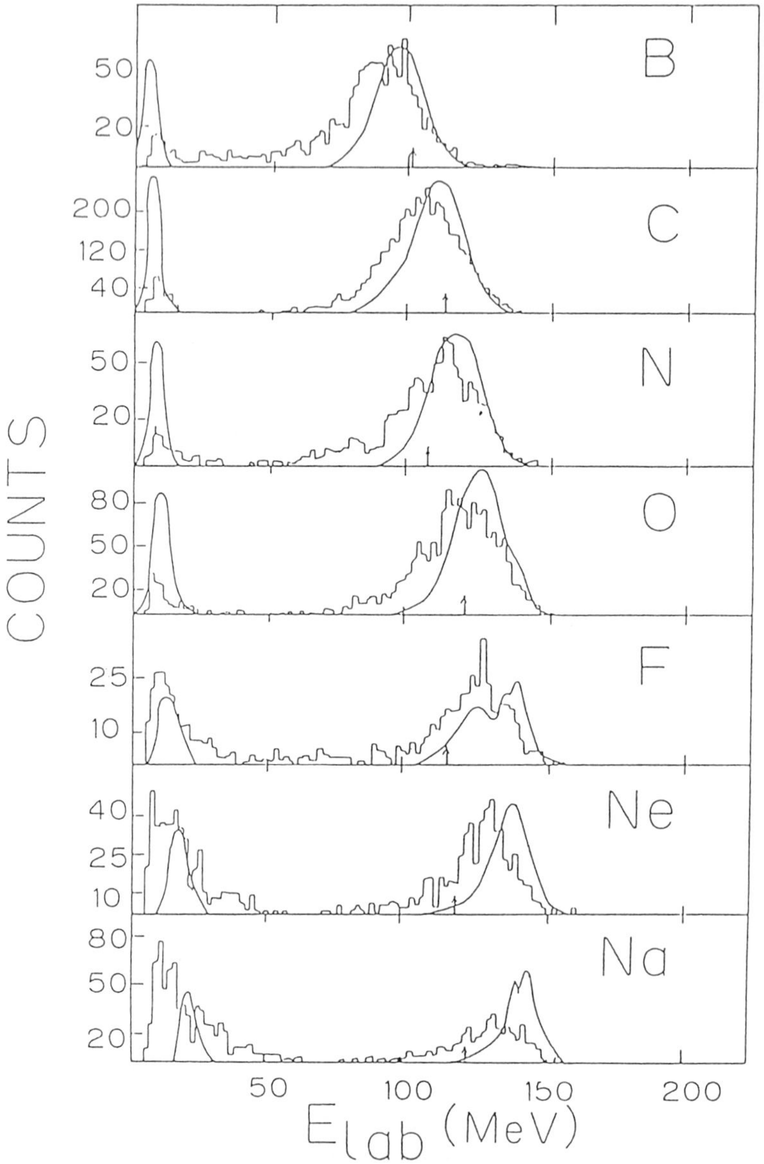

Fig.5 Comparisons of calculated kinetic energy distributions of fission fragments in the Lab-system by EHF-method (solid line) with the observed ones of $^{35}Cl +^{12} C$ reaction at $E_{Lab} = 200$MeV (histograms) at $\theta_{Lab} = 12^{\circ}$.

channels as mentioned above, the $^{23}Na + ^{24}Mg$ channel has larger fission-like cross section than $^{35}Cl + ^{12}C$ channel (see the insets of Fig.4), but the calculated partial wave dependence of the competition between fission-like and evaporation are completely same with both channels.

In Fig.5, the calculated kinetic energy distributions for the case of $^{35}Cl + ^{12}C$ reaction at $E_{Lab} = 200$MeV are shown for the lighter fission-like fragments. The calculated kinetic energy distributions in the CM-system are transformed to the laboratory system by assuming that the angular distribution of decaying fragment in the CM-system has the $1/\sin(\theta)$ dependence. By comparing with the calculated and the measured kinetic energy distributions, the calculated results reproduce well the observed ones. Then we can conclude that the observed fission-like fragments are the statistical binary decay from the compound nucleus.

4. Results and discussions for ^{111}In compound nucleus

Because the general introduction and aim of this study on the reaction $^{84}Kr + ^{27}Al$ at $E_{Lab} = 890$MeV have been already presented by Heusch [1] and by Yuasa-Nakagawa [2] including the essential parts of the calculated results by EHF-method at this conference, here some complementary interpretations are shown on the calculated results.

From the measured charge distribution and the evaporation cross section (see Fig.4 of Ref.[2]), the critical angular momentum are deduced to be $J_{cr} \simeq 83\hbar$ for the formation of the total fusion cross section. It is very interesting experimental result that this obtained J_{cr} exceeds the limit of the Rotating Liquid Model, but our fusion model predicts the almost same value of the experimentally obtained one (see Fig.5of Ref.[2]). Then we use the critical angular momentum $J_{cr} = 83\hbar$ for the total fusion cross section in this calculation of EHF-method. The used parameters in this calculation are the almost same that of the case of the lighter mass system such as ^{47}V compound nucleus, but the static quadrupole deformation is introduced to the binary fragments at scission point with the collinear configuration in the case that the fragments excite statistically.

On the first step EHF-calculation.

Before showing the results of the multi-step cascade calculation, at first the calculated results of the first step are shown. The competition between fission-like and evaporation production how depend to the partial wave is shown in Fig.6. In this calculation the cross sections with $Z > 5$ are included in the fission-like production. The calculated cross section of evaporation residues is about $500mb$ in the first-step calculation and the onset partial wave for fission-like production is about $50\hbar$. Then in the partial wave region $J > 60\hbar$ the fission-like production becomes dominant than the light particle productions.

Because the formed compound nucleus ^{111}In has the high excitation energy $E^* =$ 216 MeV in this experiment, it is easily expected that the residue nuclei of particle emission and fission-like fragments has the highly excited energy distribution at the first-step decay from the compound nucleus. In Fig.7(a) the calculated excitation energy distribution is shown for the residue nuclei of neutron and it can be understood that the residue nuclei have highly excited energy distribution. In the insets of the Fig.7(a) the calculated spin distribution is presented. As can be expected from the Fig.6, the spin distribution of the residue nuclei are almost same that of the spin distribution of the evaporation production because the relative motion between light particle and residue nuclei can not take out the large angular momentum. Then multi-step cascade decays from highly excited residue nuclei occur sequentially. Because the spin distributions of

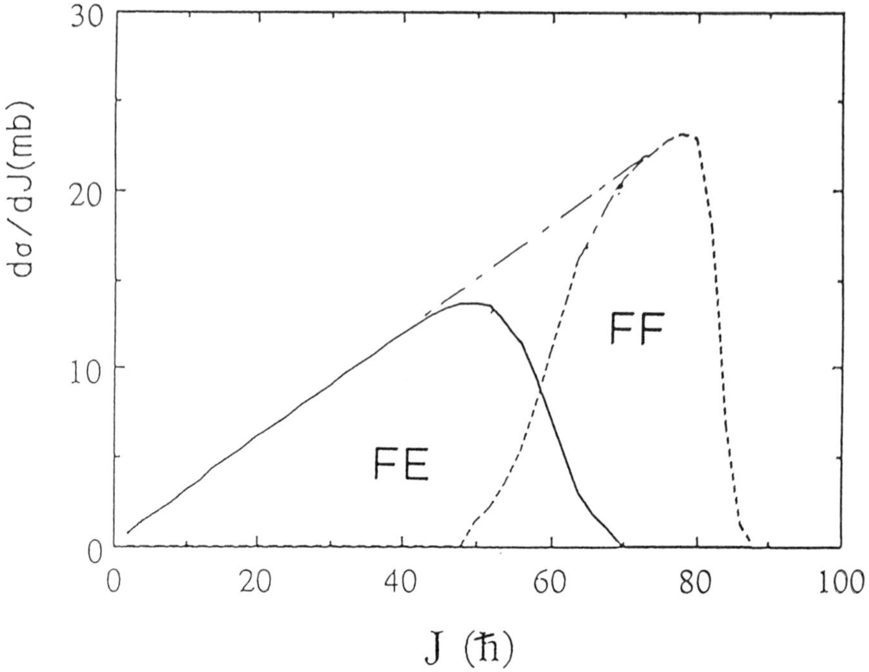

Fig.6 Partial wave dependence of fission (dashed line) and evaporation cross section (solid line) obtained by EHF-method calculation for the $^{84}Kr + ^{27}Al$ reaction at $E_{Lab} = 890$MeV. Total fusion cross section is shown by half-dashed line with $J_{cr} = 83\hbar$.

the residue nuclei have high spin one but not so high for fission-like production as can be expected from Fig.6, it is expected to be possible that the complex fragments with light mass such as $Z < 10$ may be produced in the process step of the multi-step cascade which start from the first step residue nuclei. In fact a large part of the enhancement of the light fragment cross section from the first step one to final step are accumulated through this multi-step cascade decay as shown in Fig.8.

On the other hand the fission-like fragments with heavier mass such as $Z < 30$ have the excitation energy and spin distributions as shown in Fig.7(b). The reason why the spin values of the fission-like fragments are not so large is that the relative motion between the fission-like fragments take out the large angular momentum. The energy distribution are well determined by the barrier height of scission point and the equipartition of energy between the two fragments. The obtained spin and energy distributions of the fission-like fragments are almost similar to the partial wave distribution and energy of the ^{47}V nucleus as presented above (see Fig.3 and 4). Then these fission-like fragments also decay by emitting the light particles, but the light fragments production is not important in these cascade decay. Of course because the produced light fragments such as $Z \simeq 10$ at the first step excite to somewhat high spin around $J \simeq 10\hbar$

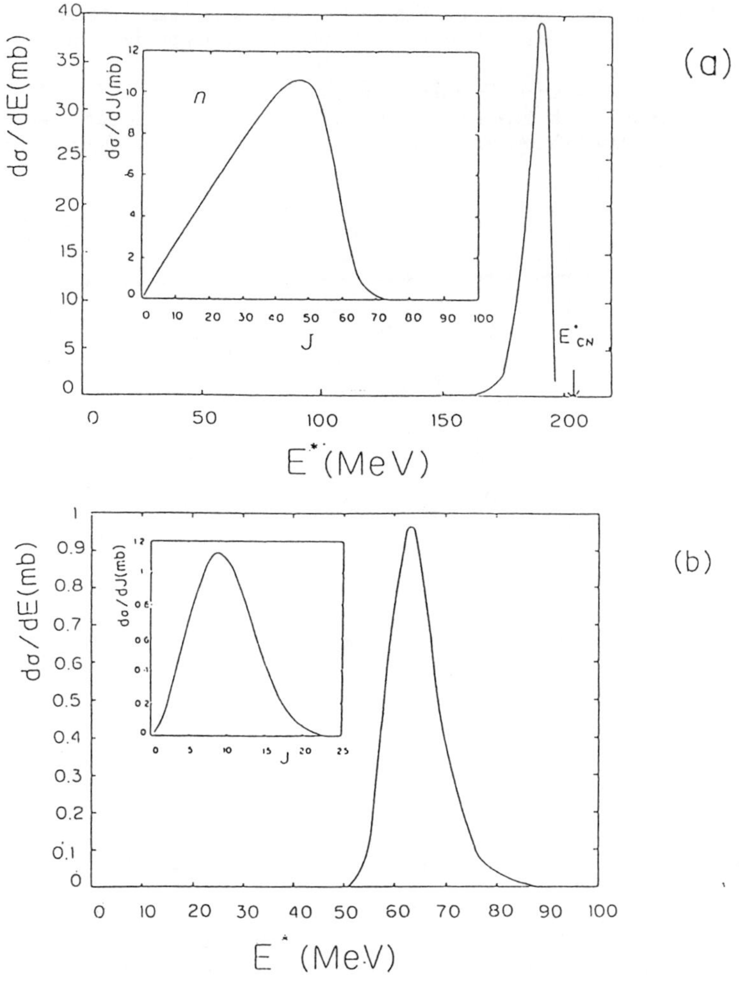

Fig.7 Excitation energy distributions of the residue nuclei of light particle emissions and fission fragment obtained by the first step EHF-method calculation for the $^{84}Kr + ^{27}Al$ reaction at $E_{Lab} = 890$MeV. The inset are the spin distribution of the first chance emission and fission fragment. (a) neutron emission, (b) fission fragment of $Z = 23, N = 29$.

and higher excitation energy around $E^* \simeq 30$ MeV, these light fragments also decay by emitting the light particles. Therefore, these residue nuclei started from the produced light fragments at first step shift down to the the nuclei with smaller charge. This may be one of the possible interpretation of the complex fragments production.

Effect of the multi step EHF-calculation.
Next the results of the final step obtained by the full cascade calculation are shown. In Fig.8 the calculated charge distributions are shown by comparing the measured one [2].

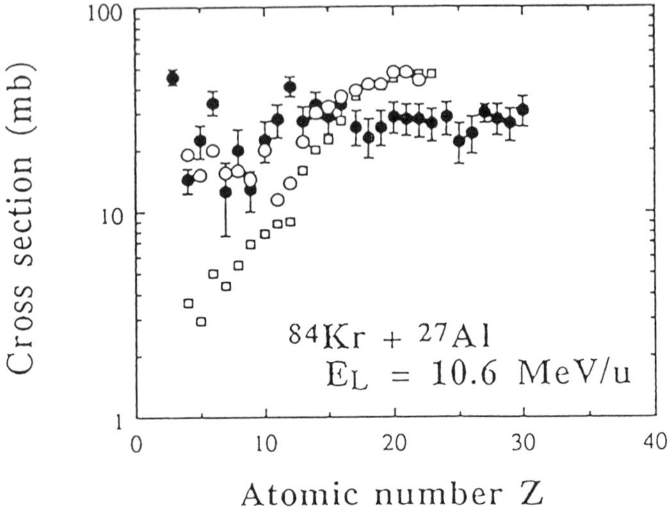

Fig.8 Comparison of the measured Z-distribution (closed circles) [2] and calculated Z-distribution obtained by EHF-method calculation for the $^{84}Kr + ^{27}Al$ reaction at $E_{Lab} = 890$MeV. The open squares show the results of first-step calculation and open circles the results of multi-step cascade calculation.

The large enhancements from the first step distribution in the region of light fragments $Z < 10$ are obtained. The main origin of the enhancement around $Z \simeq 10$ is the light particle cascade decay of fission-like fragments with a few larger mass. On the other hand the enhancements in the region $Z < 7$ come from the multi-step cascade decay of the evaporation residues in the first step. Because the calculated charge distribution in the region of symmetric fission is somewhat larger than the measured one, the values of the calculated cross sections in the lighter fragments become somewhat small by comparing the measured ones.

In Fig.9(a) and (b), the calculated isotope mass distributions for the case of Si and Ca isotope are shown comparing with the observed ones, respectively. As can be seen in the observed isotope mass distributions (see Fig.11 of Ref.[2]), the distribution of each isotope has the peak around the nucleus of the beta-stable-line. The peak of the calculated distributions obtained in the first step EHF-calculation deviates to the heavier part by two or three mass units than the observed one. However, as we have already discussed on the spin and energy distribution of fission-like fragment at first-

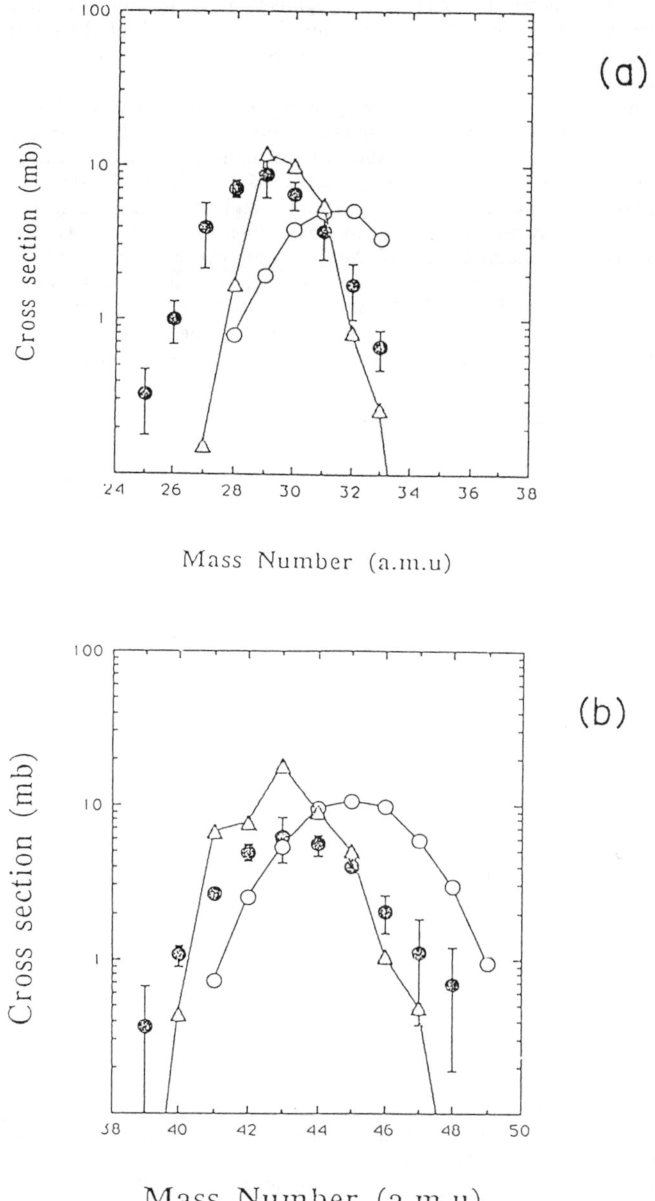

Fig.9 Comparison of the measured isotope mass-distribution (close cles) [2] and calculated distribution obtained by EHF-method calculatic the $^{84}Kr + ^{27}Al$ reaction at $E_{Lab} = 890$MeV. The open circles show the r of first-step calculation and open triangles the results of multi-step cascad culation. (a) shows Si isotope (b) Ca isotope.

step calculation shown in Fig.7(b), the fragments decay by emitting the particle. Then as a consequence of the multi-step cascade decay, the peak of the distribution of the first-step shifts down and the calculated distribution of the final-step reproduces well the observed one.

Finally the calculated kinetic energy distributions for the $Z = 20$ fragment is shown in Fig.10 by comparing with the measured one. The kinetic energy distribution is calculated by taking into account the effect of multi-step cascade decay of fission-like fragments. At first the kinetic energy distribution in the CM-system are transformed to the laboratory system with the laboratory angle θ_L by assuming that the angular distribution of decaying fragment in the CM-system has the $1/\sin(\theta)$ dependence. Next it is assumed that the velocities of the fragments conserve in the process of the light particle cascade decay. Then the kinetic energy distribution are evaluated as shown in Fig.10. By comparing with the thus calculated and the measured kinetic energy distributions, the calculated results reproduce well the observed one except that the width of the calculated one is somewhat narrower than the measured one.

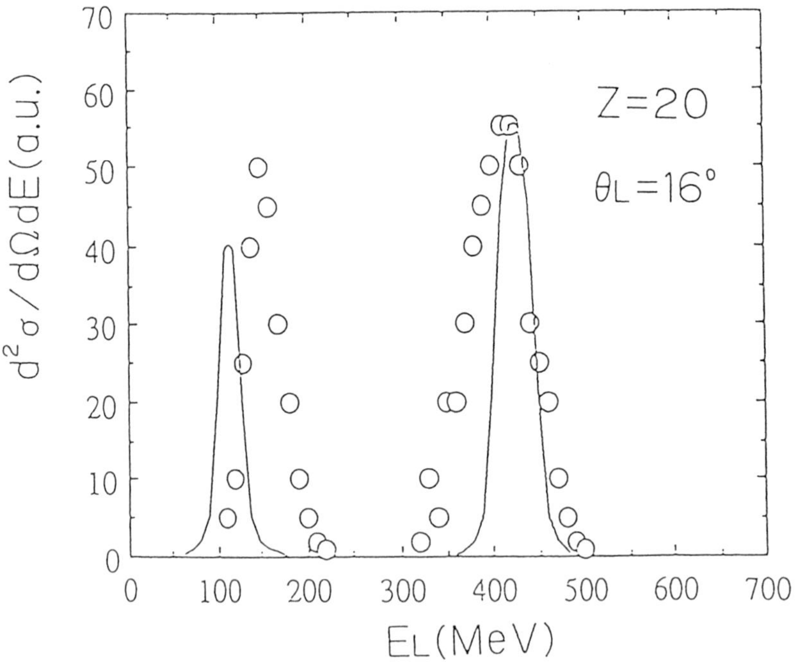

Fig.10 Comparison of the measured kinetic energy distribution (open circles) [2] and calculated distribution (solid line) obtained by EHF-method calculation for the $^{84}Kr + ^{27}Al$ reaction at $E_{Lab} = 890$MeV at $\theta_L = 16°$. The calculated kinetic energy distribution is for the fragments with $Z = 20$ at final step after the sequential multi-step cascade calculation from all available fragments.

5. Conclusions.

It is shown that the observed quantities of the complex fragment productions (fission-like productions) and the evaporation residues are well reproduced by the *EHF-method* for both $^{35}Cl +^{12} C$ and $^{23}Na +^{24} Mg$ reactions. The essential point of the EHF-method is to treat the light-particle emission and heavy fragment production in equivalent statistical way and practically to utilize fully the effect of the observed binding energy in the calculation. Then it is expected that the decaying compound nucleus has to be represented by a statistical ensemble of the binary combination of almost individual nuclei. It is very important that the significantly large amount of cross section of the fission-like fragment production really exist as a part of complete fusion cross section in the lighter mass system. Then in order to make clear which theoretical model is valid, it is hoped that the experimental studies on the complex fragment production will be performed more systematically at higher energy region in the lighter mass system.

On the complex fragment production from the ^{111}In compound nucleus formed by the reaction $^{84}Kr +^{27} Al$ at E_{Lab} = 890MeV, the EHF-method was also applied and the processes of the multi-step cascade decay of the fission-like fragments and the evaporation residues are investigated by referring the statistical binary decay of the ^{47}V compound nucleus. It was shown that the significant enhancement from the first step calculation of the intermediate mass production such as $Z < 10$ comes from not only the sequential multi-step cascade decay of the fission-like production but also the evaporation residues. The isotope mass distributions which will give the essential criterion for the study on the reaction mechanism are well reproduced by the EHF-method including the sequential multi-step cascade decay of the fission-like production.

ACKNOWLEDGMENTS

One of us(T.M.) would like to acknowledge Mr. T.Takizawa for his kind assistance. The whole calculations of this study are performed by the use of computers H260D of Shinshu University, M880 of University of Tokyo and F780 of RIKEN.

REFERENCES.

[1] B.Heusch, contribution to this conference.

[2] K.Yuasa-Nakagawa et al., contribution to this conference, and references therein.

[3] G.Guillaume, et al., Phys. Rev., C26(1982) 2458.

[4] J.P.Coffin, et al., Phys. Rev., C30 (1984) 539.

[5] S.J.Sanders, et al., Phys. Rev., C40 (1989) 2091.

[6] L.G.Moretto, Nucl. Phys., A247 (1975) 211.

[7] B.Shivakumar, et at., Phys. Rev., C35 (1987) 1730.

[8] C.Beck, et al., Z. Phys. A334 (1989) 521.

[9] C.Beck,et al., Proc. of the Workshop on the Interface between Nuclear Structure and Heavy-Ion Reaction Dynamics, Notre Dame, May, 1990; Inst. Phys. Conf. Ser. $n°$ 109 (1990) 213.

[10] T.Matsuse, Proc. of the Tsukuba Intern. Symp. on Heavy Ion Fusion Reactions, edited by K.Furuno and T.Kishimoto (Tsukuba,Japan,1984) 113.

[11] S.M.Lee, W.Yokota and T.Matsuse, Proc. of the Symp. on the Many Factes of Heavy Ion

Fusion Recations, (Argonne 1986) 63.
[12] T.Matsuse, Proc. of International Symposium on Heavy Ion Physics and Its Application, Lanzhou, China, Oct 1990, edited by W.Q.Shen, Y.X.Luo and J.Y.Liu (World Scientific) 95, and references therein.
[13] B.Djerroud et al., Proc., of 6th Intern. Conf. on Nuclear Reaction Mechanisms, Varenna, 10-15 June 1991.
[14] S.M.Lee et al., Nucl. Phys. A519 (1990) 221c. and Proc. of International Symposium on Heavy Ion Physics and Its Application, Lanzhou, China, Oct 1990, edited by W.Q.Shen, Y.X.Luo and J.Y.Liu (World Scientific) 401.
[15] S.M.Lee, T.Nakagawa and T.Matsuse, Proc. of the Tsukuba Intern. Symp. (1984),p.62.
[16] A.Bohr and B.R.Mottelson, "Nuclear Structure Vol.1"
[17] A.H.Wapstra and G.Audi, Nucl. Phys., A432(1985) 1.
[18] H.J. Krappe and J.R. Nix, Proc. of the Third Int. I.E.A. Symp. on the Physics and Chemistry of Fission,(Vinna 1973),Vol.1,p.159

DIRECT REACTION APPROACHES AND BEYOND

A UNIFIED APPROACH FOR LOW-ENERGY NUCLEAR REACTIONS

B. T. Kim
Department of Physics, Sung Kyun Kwan University, Suwon 440-746, Korea

T. Udagawa
Department of Physics, University of Texas, Austin, Texas 78712, U.S.A

ABSTRACT

We investigate the dynamic polarization potential(DPP) due to the coupling to collective states, which is an important element in a unified approach to describe all the different types of nuclear reaction processes. The construction of the nonlocal and ℓ-dependent DPP is based on the projection operator formalism of Feshbach. Dispersion relations between real and imaginary parts are discussed. We present an approach that can calculate the elastic scattering, direct reaction, and fusion cross sections simultaneously with a single DPP. For this purpose, a rapid method of solving the nonlocal Schrödinger equation is first developed to obtain the *exact* solution. The separation of the DPP into direct reaction and fusion parts is then formulated. We show that such a separation made phenomenologically in a local space nicely explains the observed data. Finally the localization of the nonlocal DPP is discussed in order to justify the phenomenological local optical model potential.

1. INTRODUCTION

The major goal of low-energy nuclear reaction theory is to describe all the different types of nuclear reactions based on a *single* unified approach. Reactions between two nuclei are generally characterized by the impact parameters. For a large impact parameter the elastic scattering is dominant and is well described by the optical model. Inelastic scattering and few-nucleon transfer reactions occur in the grazing collision. Here the multi-step direct reaction theories, like the distorted wave Born approximation, coupled-channels Born approximation, coupled-reaction channels and so on, have explained nicely the observed peripheral phenomena. In the almost head-on collisions, a thorough amalgamation of the two-colliding systems, usually referred to as fusion, is a dominant process. An approach has been suggested[1,2] to interpret the general behavior of fusion based on the coupled-channels(CC) direct reaction theories[3], in which fusion is proceeded through the formation of a compound nucleus.

With this CC formalism, one could explain, in principle, the low energy reaction phenomena on the same footing[4,5]. In fact, such attempts have recently been made in a limited coupling scheme to the low lying excited states for several systems, and explained reasonably well the observed elastic scattering, inelastic scattering, few nucleon transfer reactions, and fusion data. However, the calculations are quite involved when the dimension of the CC equations becomes larger. In addition to this, there are some other difficulties. The CC potential associated with the CC space is not well known. All coupling information, mostly unknown for most of nuclei, should be input to the calculation. Studies concerning the reduction to a one-channel description have thus attracted interest. The dynamical polarization potential (DPP), which describes the effects of coupling to the non-elastic channels, represents the potential for such a one-channel formalism.

In this paper, we describe the formal background of the DPP following the projection operator method of Feshbach[6], and dispersion relations in Sec. 2. We present a method to calculate the elastic scattering, direct reaction, and fusion cross sections in a model with the DPP in Sec. 3. A method to solve the nonlocal Schrödinger equations is introduced, and the separation of the DPP into fusion and direct reaction (DR) parts is formulated. As an example, we present a phenomenological analysis of elastic scattering, direct reactions, and fusion data for the $^{16}O+^{208}Pb$ system near the Coulomb barrier energy. Sec. 4 is devoted to the nonlocality problem. We present several localization methods and compare a localized DPP with the phenomenological local optical model potential.

2. DYNAMIC POLARIZATION POTENTIAL

Here we briefly introduce the DPP in the Feshbach formalism. The dispersion relation between the real and imaginary parts of the potential, and its consequences are also mentioned.

2.1. Construction of DPP

The present study is based on the Feshbach formalism[6], where the Schrödinger equation for the Hamiltonian H becomes two coupled equations,

$$(E - \mathcal{H}_{\mathcal{PP}})\mathcal{P}\Psi = \mathcal{H}_{\mathcal{PQ}}\mathcal{Q}\Psi, \tag{1}$$

$$(E - \mathcal{H}_{\mathcal{QQ}})\mathcal{Q}\Psi = \mathcal{H}_{\mathcal{QP}}\mathcal{P}\Psi. \tag{2}$$

Here Ψ is the total wave function of the original Hamiltonian H, and

$$\mathcal{H}_{\mathcal{PP}} = \mathcal{P}H\mathcal{P}, \quad \mathcal{H}_{\mathcal{PQ}} = \mathcal{P}H\mathcal{Q}. \tag{3}$$

In Eq. (3), $\mathcal{P}$ is the projection operator onto the model space spanned by the direct reaction channel states. $\mathcal{Q} = 1 - \mathcal{P}$ is the projection operator onto the rest of the space. It includes open DR channels as well as all the closed channels.

$\mathcal{H}_{\mathcal{PP}}$ is responsible for reactions leading to the DR channels, while $\mathcal{H}_{\mathcal{PQ}}$ couples $\mathcal{Q}\Psi$ to $\mathcal{P}\Psi$.

By eliminating $\mathcal{Q}\Psi$ in Eqs. (1) and (2), the effective Hamiltonian is obtained as,

$$H^{eff} = \mathcal{H}_{\mathcal{PP}} + \mathcal{H}_{\mathcal{PQ}} \frac{1}{E - \mathcal{H}_{\mathcal{QQ}} + i\eta} \mathcal{H}_{\mathcal{QP}}. \tag{4}$$

The Hamiltonian H^{eff} given by Eq. (4) defines the coupled-channels or coupled-reaction-channel problem. It is easy, for instance, to derive the CC or CRC equations from the Schrödinger equation for H^5. If we choose $\mathcal{P}\Psi$ as the elastic channel, then H^{eff} satisfies the optical model Schrödinger equation

$$(E - H^{eff})\chi_0^{(+)} = 0, \tag{5}$$

where $\chi_0^{(+)}$ is the optical model distorted wave function. The effective optical model Hamiltonian H^{eff} becomes

$$H^{eff} = \mathcal{H}_{00} + \mathcal{H}_{0\mathcal{Q}} \frac{1}{E - \mathcal{H}_{\mathcal{QQ}} + i\eta} \mathcal{H}_{\mathcal{Q}0}. \tag{6}$$

The first term on the right hand side (RHS) of Eq.(6) is the sum of a kinetic energy operator and a bare potential which account for the mean field, and can be calculated from the Hartree-Fock formalism. The second term is called the dynamic polarization potential, ("*dynamic*" because it arises from collisions, "*polarization*" because it represents the distortion of the internal states due to the mutual interaction.) which is the subject of this lecture.

The DPP is absorptive and thus complex since $\mathcal{Q}\Psi$ has open channels. The coupling induces a loss of flux in the elastic channel to $\mathcal{Q}\Psi$. Mathematically, the limit of the Green's function

$$\lim_{\eta \to 0} \frac{1}{E - \mathcal{H}_{\mathcal{QQ}} + i\eta} = \mathbf{P}(\frac{1}{E - \mathcal{H}_{\mathcal{QQ}}}) - i\pi\delta(E - \mathcal{H}_{\mathcal{QQ}}), \tag{7}$$

where $\mathbf{P}$ stands for principal value, is complex. Even if there are no open channels in the $\mathcal{Q}$-space, averaging the fluctuations over energy, which are associated with quasi-bound states due to couplings to the bound states of $\mathcal{H}_{\mathcal{QQ}}$, gives a complex DPP. This absorption is to be identified with the loss of flux that goes into compound-nucleus formation.

Furthermore the DPP is nonlocal and intrinsically energy-dependent due to the Green's function in Eq. (6). Also it becomes ℓ-dependent as can be seen by a partial wave expansion of the Green's function. The energy dependence of the DPP arises not only from the Green's function but also from the dispersion relations between the real and imaginary parts which are described in the next subsection, and also from the localization of the nonlocal potential when the local potential is adopted. (See Sec. 4.1)

We now construct the DPP more explicitly in the partial wave expansion. For simplicity, we take only two channels, one for elastic, and another for an inelastic channel. We further assume that the spins of both projectile and target are zero and the diagonal and non-diagonal (coupling) potentials are spherically symmetric. The CC equations Eqs. (1) and (2) for a total angular momentum ℓ may then be expressed as

$$(E - H_\ell)|\chi_\ell >= \sum_{\ell'} v_{\ell\ell'}|\omega_{\ell'} >, \tag{8}$$

$$(E - H_{\ell'})|\omega_{\ell'} >= v_{\ell'\ell}|\chi_\ell >, \tag{9}$$

where χ_ℓ and ω_ℓ are the radial partial wave functions in the elastic and inelastic channels, respectively, and $v_{\ell\ell'}$ is the coupling potential. The Hamiltonian H_ℓ is expressed as

$$H_\ell = -\frac{\hbar^2}{2\mu}(\frac{d^2}{dr^2} - \frac{\ell(\ell+1)}{r^2}) + V_{00}(r) + v_{\ell\ell}(r), \tag{10}$$

where $V_{00}(r)$ is the bare potential. The reduced optical model Schrödinger equation may again be obtained by eliminating $\omega_{\ell'}$ in Eqs. (8) and (9). The resulting equation reads

$$(E - H_\ell)|\chi_\ell >= U_\ell|\chi_\ell >, \tag{11}$$

where the nonlocal DPP, $U_\ell(r, r')$ is given by

$$U_\ell(r, r') =< r|U|r' >= \sum_{\ell'} v_{\ell\ell'}(r)G_{\ell'}(r, r')v_{\ell'\ell}(r'). \tag{12}$$

The partial wave optical model Green's function $G_\ell(r, r')$ in Eq. (12) satisfies

$$(E - H_\ell)G_\ell(r, r') = \delta(r - r'). \tag{13}$$

The explicit form of the radial Green's function may then be given as

$$G_\ell(r, r') = \frac{2m}{\hbar^2}\frac{f_\ell(r_<)h_\ell(r_>)}{W}, \tag{14}$$

where $f_\ell(r)$ and $h_\ell(r)$ are the regular and irregular solutions of H_ℓ, respectively, while W is the Wronskian. Therefore the DPP can easily be constructed in this way, once the coupling potentials are given.

As an example, we calculated the nonlocal DPP for the $^{16}O+^{208}Pb$ system at an incident ^{16}O energy of E_{lab}=80 MeV. We consider coupling of the ground state to the 2.61 MeV 3^-, 3.20 MeV 5^-, and 4.07 MeV 2^+ states in ^{208}Pb, and to the 6.13 MeV 3^- state in ^{16}O. The coupling potential $v_{\ell\ell'}(r)$ is chosen as the collective form factor. The imaginary part of the $\ell = 10$ polarization potential for this system is shown in Fig. 1a. We clearly see that $U_\ell(r, r')$ is highly nonlocal. We will discuss this problem of nonlocality explicitly in Sec. 4.

2.2. Dispersion Relations

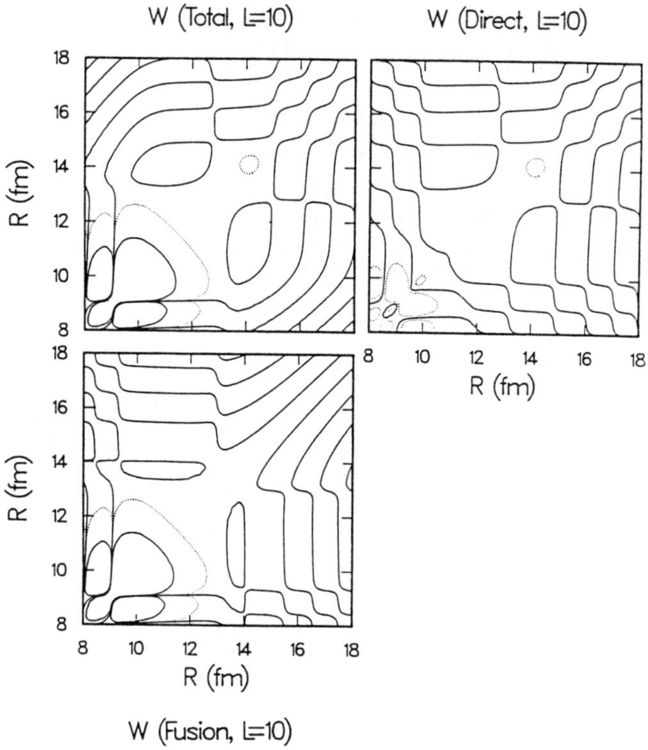

Figure 1: The imaginary part(a) of $\ell = 10$ DPP for the $O + Pb$ system at E_{lab}=80 MeV. The imaginary part is further decomposed into fusion(b) and direct reaction(c) parts.

The dispersion relation follows immediately from Eq. (6) if the eigenstates of $\mathcal{H}_{QQ}$ are given. Let us assume $\mathcal{H}_{QQ}$ has discrete states ϕ_α with eigenvalues of E_α, and continuum states $\phi(\epsilon, \alpha)$ with eigenvalues of ϵ, where α is introduced to classify all additional quantum numbers. By substituting these states into $\mathcal{H}_{QQ}$ in Eq. (6), we obtain the following dispersion relation,

$$ReH^{eff} = H_{00} + \sum_\alpha \frac{\mathcal{H}_{0Q}|\phi_\alpha >< \phi_\alpha|\mathcal{H}_{Q0}}{E - E_\alpha} - \frac{1}{\pi}\mathbf{P}\int d\epsilon \frac{ImH^{eff}}{E - \epsilon}. \qquad (15)$$

What it means is that the real and imaginary parts of the DPP are not independent. This relation also holds when the potential is decomposed into partial waves or orthogonal sets. The dispersion relations are connected to *causality*, which means in the scattering theory that the scattered wave cannot leave the target before the incident wave arrives. The Fourier transformation of the Green's function to the time domain should have $G(t - t') = 0$ for $t < t'$.

The elastic scattering data accumulated have shown that the imaginary part of the optical potential diminishes quite rapidly as the incident energy becomes lower than the Coulomb barrier. This can be understood from the fact that fewer reaction channels are open when the beam energy is lower. The origin of this energy dependence is obviously not from the intrinsic Green's functions, nor from nonlocality. Furthermore, when the optical potential at a fixed point around the strong absorption radius is plotted as a function of energy, the real depth of the potential is shown to be a bell shape, peaked around the Coulomb barrier. This is because the coupling of the elastic channel to other reaction channels gives rise to the polarization effect.

Recently, Nagarajan et al.[7] suggested a local form of the dispersion relation for the two colliding nuclei as,

$$V(r,E) = V(r,E_s) + \frac{E-E_s}{\pi}\mathbf{P}\int d\epsilon \frac{W(r,\epsilon)}{(\epsilon-E)(\epsilon-E_s)}, \tag{16}$$

where V and W are the real and imaginary parts of optical potential, respectively. The first term of the RHS, $V(r, E_s)$ is the sum of the first and second terms of Eq. (15) and E_s is the energy where the normalization is done to give a correct magnitude of the potential depth. They showed that an observed bell-shaped real potential at a fixed point is a consequence of the dispersion relation, when the imaginary part suddenly changes near the Coulomb barrier.

The sudden increase in the real depth of the potential at near-barrier energies allows for a lowering of the barrier, and enhances the penetrability of the incident wave through the barrier. It can account for anomalously large fusion cross sections recently observed in subbarrier heavy-ion fusion reactions. Therefore the dispersion relation plays a crucial role in explaining not only elastic scattering but also fusion near the barrier.

3. ELASTIC SCATTERING, DIRECT REACTION, AND FUSION

3.1. Decomposition of DPP into DR and fusion parts

We first show that the imaginary part of the DPP can be separated into direct reaction (DR) and fusion parts. Let us denote the imaginary part of $U_\ell(r, r')$ of Eq. (11) as $W_\ell(r, r')$, i.e.,

$$W_\ell(r,r') = Im U_\ell(r,r'). \tag{17}$$

$W_\ell(r, r')$ can then be decomposed into fusion and DR parts, $W_{F;\ell}(r, r')$ and $W_{D;\ell}(r, r')$, respectively, as

$$W_\ell(r,r') = W_{F;\ell}(r,r') + W_{D;\ell}(r,r'). \tag{18}$$

We use for this purpose the identity,

$$Im G_\ell = G_\ell^{+}[Im U_\ell]G_\ell + \Omega_\ell^{+}[Im g_\ell]\Omega_\ell, \tag{19}$$

where g_ℓ is the free Green's function, while Ω_ℓ is the Möller wave operator which generates the distorted wave function upon operating on the plane wave function,

$$\Omega_\ell = 1 + U_\ell G_\ell. \tag{20}$$

By plugging Eqs. (18), (19) and (20) into Eq. (11), $W_{F;\ell}(r,r')$ and $W_{D;\ell}(r,r')$ can be expressed as

$$W_{F;\ell}(r,r') = \sum_{\ell'}[w_{\ell\ell'}G_{\ell'}v_{\ell'\ell} + v^+_{\ell\ell'}G^+_{\ell'}w_{\ell'\ell} + v^+_{\ell\ell'}G^+_{\ell'}W_{\ell'\ell'}G_{\ell'}v_{\ell'\ell}], \tag{21}$$

$$W_{D;\ell}(r,r') = \sum_{\ell'} v^+_{\ell\ell'}\Omega^+_{\ell'}[Img_{\ell'}]\Omega_{\ell'}v_{\ell'\ell}, \tag{22}$$

where $w_{\ell\ell'}$ and $W_{\ell\ell}$ are the imaginary parts of $v_{\ell\ell'}$ and H_ℓ of Eq. (10), respectively. The first two terms on the RHS of Eq. (21) describes the absorption occurring during transitions, and vanishes when the real coupling form factors are used. The last one represents absorption into the compound nucleus from the elastic and nonelastic channels. $W_{D;\ell}$ of Eq. (22) is the part of the imaginary potential that accounts for the direct reaction.

We calculated the nonlocal $W_{F;\ell}(r,r')$ and $W_{D;\ell}(r,r')$ for the $^{16}O+^{208}Pb$ system at E_{lab}=80 MeV. The same coupling scheme as the example in Sec. 2.1 was used. The fusion and DR parts of the imaginary potential are displayed in Fig. 1b and 1c. Again they are highly nonlocal. The shape of W in the region of radial distances smaller than the strong absorption radius (about 12.4 fm in this case) is similar to that of W_F, while it looks like W_D in the outer region and decreases rapidly as the radial distance gets larger. This implies that fusion takes place in the inner region and the direct reaction in the nuclear surface region.

3.2. Elastic, Direct Reaction, and Fusion Cross Sections

The basic equation, Eq. (11) becomes a nonlocal integro-differential equation in the r-representation given as

$$(E - H_\ell)\chi_\ell(r) = \int U(r,r')\chi_\ell(r')dr'. \tag{23}$$

Since the celebrated work of Perey and Buck (PB) some 30 years ago[8], little progress has been made in improving the method for solving this equation. PB solved the equation by an iteration method, but convergence depended very critically on the particular local potential $U_0(r)$ used for generating the initial trial solution. The iteration often diverges, even if U_0 differs very slightly from the optimal potential suggested by PB. Because of this, the PB method has not been widely used.

Very recently, Kim and Udagawa[9] presented a new, rapid, and reliable method for solving the nonlocal optical model Schrödinger equation. We briefly explain this method. Henceforth for brevity, we shall write all equations in terms of bra and ket vectors and operators. Eq. (23) then becomes

$$(E - H_\ell)|\chi_\ell >= U|\chi_\ell > . \tag{24}$$

Eq. (24) can then be transformed into the following integral equation

$$|\chi_\ell >= |\chi_\ell^{(0)} > +G_\ell U|\chi_\ell >, \tag{25}$$

where $|\chi_\ell^{(0)} >$ is the homogeneous solution of Eq. (24) with the same outgoing wave boundary condition, and G_ℓ is the Green's function of Eq. (14).

Now let us introduce an auxiliary state function $|\lambda_\ell >$ defined as

$$|\lambda_\ell >= U|\chi_\ell >, \tag{26}$$

Multiplying both sides of Eq. (25) by U, one obtains an integral equation for $|\lambda_\ell >$,

$$|\lambda_\ell >= |\rho_\ell > +UG_\ell|\lambda_\ell >, \qquad (|\rho_\ell >\equiv U|\chi_\ell^{(0)} >). \tag{27}$$

The above Eq. (27) is the equation that we propose to solve. Since both $|\chi_\ell^{(0)} >$ and G_ℓ are defined in terms of the local potential H_ℓ, they can be calculated without any problem. This means that once $|\lambda_\ell >$ is obtained, the optical model wave function $|\chi_\ell >$ can easily be obtained from Eq. (25) with the help of Eq. (26).

In solving Eq. (27), we use the Lanczos method[10]. It is worth noting that the application of the Lanczos method for solving Eq. (27) is possible, since $|\lambda_\ell >$ is a bound state vector, as seen from the fact that it is defined by multiplying $|\chi_\ell >$ by a localized operator U. (See Eq. (27)). It is thus possible to expand $|\lambda_\ell >$ in terms of an orthonormal set of vectors, as will be seen below in Eq. (31). This is not the case for $|\chi_\ell >$ in Eq. (25), since $|\chi_\ell >$ is not bounded.

We first expand $|\lambda_\ell >$ in terms of the bi-orthonormal set of vectors $|D_i >$ with i=*0,1,....,N* that are generated according to

$$|D_0 >= \frac{1}{N_0}|\rho_\ell >, \tag{28}$$

$$|D_{i+1} >= \frac{1}{N_{i+1}}[UG_\ell|D_i > - \sum_{j=0}^{i}|D_j > \alpha_{ji}], \tag{29}$$

with

$$\alpha_{ji} = \begin{cases} < \tilde{D}_j|UG_\ell|D_i > & \text{if } j \leq i+1 \\ 0 & \text{if } j > i+1. \end{cases} \tag{30}$$

N_i in the above equations is the normalization constant determined from the condition $< \tilde{D}_i|D_i >= 1$, $|\tilde{D}_i >$ being the conjugate state to $|D_i >$. The coefficients α_{ji} given by Eq. (30) are those determined from the usual Schmidt orthonormalization procedure. We now write $|\lambda_\ell>$ as

$$|\lambda_\ell>=\sum_{i=0}^{N} C_i|D_i>, \tag{31}$$

where C_i are the expansion coefficients. Inserting Eq. (31) into Eq. (27), one easily derives a set of inhomogeneous linear equations for the expansion coefficients C_i, i.e.,

$$\sum_j(\delta_{ij}-\alpha_{ij})C_j = N_0\delta_{0i}. \tag{32}$$

The values of C_i are then determined by solving Eq. (32). Note that Eq. (32) can be solved rather easily, because $\alpha_{ji}=0$ for $j>i+1$. In addition, the value of N, i.e., the number of the basic functions D_i, can be chosen as a small number. This greatly helps to make the actual numerical calculations very rapid.

This method was tested for the case of the n+^{208}Pb system with incident energies E_n=7 and 14.5 MeV. The results showed that N=3 was enough to obtain the solution within an accuracy of 0.1%. Note that if one uses the PB method, one needs to do about 10 iterations in order to obtain the solution with the same accuracy. In the case of a highly nonlocal DPP constructed in the previous section, a good agreement with the CC result is achieved with N=8. The larger N-value may be ascribed to the complex nature of nonlocality of the DPP.

Once the optical model wave function is obtained with the method described above, the elastic scattering amplitude can be easily calculated from the outgoing wave boundary condition, and the differential elastic cross section immediately follows.

The total direct reaction and fusion cross sections can be obtained in the following way. The divergence of the current density is given[11] as,

$$div\ \mathbf{j}_\ell = \frac{2}{\hbar} Im(\chi_\ell^{(+)*}(r)U_\ell(r,r')\chi_\ell^{(+)}(r')). \tag{33}$$

The reaction cross section can be obtained by integrating the divergence of the current density all over space and summing over all the partial waves.

$$\sigma_R = \frac{\pi}{k^2}\sum_\ell(2\ell+1)T_{R;\ell}, \tag{34}$$

where the transmission coefficients $T_{R;\ell}$ can be written as

$$T_{R;\ell} = \frac{8}{\hbar v}\int dr dr' \chi_\ell^{(+)*}(r)W_\ell(r,r')\chi_\ell^{(+)}(r'). \tag{35}$$

Here v is the relative velocity in the incident channel. If W is decomposed into W_F and W_D, the fusion and direct reaction cross sections, σ_F and σ_D, respectively, can be expressed exactly the same as in Eq. (34), but with the corresponding transmission coefficients,

$$T_{F;\ell} = \frac{8}{\hbar v} \int dr dr' \chi_\ell^{(+)*}(r) W_{F;\ell}(r,r') \chi_\ell^{(+)}(r'), \tag{36}$$

$$T_{D;\ell} = \frac{8}{\hbar v} \int dr dr' \chi_\ell^{(+)*}(r) W_{D;\ell}(r,r') \chi_\ell^{(+)}(r'). \tag{37}$$

Thus all the elastic scattering, direct reaction, fusion cross sections are given in a model with a DPP.

3.3. A Phenomenological Analysis with a Local DPP

We have shown that the DPP can account for all the reaction phenomena if it is properly constructed, including all the nuclear structure information. Unfortunately, it is impossible to make such a construction of the DPP at this stage. In any nuclear reaction calculation, the phenomenological optical model Hamiltonians have thus customarily been used. In the present study, following this usual procedure, we make[12] a simultaneous χ^2-analysis of elastic scattering, DR, and fusion cross sections for the ^{16}O+^{208}Pb system at near- and sub-barrier energies and determine a DPP that satisfies the dispersion relation over all space.

We assume in this analysis the conventional local optical model Hamiltonian,

$$H = H_a + T_a + U_a, \tag{38}$$

satisfying the properties of the DPP we have previously described, namely H satisfies a dispersion relation, and W can be separated into direct reaction and fusion parts. In Eq. (31), H_a, T_a and U_a are the intrinsic Hamiltonian, kinetic energy operator for the relative motion between the projectile a and target A, and the so-called optical model potential, respectively. The optical model potential U_a to be used is local, and is given as

$$U_a(r) = U_C(r) - [V_0(r) + V(r) + iW(r)], \tag{39}$$

where $U_C(r)$ is the Coulomb potential and $V_0(r)$ is the bare potential[13,14] that corresponds to the Hartree-Fock potential[15], Note that $V_0(r)$ may have an energy dependence coming from the nonlocality of the exchange contribution. We ignore such effects, however, in the present study, as they are expected to be small for heavy-ion scattering[16]. The quantities V and W in Eq. (39) are, respectively, the real and imaginary parts of the DPP originating from couplings to the reaction channels. They are assumed to have a volume-type fusion and surface-derivative-type DR parts. Explicitly, $V_0(r)$, $V(r)$ and $W(r)$ are given, respectively, as

$$V_0(r) = V_0 f(X_0), \tag{40}$$

$$V(r) = V_F(r) + V_D(r) = V_F f(X_F) + 4V_D a_D \frac{df(X_D)}{dR_D}, \tag{41}$$

$$W(r) = W_F(r) + W_D(r) = W_F f(X_F) + 4W_D a_D \frac{df(X_D)}{dR_D}, \tag{42}$$

where $f(X_i) = [1 + exp(X_i)]^{-1}$ with $X_i = (r - R_i)/a_i$ (i=0,F and D) is the usual Woods-Saxon function.

The optical potential given above involves eleven parameters, excluding the Coulomb radius parameter in U_C, which we fix as $r_c = 1.25$ fm. We have attempted to fix these parameters by a χ^2-analysis. It is, however, impossible to determine all eleven parameters uniquely from a χ^2-analysis; we thus fix *a priori* some of the parameters by taking their values from the literature. Following Refs. 13 and 14, we fixed the parameters of the bare potential $V_0(r)$ as

$$V_0 = 60.4\ MeV, \qquad r_0 = 1.176\ fm, \qquad a_0 = 0.658\ fm. \tag{43}$$

With these values, $V_0(r)$ has values very similar to those of the folding potential at $r \approx R_A$. Further, we fixed all the geometrical parameters in the local DPP $(V(r) + iW(r))$ as

$$r_F = 1.40\ fm, \qquad a_F = 0.25\ fm, \tag{44}$$

$$r_D = 1.50\ fm, \qquad a_D = 0.45\ fm. \tag{45}$$

These values are chosen as the best possible values, taking into consideration results of studies reported in Refs. 13. Note, however, that these values are still not unique, and this non-uniqueness might introduce a certain ambiguity in the present study.

Once the geometrical parameters are fixed in this way, the dispersion relation is reduced to one for the strength parameters V_i and W_i ($i = F$ and D). The relation now reads[7]

$$V_i(E) = V_i(E_s) + \frac{E - E_s}{\pi} \mathbf{P} \int_0^\infty dE' \frac{W_i(E')}{(E' - E_s)(E' - E)}, \tag{46}$$

where $V_i(E_s)$ is the value of the potential at a reference energy $E = E_s$. Using the above relation, it is possible to evaluate values of V_F and V_D, once the values of $W_F(E)$ and $W_D(E)$ are known.

In an attempt to obtain the values of $W_F(E)$ and $W_D(E)$, we first carried out a χ^2-analysis, treating the four strength parameters V_F, V_D, W_F and W_D as adjustable parameters. We took into account all the data[17,18] available for incident energies between E_{lab}=78 MeV and 104 MeV. We included the total DR cross sections, when available, in the χ^2-analysis.

The values of the parameters determined in this way showed, as expected, a considerable fluctuation; i.e., they changed appreciably from one energy point

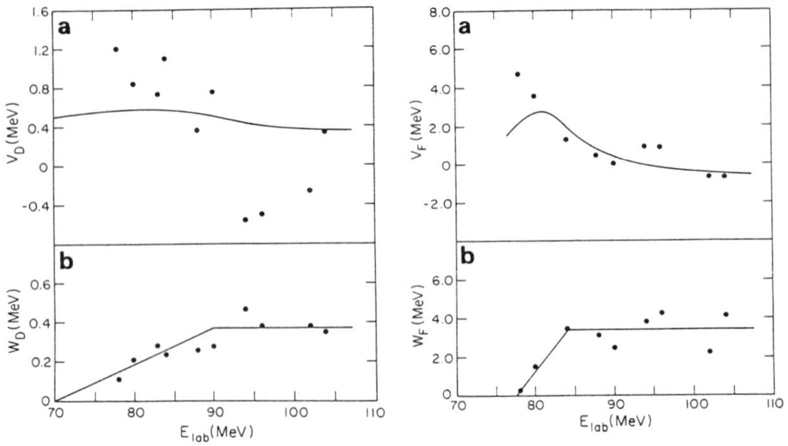

Figure 2: Strength parameters $V_D(E)$ and $W_D(E)$ of the direct reaction potential(a) as functions of E. The dots are those extracted from the χ^2-analyses, while the full lines drawn for W_D and $V_D(E)$ are the predictions of Eq. (47) and Eq. (46) using (47), respectively. (b) is the same as (a) but for the fusion potential. The full lines drawn for W_F and $V_F(E)$ are the predictions of Eq. (48) and Eq. (46) using (48), respectively.

to another, excepting those of W_D. For this exceptional case, the values were determined as a fairly smooth function of E. To illustrate this and also the fluctuation mentioned above, we show in Fig. 2a the values of V_D and W_D. As seen, the V_D-values fluctuate fairly strongly, while those of W_D behave rather smoothly as a function of E. The later W_D-values can be well represented by the following function of E (in units of MeV):

$$W_D(E) = \begin{cases} 0 & \text{for E} \leq 70 \\ 0.0185(E-70) & \text{for } 70 < \text{E} \leq 90 \\ 0.37 & \text{for } 90 \leq \text{ E} \end{cases} \tag{47}$$

The solid line shown in 2a is W_D predicted by the above Eq. (47). As seen, the line fits the empirical values rather well.

The fact that W_D could be fixed as a smooth function of E indicates that these values are reliable. There is a reason that W_D can be determined unambiguously, and hence appear as a smooth function of E. The reason lies in the fact that W_D is the dominant absorptive term in the peripheral region. Therefore, the elastic scattering cross section is quite sensitive to the value of

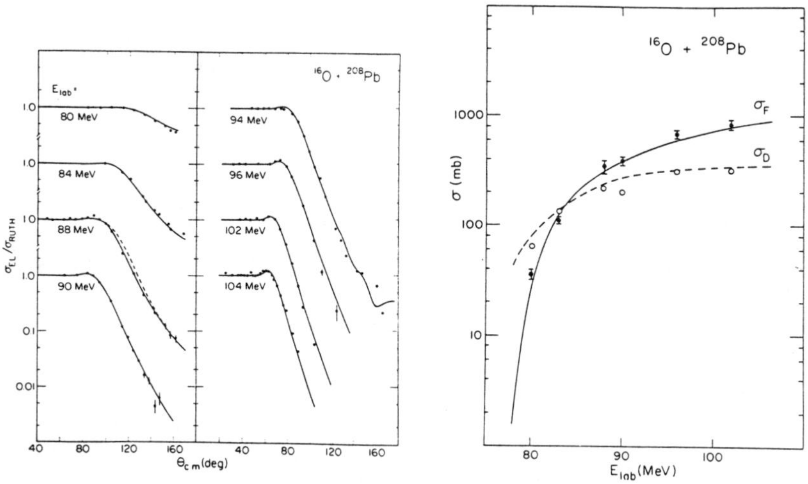

Figure 3: Comparison of the calculated ratio R of the elastic scattering cross sections to Rutherford cross sections(a), and the calculated direct reaction, fusion, and total reaction cross sections(b) using our final optical potential for the $O + Pb$ system, with the experimental data. The broken line shown for E_{lab}=88 MeV is R calculated by neglecting $V_D(r)$. The data are taken from Refs. 17 and 18.

W_D. This is not the case for V_D; even in the peripheral region; V_D is generally smaller than the bare potential $V_0(r)$, making it difficult to determine its value unambiguously. The fluctuation seen in Fig. 2a may be understood to arise from this difficulty.

Since a reliable value of W_D is now available, one can generate better values of V_D by using the dispersion relation (46). In doing this, we need to know one more parameter, i.e., the value of V_D at $E = E_s$. We may fix this $V_D(E_s)$ by fitting the average of the resultant V_D to that of the empirically determined V_D. The solid curve shown in Fig. 2a shows the V_D-values thus calculated. The $V_D(E_s)$-value used is $V_D(E_s)$=0.5 MeV at E_s=90 MeV. In what follows, we shall use thus calculated $V_D(E)$ as our final $V_D(E)$.

The second χ^2 analysis was then carried out by using the above values for V_D and W_D, and by treating V_F and W_F as adjustable parameters. Fig. 2b shows the values of V_F and W_F thus determined. As seen, both are now determined as fairly smooth functions of E. The W_F-values can now be well represented (in units of MeV) as

$$W_F(E) = \begin{cases} 0 & \text{for E} \leq 77.5 \\ 0.52(E-77.5) & \text{for } 77.5 < \text{E} \leq 84 \\ 3.4 & \text{for } 84 \leq \text{ E.} \end{cases} \tag{48}$$

The solid curve shown in Fig. 2b is the prediction of Eq. (48). The curve fits the empirical values reasonably well. We then calculated values of V_F by using Eq. (46) and (48). The constant value involved was chosen as $V_F(E_s)$=1.8 MeV at E_s=84 MeV. The solid line shown in Fig. 2b is the prediction. As seen, the predicted V_F-values again agree reasonably well with the empirically determined values, except at the lowest energy point (E=78 MeV). The discrepancy seen there, however, should not be taken too seriously, since the data are rather insensitive to the V_F-values at such a low energy. Except for this point, the empirically determined V_F and W_F are seen to satisfy the dispersion relation fairly well.

This result then suggests that we may use, as our final potential, the one with W_F and W_D given, respectively, by Eqs. (48) and (47), and also V_F and V_D generated from the dispersion relation as done above. This potential fully satisfies the dispersion relation. It can also nicely reproduce the data. This is seen in Fig. 3, where we have presented $R = \sigma_{EL}/\sigma_{RUTH}$ (where σ_{RUTH} is the Rutherford cross section), σ_F and σ_D, calculated by using our final potential. The results are also compared with the experimental data. As seen, all experimental R, σ_F and σ_D results are well reproduced by the calculations.

4. LOCALIZATION OF THE NONLOCAL DPP

Even though the DPP is formulated to be *nonlocal*, the phenomenological *local* optical potential which satisfies the dispersion relations, and is decomposed into fusion and direct reaction parts is shown to explain the observed low-energy nuclear reaction phenomena; elastic scattering, direct reaction, and fusion. This implies that there always exists an equivalent local potential to the nonlocal DPP, or that the DPP itself is local(or that nonlocality of the DPP is small in reality.). Another implication is that we may understand in a simple potential term where the reaction occurs. For example, if $W_F^{OM}(r)$ is proved to actually represent $W_F^{DPP}(r,r')$, then the place where fusion takes place, which is one of the problems in the heavy-ion fusion reactions, can be known. None of these questions have been clearly answered! The first thing to do in order to answer these questions would be to construct the nonlocal DPP in the large CC space including all coupling information, and to see whether it becomes local or not when the CC space gets larger. However, with regard to the present status in nuclear physics, this is a rather difficult task. Another test would be to find an appropriate localization procedure in which the physics remains the same. In this section, we introduce several known localization procedures, apply them to the localization of the DPP, and discuss the results.

4.1. Methods for Localization

The problem we encounter is to find a $U_\ell^L(r)$ which contains all the same physical properties as that of $U_\ell^N(r,r')$. They are related through nonlocal and local Schrödinger equations as,

$$(E-H_\ell)\chi_\ell^N(r)=\int dr' U_\ell^N(r,r')\chi_\ell^N(r'), \tag{49}$$

$$(E-H_\ell)\chi_\ell^L(r)=U_\ell^L(r,)\chi_\ell^L(r). \tag{50}$$

We introduce several approximation methods here.

1) Trivially equivalent local potential.

Franey and Ellis[19] defined the trivially equivalent local potential(TELP) as

$$U_\ell^L(r)=\frac{1}{\chi_\ell}\int dr' U_\ell^N(r,r')\chi_\ell^N(r'), \tag{51}$$

by setting, in Eqs. (49) and (50),

$$\chi_\ell^L(r)=\chi_\ell^N(r)=\chi_\ell(r) \tag{52}$$

Because $\chi_\ell^L=\chi_\ell^N$, this potential always guarantees the elastic scattering. Apart from the ℓ-independent phenomenological optical potential, it is ℓ-dependent. Because of χ_ℓ in the denominator, this potential varies rapidly where $|\chi_\ell|$ is small.

This localized local potential has an unphysical contribution; the imaginary part of $U^L(r)$ has a contribution from $ReU^N(r,r')$, i.e.,

$$\begin{aligned} ImU_\ell^L(r) &= \tfrac{1}{|\chi_\ell|^2} Im \textstyle\int dr' \chi_\ell^*(r) U_\ell^N(r,r')\chi_\ell^N(r') \\ &= \tfrac{1}{|\chi_\ell|^2} Re \textstyle\int dr' \chi_\ell^*(r)[ImU_\ell^N(r,r')]\chi_\ell^N(r') \\ &+ \tfrac{1}{|\chi_\ell|^2} Im \textstyle\int dr' \chi_\ell^*(r)[ReU_\ell^N(r,r')]\chi_\ell^N(r'). \end{aligned} \tag{53}$$

A similar contribution in $ReU^L(r)$ comes from $ImU^N(r,r')$. This unphysical term is quite significant, and plays an important role in constructing the distorted waves in the elastic channel, and thus in accounting for the elastic scattering when nonlocality becomes large. Of course, it does not affect the direct reaction and fusion cross sections, since the integration of this term over all space vanishes. Fig. 4 displays each term in Eq. (53) in the case of the ^{16}O + ^{208}Pb system at E_{lab}=80 MeV, where nonlocality is very large (See Fig. 1). The solid curve is the resultant imaginary part of the $\ell=10$ DPP, while the dashed and dotted curves represent contribution from $ImU^N(r,r')$ and $ReU^N(r,r')$ (fictitious) parts, respectively.

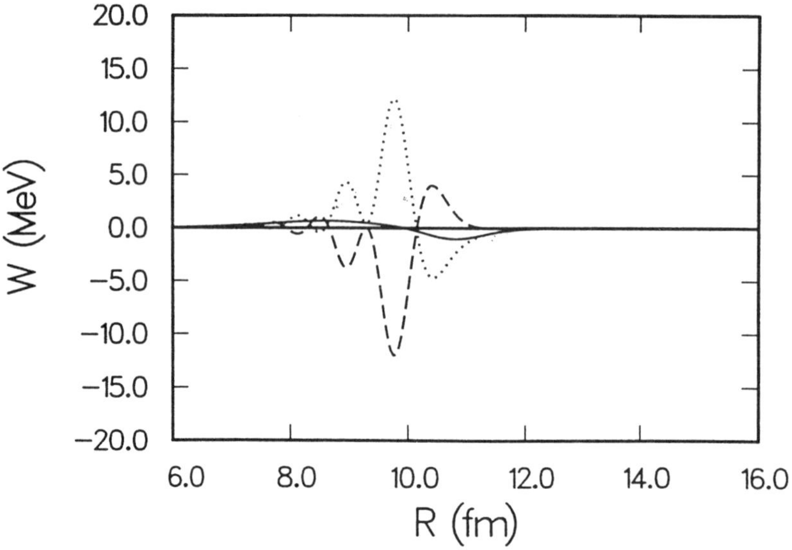

Figure 4: Imaginary part(solid) of the potential obtained from a TELP localization method. The dashed curve denotes contribution from the imaginary part of $U^N(r, r')$, while the dotted one from the real part.

2) Taylor expansion method

Perey and Buck[8] assumed a separable form of a nonlocal potential with a Woods-Saxon type local part, $U^{N,local}$, and a Gaussian type nonlocal part. They defined $\beta \mathbf{s} = \mathbf{r} - \mathbf{r}'$ where β is a nonlocality parameter, expanded U^N and χ^N in Taylor series, and performed an integration over $\mathbf{s}$. By assuming again $\chi_\ell^L = \chi_\ell^N$, and by neglecting the surface effect of the local part, $U^{N,local}$, they could relate U^L to $U^{N,local}$ as

$$U^L(\mathbf{r})exp[\frac{m\beta^2}{2\hbar^2}(E - U^L(\mathbf{r})] = U^{N,local}(\mathbf{r}) \tag{54}$$

This method only works well when the nonlocality of the potential has a Gaussian form, otherwise integration is difficult to perform and it cannot be placed into a simple analytic form.

3) Fourier transformation method

Following the famous Perey-Saxon's local energy approximation[20] to nonlocality through the Fourier transform, Madsen and Osterfeld[21] recently developed a simple Fourier transformation technique for the localization of the imaginary optical potential. They defined coordinates $s = r - r'$ and $\frac{1}{2}(r + r') = r + \frac{1}{2}s$, and again assumed $\chi_\ell^L = \chi_\ell^N$. The Perey-Saxon's local energy approximation is

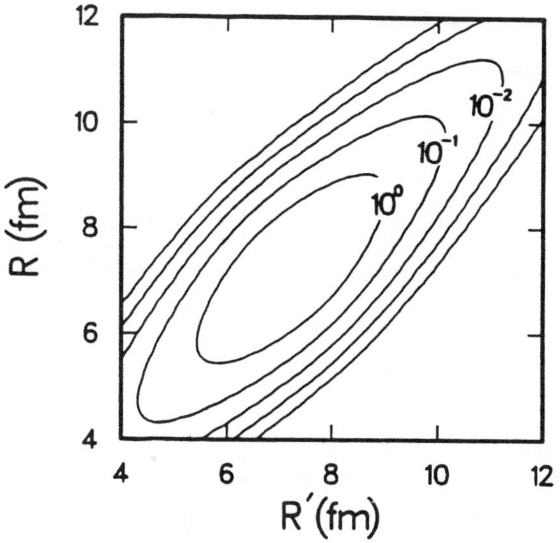

Figure 5: Imaginary part of Perey-Buck nonlocal potential for the $n+Pb$ system at $E_n = 7$ MeV.

then applied to the radial nonlocal potential as

$$\begin{aligned} U_\ell^L(r)\chi_\ell^L(r) &= \int dr' U_\ell^N(r,r')\chi_\ell^N(r') \\ &\approx \int ds U_\ell^N(r, r-\tfrac{s}{2})\chi_\ell^L(r-\tfrac{s}{2}) \\ &\approx 2\pi \mathcal{U}_\ell^N(r,\kappa)\chi_\ell^L(r). \end{aligned} \quad (55)$$

where $\mathcal{U}_\ell^N(r,\kappa)$ is a simple Fourier transformation of $U_\ell^N(r, r-\frac{s}{2})$

$$\mathcal{U}_\ell^N(r,\kappa) = \frac{1}{2\pi}\int_{-\infty}^{\infty} ds U_\ell^N(r,s)e^{-i\kappa s} \quad (56)$$

where κ is chosen equal to the local wave number

$$\kappa = \kappa_\ell = [\frac{2m}{\hbar^2}(E - V(r)) - \frac{\ell(\ell+1)}{r^2}]^{1/2} \quad (57)$$

for a given ℓ. It should be noted that this approximation induces local momentum (or energy) dependence in the localized potential and in addition there is another energy dependence of the local optical potential due to localization.

4) Comparisons in the Perey-Buck nonlocal potential

We now compare these three localization methods to the Perey-Buck nonlocal potential[8], which consists of a Woods-Saxon type local part $U^{N,local}$ and a Gaussian type nonlocal part H as,

$$U^N(\mathbf{r},\mathbf{r}') = U^{N,local}((\mathbf{r}+\mathbf{r}')/2)H(|\mathbf{r}-\mathbf{r}'|), \quad (58)$$

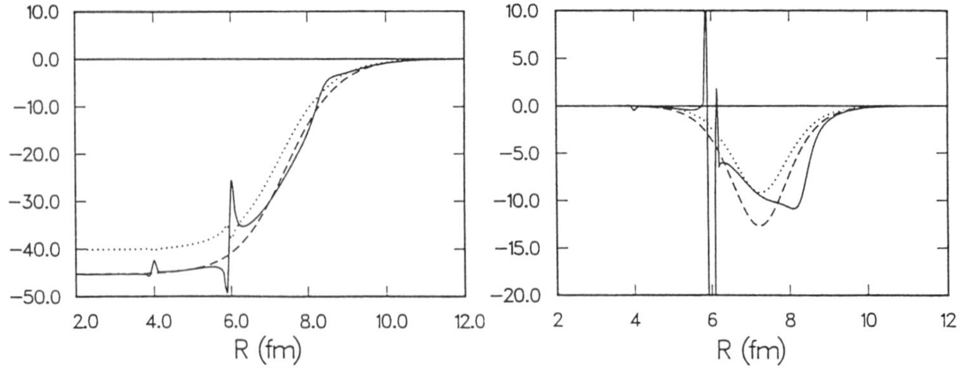

Figure 6: Localized real(a) and imaginary(b) parts of Perey-Buck nonlocal potential for the $n+Pb$ system at $E_n = 7$ MeV obtained from TELP (solid), Taylor expansion (dashed) and Fourier transformation (dotted) methods.

$$H(|\mathbf{r}-\mathbf{r}'|) = \frac{1}{(\sqrt{\pi}\beta)^3} exp[-(\frac{\mathbf{r}-\mathbf{r}'}{\beta})^2]. \tag{59}$$

The radial part of the nonlocal potential in Eq. (23) for this case then becomes

$$U_\ell^N(r,r') = U^{N,local}((r+r')/2)h_\ell(r,r'), \tag{60}$$

and

$$h_\ell(r,r') = \frac{1}{(\sqrt{\pi}\beta)} exp[-(\frac{r^2+r'^2}{\beta^2})](2i^\ell z j_\ell(-iz)) \tag{61}$$

where $z = \frac{2rr'}{\beta^2}$ and $j_\ell(z)$ is a spherical Bessel function. Fig. 5 displays the imaginary part ($\ell = 0$) of the nonlocal potential of which $U^{N,local}$ has a surface type(derivative of the Woods-Saxon).

The calculated real(a) and imaginary(b) parts of the localized DPP from the three methods are shown in Fig. 6. The solid, dashed, and dotted curves denote the results of the TELP method, Taylor expansion method, and Fourier transformation technique, respectively. The potentials look similar to each other as seen in Fig. 6, but generate quite different elastic differential cross sections as described below. The elastic scattering cross sections for the $n+{}^{208}Pb$ at $E_n = 7$ MeV by using potentials obtained from the corresponding methods are displayed in Fig. 7. The *exact* differential cross section from the nonlocal calculation of Sec. 3.2 coincides with the TELP result. The elastic results, except for that of TELP, are unfortunately quite different from the nonlocal one and it becomes even more severe as the nonlocality gets larger. From now on, we will use the TELP method which reproduces the correct elastic scattering cross sections even though it has itself drawbacks as mentioned before.

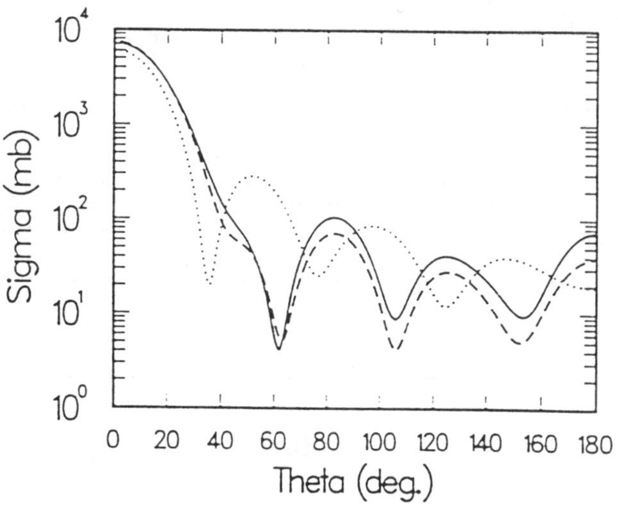

Figure 7: Elastic differential cross sections for the $n + Pb$ system at $E_n = 7$ MeV by using potentials obtained from TELP (solid), Taylor expansion (dashed) and Fourier transformation (dotted) methods. The exact cross section coincides with the solid curve.

4.2. Localized W_F and W_D

In Fig. 8, we show the localized $W(r, r')$, $W_F(r, r')$, and $W_D(r, r')$ generated for the ^{16}O + ^{208}Pb system at E_{lab}=80 MeV as solid, dashed, and dotted curves, respectively. The scale in the figure at greater radial distances than 11 fm is magnified. The construction of the DPP is the same as in Sec. 2.1. The localization was carried out for $\ell = 10$. Since only the imaginary parts of the nonlocal potential are localized, the unphysical parts do not contribute. The localized W is well separated into W_F and W_D parts; the inner part of W coincides with W_F, while the outer part with W_D. A similar separation is also seen for different partial waves. It is worth noting that W_F extends out to about 12 fm, which is definitely outside the Coulomb barrier, supporting the direct reaction approach to fusion[1,4].

4.3. Comparison of a localized DPP with the phenomenological OM potential

We assume that there exists an equivalent local potential to the nonlocal DPP, and then derive this local potential by using the TELP method. This local potential is then compared to a phenomenological optical model potential. Since it is impossible to construct a nonlocal DPP including all the open and

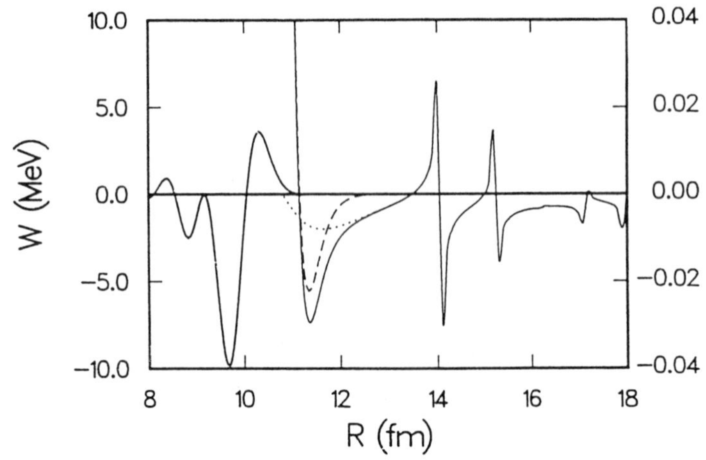

Figure 8: Separation of a localized W(solid) into W_F(dashed) and W_D(dotted).

closed channels formed by the two colliding nuclei at this stage, we restrict ourself to a limited CC space, perform CC calculations in that space, and treat the CC results as a data set. We now obtain a phenomenological ℓ-independent OM potential by performing χ^2- fitting to CC results. On the other hand, we construct the nonlocal DPP in this limited CC space, and localize it by using the TELP method. The approximated ℓ-independent local potential is then obtained from the ℓ-dependent localized potential as follows[22],

$$U(r) = \frac{\sum_\ell (2\ell+1) T_\ell U_\ell(r)}{\sum_\ell (2\ell+1) T_\ell} \tag{62}$$

where T_ℓ is the transmission coefficient of Eq. (35). We then compare the two potentials with each other.

For this purpose, we choose the ^{16}O + ^{40}Ca system at E_{lab}=60 MeV, of which the CC potential is known[23]. The same CC space is chosen as that of Ref. 23, i.e., coupling of the ground state to the 3.90 MeV 2^+, 3.74 MeV 3^-, and 4.49 MeV 5^- states in ^{40}Ca. The coupling strengths are also obtained from Ref. 23. In Fig. 9, the solid curve is the resultant localized potential, while the dashed one the phenomenological potential obtained by a χ^2- fitting to the CC results. The dotted curve is the imaginary part of CC potential used in the calculation. Except for values around 8 fm, the two potentials agree nicely. However, we cannot conclude that an equivalent local potential to the nonlocal one always exists, because the TELP method itself has some drawbacks as mentioned before. A better localization procedure should be made for a highly nonlocal potential.

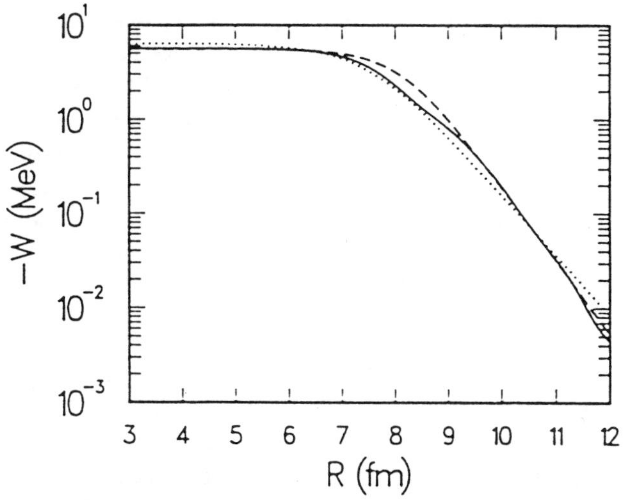

Figure 9: Comparison of $W^{localizedDPP}$(solid) with W^{OM}(dashed) for the $O+Ca$ system at E_{lab}=60 MeV. The dotted curve is the imaginary part of CC potential. The coupling scheme is mentioned in the text.

5. SUMMARY

We have studied the dynamic polarization potential due to the coupling to collective states, based on the projection operator formalism of Feshbach. It turns out to be nonlocal and ℓ-dependent. Dispersion relations between the real and imaginary parts were discussed.

We have presented how to calculate the elastic, direct reaction, and fusion cross sections simultaneously with a single DPP. For this purpose, a rapid way to solve the nonlocal Schrödinger equation was developed to obtain the *exact* solution. The separation of the DPP into the direct reaction and fusion parts was then formulated. We showed that such a separation phenomenologically made in a local space explains well the observed data.

In order to justify the phenomenological local optical model potential, localization procedures of the nonlocal DPP were discussed. A few attempts with the trivially equivalent local potential method were made to account for the phenomenological fusion potential which has a large fusion radius. It turns out that the localized fusion potential indeed exists out to a large radial distance, supporting the direct reaction approach to fusion. In a limited coupled-channels space, we constructed both a localized polarization potential and a phenomenological potential which fits the coupled-channels results. We found that these two potentials agree reasonably well with each other. Nevertheless, there is still room to justify phenomenological local optical potential with a dynamic polarization potential. It is too early to give a definite answer whether or not the equivalent local potential to the nonlocal one always exists. A better localization procedure should be made for a highly nonlocal potential.

We thank Dr. D. Knobles for a careful reading of the manuscript. This work was supported in part by the Ministry of Education, through the Basic Science Institute Program, 1990, and the Center for Theoretical Physics, Seoul National University, through the SRC program and in part by the U. S. Department of Energy.

REFERENCES

1. T.Udagawa, B. T. Kim, and T. Tamura, Phys. Rev. **C32**, 124 (1985).
2. G. R. Satchler, M. A. Nagarajan, J. S. Lilley, and I. J. Thompson, Ann. Phys. **178**, 110 (1987).
3. G. R. Satchler, *Direct reaction theories*, (Oxford, London, 1984).
4. B. T. Kim, in *Proc. of the 1st Sorak symposium on nuclear physics, Sorak, 1988*, ed. D. P. Min, p.235.
5. B. T. Kim, in *Proc. of the RIKEN symposium on nuclear collective motion and nuclear reaction dynamics, Wako, Japan, 1989* ed. K. I. Kubo et al., (World-Scientific, Singapore, 1990) p.249.
6. H. Feshbach, Ann. Phys. **5**, 357 (1958), ibid, **19**, 287 (1962).
7. M. A. Nagarajan, C. Mahaux and G. R. Satchler, Phys. Rev. Lett. **54**, 1136 (1985).
8. F. Perey and B. Buck, Nucl. Phys. **32**, 353 (1962).
9. B. T. Kim and T. Udagawa, Phys. Rev. C **42**, 1147 (1990).
10. R. R. Whitehead, A. Watt, B. J. Cole, and I. Morrison, Adv. Nucl. Phys. **9**, 123 (1977).
11. G. R. Satchler, Phys. Rev. C **32**, 2203 (1985).
12. B. T. Kim, M. Naito, and T. Udagawa, Phys. Lett. **237B**, 19 (1990).
13. M. J. Rhoades-Brown and M. Prakash, Phys. Rev. Lett. **53, 333** (1984).
14. B. T. Kim, H. C. Kim and K. E. Park, Phys. Rev. C **37**, 998 (1988).
15. C. Mahaux, H. Ngo and G. R. Satchler, Nucl. Phys. **A456**, 134 (1986); M. A. Nagarajan, C. Mahaux and G. R. Satchler, Phys. Rev. Lett. **54**, 1136 (1985).
16. G. R. Satchler and W. G. Love, Phys. Rep. **55**, 183 (1979).
17. F. Videvaek *et al.*, Phys. Rev. C **15**, 954 (1977).
18. E. Valgaris, L. Gradzins, S. G. Steadman and R. Ledoux, Phys. Rev. C **33**, 2017 (1986).
19. M. A. Franey and P. J. Ellis, Phys. Rev. C **23**, 787 (1981).
20. F. G. Perey and D. S. Saxon, Phys. Lett. **10**, 107 (1964).
21. V. A. Madsen and F. Osterfeld, Phys. Rev. C **39**, 1215 (1989).
22. I. J. Thompson, M. A. Nagarajan, J. S. Lilley and M. J. Smithson, Nucl. Phys. A **505**, 84 (1989).
23. K. I. Kubo and P. E. Hodgson, Nucl. Phys. **A366**, 320 (1981).

INELASTIC BREAKUP IN INCLUSIVE CONTINUUM SPECTRA.

A.Bonaccorso
Istituto Nazionale di Fisica Nucleare, 56100 Pisa,Italy

D.M.Brink
Department of Theoretical Physics, Oxford OX1 3NP, UK.

ABSTRACT

In this paper we discuss properties of transfer reactions which are important for understanding the dynamics of heavy-ion grazing collisions. In particular we stress the evolution of the reaction mechanism as a function of the incident beam energy and discuss the varius components of a continuum energy spectrum.

INTRODUCTION

In a heavy-ion grazing collision transfer reactions play a major role as a source of dissipation of energy therefore it is very important to understand the dynamics of such reactions and the evolution with energy.

Transfer can occur both from the projectile to the target and vice versa but in this paper we will always refer to the first case and in particular we will discuss one neutron transfer to continuum states according to the formalism developped in refs.1, 2 and 3. The discussion about inelastic breakup is instead new.

When the incident energy per nucleon is less than the average bound energy of a nucleon in a nucleus ($\sim 10MeV$) transfer to a final bound state is the leading mechanism. At higher energy the cross section for transfer to final bound states decreases exponentially as it has been shown both theoretically [(4,5)] and experimentally [(6,7)]. When the incident energy per nucleon increases above the average bound state energy then transfer to a continuum final state dominates.

A continuum final state is a scattering state of positive energy of the transferred neutron in the field of the target. There are two main mechanisms that can produce a continuum final state. One is due to direct type of reactions like elastic and inelastic breakup. Inelastic breakup means that the target can be excited as a result of the final state interaction of the neutron with the target potential while in the elastic breakup case there is no final state interaction and the target potential acts only as a disturbance which allows the neutron to breakup from the projectile. The other mechanism is of compound type in which the transferred neutron goes into a single particle resonance state of the target or it forms a compound nucleus.

An experimental inclusive spectrum is due to the sum of direct and compound type of reactions and it is useful for obtaining the amount of the total reaction cross section which is due to transfer. However to understand the dynamics of the reaction it is more interesting to disentangle the various components.

Direct type of reactions like elastic and inelastic breakup can be studied by coincidence experiments in which the light particle emitted by the projectile is detected in coincidence with the projectile like fragment. Direct reactions have forward peaked angular distributions. Compound nucleus reactions can be studied by detecting the particles which decay from the compound nucleus. Such a decay is instead isotropic and therefore by detecting light particles at backward angles one should be able to get the compound component of the inclusive spectra.

From the theoretical point of view it is possible to study transfer reactions both to bound as well as to unbound continuum states by using a semiclassical method described in refs.(1-4). It is also possible to distinguish the various types of reactions contributing to continuum spectra using a Green's function formalism described in ref. (2).

In order to compare experimental spectra with calculations we discussed in ref.(2) the contributions from elastic breakup and from absorption. These two components can be separated in the framework of an optical model treatment of the final state interaction of the transferred neutron with the target. The absorption part of the spectra is due at least to two different types of reactions. One is transfer to single particle or compound nucleus resonance states of the target the other is inelastic breakup. The intersting point is that while inelastic breakup from a theoretical point of view belongs to the class of absorptive effects from an experimental point of view it can be studied only in a coincidence experiment which would give a spectrum that is the sum of elastic and inelastic breakup. This subtle difference in the way inelastic breakup can be studied experimentally and theoretically has been the origin of some confusion in the past and it is very useful to clarify it if one is looking for example for optimum transfer conditions or peak position of the spectra.

ABSORPTION vs. ELASTIC BREAKUP

In refs (1,2,3) we used the following formula which gives the transfer probability from an initial bound state with energy $\varepsilon_i < 0$ to a final continuum state with energy $\varepsilon_f > 0$

$$\begin{aligned}\frac{dP_t}{d\varepsilon_f} &= \Sigma_{l_f} < |1 - S_{l_f}|^2 > B(l_f, l_i) \\ &= \Sigma_{l_f}(|1 - < S_{l_f} > |^2 + T_{l_f})B(l_f, l_i) \qquad (1)\end{aligned}$$

where

$$T_{l_f} = 1 - | < S_{l_f} > |^2$$

is the optical model compound nucleus probability and $|1- < S_{l_f} > |^2$ is the shape elastic probability and

$$B(l_f, l_i) = \frac{1}{4}\left(\frac{\hbar}{mv}\right)^2 \frac{m}{\hbar^2 k_f} \frac{e^{-2\eta s}}{\eta d} |C_i|^2 e^{-2\eta R_i} P_{l_i}(X_i)(2l_f + 1) P_{l_f}(X_f) e^{-2\eta R_f} \tag{2}$$

where R_i and R_f are the radii of the initial and final nuclei and $d = R_i + R_f + s$ is the distance of closest approach between the trajectories of their centers.

The quantity η in eq.(2) is defined by

$$\eta^2 = k_1^2 + \gamma_i^2 = k_2^2 - k_f^2 \tag{3}$$

where

$$\gamma_i^2 = -\frac{2m\varepsilon_i}{\hbar^2} \qquad \text{and} \qquad k_f^2 = \frac{2m\varepsilon_f}{\hbar^2} \tag{4}$$

and $\hbar k_1$ and $\hbar k_2$ are the z-components of the momentum of the transferred neutron relative to the first and second nucleus. They are given by

$$\hbar k_1 = -\frac{(Q + \frac{1}{2}mv^2)}{v}, \qquad \hbar k_2 = -\frac{(Q - \frac{1}{2}mv^2)}{v} \tag{5}$$

where Q=$\varepsilon_i - \varepsilon_f$ is the Q-value of the reaction. In the case of transfer to the continuum $Q < 0$. The coefficient C_i is the asymptotic normalization constant of the initial state wave function , v is the velocity of relative motion at the distance d defined as

$$v^2 = \frac{2}{\mu}(E_{CM} - V_{CB})$$

where E_{CM} is the center of mass energy, V_{CB} is the Coulomb barrier at d and μ is the reduced mass of projectile and target. We choose the nucleus-nucleus orbit of relative motion to be a straight line trajectory along the z-direction with constant speed v. P_{l_i} and P_{l_f} are Legendre polynomials of argument

$$X_i = 1 + 2\frac{k_1^2}{\gamma_i^2} \qquad \text{and} \qquad X_f = 2\frac{k_2^2}{k_f^2} - 1 \tag{6}$$

S_{l_f} is the S-matrix which describes the scattering of the neutron on the target. Equation (1) contains an average over the resonances. This is valid because $B(l_f, l_i)$ has a smooth dependence on the final energy.

An approximate formula for the total transfer cross section can be obtained from eq.(1) by integrating over the impact parameters of the nucleus-nucleus relative motion as in Bonaccorso,Brink and Lo Monaco [(4)]

$$\frac{d\sigma}{d\varepsilon_f} = 2\pi \int_0^\infty \frac{dP_t}{d\varepsilon_f} P_{el}\, b\, db$$
$$\approx \pi \frac{R_s - a_c}{\eta} \frac{dP_t}{d\varepsilon_f}(R_s) \tag{7}$$

where $dP_t/d\varepsilon_f$ is given by eq.(1) with $d = R_s$ and a_c is the Coulomb length parameter. The nucleus-nucleus probability of elastic scattering [4] P_{el} was given by the sharp cut-off model with strong absorption radius R_s . Experimental results have shown that in the kind of reactions we are discussing here the ejectile is scattered at very forward angles and that the spectra show little dependence on the scattering angles[8]. Therefore we will calculate some spectra according to eq. (7) which gives the total transfer cross section integrated over angles.

In ref.(2) we showed also that closed formulae for the elastic breakup and the absorption could be obtainedby calculating the S-matrix in eq.(1) in Born approximation.

The formulae were

$$\frac{dP_{breakup}}{d\varepsilon_f} = \frac{m}{\hbar^2 k_f} \frac{1}{2\pi v^2 \hbar^2} \int d\mathbf{b}\, |\tilde{V}_2(\mathbf{b}, k_f - k_2)\tilde{\psi}_1(\mathbf{b}, k_1)|^2 \qquad (8).$$

$$\frac{dP_{abs}}{d\varepsilon_f} = \frac{1}{\pi \hbar^2 v^2} Im \int_{-\infty}^{+\infty} d^3\mathbf{r}_2 |\tilde{\psi}_1(x, y, k_1)|^2 \; V_2(\mathbf{r}_2) \qquad (9)$$

Born approximation is valid when the interaction with the potential is not too strong for examples in cases in which the neutron feels only the tail of the potential. This is what happens for the breakup and in fact we showed in ref.(2) that the first term in the RHS of eq.(1) and eq.(8) are equivalent .

INELASTIC BREAKUP

The elastic breakup formula eq.(8) can be generalized to include cases in which the neutron breaks up from the projectile but at the same time the target is left in an excited state. The inelastc breakup formula is given by[9]

$$\frac{dP_{cont}}{d\epsilon_f} = \frac{1}{v\hbar} \frac{1}{2\pi v^2 \hbar^2} \sum_{\beta \neq \alpha} \int d\mathbf{b}\, |\tilde{V}_{\beta\alpha}(\mathbf{b}, k_\beta - k_2)\tilde{\psi}_1(\mathbf{b}, k_1)|^2. \qquad (10)$$

Where α is the target ground state and β are excited states. $\tilde{V}_{\beta\alpha}$ is the one dimensional Fourier transform of a matrix element of the neutron-nucleus interaction between the ground state and excited states of the target.

Eq.(10) is interesting but it is very difficult to calculate because it contains a sum of matrix elements of the neutron-nucleus potential between the ground state and excited states of the target. However it can be shown [9] that eq.(10) is equivalent to eq.(9) which gives the absorption in Born approximation. The proof uses a microscopic formula for the imaginary part of the optical potential obtained by Bonaccorso and Brink in 1982[10]. In ref.(10) we first showed how a local phenomenological optical potential like $V_2(\mathbf{r_2})$ in eq.(9) could be equivalent to Feshbach potential ΔV, then obtained the following formula for the Wigner transform of ΔV

$$Im \int_{-\infty}^{\infty} \Delta\tilde{V}(r,p)dz = -\frac{1}{2v\hbar} \sum_{\beta\neq\alpha} \left| \int_{-\infty}^{\infty} dz \; e^{(i(k_z-k_2)z)} < \alpha|V(b,k_z-k_2,\xi)|\beta > \right|^2$$
$$= -\frac{1}{2v\hbar} \sum_{\beta\neq\alpha} \left| \tilde{V}_{\beta\alpha}(b,k_z-k_2) \right|^2 \qquad (11)$$

k_z is the momentum along the direction of relative motion and it is the approximated as

$$k_z \approx k_\beta = \sqrt{2m\frac{(\varepsilon_f-\varepsilon_\beta)}{\hbar}}$$

where ε_f is the neutron final energy relative to the target and ε_β is the target excitation energy and

$$k_2 = \frac{\varepsilon_f + \varepsilon_\beta - \varepsilon_i + \frac{1}{2}mv^2}{\hbar v}$$

In the above formula the z-integral gives the Fourier transform of a matrix element $V_{\beta\alpha}$ of the neutron-nucleus interaction between the ground state and excited state of the target. Using eq.(11) in eq.(10) one obtains eq.(9) which therefore can be used to calculate the inelastic breakup component of an inclusive spectrum and in the following we will show some examples. The physical interpretation of the equivalence of eqs.(9) and (10) is the following. From the absorption formula (9) calculated in Born approximation one can see that the integrand is different from zero only in a limited region of space between the two nuclei where the surface of the target potential overlaps with the initial state wave function. This means that resonance effects are neglected since they are due instead to a strong interaction of the neutron with the inside of the potential. Using eq.(9) to calculate the absorption corresponds to consider that part of the optical potential that describes direct inelastic excitations of the target only.

ENERGY SPECTRA

In figs.(1), (2) ,(3) we give some examples of spectra calculated with the formulae given in the previous section. Spin was taken into account according to the Appendix to ref.(3). Figs.(1), (2) are for the reaction $^{208}Pb(^{20}Ne,^{19}Ne)^{209}Pb$ at $E_{inc} = 30,40MeV/u$ respectively while fig.(3) is for the reaction $^{208}Pb(^{14}N,^{13}N)^{209}Pb$ at $E_{inc} = 60MeV/u$. Numerical parameters are the same as in ref.(3) a part from the strong absorption radii for the ^{20}Ne induced reactions which are given in the figure captions. In all the figures the dot curves represent the elastic breakup while the close dotted curves represent the inelastic breakup and the dotdashed curves represent the sum of elastic and inelastic breakup. This is the spectrum that should be seen in a coincidence experiment between the projectile like fragment and the light particles emitted in forward direction. The dashed curves give the absorption which is due to compound nucleus effects and inelastic breakup. The compound nucleus effects can be isolated by subtacting the inelastic breakup. In figs.(2), (3) this has been done and the results are given by the lower full curves. The upper full curves represent instead the sum af all reactions (elastic and inelastic breakup and compound nucleus) and correspond to the results of singles experiments.

As we discussed in ref.(3) the elastic breakup spectrum has a peak at

$$\varepsilon_f \sim \frac{1}{2}mv^2 \tag{12}$$

while the absorption and then the inelastic breakup which is part of it has a peak at

$$\varepsilon_f \sim \frac{1}{2}mv^2 - |\varepsilon_i| \tag{13}$$

where ε_i is the initial bound state energy. When the incident energy per nucleon is much larger than ε_i the two above conditions give the same peak energy. An experimental breakup spectrum giving the sum of elastic and inelastic breakup should peak at a final energy intermediate between (12) and (13). It is interesting to note that the elastic breakup is only a small part of the inclusive spectrum at low incident energies while at high incident energy (cf. fig.(3)) its magnitude is comparable to the absorption .

The second term of eq.(1) gives the absorption spectrum which is represented by the dashed lines in the figures. Subtracting from it the inelastic breakup obtained from eq.(9) gives the resonant component of the inclusive spectrum which corresponds to the neutron beeing absorpted into single particle or compound nucleus states of the target. This component shows a trend opposite to the breakup. It is in fact dominant at low incident energies because matching conditions tend to favour the population of single particle resonance states. These states are broad and overlap therefore in an experimental spectrum one sees bumps in which many single particle states contribute. We will call this component the transfer component and notice that the structure of the bumps depends a lot on the resonance states available in the target and also on the momentum distribution of the neutron in the initial state as discussed in ref.(3).

Comparing figs.(1), (2), (3) it is interesting to note that there is a definite evolution in the reaction mechanism when the incident energy increases. At low incident energy , fig.(1) , matching conditions favour transfer to the available low lying single particle resonance states of the target. Both the elastic and inelastic breakup give a small contribution. As the incident energy increases transfer and breakup become comparable while at high incident energy, fig.(3) , the breakup component , dotdashed curve , is larger than the transfer component, lower solid curve . However it is very interesting to note that even at such high energy transfer still gives a large contribution to the inclusive spectrum and the dynamics of the reaction is very similar to low energy dynamics. It has often been suggested that at incident energies close to the Fermi energy ($\sim 40 MeV/u$)

there should be a change in the dynamics of the nuclear reactions from a mean field dynamics to a nucleon-nucleon scattering dominated dynamics. However both experimental[8,11,12] and theoretical results discussed in this paper seem to suggest that such a change occurs at higher energies, at least higher than $60MeV/u$.

CONCLUSIONS

In this paper we have shown that in heavy ion grazing collisions there is an evolution in the transfer mechanism when the incident energy increases. Up to incident energies of the order of $10MeV$ bound state transfer dominates. Increasing the energy transfer populates single particle low lying resonance states which are in the continuum. There is also a small breakup component. At $40MeV/u$ transfer is still important and one sees the effect of high angular momentum states of the target in the second bump of fig.(2). Elastic and inelastic breakup are comparable in magnitude to the transfer. At $60MeV/u$ breakup is the dominant mechanism and in the transfer component single particle effects are very small. At this energy however the interaction with the target mean field is still responsabile for the dynamics of the reaction. Increasing the energy one expects the compound nucleus component to disappear and the maxima of the spectra will correspond to the most favourite final energy for breakup given by eq.(12).

REFERENCES

1) A.Bonaccorso and D.M.Brink, Phys. Rev. **C38**,1776 (1988).
2) A.Bonaccorso and D.M.Brink, Phys.Rev. **C43**, 299 (1991).
3) A.Bonaccorso and D.M.Brink, IFUP-TH 5/91 preprint, submitted to Phys.Rev.C.
4) A.Bonaccorso,D.M.Brink and L.Lo Monaco,J.Phys.G **13**,1407 (1987).
5) W.von Oertzen, Phys.Lett. **B115** 95 (1985).
6) T.Kurihara and M.Sakai, Phys.Lett. **133B**, 157 (1983).
7) J.S.Winfield, Sam M.Austin, G.M.Crawley, C.Djalali, C.A.Ogilvie, R.J.Smith, Ziping Chen and M.Torres, Phys.Lett. **203B**, 345 (1988).
8) S.Fortier, S.Gales, S.M.Austin, W.Benenson, G.M.Crawley, C.Djalai, J.S.Winfield and G.Yoo, Phys.Rev.**C41**,2689 (1990) .
9) A.Bonaccorso and D.M.Brink, IFUP-TH 22/91 preprint.
10) A.Bonaccorso and D.M.Brink,Nucl.Phys. **A384**, 161 (1982).
11) D.Beaumel, Y.Blumenfeld, Ph.Chomaz, N.Frascaria , J.P.Garron, J.C.Roynette , T.Suomijarvi, J.Barrette, B.Berthier, B.Fernandez, J.Gastebois, W.Mittig, 1987, Annual Report Orsay, pag. 41 .
12) W.Lahmer, W.von Oertzen, A.Miczaika, H.G.Bohlen, W.Weller, R.Glasow, R.Grzonka, R.Santo, Y.Blumenfeld, N.Frascaria, J.P.Garron, J.C.Jacmart and J.C.Roynette ,Z.Phys.- Atomic Nuclei **A337**, 425 (1990).

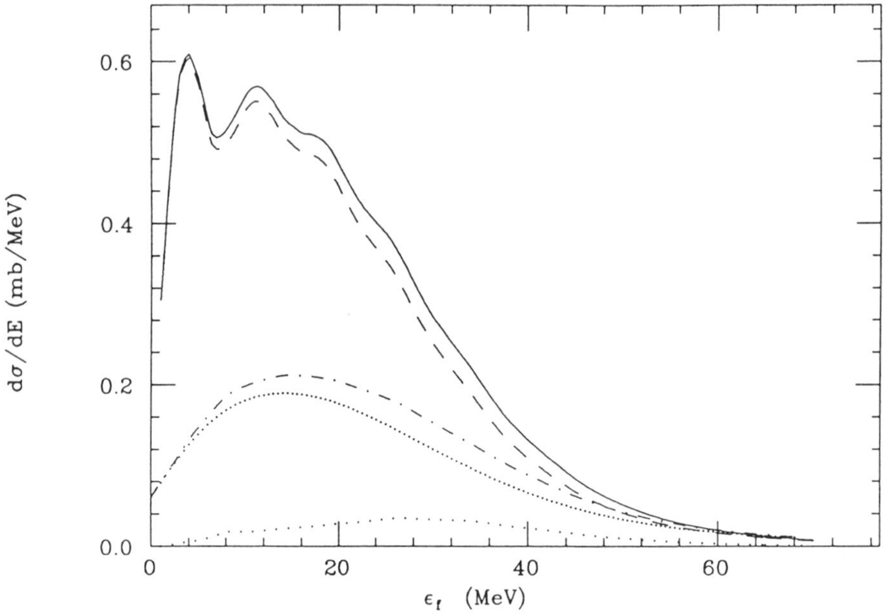

Fig.1 Spectrum of the reaction $^{208}Pb(^{20}Ne,^{19}Ne)^{209}Pb$ at $E_{inc} = 30MeV/u$, with $R_s = 11.75fm$, cf. ref.(8) for a similar experimental spectrum. In all figures the dot curve gives the elastic breakup while the close dotted curve gives the inelastic breakup, the dotdashed curve gives the total breakup, the dashed curve gives the absorption, the lower solid curve gives the resonant transfer obtained subtracting the inelastic breakup from the absorption while the upper solid curve is the sum of absorption and elastic breakup and corresponds to an inclusive experimental spectrum.

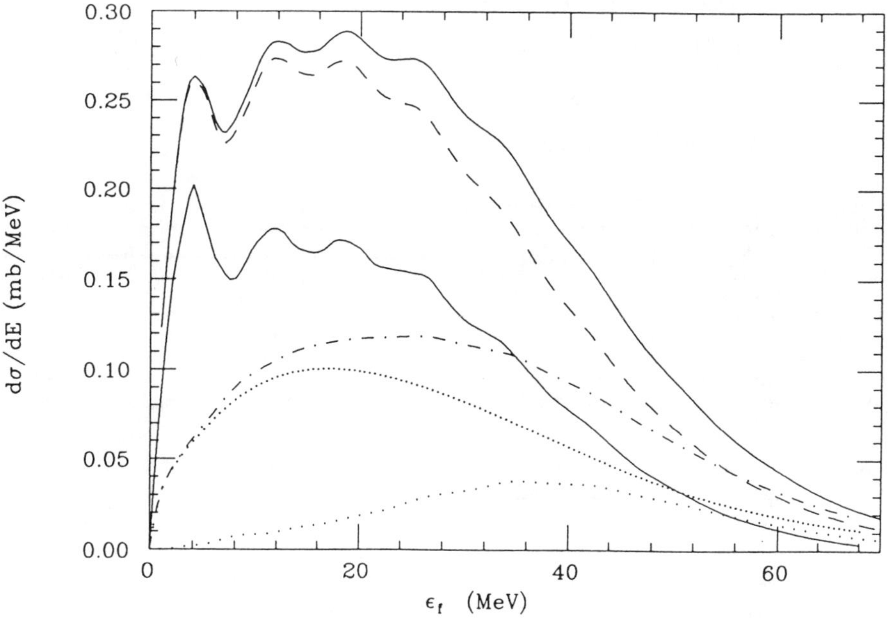

Fig.2 Spectrum of the reaction $^{208}Pb(^{20}Ne,^{19}Ne)^{209}Pb$ at $E_{inc} = 40MeV/u$, with $R_s = 11.92fm$, cf. ref.(11) for a similar experimental spectrum.

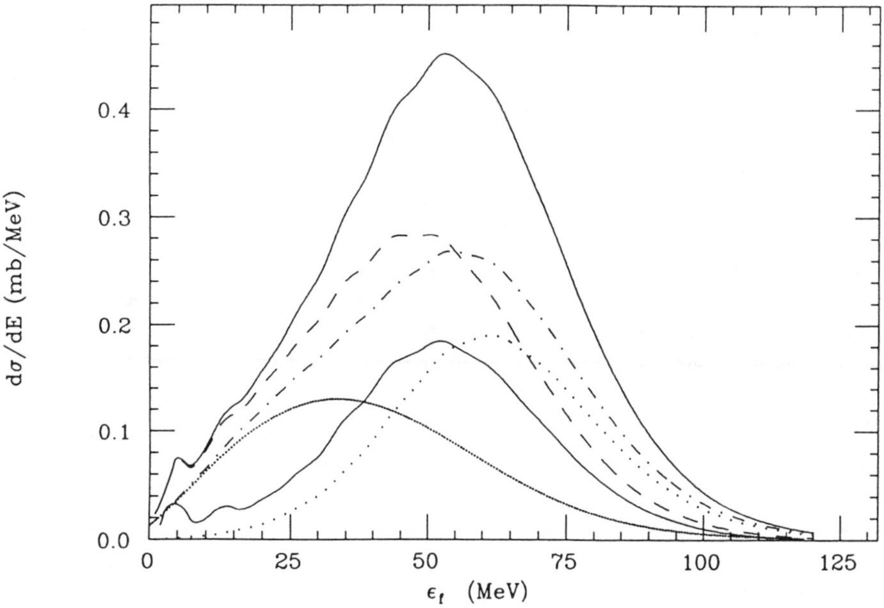

Fig.3 Spectrum of the reaction $^{208}Pb(^{14}N,^{13}N)^{209}Pb$ at $E_{inc} = 60MeV/u$,cf. ref.(12) for a similar experimental spectrum.

BOLTZMANN DYNAMICS AND
CLUSTERIZATION IN LIGHT NUCLEI

G.A.Lobov

ITEP -USSR II7259 Moscow

The process of hadron-nucleus interaction associated with the knockout of considerable number of nucleons from the nucleus is investigated in the framework of intra-nuclei cascade model. It is shown that final states of target nucleus corresponding to the knockout of all possible numbers of nucleons form a continuous spectrum. Obtained results are discussed in connection with the problem of possible existence of multi-nucleon clusters in the nuclei. It is shown that intra-nuclear cascade model allows great occurences of exited nuclear states corresponding to the knockout of α -clusters in hadron-nucleus scattering process without the assumption of their existence in the nucleus.

Introduction

Deep-splitting reactions of nuclei in hadron-nucleus collisions pertain to the phenomena which are incompletely investigated and have an ambiguous interpretation. One of investigation methods relevant for these reactions is instantaneous nuclear γ -radiation spectroscopy which makes it possible to identify a wide spectrum of residual nuclei produced in deep-splitting reactions by the energy of registered γ -transitions.

A great amount of experiments performed with π -mesons are connected with the investigation of pion-nucleus interaction mechanism in deep-splitting reactions. In the series of works the conclusion is made that the interaction is associated with the absorption of slow π -meson by the nucleus [1,2] . At greater π -meson energies the pion collision process with the production of exited nuclear states becomes significant [3] .

The mechanism of proton-nucleus interaction in deep-splitting reactions is rather poor investigated. At proton energies up to 500 MeV the basic mechanisms of exitation energy transfer to the target nucleus are inelastic scattering with giant resonance exitation or quasi-elastic scattering [4] . At greater energies of striking protons the interaction process becomes of cascade type. Both primary and secondary particles created in collision process contribute to the production of exited nuclei. Absorption of slow pions created in inelastic scattering of cascade particles by nucleons in nuclei, by nucleon ~~aggregates~~ associations with its subsequent breakdown and knockout of nucleons from the nucleus becomes substantial.

Many works deal with investigation of exited nucleus decay channels associated with the emission of α-particle [5,6] . This kind of channels is selected by the detection of intense γ-transitions of residual nuclei in instantaneous γ-quanta spectra. Creation of these nuclei might correspond to the emission of one or several α-particles from the target nucleus in hadron-nucleus interaction process. This effect is especially noticeable for even-even nuclei. Obviously, in order to give an adequate interpretation of the phenomenon mentioned above one has to found to what extent the creation of residual exited nuclei is connected with the emission of α-particles rather than with that of single nucleons. Correlated measurements of instantaneous γ-quanta and secondary charged particle spectra are necessary for this purpose [7] .

Thus, the use of $h\gamma$-coincidences in experiments as the method of nucleus splitting in hadron-nucleus collisions is of great interest [5] . In this method the scattered hadron h' is registered by coincidence with γ-quantum emitted by exited daughter nucleus B^* produced by splitting of target nucleus A in the process

$$h+A \longrightarrow h'+B^*+X, \quad B^* \longrightarrow B_0+\gamma \qquad \text{(I)}$$

where B_0 is the ground state of the nucleus B. The method makes it possible to identify the splitting channel-daughter nucleus - B_0 and to measure reaction cross sections for selected channel as a function of kinematic variables. Thus, it has been shown by investigation of the reaction $^{16}O(\pi,\pi X)^{12}C^*(2^+)$ [6] that the following three mechanisms are the most important in production of residual nuclei: direct quasi-elastic knockout of α-clusters, knockout of α-cluster followed by its decay, the knockout of nucleon followed by the emission of three-nucleon aggregate by the residuum.

In the work [8] the elastic cross sections for protons and inelastic cross sections for protons and pions at the energy I GeV scattered by the nucleus ^{16}O have been calculated by Glauber theory. In the calculations the nucleus ^{16}O was assumed to consist of four α-clusters positioned in the vertexes of a perfect tetrahedron in equilibrium state. In the work [8] a good description of experimental data has been reached for transfered momenta up to inelastic scattering maximum. In particular, in the framework of cluster model nuclear state spectra of ^{16}O [9] and ^{40}Ca [IO] appeared as a result of vibrations of α clusters in oscillator potentials near equilibrium and rotational positions of tetrahedron (and near more complicated configurations for heavy nuclei) are well described. Besides that, the radiation transition lifetimes of the lowest nuclear exited states can also be well described in the framework of α -cluster model [9, 10] . One should attribute to the shortcomings of cluster model the fact that still it is incapable for the explanation in what way the nuclear forces lead to tetrahedron symmetry group in the nucleus ^{16}O and to more complicated configurations in other α - cluster nuclei. This problem still remains open to rather substantial extent. The problem of theoretical justification of α -cluster model for oxygen nucleus was discussed, in particular, in [9a] . In this work an attempt to connect the tetrahedron symmetry of ^{16}O nucleus with hadron quark structure has been undertaken.

In this work in the frame of intra-nuclear cascade model characteristics of the process (I) observed experimentally are investigated, namely, the characteristic distributions of nuclear target final states and secondary hadrons. In Chap.I the justification of the model is presented. In Chap.2 the model under consideration is described and the scheme of calculations is given. In Chap.3 the results of numerical calculations are given and the comparison with experimental data is made. The main conclusions of the work are given in the last Chapter.

I. Justification of the model

In view of the choice of intra-nuclear cascade model as the calculation model some principal points concerning its place among other nuclear models are worth mentioning. The problem of intra-nuclear cascade model justification is connected with the possibility of passing from dynamic Schrödinger equations (or Liouville equations) to Boltzmann kinetic equations.

We will describe the nuclear system by the Hamiltonian $\hat{H}=\hat{H}_0+\hat{V}$ where $\hat{H}_0$ corresponds to non-interacting nuclear components. The representation is chosen in which unperturbed Hamiltonian is diagonal,

$$\hat{H}_0 |Ea\rangle = E|Ea\rangle$$

Other subscripts describing the state of the system are meant by a. Consider the perturbation $\hat{V}$ to be small enough (see below) to neglect the change in the energy of the system. Then, using statistical mechanics methods one can show that diagonal elements of density matrix satisfy the master equation [II, I2]

$$\partial P(E,a,t)/\partial t = \sum_{a \neq a'} \left[\alpha(Ea,Ea')P(E,a',t) - \alpha(Ea',Ea)P(E,a,t) \right] \quad (2)$$

Transition probability per unit time that conserves the energy is defined in the first order of perturbation theory as

$$\alpha(Ea,Ea') = \frac{2\pi}{\hbar} \left| \langle Ea|\hat{V}|Ea'\rangle \right|^2 \omega_a(E) \quad (3)$$

where $\omega_a(E)$ is the density of final states of the system. Equation (2) can be described in the approximation of short memory time of the system as compared to the characteristic transition time $\tau \sim \hbar/\alpha(Ea,Ea')$. Due to the last condition the random process described by the equation (2) is Markovian process. In [I3, I4] it has been shown that type (2) equations are valid for the description of finite nuclear system. The equation (2) is a mathematical basis for intra-nuclear cascade model. Transforming (2) into occupation number representation in case of dilute matter (or short-range potential) and taking into account pair collisions only we ontain Uling-Ulinbeck modified equation of quantum statistics [I5] that turns into corresponding classical Boltzmann kinetic equation describing the transport process in the medium for steady flux of incident particles. Monte-Carlo realization of cascade part can be interpreted as a convenient method of numerical solution of these very complicated kinetic equations.

It is notewothy that existing methods of analytical solution of kinetic equations make it possible to calculate the cascade part of the process in those energy range where pion production is still negligible [I5, I6] .

Thus, intra-nuclear cascade model based on master equation (2) describes non-equilibrium processes occuring in the nucleus and contains a number of general physical assumptions. The main assumptions are "weak bond" approximation, i.e. the small value of perturbation $\hat{V}$, and allowance for two-particle interaction only.

A problem of intra-nuclear cascade model justification based on quantum theory of multiple scattering has been considered in details in [I5, I7] . In the work [I7] , in particular, the following validity conditions for the model have been obtained in case of multiple scattering:

I) nucleon-nucleon interaction radius is much smaller than mean free path of the particle in the nucleus (gas approximation);

2) wavelength of striking particle is much smaller than mean free path;

3) the energy transfered to the nucleons of the nucleus from the striking particle is much greater than mean kinetic and potential energies of the particles in the nucleus (quasi-free scattering approximation);

4)full system approximation is used which is necessary when summing the cross sections of quantum mechanical process over all final target states corresponding to the flowout of quasi-elastic particle.

Therefore, intra-nuclear cascade model algorithm is that of solution for the equation chain that is deduced from kinetic balance equations for multi-particle nucleon distribution functions in the approximations mentioned above. If these conditions are satisfied, intra-nuclear cascade model is no less rigorous physical model for nuclear reaction description than Glauber theory, optical model or distorted wave model and its various modifications.

2. Calculation scheme

Structure scheme of intra-nuclear cascade model HAMLET [18] is briefly described below. Simulation procedure for the interaction of cascade particles with nucleons is organized in this scheme as follows. The calculations imply that nuclear matter consists of three components, namely, of neutron, proton and deuteron components. Introduction of deuteron component allows for pion absorption by nuclear two-nucleon association. Each component mentioned above was described by its own density, Fermi momenta distribution for nucleons that form it and the cross section of its interaction with the particle passing through it. Nuclear zone model was used. For every zone the characteristics of nuclear components are constant in absence of trailing effect. Knowing kinetic energy and the type of interacting particle, Fermi momentum of its partner was simulated and interaction probability for trajectory length R (from the point of particle location to the point where it leaves a given zone) were determined for each component of nuclear matter. If the interaction with the i-th medium does not take place, the partner of corresponding component is considered to fall out of"the game", i.e. arbitrary free path of the particle L_i is set to be equal to the distance from the next zone R. Otherwise the least of the obtained values L_i , $i = 1,3$ is chosen and simulation of elementary interaction of the particle with the i-th medium partner is performed.

To allow for Pauli principle the energy of nucleons created during the interaction was claimed to exceed Fermi energies in interaction zone. If this claim is satisfied, procedure of interaction simulation stops, otherwise the least of the rest values $\{L_i\}$ is chosen and the interaction is simulated again, but now with a new partner. Finally, as soon as all L_i happen to be equal to R, it occuring either in the beginning of path simulation or after its rejection by Pauli principle, the particle is considered to fly without interaction to the next zone along the vector $\mathbf{R}$. This procedure differs from the versions described in monograph [19] and seems to be more adequately related to simulated physical process. Input of experimental information program on elementary hadron-nucleon interactions into data bank made it possible to realize simulation procedure of interaction type, number and charges of secondary particles sufficiently simply and correctly.

In the work [18] the comparison of experimental data and theoretical results calculated by HAMLET was made by various characteristics (angular distributions, secondary particle spectra, multiplicity and others) of inelastic nucleon-nucleus scattering. In particular, this comparison shows that proposed version of intra-nuclear cascade model matches well experimental data on fast nucleon output in nucleon-nucleus collisions in the range of initial energies from 0.45 to 3 GeV both for light and heavy nuclei.

3. Results of numerical calculations

In this work the production of residual nuclei connected with the emission of single nucleons was under investigation. Interaction of striking hadrons with the nucleus was treated like a cascade process, in which not only primary hadrons but also secondary particles created both in elastic and inelastic hadron-quasi-free nuclear nucleon scattering contribute to the production of exited nuclei. The whole spectrum of target nucleus final states was investigated that corresponds to the outflow of all possible numbers of created particles. In this respect the problem is in consistence with the full system approximation which is used in summing of quantum mechanical process cross section over all final states.

Nucleons that flow out of the nucleus are produced both in sequential collisions of striking hadron with secondary cascade particles and in two-nucleus ~~aggregate~~ association decay in the process of secondary slow pion absorption. Flying out nucleons can be treated like a system consisting of k nucleons which corresponds to the creation of residual exited nucleus with atomic number $(A-k)$.

In fig.I distribution over charge Z for the system consisting of k nucleons (in particular, k = 4, 8, I2) emitted upon the interaction of protons with energy 2.9 GeV with nuclei ^{40}Ca in (I)-type process obtained in calculations using intra-nuclear cascade model is shown. Distributions in fig.I are normalized to unity. Slight charge asymmetry of the distribution is due to the charge of striking proton.

It should be stressed that great probabilities of nucleon aggregate emission observed experimentally by registration of exited residual nucleus B^* were obtained in our calculations without any special suggestions concerning symmetry group of target nucleus ^{40}Ca . Distributions over number of nucleons k emitted in this process are shown in fig.2. Statistical accuracy of the calculations is $\sim$ 5% for the cases $k < 12$ and $\sim$ 20% for greater k. Experimental results obtained in ITEP [20] are shown by dots in fig.2. In the work [20] intermediate states in respect of the number of emitted nucleons different from states related to the knockout of definite number of α– clusters were not observed that is connected with experimental method. Comparison of theoretical and experimental results shows that the description of the process under consideration based on intra-nuclear cascade model looks quite satisfactory. Exceeding of calculated dots over experimental dots in the range $k < 16$ where stasistical accuracy of the results is sufficiently high is due to the summation over all final states of corresponding exited residual nucleus $(A-k)$. On the contrary, only definite final states corresponding to γ - transitions of those residual nuclei which formation might correspond to the emission of one or some α - particles from the target nucleus are investigated in experiment. It follows from the results shown in fig.2 that creation of final states of residual nuclei corresponding to the emission of one, two or more α -clusters in type (I) scattering process has noticeable occurence.

It can be also seen in fig.2 that highly exited nuclear states appearing by emission of definite number of nucleons form a continuous spectrum and the states corresponding to the flowout of α -clusters (k = 4, 8, I2) are by no means distinguished. This result is supported by experimental investigations of nuclear state spectrum for the states produced in the interaction of protons with momentum 9 Gev/c with copper and tantalum nuclei [2I, 22] . Investigation of xenon and krypton nucleus fragmentation upon its irradiation by the beam of protons with momenta from 80 to 350 GeV/c also indicates that fragment A_f distribution in the range $6 < A_f < 30$ is described by power function $A_f^{-2,6}$ [23] .

Angular distribution of pions in the process

$$\pi + {}^{28}Si \rightarrow \pi + {}^{24}Mg + X$$

is shown in fig.3.

Here pions with momenta above 200 MeV/c are analysed that corresponds to the conditions of experiment which is presently carried out in ITEP. Statical error in one histogram channel is $\sim 30\%$, therefore the curve has been drawn through the dots in fig.3 for the purpose of visual demonstration. Pion angular distribution in fig.3 is practically isotropic in the range where scattering angles are less than 50°. This result is typical for the multiple re-scattering process described in intra-nuclear cascade model. In case of pion scattering by α -cluster of target nucleus associated with its subsequent emission a peak has to be observed in forward angular distribution in the range of angles less than 20°. For this reason experimental investigation of angular pion distribution in the range of small scattering angles is of great interest.

Distribution over number of nucleons emitted in the interaction of pions with momentum 2 GeV/c with nuclei ${}^{28}Si$ calculated by HAMLET is shown in fig.4. It follows from calculation results that the dependence of cross section on k is exponential with reasonably good accuracy.

Conclusion

Rather great number of works is devoted to the study of decay channels of exited nuclei produced in hadron-nucleus collisions with α -particle emission. The reason for the selection of these channels is the presence of instantaneous γ -quanta of intence γ -transitions of those residual nuclei which production might correspond to the emission of a single α -particle or some α -particles.

In this work using intra-nuclear cascade model which we have realized in HAMLET program the characteristics of high energy hadron -nuclei scattering process associated with the emission of α -particle that are observed experimentally have been first calculated. Occurences of these states proved to be high without any additional suggestion concerning clusterization inside the target nucleus. Assumptions on the structure of target nucleus accepted in this work are common for intra-nuclear cascade model. It turns out that the cross section of target nucleus highly exited state production can be described as the process of single nucleon knockout.

Knockout of nucleons from the nucleus at mean energies is of cascade type. Not only primary but secondary particles as well contribute to exited nuclei creation. It has been shown that final states of target nucleus form a continuous spectrum with respect to the number of knocked out nucleons. The measurement of production cross section of exited states different from the states corresponding to α -particle emission would be important.

It is clear that correlated measurements of instantaneous γ -quanta spectra characterizing final states of target nucleus and spectra of secondary hadrons are necessary for unambiguous interpretation of considered effect. Such experiments make it possible to study the characteristics of particles emitted in selected channels of deep-splitting reaction and to investigate hadron-nucleus interaction mechanisms more completely. In this work the results of calculation carried out in the frame of intra-nuclear cascade model are given for angular distribution of π -mesons created in the reaction $^{28}Si(\pi,\pi X)^{24}Mg$ Obtained angular distribution differs significantly from that expected in α -cluster model in which the existence of α -cluster in the nucleus is assumed a priori. Therefore experimental research of pion angular distribution in reactions of this kind is of considerable interest.

References

I. Gassanou Y. et al // Phys.Rev. 1977. V.C16. P.748.

2. Zaider M. et al // Phys.Rev. 1977. V.C16. P.2313.

3. Lind V.G. et al // Phys. Rev. Lett. 1978. V.41. P.1023.

4. Yonnet J. et al // Phys. Rev. Lett. 1978. V.40. P.164.

5. Kirpichnikov I.V. et al // Nucl.Phys. 1983. V.A392. P.352.

6. Vasenko A.A et al. Preprint ITEP-185. Moscow, 1987 (in Russian).

7. Batist L.H. et al Preprint LINP-606. Leningrad, 1980 (in Russian).

8. Karmanov V.A. // Nuclear Physics. 1982. V.35. P.848 (in Russian).

9. a)Robson D. // Nucl. Phys. 1978. V.A308. P.381. b)Baldock R. // Nucl. Phys. 1984. V.A462. P.222.

10. Merchant A.C. // Phys.Rev. 1988. V.C37. P.414.

11. Bogolyubov N.N. Problems of Dynamic Theory in Statistical Physics. Moscow, Gostekhizdat, 1946 (in Russian).

12. Van Hove L. // Physica. 1955. V.21 P.517.

13. Mantzouranis G. et al // Z. Phys. 1976 V.A276. P.145.

14. Bunakov V.E. // Nuclear Physics. 1977. V.25. P.505 (in Russian).

15. Bunakov V.E. // Elementary Particles and Atomic Nuclei. 1980. V.11. P.1285 (in Russian).

16. Kazarnovskii M.V., Par'ev E.Ya. // Nuclear Physics. 1981. V.33. P.660 (in Russian).

17. Goldberger M., Watson K. Collision Theory. Moscow, Mir, 1967. Chap. II (in Russian).

18. Lobov G.A. et al Preprint ITEP-91. Moscow, 1983. Sibirtsev A.A. et al Preprint ITEP- 129. Moscow, 1985 (in Russian).

19. Barashenkov V.S., Toneev V.D. Interaction of High Energy Particles and Atomic Nuclei with Nuclei. Moscow, Atomizdat, 1972 (in Russian).

20. Kirpichnikov I.V. et al. Preprint ITEP-169. Moscow, 1976 (in Russian).

21. Kozma P., Kliman J. Preprint JIRN E1-86-606. Dubna, 1986.

22. Kozma P. et al Preprint JINR E1-87-350. Dubna, 1987.

23. Hirsh A.S. et al // Phys.Rev. 1984. V.C29. P.508.

Figure captions

Fig.I. Distributions over charge Z for the system of k nucleons (4N, 8N, I2N) emitted upon interaction of protons with energy 2.9 GeV with nuclei ^{40}Ca .

Fig.2. Distributions over number of nucleons k produced in interaction of protons with energy 2.9 GeV with nuclei ^{40}Ca . Dots are experimental results [20] ..

Fig.3. Angular distribution of π mesons with momentum above 200 MeV/c created in the reaction $\pi^+ + {}^{28}Si \rightarrow \pi + {}^{24}Mg + X$ at striking pion momentum 2 GeV/c. The curve was drawn visually.

Fig.4. Distribution over number of nucleons k emitted upon interaction of π -mesons with momentum 2 GeV/c with nuclei ^{28}Si .

FIGURES

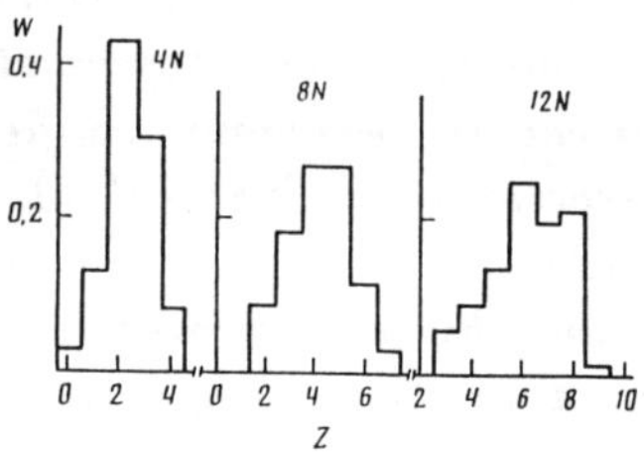

Fig.I

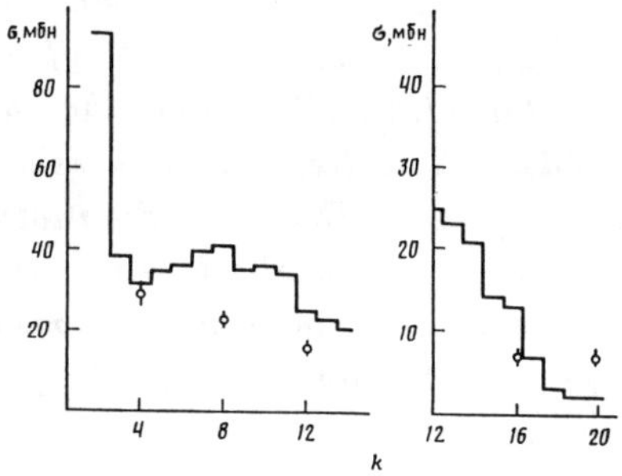

Fig.2

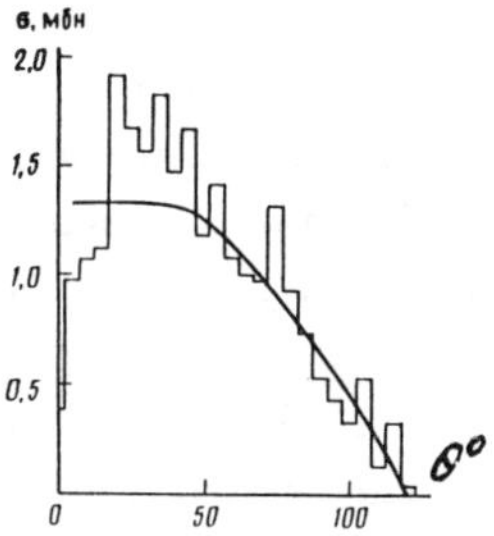

Fig.3

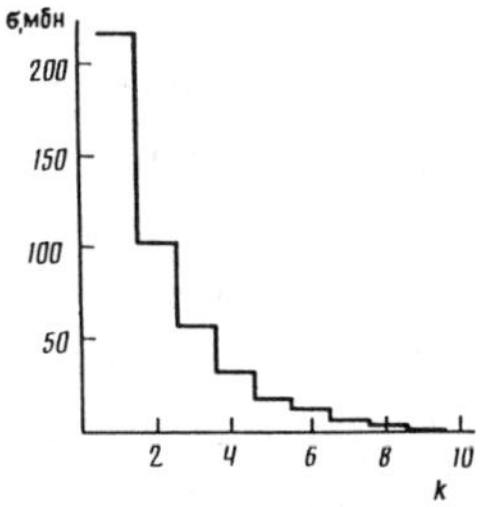

Fig.4

Chaotic scattering in heavy–ion collisions

Marcello Baldo and Andrea Rapisarda

INFN Sez. Catania, Corso Italia 57, 95129 Catania, Italy

ABSTRACT

We demonstrate that the classical dynamics of a collision between a spherical nucleus and a deformed one at energies near the Coulomb barrier can display chaotic behaviour, provided the deformed nucleus is not too heavy. The chaotic scattering is produced by sets of trajectories which are trapped in the potential pocket for a time longer than the one characteristic of a simple spherical symmetric potential scattering. At a semiclassical level this chaotic behaviour justifies the presence of Ericson's fluctuations observed in reactions like $^{28}Si + ^{24}Mg$*, while the presence of islands of regular motion embedded in the chaotic region could be the crucial point to explain the nature of the observed isolated resonances. Preliminary results of a full quantal calculation for the same system are also presented.*

INTRODUCTION

In the last few years there was a great interest in classical chaotic scattering of simple non-integrable systems with a few degrees of freedom [1]. This phenomenon has a close analogy with the chaotic dynamics encountered in bounded ones. In the simple case of potential scattering the chaotic regime is characterized by deflection functions which oscillate on any scale of the initial boundary conditions. These oscillations are associated with trajectories which remain trapped for an exceedingly long time inside the interaction region. For very long time these trajectories approach and stay close to a set of unstable trapped orbits which form a fractal set in phase space, the so-called "strange repeller", and from which unbounded orbits must finally escape. In this sense the deflection function has a fractal structure. Another feature which often appears in chaotic scattering is the occurrence of "islands" of regular scattering embedded also in the irregular region: usually these islands persist on all scales of the initial conditions. Semiclassical quantization of systems which display chaotic scattering leads to an S-matrix which belongs to the random unitary matrix ensemble[1] and it has been argued that the corresponding fluctuations in the cross sections can be a model of the Ericson's fluctuations [2,3] which are well known in nuclear reactions for many years. However it has not yet been established any general connection between the presence of a classical chaotic behaviour and the features of the corresponding quantal scattering amplitude.

In this paper, we show that in the classical treatment of a collision between a deformed and a spherical nucleus, chaotic scattering is present with all the above mentioned features, provided the following two conditions are met: i) the scattering is close to the grazing condition; ii) the ions are sufficiently light. In the present treatment, other degrees of freedom different from rotational ones are excluded and absorption is neglected. Though the classical treatment is not fully justified, it can shed light on the dynamics of ion–ion collisions and it is the basis for the semiclassical approach. A generalization is also presented to the case of a fully quantum mechanical approach to the same system.

CLASSICAL CASE

For simplicity we consider planar geometry, namely we choose the initial boundary conditions in such a way that the scattering occurs in a plane. Then, with reference to fig. 1, the system has three degrees of freedom, the relative distance r between the centers of mass of the two nuclei, the polar angle ϕ and the rotational angle θ.

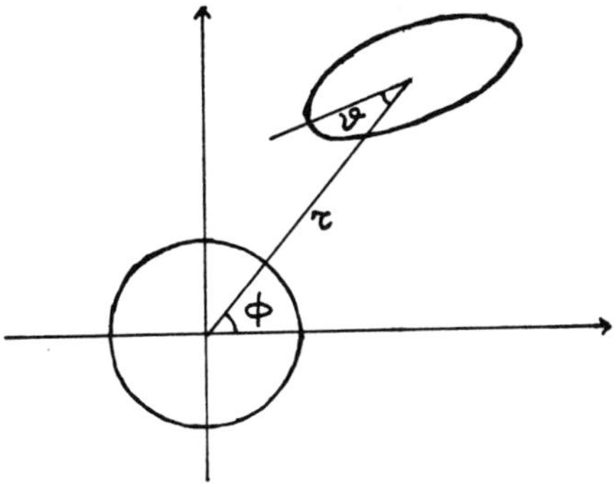

Fig. 1
Geometry of the collision and correspoding variables.

If one assumes an axially symmetric deformed nucleus, the lagrangian of the system can be written

$$\mathcal{L} = \frac{1}{2}m(\dot{r}^2 + r^2\dot{\phi}^2) + \frac{1}{2}\Im(\dot{\phi} + \dot{\theta})^2 - V(r,\theta) \qquad (1)$$

where m is the reduced mass of the two colliding nuclei, $\Im$ is the moment of inertia of the deformed nucleus, and $V(r,\theta)$ the ion–ion potential. The corresponding canonically conjugated momenta are given by

$$\begin{aligned} p_r &= m\dot{r} \\ p_\phi &= mr^2\dot{\phi} + \Im(\dot{\phi}+\dot{\theta}) \\ p_\theta &= \Im(\dot{\phi}+\dot{\theta}) \end{aligned} \tag{2}$$

where p_r is the radial orbital momentum, p_ϕ the projection of the total angular momentum along the axis perpedicular to the reaction plane, which for convenience we indicate by L_{tot}, and p_θ the angular momentum of the rotor, which we indicate by I. In terms of these conjugated momenta the Hamiltonian reads

$$\mathcal{H} = \frac{p_r^2}{2m} + \frac{(p_\phi - p_\theta)^2}{2mr^2} + \frac{p_\theta^2}{2\Im} + V(r,\theta) \tag{3}$$

and the correspoding equations of motion are

$$\begin{aligned} \dot{p}_r &= -\frac{(p_\phi - p_\theta)^2}{2mr^3} - \left(\frac{\partial V}{\partial r}\right) \\ \dot{p}_\phi &= 0 \\ \dot{p}_\theta &= -\left(\frac{\partial V}{\partial \theta}\right) \end{aligned} \tag{4}$$

Our Hamiltonian has two constants of motion, namely the total energy E_{tot} and the total angular momentum L_{tot}, as it can be seen from the second of the eqs (4). Neglecting the θ-dependence of the full ion-ion potential, the Hamiltonian is separable and thus integrable, because the internal angular momentum I and the orbital one L are conserved separately. The θ-dependence of the ion-ion potential introduces a symmetry-breaking term leading to the conservation of L_{tot} only. If the perturbation caused by the θ-dependent coupling term is small enough, the system is expected to be a mixed one[4].

Making use of the conservation of the total angular momentum L_{tot} along an axis perpendicular to the reaction plane, the effective Hamiltonian of the system can be written in general as

$$\mathcal{H} = \frac{p_r^2}{2m} + \frac{I^2}{2\Im} + \frac{(L_{tot} - I)^2}{2mr^2} + V(r,\theta) \tag{5}$$

The total ion–ion potential can be written

$$V(r,\theta) = \frac{Z_1 Z_2 e^2}{r} + \frac{Z_1 e^2 Q_o}{2r^3} P_2(cos\theta) + U_N(r,\theta) \tag{6}$$

where Z_1 and Z_2 are the charges of the two nuclei, Q_o the intrinsic quadrupole moment of the deformed nucleus and P_2 the Legendre polynomial of order 2.

The nuclear part of the ion–ion potential U_N is chosen to be the proximity one[5], which is well suited for calculating the interaction between deformed nuclei[6]. That is

$$U_N(r,\theta) = 4\pi\gamma\Re b\Phi(s(\theta)) \tag{7}$$

$$s(\theta) = \frac{r - R_1 - R_2(\theta)}{b} \tag{8}$$

$$R_2(\theta) = R_2^o(1 + \alpha_{20}Y_{20}(\theta)) \tag{9}$$

$$\Re = \frac{R_1 R_2^o}{R_1 + R_2^o}(1 - \frac{2R_1}{R_1 + R_2^o}\alpha_{20}Y_{20}(\theta)) \tag{10}$$

where Φ is the proximity function, γ is the surface tension, b = 1 fm is the diffuseness parameter, R_1 the radius of the spherical nucleus, $R_2(\theta)$ the radius which describes the quadrupole surface of the deformed nucleus and α_{20} is the deformation parameter.

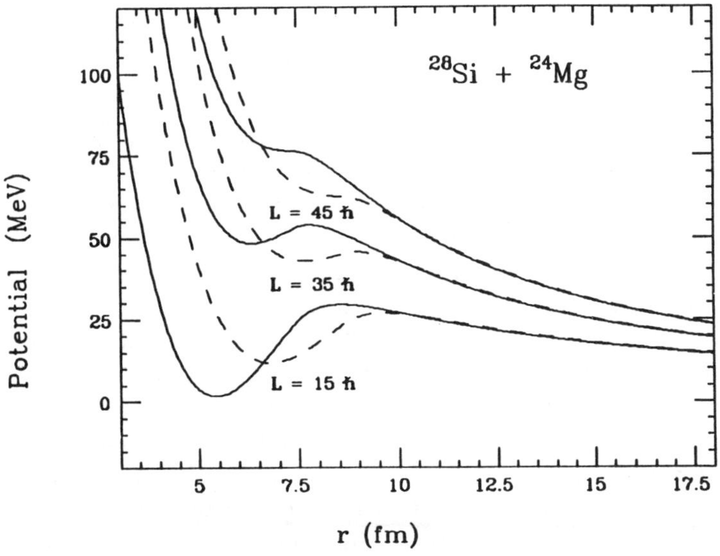

Fig. 2

Ion–ion total potential for three values of the orbital angular momenta.

As a typical example we take the system $^{28}Si + ^{24}Mg$. For simplicity we assume the ^{28}Si as spherical. The inclusion also of the deformation of the ^{28}Si nucleus would complicate the dynamics, but it would not add anything essentially new to the present qualitative discussion. The values of the different parameters[7,8] used in the calculations are reported in table 1. In fig. 2 is shown the ion-ion potential, including the centrifugal part, for three values of the relative angular momentum. The dashed curve refers to $\theta = 0^o$ while the full one refers to $\theta = 90^o$.

γ	R_1	R_2^o	α_{20}	Q_o	$\Im\hbar^{-2}$
$MeV fm^{-2}$	fm	fm		fm^2	MeV^{-1}
11.959	3.39	3.21	0.423	57	2.378

Table 1

Parameters used in the calculations

Integrating the equations of motion one gets the behaviour shown in figs. 3 and 4 for the final angular momentum of the rotor and the final scattering angle. Because of the above mentioned conservation of the total angular momentum, i.e. $L(t) + I(t) = L_{tot}$, the final scattering angle ϕ_f is obtained integrating

$$\phi_f = \int \frac{(L_{tot} - I)}{mr^2} dt. \tag{11}$$

The initial angular momentum of the rotor was $I_i = 0$. The integration was carried starting from an initial distance of $r = 18$ fm up to the internal region and then back until the distance was again 18 fm. There was also a time limit corresponding to $T_{max} = 10^4\ fm/c$, in case the trajectory was trapped inside the interaction region. Very few trajectories, however, remain trapped for so long. There are wild fluctuations in some regions due to irregular scattering and a smooth behaviour in others. Moreover there are regions of regular scattering embedded in the irregular one. The large regular region beyond 60° for $L_{tot} = 16\ \hbar$ corresponds to scattering from the tail of the potential which does not depend on the angle θ. In this region the Hamiltonian is quasi-integrable and the coupling term is nearly zero. Increasing the coupling term, the grazing condition is reached and symmetry-breaking occurs: our system shows a mixed behaviour of regular and chaotic scattering. The islands of regular motion are very sensitive to the ion-ion potential and increasing the incident energy chaotic scattering tends to disappear.

In fig. 3 the final value of I_f never reaches the extreme values $\pm I_{max} = \pm\sqrt{2\Im E_{tot}}$. In general it is limited by two smaller values fixed by the shape of the potential, which determines the apparent band of points in the figures. Those isolated points which are outside this band are not real final values of I: they in fact correspond to trajectories trapped inside the pocket of the potential because the time limit was reached. In order to observe chaotic motion, the time associated to the rotation of the deformed nucleus should be comparable to the collision time. If the moment of inertia is too big, the change in the potential due to rotation would occur in a time too long compared to the collision time and the motion would be regular everywhere. This is one reason why light systems are favoured. The other one is the fact that for these systems the quantal absorption is less than in heavier ones.

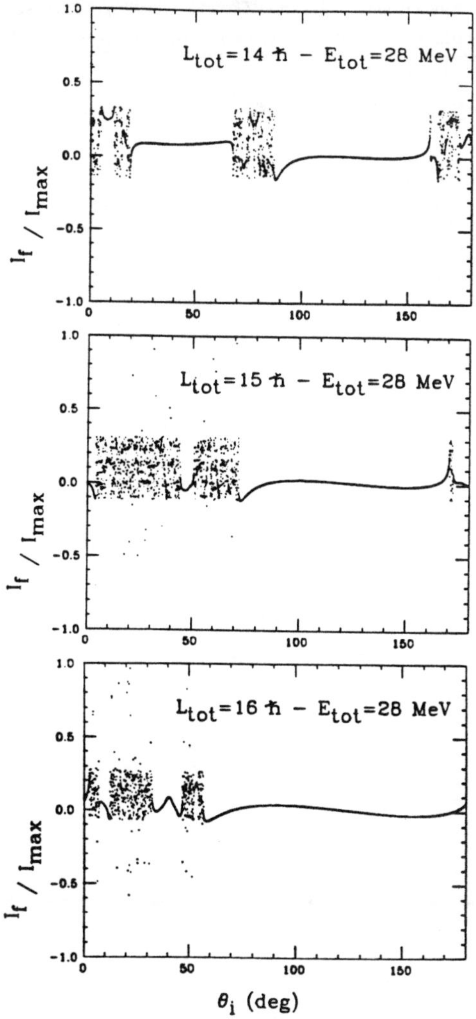

Fig. 3
Final angular momentum of the rotor as a function of its initial orientation

The simplest semiclassical quantization leads to the probability

$$P = |\sum_k \sqrt{P_k} e^{\frac{i\Phi_k}{\hbar}}|^2 \tag{12}$$

where Φ_k is essentially the action integral, including the Maslov phase[1,9] and P_k

is the classical probability for the kth trajectory which gives the final classical state of the system.This treatment gives an excellent reproduction[5] of the full quantal result in the case of pure Coulomb excitation of rotational bands, provided it is analitically extended also into the classically forbidden region. The classical probability P_k are essentially given by the Van Vleck determinants[10] if the final variables are smooth function of the initial boundary conditions. As shown in ref.[1], in the case of irregular scattering the averaged energy probability and the correlation function follow Ericson's fluctuations theory[2,3] provided the energy interval is properly chosen. In the past molecular resonances were found for light systems at energies above the Coulomb barrier[11]. The nature of these resonances has remained till now rather obscure.

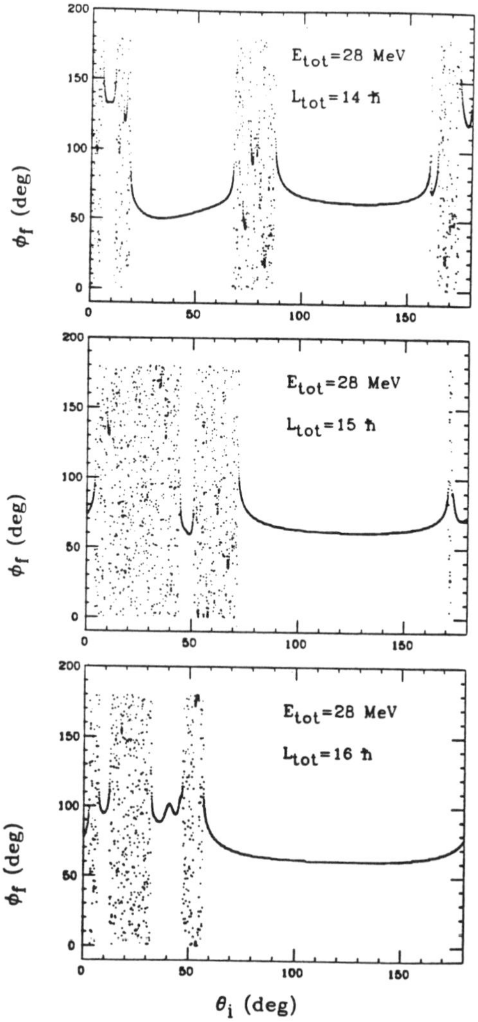

Fig. 4
Final scattering angle of the collision as a function of the rotor initial orientation.

In particular for the system $^{28}Si +^{24} Mg$ isolated resonances embedded into a statistical noise (Ericson's fluctuations) have been discovered quite recently [12,13].

Qualitatively the nature of these isolated resonances can be explained with the presence of the above discussed islands of regular trajectories inside the chaotic region. For those initial conditions which give regular scattering the system displays a quasi–integrable dynamics and the motion takes place on invariant surfaces. In the case of bounded systems the generalized Bohr–Sommerfeld quantization rules in the invariant surfaces give the semiclassical eigenenergies[9]. In the present case of unbounded scattering dynamics, similar considerations, applied to quasi–closed trajectories which explore the same phase-space region many times, should give the position of the isolated resonances. The proposed explanation is alternative to the one invoking quasi–bounded quantal states, inside the pocket of a spherical potential, which decay by tunnelling and can have also some connection with resonances due to quantal coupled-channels. However, for a quantitative analysis a careful study of the ion–ion potential and of the corresponding quantal scattering problem is required. This is beyond the aim of the present contribution.

QUANTAL CASE

The connection between classical chaotic scattering and quantal S–matrix amplitudes has been clarified till now only at the semiclassical level of approximation. To investigate the possible relevance of classical chaotic behaviour for the corresponding fully quantal scattering, we have performed pleriminary quantal calculations for the same system $^{28}Si+^{24}Mg$ for values of the conserved quantum variables equal to to the ones already considered in the classical case. Quantum mechanically the scattering process is three–dimensional, since the wave packet which describes the initial state must have a finite size, according to the uncertainity principle. To simplify the problem, which otherwise would be numerically prohibitive, we contrained the two nuclei to move in a plane. Then also in this case the coordinates r and ϕ of the orbital motion are the relevant variables and the corresponding kinetic energy of the relative motion is described by a two–dimensional Laplacian. Similarly the symmetry axis of the deformed nucleus is contrained to lie in the same reaction plane, so that the rotor is simply a one dimensional quantal rotor. The ion–ion potential is however still considered as given by eqs. (6)–(10). This model can be considered the quantal system closest to the classical one analyzed in the previuos section, and should be useful to clarify the connection between classical chaoticity and the corresponding quantal scattering problem.

Due to the conservation of total angular momentum, the orbital angular momentum L and the angular momentum I of the rotor are such that $L + I =$

L_{tot}, for a given total angular momentum L_{tot}. Then each quantal channel can be labeled by the rotor angular momentum I only. As a consequence the S–matrix can be parametrized by the total energy E, the total angular momentum L_{tot}, and the initial and final rotor angular momentum I and I' respectively

$$\hat{S} = \{S_{I,I'}(E)\} \tag{13}$$

Since in our calculations $I = 0$, the total angular momentum coincides with the initial orbital angular momentum.

In the preliminary calculations we have introduced only up to six channels in the quantal coupled equations. Results for the modulus square of the S–matrix corresponding to transitions to different final values of I' are presented in fig. 5. The results are presented as a function of energy, which is the only continuous quantum number.

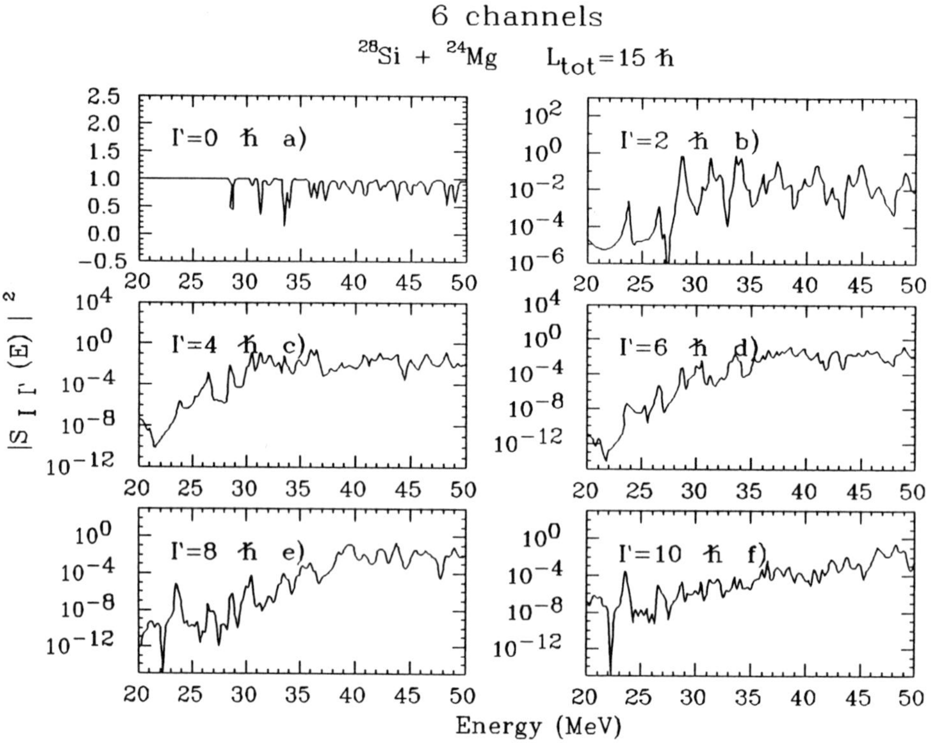

Fig. 5

Modulus square of the scattering amplitudes as a function of energy for different transitions.

The sharp structures observed in all the matrix elements are reminiscent of an irregular sequence of resonances, typical of quantal S–matrix fluctuations. It has to be noticed that the energies where these structures occur do not correspond to opening of new channels. They seem to correspond to resonances in the coupled channel system. We have verified that decresing the number of channels has the effect of rarefying the structures. Despite the limited number of channels, the results appear encouraging. The classical chaotic behavior seems to have a persistent effect on the quantal S–matrix. It would be desirable to increase the number of channels, and work in this direction is in progress.

In relation to possible relevance of our results with experimental data on heavy ion quasi–elastic scattering, it has to be noticed that in realistic cases one should consider also the absorption from the elastic channel, which could wash out the observed irregular structures. However the amount of absorption in light nuclei can be not so strong near the grazing. Furthermore it depends very much on the system, and this could indicate why the fluctuations of the cross sections, which we referred to in the preceeding section, are observed only in some systems and not in others.

In principle the phenomena here discussed for rotational degrees of freedom could also appear when considering vibrational ones. It is also interesting to investigate the connections of the arguments discussed so far with the coherent fluctuations found in more dissipative heavy-ion reactions[14].

References

[1] R. Blümel and U. Smilansky, Phys. Rev. Lett. **60** 477 (1988).
U. Smilansky, Miniworkshop on Quantum Chaos, 4-6 June 1990, preliminary notes, not published.

[2] T. Ericson, Phys. Rev. Lett. **5**, 430 (1960).

[3] D. Brink and R. Stephen, Phys. Lett. **5**, 77 (1963).

[4] I. C. Percival, Proceedings of the Royal Society of London, eds. M. V. Berry, I. C. Percival and N. O. Weiss, Princeton University Press, 1987.

[5] J. P. Blocki, J. Randrup, C. F. Tsang, and W. J. Swiatecki, Ann. of Phys. **105**, 427 (1977).

[6] R. A. Broglia and A. Winther , Heavy Ion Reactions, Lecture Notes, vol. I, Benjamin (New York) 1981.

[7] A. Bohr and B.R. Mottelson, Nuclear Structure, Ed. Benjamin II vol.

[8] E. Browne, J. M. Dairiki and R. E. Doebler, Table of Isotopes, Eds. C. M. Lederer and V. S. Shirley, Wiley-Interscience 1978 New York.

[9] M. V. Berry, Chaotic Behaviour of deterministic systems, Les Houches school, 1981, eds. G. Looss, R.H.G. Helleman and R. Stora, North-Holland.

[10] D. M. Brink, Semiclassical methods in nucleus-nucleus scattering,

Cambridge University Press, 1985.
[11] K. A. Erb and D. A. Bromley, Treatise on Heavy Ion Science, vol.3, Ed. by D. A. Bromley , Plenum Press (1984), New York.
[12] A. H. Wuosmaa, S. Saini, P. H. Kutt, S. F. Pate, R. W. Zurmühle and R. R. Betts, Phys. Rev. C **36** 1011 (1987).
[13] A. Sarma and R. Singh, Zeit. f. Phys. **337**, 23 (1990).
[14] G. Pappalardo, Nucl. Phys. A**488** 395c (1988).

INCOMPLETE DEEP INELASTIC PROCESSES IN MEDIUM MASS HEAVY ION COLLISION AT 18 AND 24 MeV/u

M. Petrovici*
Institute for Physics and Nuclear Engineering
Bucharest, Romania

ABSTRACT

Experimental evidences for an impact parameter dependence of preequilibrium processes in $^{100}Mo + ^{100}Mo$ at 18.67 MeV/u, 23.75 MeV/u and in $^{120}Sn + ^{120}Sn$ at 18.34 MeV/u are presented. An incomplete deep inelastic mechanism is proposed. Such a process explains the deviations observed at these incident energies in σ_Z^2 - TKEL/l_g (for two body) and $P_3/(P_2 + P_3)$ - TKEL (for three body) systematics. Model predictions for the preequilibrium emitted nucleons and the corresponding excitation energy per nucleon which remains in the dinuclear system are compared with the experimental data.

* Work done in collaboration with: R.Bock, R.J.Charity, R.Freifelder, A.Gobbi, N.Herrmann, K.D.Hildenbrand, H.Stelzer, J.Wessels - GSI, Darmstadt(Germany); I.Berceanu, A.Pop - IPNE, Bucharest(Romania); M.Blann - LNL, Livermore(USA); G.Casini, A.Olmi, P.R.Maurenzig, A.A.Stefanini - INFN and Univ. of Florence (Italy); M.Gnirs, D.Pelte - Univ. of Heidelberg(Germany); J.Galin, D.Guerreau, U.Jahnke - Ganil, Caen(France); J.C.Adlof, B.Bilwes, R.Bilwes, F.Rami, G.Rudolf - CRN, Strasbourg(France).

Since the heavy ion accelerators in the 10 MeV/u-100 MeV/u energy range become available, large experimental effort has been devoted for the investigation of heavy ion induced reactions in this intermediate regime.

While at energies not too far above the Coulomb barrier the general experimental aspects can be described using mean field approximation, from 10 MeV/u to 100 MeV/u transitional features become more important.

This transient regime, where the collision speed reaches or surpasses the intrinsic Fermi velocity is characteristic for an interplay between the dinuclear mean field effects and nucleon-nucleon collision processes.

As a consequence, mechanisms observed at low incident energies could still have some reminiscences and some precursors of high energy phenomena could appear at these energies. These explain the variety of the new experimental aspects emphasized in the intermediate energy regime and the necessity for more experimental information in order to aim their complete understanding.

The two dominant mechanisms at bombarding energies lower than 10-15 MeV/u, fusion and deep inelastic collisions , continue to play a role also at higher incident energies. If at low energies in both these processes, with few exceptions, the whole projectile and target interact and follow the standard dynamics, at higher incident energies part of nucleons and momentum are removed, mainly in the first stage of the collision, before the remaining heavy constituents continue a fusion or deep inelastic process. These type of processes, specific for intermediate energy regime are called incomplete.

An other class of incomplete processes are the few exceptions evidenced at low incident energies when the projectile, due to the deceleration by the nuclear friction force breaks-up into two parts corresponding to its underlying cluster structure and only one of the two fragments will follow a deep inelastic or fusion process. As a consequence, light alpha-like projectiles have been mainly used in this type of studies [1]. Such mechanisms, coined also as incomplete fusion or incomplete deep inelastic collisions are not specific for intermediate energy regime and we will not refer to them here.

Extensive studies have been devoted up to now to the incomplete fusion phenomena (IF) specific for intermediate energies[2].

In this contribution we will present experimental evidences that at 18.67 MeV/u in ^{100}Mo+ ^{100}Mo, at 18.34 MeV/u in ^{120}Sn+^{120}Sn and at 23.75 MeV/u in ^{100}Mo+^{100}Mo collisions an incomplete deep inelastic mechanism (IDIC), similar with the one observed in incomplete fusion but for the whole range of impact parameters could be advocated by the experimental trends[3].

Such a process gives the possibility to study the dynamics of preequilibrium phenomena as a function of impact parameter, from head-on to the grazing collisions. Treated in an appropriate way it can be also seen as a transient mechanism to the participant-spectator process evidenced at much larger energies.

We performed detailed studies of the main aspects related to the deep inelastic process at 15 MeV/u for many combinations using detection and identification in charge, mass and energy measurement of the secondary products.

The primary quantities obtained by a multistep iterative correction procedure for two massive fragments outgoing channels under the hypothesis of secvential evaporation process, have been proved to be obtained with good accuracy using extensive Monte-Carlo simulation technique[4].

It is evident that going at larger incident energies, the best experimental set-up would be a 4π configuration with enough granularity, based on modules of large dynamical range in energy measurement and charge and mass identification, which gives the possibility of a complete event characterization.

Different mechanisms, where two, three or four heavy fragments are produced in the exit channels and where the charge identification is not absolutely necessary, can be successfully studied using much simpler experimental devices which give the possibility to measure the velocity vectors of the reaction products. For the energy range where only a secvential evaporation process takes place between the creation of the primary fragments and the measurement of their velocity vectors, a kinematical reconstruction procedure supplies primary values for the average kinematical quantities with good accuracy, starting from the measured velocity vectors.

Starting from these arguments, we used an experimental set-up based on twelve identical large area position sensitive parallel plate avalanche counters, arranged in three groups and mounted in an axially symmetric configuration around the beam axis. The detector groups match each to other in a way which covered about 75% of the forward hemisphere.

Two experiments were performed using symmetric projectile-target combinations.

The first experiment has been done at GSI-Darmstadt using beams of ^{100}Mo and ^{120}Sn at energies of 18.7 and 18.3 MeV/u, respectively, supplied by the UNILAC accelerator.

The second experiment was performed with exactly the same experimental set-up at Caen using a ^{100}Mo beam of 23.7 MeV/u delivered by the GANIL accelerator[5,6].

Due to reasons which will become more transparent later, we prefer to analyse the velocity diagrams (parallel versus perpendicular component) for different mass asymmetries of the outgoing two massive fragments (η = 0.0±0.02; 0.5±0.02; 0.8±0.02) for the three measured cases[7]. The results are presented in Fig.1. The full range of velocities from elastic , quasielastic to completly damped events, representative for a dissipative process, is present in the velocity diagrams corresponding to η = 0.0±0.02. The velocity diagrams for η = 0.5±0.02 and η = 0.8±0.02 correspond already to completely damped events, the corresponding velocity rings are centered in the center of mass velocity vector and are quite isotropic populated for both measured systems at 18 MeV/u.

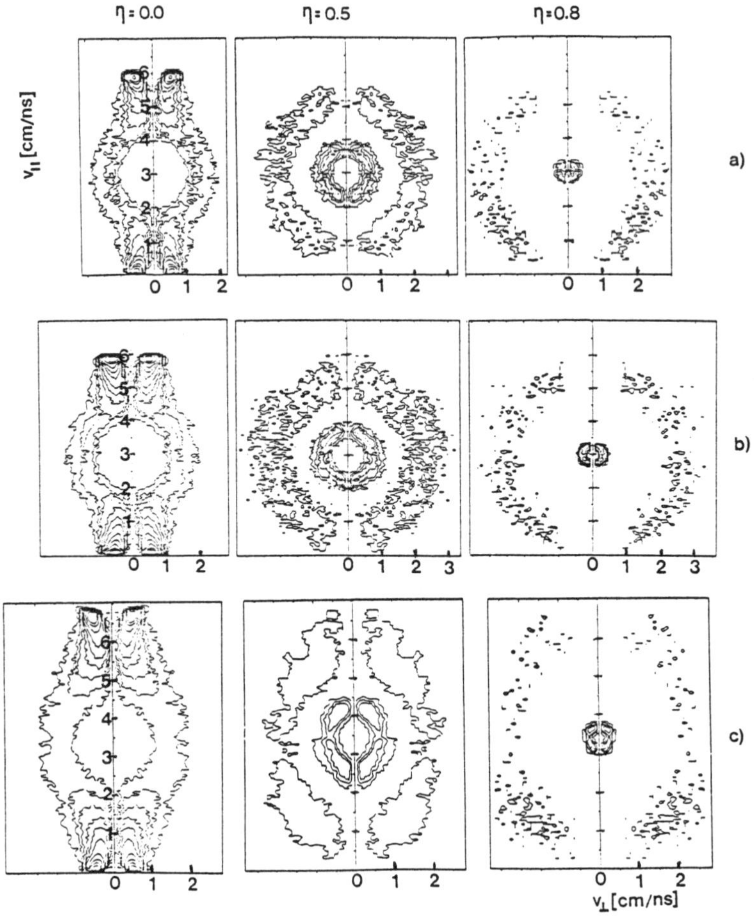

Fig. 1 Velocity diagram representation for:

a) ^{100}Mo + ^{100}Mo at 18.67 MeV/u
b) ^{120}Sn + ^{120}Sn at 18.34 MeV/u
c) ^{100}Mo + ^{100}Mo at 23.75 Mev/u

At the largest incident energy measured in this experiment (^{100}Mo + ^{100}Mo at 23.75 MeV/u) a sizable anisotropy appears for η =0.5 and 0.8. This could be a consequence of the interaction time which is so short that even for the largest energy losses, the system has not enough time to rotate in order to populate with equal probability all angles in the CM-system. Arguments in favour of this hypothesis will come later on.

A velocity diagram representation is not affected by the kinematical reconstruction procedure used in order to obtain other kinematical quatities and clearly shows that the main features of a deep inelastic collision are still present at these incident energies. Using a collision between identical nuclei, possible preequilibrium emission of nucleons will not change the mean value of the center of mass velocity vector [8]. Statistical effects could introduce only an extra smearing around the mean value which may be the origin of the ellipsoidal shape evidenced in the velocity diagrams for the 23.75 MeV/u case.

A kinematical reconstruction procedure [9] starts under the hypothesis that the heavy detected fragments originate from a dinuclear system which conserves the mass and momentum of the incident channel. This is the case at lower incident energies where this method proved its wide applicability. At higher incident energies (E/A ≥ 15 MeV/u) , as soon as the preequilibrium processes start to appear, they remove part of

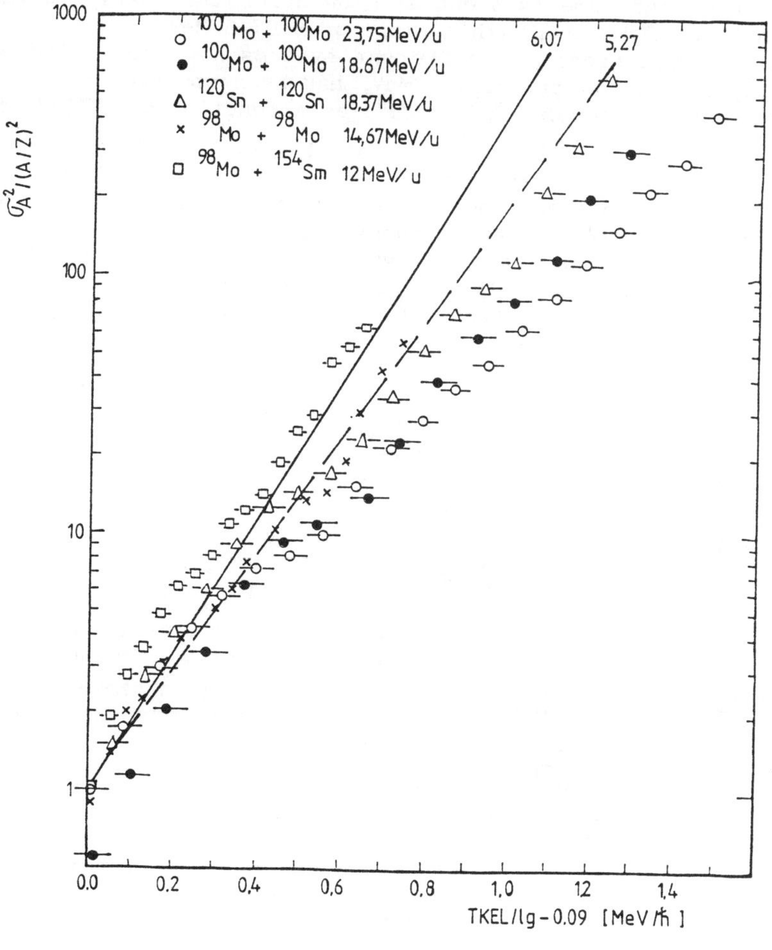

Fig. 2. Experimental charge variances σ_Z^2 as a function of TKEL/lg

nucleons and momentum, most of them in the very first stage of the collision. In such cases, kinematical reconstruction procedure , based only on the velocity vectors of the heavy fragments could propagate distortions on the reconstructed observables. These distortions combined with the hypothesis that the whole missing energy from the exit channel relative to the entrance one is transfered to the internal excition of the interacting system, could bring important deviations from the systematics observed at lower incident energies.

For deep inelastic collisions a $\sigma_Z^2(\sigma_A^2)$ systematics as a function af TKEL/lg has been found to have a wide applicability[10,11,4].

The first systematics was found using experimental data obtained at lower incident energies (< 10 MeV/u)[10]. All the systems shown the same exponential growth for the charge variance as a function of TKEL/lg with a slope of 6.07 ħ/MeV.

At 15 MeV/u where six different systems have been studied[4], a slightly smaller slope, 5.27 ħ/MeV, has been observed.

The charge variance ($\sigma_A^2/(A/Z)^2$) as a function of TKEL/lg for ^{100}Mo + ^{100}Mo and ^{120}Sm + ^{120}Sm at higher incident energies measured in our last experiment are presented in Fig.2. The straight full and dashed lines correspond to the systematics found at energies < 10 MeV/u[10] and respectively 15 MeV/u[4].

Strong deviations from the systematics can be observed. The deviations are larger at higher incident energy and larger energy losses (smaller impact parameters). In the same figure are

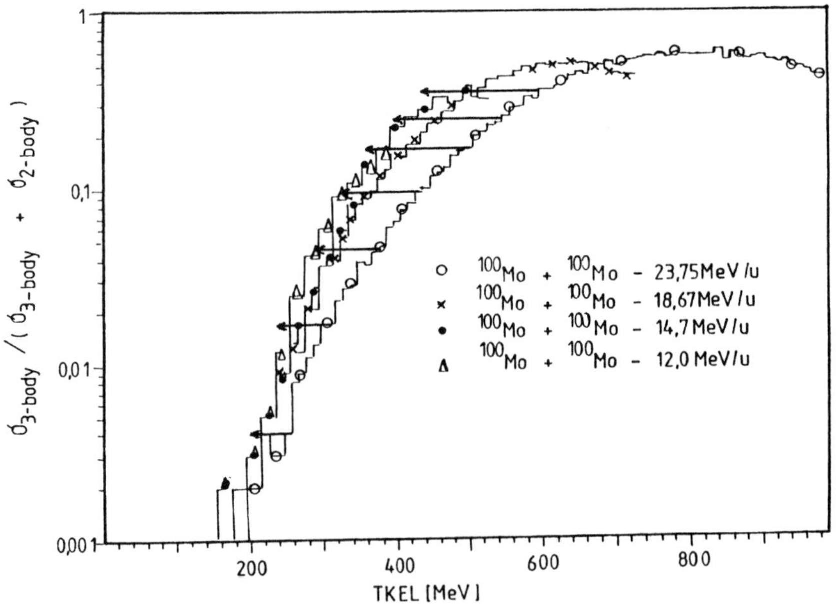

Fig.3. Ratio of 3-body (P3) to the sum of 2- and 3-body cross sections at four bombarding energies.

represented the experimental data for $^{98}Mo + ^{98}Mo$ at 15 MeV/u [4] and $^{98}Mo + ^{154}Sm$ at 12 MeV/u [12].

Fig.3 shows the three body probability (P_3) for $^{100}Mo + ^{100}Mo$ collision as a function of TKEL at four incident energies 12., 14.7, 18.67 and 23.75 MeV/u [6,13].

If at 18 MeV/u the deviation from the trends observed at lower incident energies is not too striking, at 24 MeV/u it cannot be neglected any longer. If the main bulk of the three body events could be interpreted as the result of a two step process, a deep inelastic interaction followed by a fission step of one of the final fragments, then for a given TKEL the same P_3 probability should be observed

Dynamical effects in the entrance channel would enhance the probability for three or four body events at the same TKEL for higher incident energies. The experimental trends do not support such a mechanism.

The deviations from the systematics observed at lower incident energies (< 10 MeV/u) can be due to different processes which start to be important in the intermediate energy regime.

These processes can be classified as follows:

a.- entrance channel effects - very fast processes as preequilibrium particle emission

b.- effects which take place during the evolution time as a consequence of the high excitation energy of the interacting system

c.- exit channel effects - clusters or heavy fragments emission or the onset of the multifragmentation process.

Fig.4 shows the multiplicity spectra of the fragments detected by our experimental set-up for the measured systems, normalized with the corresponding factors obtained from a comparison of the measured elastic events and Rutherford scattering cross section. They are not corrected for the detector geometry and threshold effects.

The shape of the spectra are very similar presenting a continuous rapid decrease with increasing multiplicity. This shows that processes from class c can not account for the above mentioned experimental deviations.

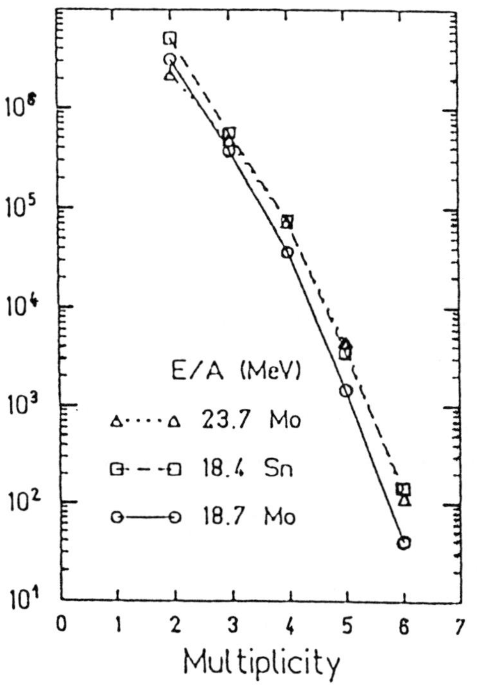

Fig.4. The multiplicity spectra of detected fragments.

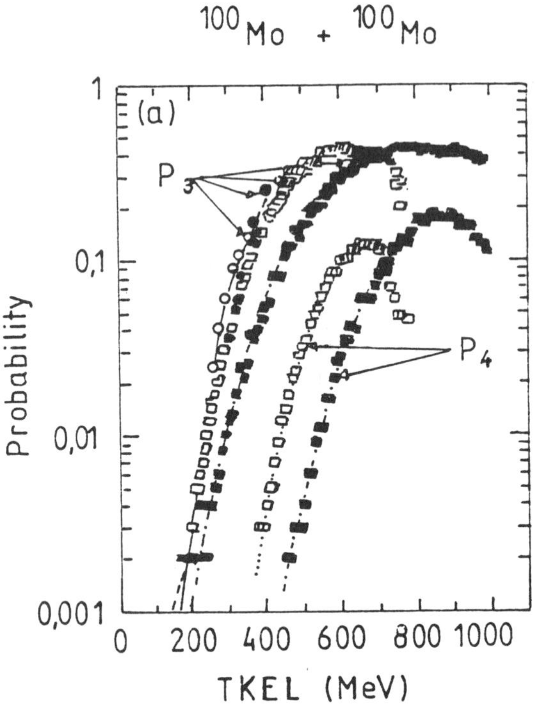

Fig. 5. 3- and 4-body probabilities as a function of TKEL.

This conclusion is supported also by the experimental trends observed for the four body probabilities as a function of TKEL. As can be seen in Fig. 5, the two curves of the P_4 probability for ^{100}Mo at 18.7 MeV/u and 23.75 MeV/u are displaced relative to each other very similar with the P_3 representations for the same energies.

We will analyse now in which extend the effects of class b) could explain the experimental trends. In order to investigate this in a more quantitative way, we compare theoretical predictions for deep inelastic interaction time with the statistical model decay time for the light particle emission.

Fig. 6 shows the estimates of these times and their dependence on TKEL. Under the hypothesis of a standard deep inelastic mechanism also at higher incident energies, we used a phenomenological model[14] for interaction time calculations. The results are represented in Fig. 6 by continuous lines for four incident energies. The average emmission time for the first particle has been estimated using the statistical model under the assumption that the whole energy transfered to the interacting system as internal excitation reaches instantaneously the thermal equilibrium[15]
The predicted time for the first particle emission is represented in Fig. 6 by a dotted line. As can be expected, this time decreases very rapidly with increasing TKEL or the nuclear temperature. Comparing now the two times, interaction and particle emission times, can be seen that for peripheral collisions (small TKEL) the interaction time is much shorter. For the events originating from this region, only preequilibrium or sequential evaporation processes are important. For larger energy losses or more central collisions, the decay time for particle emission of the interacting system becomes shorter. Here a considerable amount of excitation energy may be lost during the interaction time by evaporation processes.

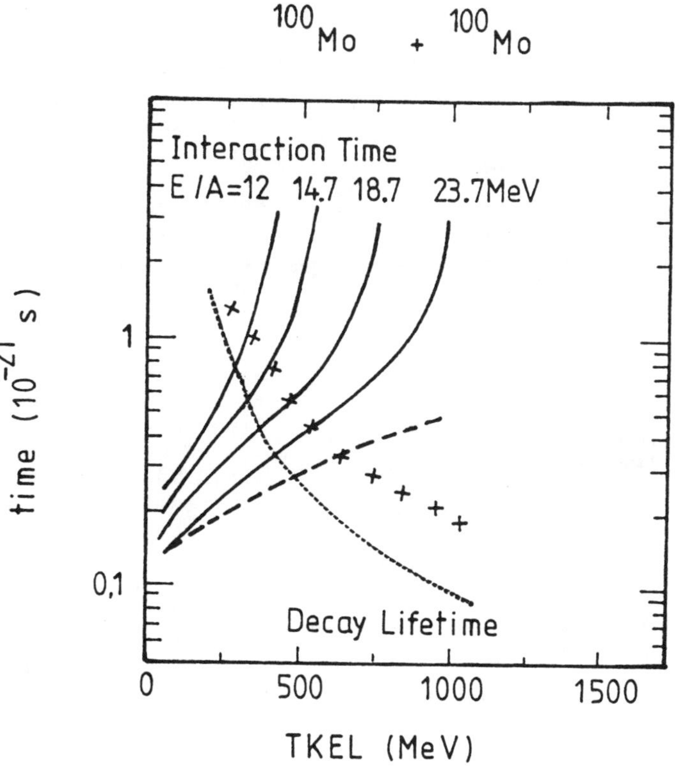

Fig. 6. Predictions for the interaction time of a deep inelastic process as a function of TKEL (continous lines) and the lifetime for statistical light particle emission from the interacting system (dotted line).

Fig. 7 presents the results for three and four body probability as a function of the true excitation energy which remains in the interacting system at the separation moment. As can be seen, such a correction removes the saturation and even a decreasing yield for three and four body events for large energy losses. The observed deviations discussed above at larger incident energies are not affected.

Let us analyse now in which extend the experimental trends can be attributed to the processes of class a). At higher incidents energies (≥ 15 MeV/u), as soon as preequilibrium processes start to appear, they remove part of nucleons and momentum, most of them in the very first stage of the collision. Such phenomena take place for all impact parameters. If for the central collisions they continue with a fusion process, for more peripheric ones, a deep inelastic will take place in the second step of the collision.

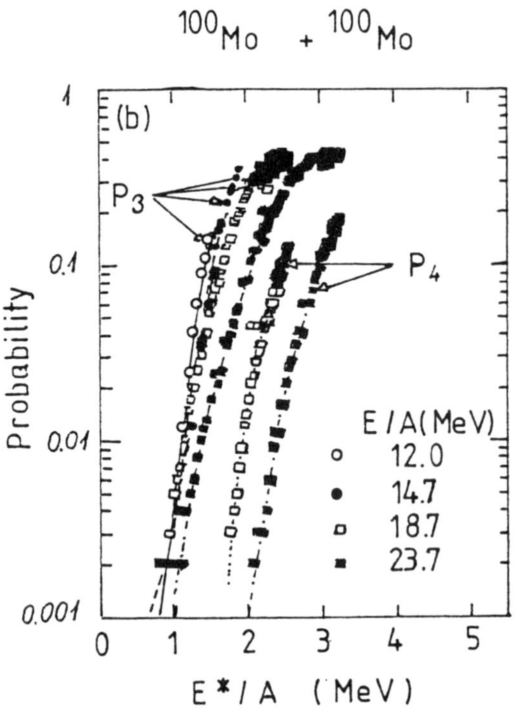

Fig. 7. As Fig. 5 when the TKEL scale was corrected by the energy removed through statistical emission process.

In this situation, it seems natural to baptize them as incomplete deep inelastic collision[3].

We would like to stress here two aspects :

i) from the observed experimental deviations is difficult to disentangle between the two pre-equilibrium contributions: the number of particles and their velocity.

ii) a two step mechanism, where the preequilibrium step is followed by a deep inelastic process is an aproximative scenario.

In reality, during the interaction, the two processes coexist with a relative contribution very well aproximated by a two step process.

In the remaining part of this talk, the consequences of such a process on the kinematical reconstructed quantities will be analysed and comparison with model predictions will be done.

A kinematical reconstruction procedure which starts with the measured velocity vectors and conserves the initial mass and momentum, gives a wrong final total kinetic energy value (TKE^f_w) in the presence of preequilibrium processes. Between the average value of the kinematical reconstructed TKE^f_w and the real one TKE^f_r the following relation can be found [16] :

$$TKE^f_w = \frac{A_T + A_P}{A_T + A_P - N_e} TKE^f_r \qquad (1)$$

A_T, A_P and N_e are the target mass, projectile mass and the number of preequilibrium emitted nucleons, respectively.

As a consequence, $TKEL^f_w = TKE^i - TKE^f_w$ and

$$TKEL^f_r = \varepsilon^* (A_T + A_P - N_e) \qquad (2)$$

where ε^* is the excitation energy per nucleon which remains in the dinuclear system after the first stage of a preequilibrium emission.

A quantity which is not affected in the average by the

reconstruction method, is the mass asymmetry in the outgoing channels :

$$\eta = \frac{A_1 - A_2}{A_1 + A_2} = \frac{A_1/A_2 - 1}{A_1/A_2 + 1} = \frac{v_2^{cm} / v_1^{cm} - 1}{v_2^{cm} / v_2^{cm} + 1} \qquad (3)$$

As the center of mass velocity vector is not modified in the average by the preequilibrium emission for a symmetric system, v_1^{cm}/v_2^{cm} is a correct value obtained from the corresponding measured quantities , v_1^{lab}, v_2^{lab}. As a consequence:

$$(\sigma_A^2)^w = \left(\frac{A_T + A_P}{A_T + A_P - N_e} \right)^2 \times (\sigma_A^2)^r \qquad (4)$$

$(\sigma_A^2)^w$ and $(\sigma_A^2)^r$ label the variances of the mass distributions obtained via a kinematical reconstruction procedure and real one, respectively.

If a two step mechanism is a good approximation for IDIC than the second step should follow the σ_Z^2 versus $TKEL^r/l_g$ systematics observed at lower incident energies. Taking N_e and ε^* as parameters and trying to adjust them in order to restore the slope of 6.07 ħ/MeV in a representation where the two variables, σ_Z^2 and TKEL are corrected using the relations (4) and (2) it is possible to obtain information on the ε^* and N_e quantities which characterize the first step of the process.

In order to apply the same recipe at the same level of accuracy for P_3 and P_4 systematics as a function of TKEL, the effect of decreasing mass of the heavy partners (as a consequence of preequilibrium emission) on the sequential fission probability and on the TKEL correction offset[6] has to be analysed in more details.

Let us see now in which extend some model predictions for preequilibrium emission can be used in order to estimate the effect of the first step on a deep inelastic process and compare them with the experimental results.

Recently, a theoretical appproach for heavy ion collisions at intermediate bombarding energies has been developed[17] . This model explained very well the main facts observed in incomplete fusion experiments for asymmetric systems. As far as the preequilibrium emission concerns, the authors used the master equation approach of Harp and Miller and parametrized afterwards the number of emitted nucleons N_e and the excitation energy per nucleon ε^* of the residual nucleus at the end of the preequilibrium phase by the following expressions:

$$N_e(t_1) = 0.96\mu^{5/4} \left[1 - \exp\left(-\frac{\varepsilon}{0.736 A_T}\right)\right] \qquad (5)$$

and

$$\varepsilon^*(t_1) = 6.8 A_T^{-1/3} \mu^{-2/3} \left[1 - \exp\left(-\frac{\varepsilon}{350\mu^{-1/3}}\right)\right] \qquad (6)$$

ε is the incident energy per nucleon, μ is the reduced mass of the incident channel, A_T is the target mass. The quantitative estimates for linear momentum transfer , percent energy loss and the number of preequilibrium emitted nucleons, obtained using these expressions are in a very good agreement with the experimental systematics[2] and the predictions of other theoretical calculations[18,19]

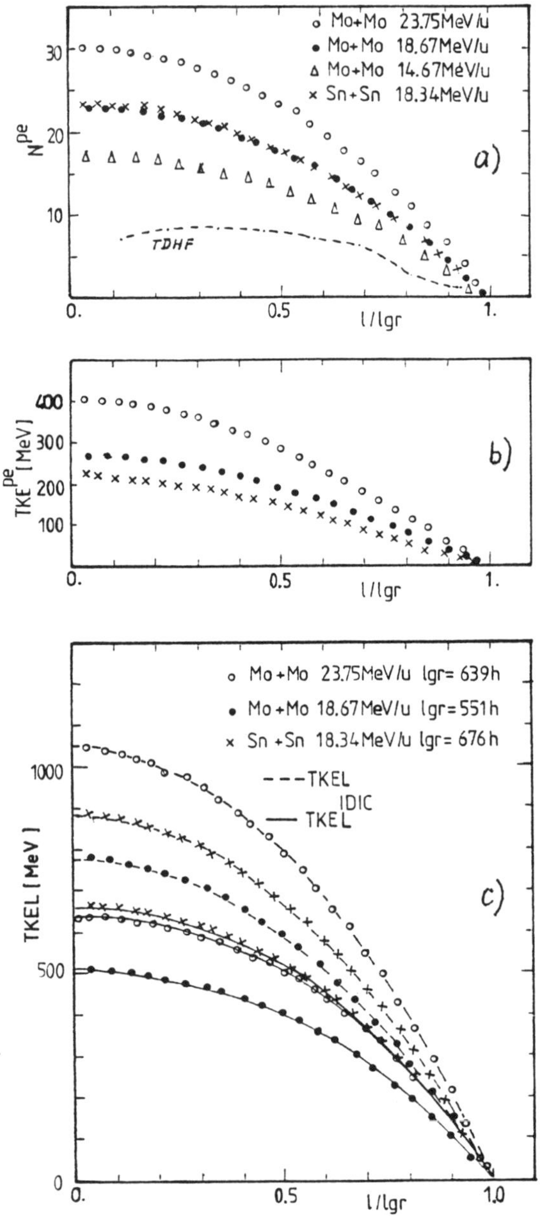

Fig. 8. The results of model predictions (see text)

The main argument to extrapolate this parametrisation to the symmetric case is based on the fact that the expression (5) gives a total multiplicity comparable with the one predicted by an extended nucleon exchange transport model[19].

In order to take into account the impact parameter dependence of such processes, a crude approximation would be to include it in ε. In this situation, the extention of the above parametrization to incomplete deep inelastic collisions can be made if we take into account only the available radial kinetic energy per nucleon for a given impact parameter.

In the present application ε has been calculated as:

$$\varepsilon = (E_{cm} - V_l - V_c)/\mu \tag{7}$$

With such a hypothesis we can apply the above equations in order to find the number of preequilibrium emitted nucleons and corresponding excitation energy per nucleon which remains in the nuclear system after the preequilibrium phase.

Fig. 8 a, b and c represent the results of such calculations for preequilibrium nucleon multiplicity as a function of angular momentum, the total energy removed by such a process and the difference between the total kinetic energy loss and the energy loss associated with the IDIC, respectively, for the systems measured by us. For the preequilibrium emitted particles are included also the estimates of TDHF calculation for $^{92}Mo + {}^{92}Mo$ at 14.6 MeV/u[4]. With these estimates we corrected the curves for the interaction time of the dinuclear system and decay time for light particle emission; the results are presented in Fig. 6 by dashed line and crosses, respectively. A dramatic decrease for the life time of the IDIC process can be observed. This is in agreement with the experimentaly observed anisotropy in the $V_{\parallel} - V_{\perp}$ representation at higher incident energy. One can observe also that the evaporation processe during the interaction time is still possible but for TKEL larger than 600 MeV.

Let us see now which is the effect of the correction applied to $(\sigma_Z^2)^W$ and $TKEL^W$ for the experimental points presented in Fig. 2

Fig. 9 presents the results of this correction. The agreement between the systematics obtained at lower incident energies and the experimental results, corrected by the assumption of an incomplete deep inelastic process is quite satisfactory. In Fig. 3 the thick arrows represent the corrected experimental points for 23.75 MeV/u. An agreement with the experimental results obtained at 12 MeV/u is observed.

A similar mechanism was recently proposed in order to explain the experimental trends observed in $^{40}Ar + {}^{108}Ag$ reaction at 27 MeV/u[20]

A semiclasical calculation based on the Landau - Vlasov equation and using a Zamick force reproduced the experimental observations for the above reaction. These calculation predicted about twelve nucleons rapidly ejected for central collisions. For a heavier projectile and a symmetric system, the number of the preequilibrium emitted nucleons is expected to increase.

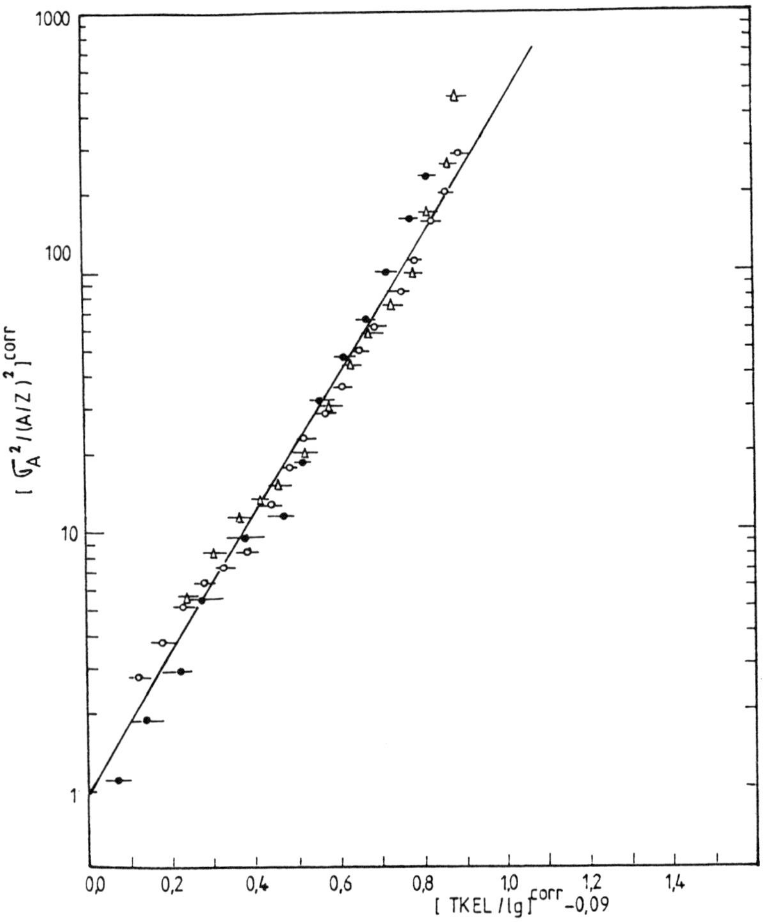

Fig. 9. Corrected experimental points represented in Fig. 2 using the procedure presented in this paper.

More realistic estimates of the preequilibrium step , using Boltzmann master equation where the number of excitons are taken proportional to the number of nucleons contained in the overlap region, are in progress.

Fig. 10 shows some preliminary results. The predictions of the above parametrization which seems to be in agreement with the experiment, are reproduced by our calculations using the same scaling factor between the number of nucleons in the overlap region and the number of excitons for a given incident energy.

These factors are 0.3, 0.4 and 0.5 for ^{100}Mo + ^{100}Mo case at 14.67 MeV/u, 18.67 MeV/u and 23.75 MeV/u, respectively. For ^{120}Sn + ^{120}Sn at 18.34 the scaling factor is 0.35.

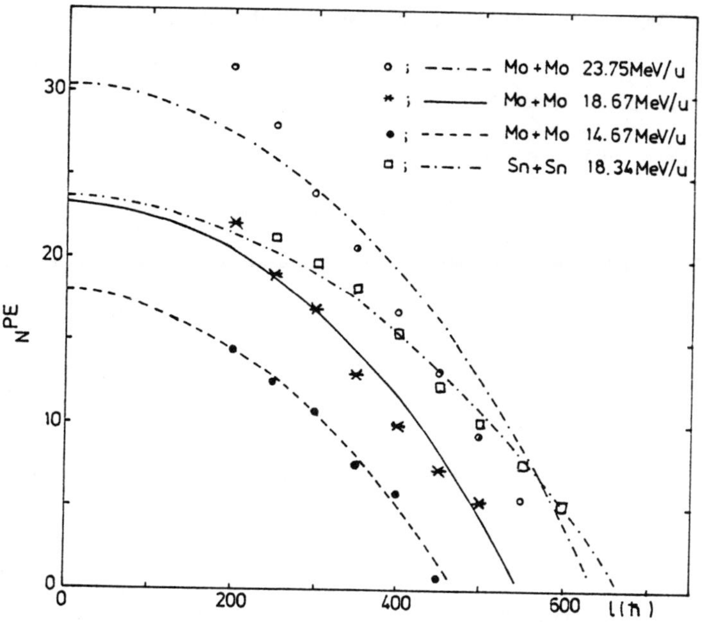

Fig. 10. Predictions of a model based on Boltzmann master equation when the number of excitons are taken propotional to the number of nucleons contained in the overlap region.

To summarize, in our recent studies of $^{100}Mo + ^{100}Mo$ and $^{120}Sn + ^{120}Sn$ collisions at 18 MeV/u and 24 MeV/u, strong deviations from the systematics obtained at lower incident energies are observed. These deviations are larger at higher incident energy per nucleon. A similar mechanism to the incomplete fusion process but for noncentral collisions - an incomplete deep inelastic collision (IDIC) process has been supposed to account for the observed discrepances. Such a mechanism gives the possibility to study the dynamics of the preequilibrium phenomena as a function of impact parameter and presents a challanging testing ground for theoretical models.

REFERENCES

1. P.L.Gonthier et al., Phys. Rev. C35, 1946 (1987)
W. Terlau at al.,Z. Phys A330, 303(1988) and refs. therein

2. V.E.Viola, Nucl. Phys. A471, 53c (1987) and refs. therein
T.Péter and B.Tamain, Proc. Riken - IN2P3 Symposium on Heavy Ion Collisions, Shimoda, October (1987) and refs. therein

3. M.Petrovici et al., IPNE Annual Report 1998, p.100

4. M.Petrovici et al., Nucl. Phys. A477, 277 (1988)

5. R.Freifelder et al., GSI Annual Report 1988, p.64

6. R.J.Charity et al., Preprint DFF 123/12/90 University of Florence, submited for publication

7. M.Petrovici et al., GSI Annual Report 1988, p.64

8. M.Morgenstern et al., Phys. Rev. Lett. 52, 1104 (1984)

9. G.Casini et al., Nucl. Instr. and Meth. A277, 445 (1989)

10. H.J.Wollersheim et al., Phys. Rev. C25, 338 (1982)

11. S.Gralla et al., Phys. Rev. Lett 54, 1898 (1985)

12. M.Petrovici et al., Nucl. Phys A387, 313c (1982)

13. P.R.Maurenzig et al., GSI Annual Report 1988, p.63

14. G.Wolschin and W.Nörenberg, Z. Phys. A284, 209 (1978)

15. R.J.Charity et al., Symp. on Nuclear Dynamics and Nuclear Disassembly, Dallas, Texas 10 -14 April 1989

16. M.Petrovici, Invited Talk at the International School Seminar on Heavy Ion Physics, Dubna, USSR, October 3-12 1989

17. C.Cerruti et al., Nucl. Phys. A476, 74 (1988)

18. M.Blann, Phys. Rev. C31, 1245 (1985)

19. J.Randrup and R.Vandenbosch, Nucl. Phys. A474, 219 (1987)

20. M.F.Rivet et al., Phys. Lett. 215B,55 (1988)

FISSION DECAY AND COLLECTIVE DYNAMICS OF HOT NUCLEI

INTERMEDIATE-MASS FRAGMENT EMISSION: PROBING NUCLEAR DYNAMICS AT HIGH EXCITATION ENERGIES*

V.E. Viola and K. Kwiatkowski
Indiana University, Bloomington, IN 47405

ABSTRACT

The emission of intermediate-mass fragments (IMFs) from highly excited nuclei is examined as a function of projectile energy. Near threshold the principal IMF sources are (1) the statistical decay of equilibrated composite systems formed in central collisions and (2) peripheral processes which form projectile-like ejectiles. In the vicinity of the Fermi energy, nonequilibrium IMF emission, presumably from intermediate impact-parameter collisions, becomes a third important mechanism. At total bombarding energies in excess of 1 GeV, multifragment emission is also observed. In this report we investigate the evolution of these mechanisms as a function of beam energy in both inclusive and exclusive studies of ^{14}N-induced reactions below E/A = 100 MeV and ^{3}He-induced reactions up to E/A = 1.2 GeV.

INTRODUCTION

Accounting for the emission of intermediate-mass fragments from highly excited nuclei poses a severe challenge for any effort to formulate a unified picture of nuclear dynamics. These ejectiles, or IMFs, are usually defined as fragments with atomic numbers $3 \leq Z \leq 15$; i.e. heavier than light-charged particles (LCP=H and He) but lighter than fission fragments. IMFs were initially observed over 30 years ago in radiochemical studies of proton- and alpha-particle-induced reactions on heavy target nuclei.[1-3] These early measurements were especially important in demonstrating the strong sensitivity of IMF emission to bombarding energy up to energies of several GeV -- thus associating these fragments with the disintegration of highly excited nuclei.[4,5] In addition, these results

*Work supported by the US Department of Energy, grant no. DE-FG02-88ER.40404.A000.

stimulated many early discussions relating to the question of energy dissipation on time scales comparable to the collision time, including the possible role of subnucleonic degrees of freedom and, when relativistic heavy ions became available, density compressions. The concept of multifragmentation -- or the instantaneous breakup of nuclei -- also emerged in response to these early studies.

With recent developments in accelerator technology, detector systems and computing power, both the experimental and theoretical situation with respect to our understanding of IMF emission have improved markedly. We now believe that a broad continuum of reaction phenomena contribute to the observed IMF spectra -- evolving due to the interplay of impact parameter, reaction time and projectile energy. However, as will be apparent in much of the following, our understanding of IMF emission remains for the most part at the qualitative level. Theorists face a formidable task in assimilating this spectrum of collective processes into any unified theory of nuclear dynamics. Fortunately, impressive efforts in this direction are underway.[6-12]

In the following section we discuss the various mechanisms responsible for IMF emission, along with some of the relevant experimental signatures of each. This is followed by a discussion of the evolution of IMF formation mechanisms with bombarding energy, illustrated by recent systematic studies of the $^{14}N + {}^{nat}Ag$, ^{197}Au and $^{3}He + {}^{nat}Ag$ reactions. The fourth section of this paper deals with multifragment emission, emphasizing the results of measurements on the $^{3}He + {}^{nat}Ag$ system up to 3.6 GeV. In the final section we attempt to summarize the present state of our experimental knowledge concerning IMF emission.

IMF SOURCES

Because the study of IMFs requires beams of intermediate-to-high energy projectiles, experimental investigations are complicated by the multibody nature of such reactions. This introduces the attendant difficulties of defining the thermal and kinematic properties of the emitting source, which in turn necessitates the implementation of increasingly complex coincidence experiments as the projectile energy increases. Studies of this nature have identified several different mechanisms which contribute to the observed IMF spectra. For mass-asymmetric collision partners in normal kinematics, these can be generalized as follows: (1) projectile remnants from peripheral interactions, (2) nonequilibrium emission from the projectile-target composite system, (3) statistical decay of equilibrated composite nuclei, and (4) multifragment emission. We stress that although each of these mechanisms can be isolated under specific experimental conditions, they represent a continuum of impact-parameter- and bombarding-energy-dependent events, all of which may co-exist in a given system. For the case of collisions between nearly symmetric heavy nuclei, mechanisms (1) and (3) must obviously merge; at the other mass-asymmetry extreme, for H- and He-induced reactions, the situation is simplified since (1) is not present.

The projectile-like component (for $A_p \lesssim 40$) is composed of several contributions, including inelastic scattering, transfer reactions, damped collisions and projectile fragmentation.[13-14] It is primarily important near the grazing angle and presumably involves short interaction times. The salient properties of this component are illustrated in Fig. 1(a), where we show spectra for Z=4,6 and 8 fragments observed at 15° in the E/A = 20 MeV $^{16}O + ^{238}U$ reaction in coincidence with angle-correlated fission fragments.[15,16] The fission folding angle serves as a measure of the linear momentum transfer (LMT) from the beam to the target. Correlation angles near 180° signify small LMT, or weak target-projectile interactions, while smaller folding angles indicate increasingly strong interactions. On the left side of Fig. 1 the inclusive correlation function for all fission events is compared with correlated events in coincidence with IMFs. For the data at 15° the average longitudinal momentum transfer relative to the beam momentum is small, $\langle p_\parallel \rangle / p_{beam} = 0.10 \pm 0.10$. The corresponding IMF energy spectra (given in the frame on the right) are peaked near the E/A value of the beam; in addition, the mass distribution is strongly peaked near the projectile mass. All of these facts support an origin in peripheral interactions. Although the projectile-like component is of considerable interest in its own right, the focus of this paper will be on IMF emission from heavy systems.

The nonequilibrium source becomes important as the projectile velocity approaches the Fermi velocity and resembles in many respects pre-equilibrium nucleon emission. It is characterized by forward-peaked angular distributions and is usually accompanied by pre-equilibrium nucleon emission. Whereas spectra for the peripheral component is strongly dependent on the energy, mass and charge of the projectile, the nonequilibrium component is not significantly influenced by projectile properties. This is supported by the very similar spectra for IMFs from both light- and heavy-ion-induced reactions at intermediate angles, where this component is most readily observed.[20,21] In Fig. 1(b) the angular correlation and spectra are shown for IMFs measured at 51° in the E/A = 35 MeV $^{14}N + ^{232}Th$ reaction.[22] In this case a strong projectile-target interaction is found, $\langle p_\parallel \rangle / p_{beam} = 0.72 \pm 0.03$, indicating that most of the beam momentum has been transferred to the composite system prior to IMF emission. The corresponding kinetic energy spectra are distinctly different in

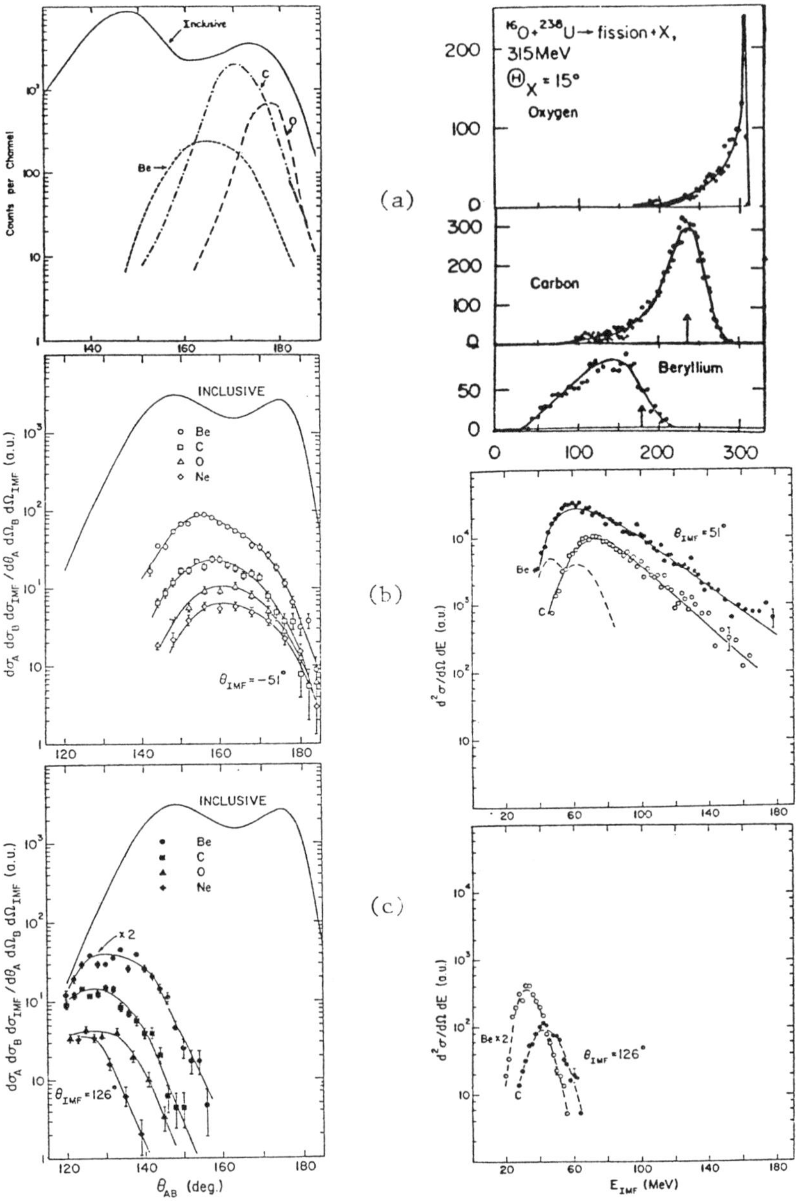

Figure 1. Left column: Fission-fission angular correlations in coincidence with IMFs compared with inclusive fission-fission results. Right column: Energy spectra for IMFs in coincidence with correlated fission fragments. Top: IMFs at 15° from E/A = 20 MeV $^{16}O + ^{238}U$; Middle: IMFs at 51° from E/A = 35 MeV $^{14}N + ^{232}Th$, and bottom: IMFs at 126° from E/A = 35 MeV $^{14}N + ^{232}Th$.

character relative to the peripheral component, being peaked near the Coulomb energy for touching spheres; similarly the charge and mass distributions follow a power law distribution, $\sigma(Z,A) \propto (Z,A)^{-\tau}$ instead of being peaked near the projectile Z and A. The fact that the spectral slope temperatures are ~3 times greater than expected for a fully equilibrated system, combined with the forward-peaked character of the angular distributions, gives rise to the concept of a nonequilibrium process in which IMFs are emitted from a localized region of excitation in the composite system. Thus, we appear to be observing a fast collective process in which the reaction dynamics and decay of the system are closely coupled. At some forward angle, both the peripheral and nonequilibrium components must overlap so as to become indistinguishable.[23]

Statistical decay of equilibrated heavy residues constitutes the simplest, and probably best understood, mechanism of IMF emission. This component, which resembles proton and ^4He evaporation in character, is most readily observed near the threshold for IMF emission[24] ($E^* \approx 50$-100 MeV) or at backward angles[20] at higher bombarding energies. For example, in Fig. 1(c), IMFs detected at 126° are found to involve large LMT values, $\langle p_\parallel \rangle / p_{beam} = 0.95 \pm 0.05$, corresponding to emission from a fusion-like source with nearly the maximum available excitation energy. The corresponding spectral slope temperatures are consistent with expectations for evaporation from such a source. This component of the spectrum is strongly sensitive to angular momentum effects[24] and exhibits corresponding forward-backward symmetric angular distributions, indicating that the emission time scale is relatively long, at least with respect to the nuclear rotation period. Although data are sparse, studies with light-ion projectiles suggest that central collisions are primarily responsible for the equilibrium component, whereas nonequilibrium ejectiles originate primarily from impact parameters intermediate between central and peripheral collisions.[25] In Fig. 2 we outline the schematic picture proposed above, along with the basic spectral features associated with each. This representation emphasizes the impact-parameter evolution of the interaction; the interplay of bombarding energy and time scales must also be folded in.

Multifragment emission, i.e. final states involving two or more IMFs, is a subject of intense current interest. Evidence for such events has been demonstrated in several experiments, extending back to very early emulsion studies.[4,5,26-31] An important question concerning multifragment decay is the distinction between sequential binary decays and instantaneous breakup of the system (or multifragmentation).[32] In the former case, the mechanism is a simple compounding of equilibrium emission probabilities

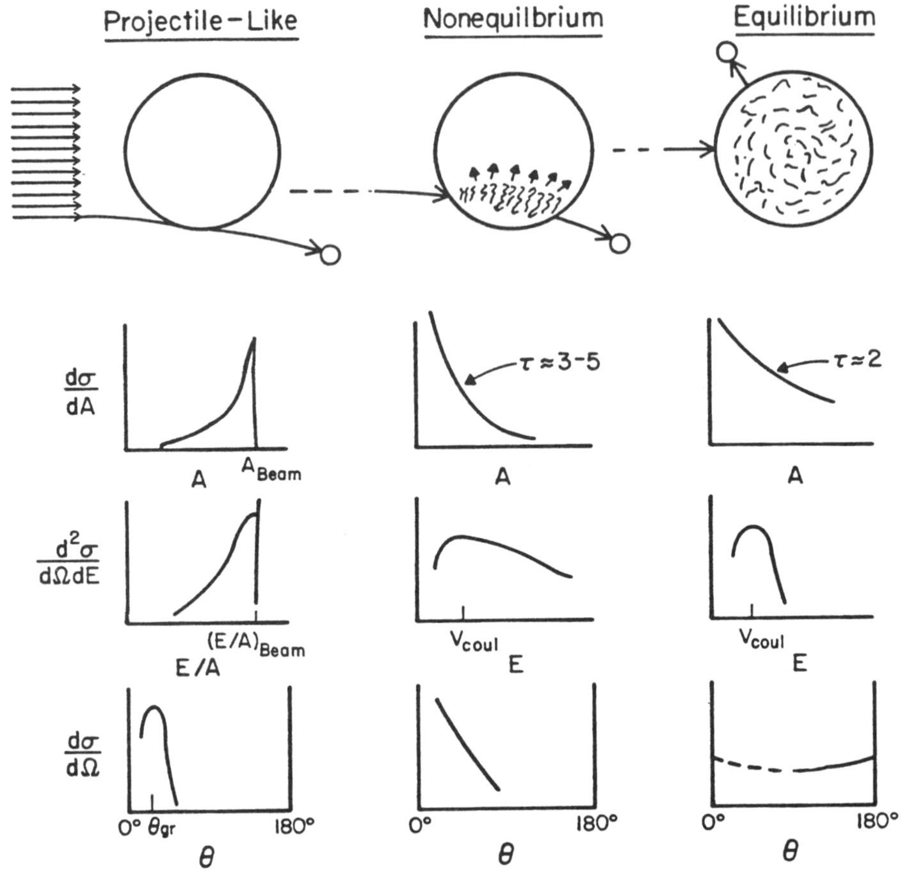

Figure 2. Schematic picture of projectile-like, nonequilibrium and equilibrium sources, along with salient experimental observables: IMF mass, kinetic energy and angular distributions.

and must certainly be present in most heavy-ion-induced reactions due to the angular-momentum enhancement of the IMF emission probability. Multifragmentation, on the other hand, represents a unique phenomenon for which clearcut experimental verification remains to be established. Hence, both mechanisms may coexist, depending on reaction parameters. Or, the difference may be semantic, given the very short lifetimes expected for systems at very high excitation energies. Clearly, a major experimental effort which employs versatile 4π detector arrays and covers the full range of target-projectile combinations will be required to sort out the mechanism(s) of multifragment emission.

SYSTEMATICS OF INCLUSIVE SPECTRA

Despite the extensive experimental effort that has gone into the study of IMF emission at intermediate energies, relatively few systematic data sets exist which trace the bombarding energy evolution of these processes from threshold to energies approaching the total nuclear binding energy. In addition to the data presented below, Trockl et al. have studied the ^{18}O + ^{nat}Ag and ^{197}Au reactions

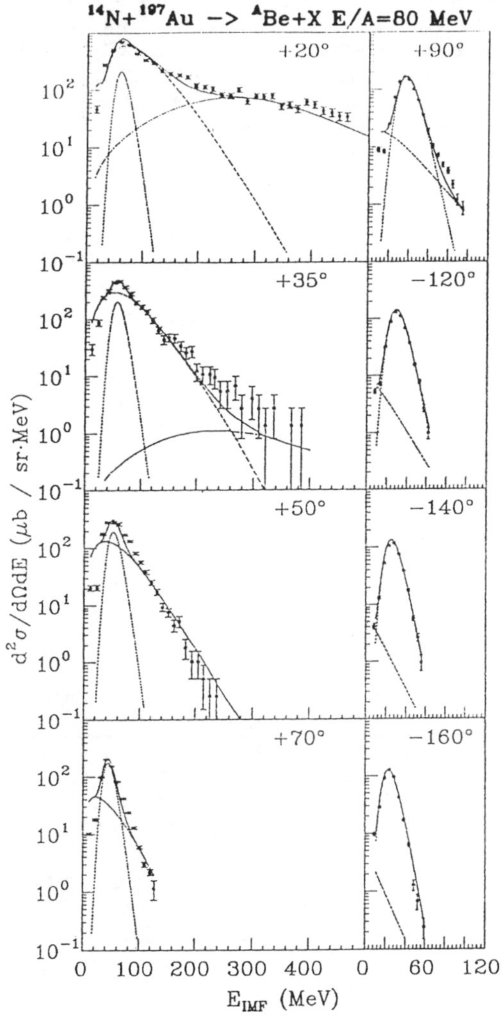

Figure 3. Energy spectra for Be fragments in the E/A = 80 MeV ^{14}N + ^{197}Au reaction. Fits are for equilibrium (dots), nonequilibrium (dashes), projectile-like (dash-dots), and sum of these sources (solid line).

at E/A = 30-84 MeV;[33] Papadakis et al. have performed measurements of the E/A = 20-60 MeV ^{20}Ne + ^{nat}Ag and ^{197}Au reactions;[34] Bowman et al. have examined the E/A = 50-100 MeV ^{139}La + ^{12}C system,[35] and the ^{12}C + ^{197}Au reaction has been measured from E/A = 30 MeV to 1.2 GeV.[36] In addition, proton-induced IMF emission has been studied on medium-mass target nuclei (Ag, Xe) from 160 MeV to 80 GeV.[37]

Here we describe a systematic set of inclusive IMF excitation function measurements for the E/A = 20-100 MeV ^{14}N + ^{nat}Ag, ^{197}Au system[21,38] and the E/A = 67 MeV to 1.2 GeV ^{3}He + ^{nat}Ag system.[39] Measurements were performed with low detector thresholds, large dynamic range ($0.4 \leq (E/A)_{IMF} \lesssim 70$ MeV), and angular coverage from 20° - 160°. The experiments were performed at the National Superconducting Cyclotron at Michigan State University, the Indiana University Cyclotron Facility and the Saturne accelerator in Saclay. These data provide a qualitative overview of the mechanisms responsible for IMF emission as the projectile energy is increased from the threshold region, across the Fermi energy and eventually approaches the total nuclear binding energy of the composite system. In addition, the results should provide a useful definition of IMF properties that are necessary for interpretation of data from more complex exclusive experiments. Preliminary reports of these results have been presented in refs. 21, 38 and 39, which contain the experimental details.

We first examine the ^{14}N-induced reactions. Representative kinetic energy spectra are shown in Fig. 3 for Be fragments emitted in the E/A = 80 MeV ^{14}N plus ^{197}Au reaction. The Be spectra are typical for fragments with Z ≤ 7 at all beam energies. In fact, one of the most striking results of the inclusive data for total energies below about 1 GeV is the relative insensitivity of the spectra at a given angle to bombarding energies above E/A = 20 MeV. The forward-angle spectra near the grazing angle exhibit a peak in the vicinity of the exit channel Coulomb barrier energy, followed by a flat tail that extends up to energies near the E/A value of the projectile. The spectral tails are identified with projectile remnants arising from peripheral interactions, as demonstrated by the coincidence experiments discussed along with Fig. 1. The projectile-like component of the fragment spectra for ejectiles with Z > 7 is not significant. In addition, the spectra of heavier fragments show little sensitivity to bombarding energy. As the emission angle increases, the projectile-like component disappears and the spectral slopes become increasingly steep, until at backward angles they are approximately equal to those expected for evaporation from an equilibrated composite system.

A three-component moving source fit to the energy spectra has been performed at all angles for the E/A = 20-100 MeV ^{14}N induced reactions. The equilibrium source has been characterized by a mean velocity, v_{EQ}; Coulomb barrier, B_{EQ}; amplification parameter, p, and temperature, T_{EQ}, according to the scission-point model of Moretto.[40] In addition, the angular distribution function was assumed to follow $w(\theta) = (1 + \alpha \cos^4\theta)$ where α is an additional fit parameter. The equilibrium component is indicated by the dotted curves in Fig. 3. The nonequilibrium component was assumed to result from a moving source with velocity v_{NEQ} emitting IMFs according to a Maxwellian energy spectrum[41] described by a slope parameter, T_{NEQ}, and a Coulomb barrier, B_{NEQ}. A c. of m. angular distribution function of the form $w(\theta) \alpha \exp(-b\theta)$ was assumed, where b is an additional fit parameter. The results of this portion of the fit are illustrated by the dashed curves in Fig. 3. Finally, the projectile remnants that arise from peripheral interactions were parameterized according to the empirical model of Kiss et al.[42] This component of the fit is shown by the dot-dashed curves in Fig. 3 and is only important for IMFs with $Z \le 7$ emitted at the most forward angles. The fits employed 14 parameters plus normalizations for each source, which were fitted simultaneously.

The solid curves in Fig. 3 represent the sum of all three components of the fit and describe the data well. In general, χ^2/DF values between 0.5 and 1.5 were obtained. The equilibrium and nonequilibrium fitting functions for Z = 3-15 fragments from both targets have been integrated over all energies and angles to yield elemental cross sections for equilibrium emission, $\sigma_{EQ}(Z)$; nonequilibrium emission, $\sigma_{NEQ}(Z)$, and the total target-like source, $\sigma_{IMF}(Z) = \sigma_{EQ}(Z) + \sigma_{NEQ}(Z)$. The projectile-like yields have been removed from these results using the fitting routine described above and are not considered further here. Fig. 4 presents the excitation functions for even-Z IMFs. Representative errors include both statistical and systematic uncertainties, as well as those associated with the moving-source fits. Odd-Z cross sections are systematically similar to their even-Z neighbors, except for Li (Z=3) which is about 2.5 - 3 times larger than for Be (Z=4). Also shown in Fig. 4 are the total IMF cross sections summed over Z = 3-15 and the reaction cross sections, σ_R, calculated according to the prescription of Kox et al.[43] This comparison indicates that IMF emission accounts for about 30% of the total cross section. The results for $\sigma_{IMF}(Z)$ are in general agreement with those of Refs. 33-36.

In addition to providing a systematic overview of IMF emission between E/A = 20 and 100 MeV bombarding energy, two distinct features of the excitation functions in Fig. 4 should be noted. Most apparent is the significant increase in $\sigma_{IMF}(Z)$ from the ^{nat}Ag target between E/A = 60 and 80 MeV. In contrast, the variation of $\sigma_{IMF}(Z)$ for the ^{197}Au target is uniform across this region. The discontinuity in

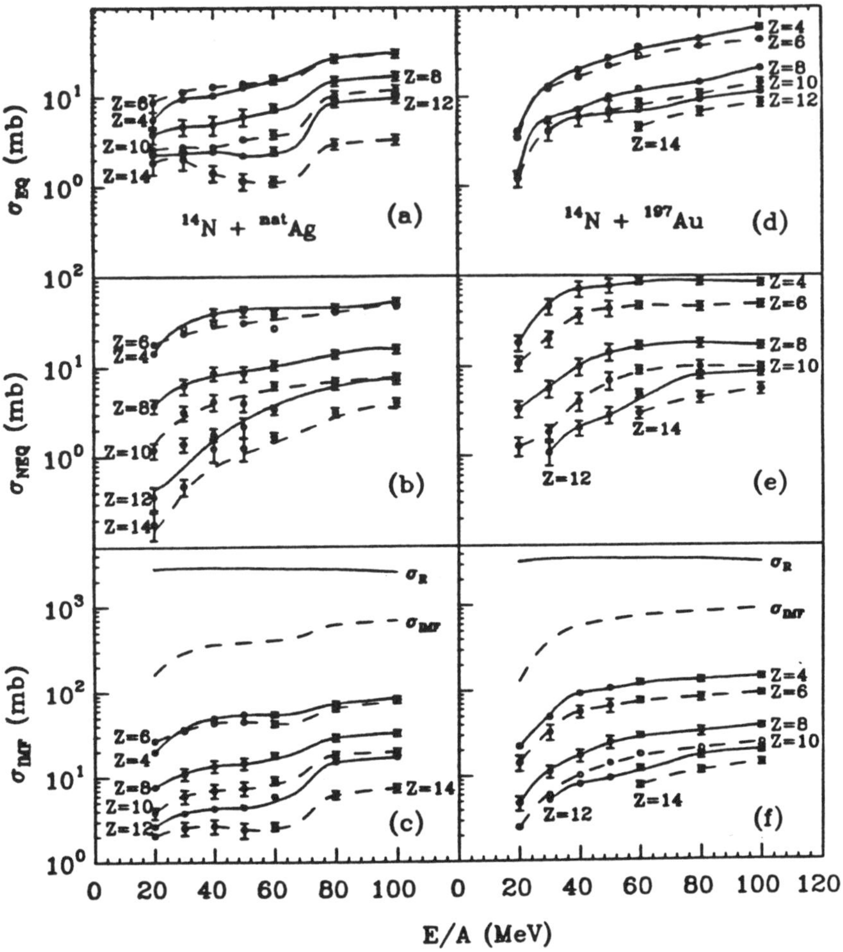

Figure 4. Excitation functions for even-Z IMFs from ^{14}N + ^{nat}Ag (left) and ^{14}N + ^{197}Au (right) for the equilibrium (top), nonequilibrium (middle) and the sum of these sources (bottom). Also shown are σ_{IMF} for each case and σ_R.

the ^{14}N + ^{nat}Ag excitation function grows with increasing fragment charge and is essentially independent of the fitting procedures employed in this analysis. On the basis of the moving-source decomposition of $\sigma_{IMF}(Z)$, the enhanced IMF yields appear to be associated with the equilibrium component. One possible explanation for this behavior may be the onset of multifragment decays at bombarding energies which approach the total nuclear binding energy. The total binding energy for the composite ^{14}N + ^{nat}Ag system is exceeded at bombarding energy of E/A ≈ 80 MeV; for ^{14}N + ^{197}Au, it is not reached until E/A ≈ 125 MeV. In this regard it should be noted that the character of the spectra at E/A = 80 and 100 MeV is consistent with that at lower energies, suggesting that if multifragment emission is responsible for the increase in cross section, sequential decay is primarily responsible.

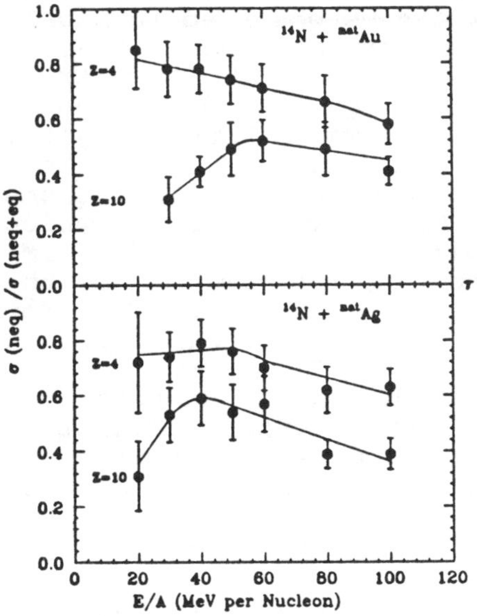

Figure 5. Ratio of $\sigma_{NEQ}(Z)$ to $\sigma_{IMF}(Z)$ for Z=4 and 10 fragments as a function of beam energy.

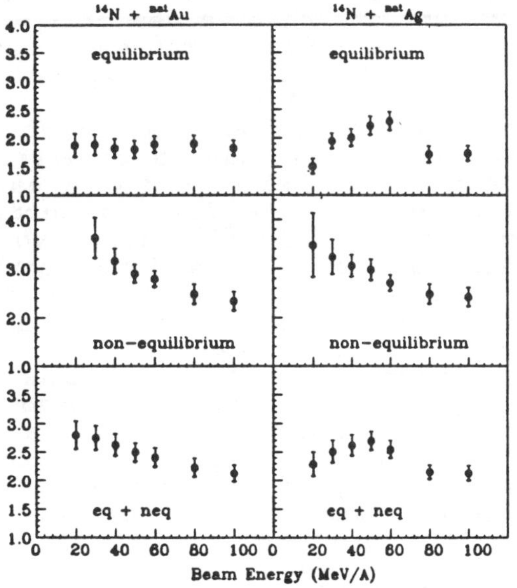

Figure 6. Power law exponent τ as a function of beam energy for ^{14}N + ^{197}Au (left) and ^{14}N + ^{nat}Ag (right) reactions. Data for equilibrium (top), nonequilibrium (middle) and target-like IMF (bottom) components.

A second important aspect of these data is the relative contribution of equilibrium and nonequilibrium IMF emission as a function of increasing projectile energy. Previously, analysis of our data below E/A = 50 MeV ^{14}N energy[21] indicated that nonequilibrium IMF emission increased rapidly with bombarding energy relative to equilibrium emission, especially for heavier fragments. This suggested that the former process might be dominant at higher energies. However, the present data do not support such a conclusion. Rather, as shown in Fig. 5, for E/A ≥ 60 MeV, the ratios of $\sigma_{NEQ}(Z)/\sigma_{IMF}(Z)$ remain approximately constant, with non-equilibrium emission favoring lower-Z fragments and equilibrium emission favoring those with higher Z.

The fragment charge distributions may also carry information relevant to the mechanisms for IMF emission.[44] The data in Fig. 4 have been analyzed in terms of a power law, $\sigma(Z) \propto Z^{-\tau}$. Values of τ obtained in this work are plotted as a function of bombarding energy in Fig. 6. For the equilibrium component, the charge distributions for the ^{197}Au target are essentially constant over the entire range of bombarding energies at $\tau = 2.0 \pm 0.1$. Thus, these data demonstrate that values of $\tau \approx 2$ can be observed in simple statistical decay processes and are not a unique signature of a nuclear liquid-gas phase transition.[45] The ^{nat}Ag target produces similar values of τ for the equilibrium component near threshold and at the highest bombarding energies. However, a maximum is observed near E/A $\approx$ 50 MeV. This behavior is believed to be associated with contributions from binary-like fission decay in these reactions which contribute significantly to the tails of the IMF distribution at the lower bombarding energies. For the nonequilibrium component, τ decreases monotonically for both targets from values near $\tau \approx 3.5$ near threshold to values near $\tau \approx 2$ at E/A = 100 MeV. This result is in agreement with previous studies that have demonstrated larger τ values for the nonequilibrium component relative to the equilibrium component at low bombarding energies.[20,33]

In order to examine the physics implications and self-consistency of the three-source fits described above, we present the systematic behavior of the velocity and temperature fit parameters, averaged over Z = 3-15 fragments, as a function of bombarding energy in Fig. 7. In calculating averages, equal weight was assigned to each Z value, and the errors give the corresponding standard deviations. For the velocity, temperature and Coulomb parameters, there is little dependence on IMF charge, which permits the use of averages for each bombarding energy and target nucleus.

The ratio of the average source velocity, $\langle v_{EQ} \rangle$, to that of the composite system formed in a complete fusion reaction, v_{CN}, is plotted as a function of projectile energy in the top-frame of Fig. 7. Near threshold, the source velocities are approximately consistent with complete fusion for both projectile-target systems. As the beam energy increases, $\langle v_{EQ} \rangle / v_{CN}$ decreases systematically below the complete fusion expectation, consistent with the corresponding increase in the probability for incomplete fusion and pre-equilibrium light-charged-particle emission prior to IMF emission. The solid curves in the top panel of Fig. 7 show the most probable velocities of fusion-like residues derived from inclusive fission-fission angular correlation studies,[46,47] corrected for target mass dependence according to the systematics of Batsch et al.[48]

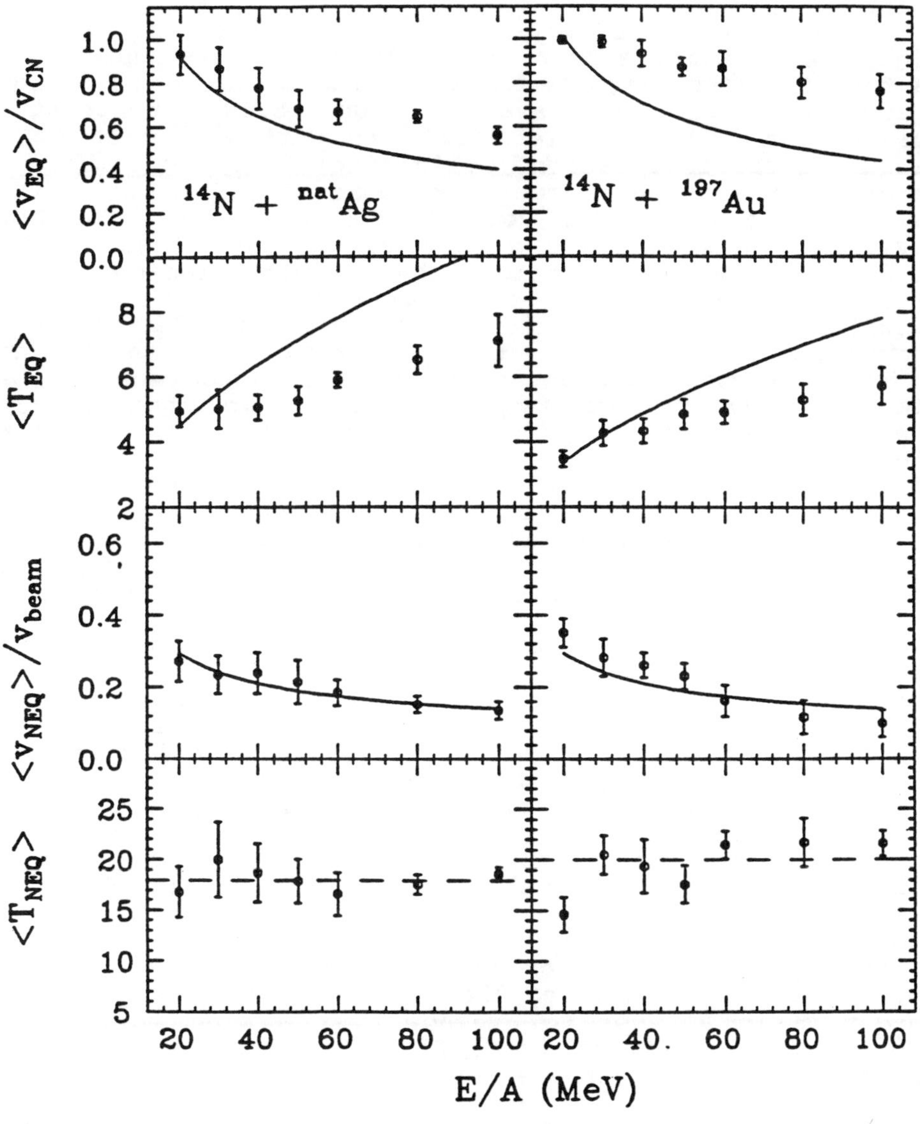

Figure 7. The dependence of fit parameters on bombarding energy for $^{14}N + {}^{nat}Ag$ (left) and $^{14}N + {}^{197}Au$ (right). Lines on plots are discussed in text.

The equilibrium IMF velocity ratios fall consistently above these systematics, implying that IMF emission is associated with higher than average linear momentum transfer;[5] i.e., hotter sources.

The second frame in Fig. 7 shows the average equilibrium temperatures, $\langle T_{EQ} \rangle$, extracted from the fits as a function of projectile energy. Also shown is the energy-dependence of the average equilibrium temperature calculated for a simple Fermi gas with a = A/10 MeV^{-1}, assuming complete fusion (solid line). At low bombarding energies the temperatures correspond to IMF emission from a nearly equilibrated complete-fusion residue for both systems. With increasing beam energy the values of $\langle T_{EQ} \rangle$ fall systematically below the complete-fusion limit, reflecting the loss in internal excitation due to non-equilibrium processes prior to IMF emission. For ^{nat}Ag this deviation grows at nearly twice the rate as for the heavier ^{197}Au nucleus, consistent with the excitation energy loss implied by the ratio of $\langle v_{EQ} \rangle / v_{CN}$. Without more quantitative information concerning the time evolution of the emitting source, one must be cautious in associating T_{EQ} with a true thermodynamic temperature. It is also noted that the Coulomb parameters for the equilibrium source are approximately equal to fission kinetic-energy systematics[49] for the $^{14}N + {}^{197}Au$ system. For $^{14}N + {}^{nat}Ag$, these values decrease from near unity at low bombarding energies to about 0.6 at the highest energy.

The ratio of the average nonequilibrium source velocity to that of the beam, $\langle v_{NEQ} \rangle / v_{beam}$, is shown in the third frame of Fig. 7 as a function of bombarding energy. For both targets the results are similar and can be described by an approximately constant velocity for the average nonequilibrium source, $\langle v_{NEQ} \rangle \approx 0.06c$ (solid line). The Coulomb parameters for this source are typically 20-30 percent those for fission energy systematics.

The slope (or effective) temperature, T_{NEQ}, results are shown in the bottom frame of Fig. 7. The values of T_{NEQ} are also found to be essentially constant with bombarding energy. For the ^{nat}Ag target, all values are consistent with $T_{NEQ} \approx 18$ MeV and for gold we find $T_{NEQ} \approx 20$ MeV. Since our model for describing the non-equilibrium component is a schematic attempt to average over a wide spectrum of source properties, it is difficult to ascribe any direct physical meaning to these constant values. However, we do note that similar behavior is observed for pion spectra, as pointed out by Suzuki.[50]

The constancy of the slope "temperatures" over this projectile energy range is in approximate agreement with earlier results,[33] but differs from those of Jacak et al.[51] where a dependence of T_{NEQ} on bombarding energy is reported. This discrepancy is most likely related to the dependence of moving-source fits on the angular range and low-energy

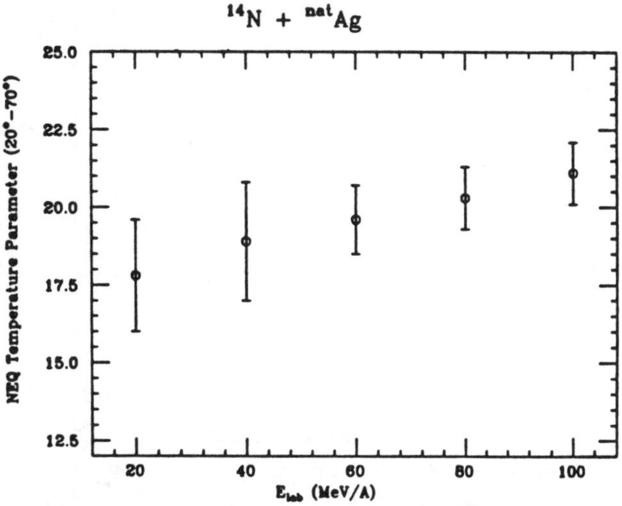

Figure 8. Values of T_{NEQ} as a function of bombarding energy when three-source fit is limited to angular range from 20° to 70°.

thresholds covered by the experiment. For example in Fig. 8 we show a fit to our data that has been constrained to the angular range 20°-70°. Analyzed in this way, one obtains a positive correlation between T_{NEQ} and bombarding energy.

At still higher bombarding energies, systematic inclusive measurements of complex fragment spectra in energetic light-ion-induced reactions have indicated important changes in the mechanism of IMF formation at total bombarding energies in excess of 1 GeV. In addition to rapidly rising cross sections which do not saturate until beam energies of 5-10 GeV are reached, distinct changes in the fragment charge distributions and energy spectra are observed in the vicinity of 1-2 GeV. In Figs. 9 and 10 we show these results for the $^3He + {}^{nat}Ag$ reaction. The power-law parameter τ, which characterizes the charge distribution for the total IMF cross section, is observed to decrease monotonically up to 1.8 GeV total bombarding energy and then remain constant at 2.0 ± 0.1 thereafter. Similar results have been reported earlier for proton-induced reactions.[37] The energy spectra show a corresponding change in spectral slopes in this energy interval, also observed by Trautman et al. in these proceedings.[31] Most noticeable are the broadening of the Coulomb peaks toward lower energies and the hardening of the high energy spectral tails. This behavior is qualitatively expected on the basis of the expanding-evaporating source model, discussed by Friedman[11] in these proceedings. The possibility that multifragment emission may account for these results is discussed in the following section.

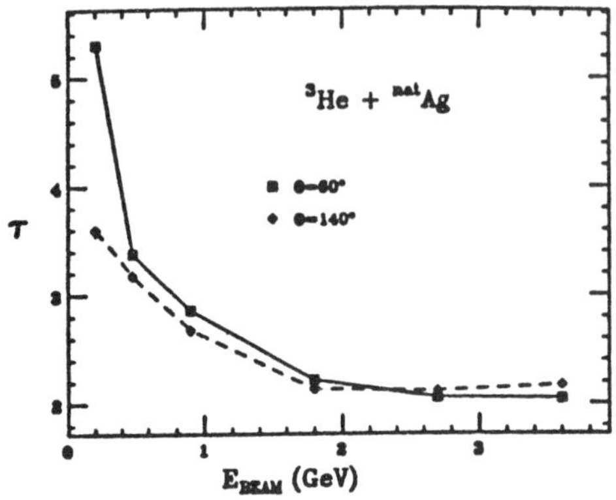

Figure 9. Power-law parameter τ for the ^{3}He + ^{nat}Ag system as a function of bombarding energy. Data are from refs. 20 and 39.

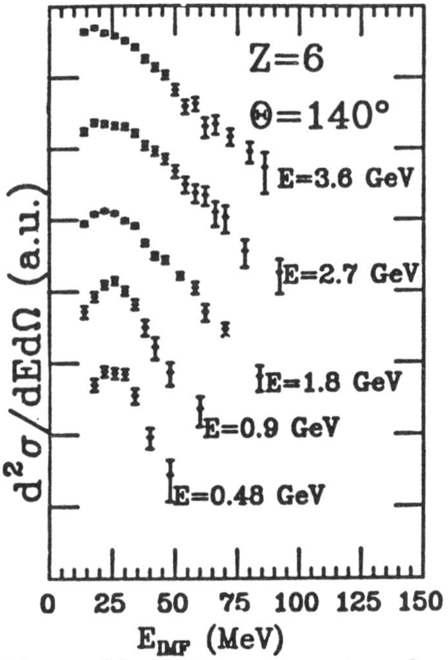

Figure 10. Energy spectra for carbon fragments at 140° from ^{3}He + ^{nat}Ag reaction at several energies.[39]

MULTIFRAGMENT EMISSION

Numerous studies have demonstrated the existence of multiple IMF emission at bombarding energies which exceed the total nuclear binding energy.[26-31] Whether the observed multifragment events are the consequence of sequential statistical binary decays or represent an "instantaneous" breakup of nuclear matter (multifragmentation), remains an important unresolved question.[32] The objective of searches for instantaneous multifragmentation is to understand the response of nuclear matter to extreme conditions of temperature and pressure, thereby shedding light on nuclear equation of state. Of related interest are questions of excitation energy deposition on a fast time scale and collective flow effects associated with transverse momentum transfer.

We next examine results of a study designed to measure multifragment emission at bombarding energies just below and well above the transition region indicated in Figs. 9 and 10, using a 36-detector array. These experiments are described in an earlier report.[52] In Fig. 11 we show the

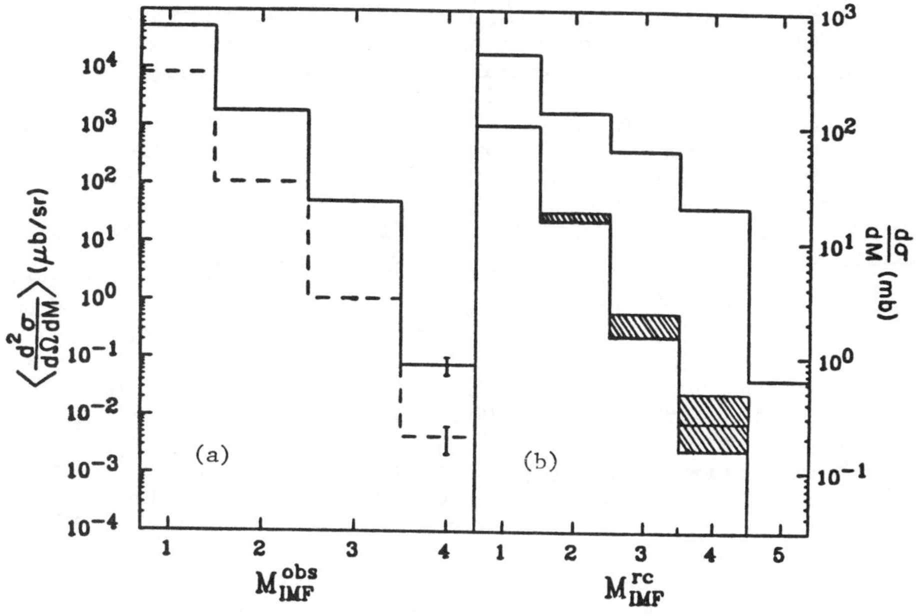

Figure 11. Left: Differential IMF multiplicities for observed multifragment events measured in the ^3He + ^{nat}Ag reaction at 900 MeV (dashed line) and 3.6 GeV (solid line). Right: Monte Carlo reconstruction of data in Fig. 11 to account for detector geometry. Upper line shows results for 3.6 GeV which assumes singles angular distribution.[39] Lower line is for 0.90 GeV; the cross-hatched area indicates effects of including angular distribution assumptions and statistics.

average differential cross sections as a function of observed coincidences (M_{IMF}^{obs}) for IMFs directly measured in this work. The range of IMF charge acceptance is Z = 3-12, and the solid angle for the full 36-detector array is 1.0 sr. At both energies the evidence for multiple fragment events is clear, even with the relatively low solid-angle coverage. In addition to the significant increase in probability for multifragment emission at the higher energy, there is a clear enhancement in the cross section as a function of M_{IMF}^{obs} at 3.6 GeV relative to 0.90 GeV. For example, the cross section for M_{IMF}^{obs} = 3 events is over 40 times larger at 3.6 GeV than at 0.90 MeV.

In an effort to assess the effects of detector geometrical acceptance as well as IMF angular distributions on the observed coincidence distributions, baseline Monte Carlo simulations have been performed. Two assumptions are included in this simulation: (1) experimental angular distributions for the inclusive IMF yields[39] and (2) an isotropic emission pattern. These values of the reconstructed coincidence distribution (M_{IMF}^{re}) are shown in Fig. 1(b) for the data at 0.9 GeV, where these two assumptions diverge most noticeably. This should provide approximately lower limits for the true multiplicity distributions. The reconstructed distributions yield minimum cross sections for multifragment events ($M \geq 2$) of about 220 mb at 3.6 GeV and 20 mb at 0.90 GeV.

To examine the possible sequential nature of the observed multifragment events, calculations were performed with the statistical decay codes EDMON[53] and GEMINI.[54] An average excitation energy of 610 MeV was employed for the post-cascade heavy fragments. This value is based on calculations with the ISABEL cascade code[55] for central collisions at 3.6 GeV within an impact parameter $b \leq 2.6$ fm, which corresponds approximately to the observed cross section for reconstructed multifragment events. Both calculations severely underestimate the multiplicity distributions with self-consistent input values for the cross section and angular momentum distributions.

In Fig. 12 we show the summed IMF charge distributions and the summed IMF kinetic energies for $M_{IMF}^{obs} \geq 2$ events. Light-charged particles are not included in these sums. At 3.6 GeV bombarding energy events are detected which account for up to 75 percent of the total available charge--even with only 8% of total solid-angle coverage. Fig. 12(a) demonstrates that the observed fragments contain significant contributions from higher-Z fragments (but $Z \leq 12$) and are not exclusively due to light elements such as Li and Be. In addition, the total observed kinetic energy per event extends up to 400 MeV, even with light-charged particles excluded from the sums. The solid line in Fig. 12(b) is the result of a calculation based upon sequential statistical emission of IMFs from an equilibrated target-like source with normal nuclear matter density, initially at a temperature of 8 MeV and corrected for detector acceptance. The simulation assumed as input experimental values for the fragment charge distributions[39] and the reconstructed multiplicity results from Fig. 11. Corrections also were included for the velocity of the emitting source, as determined by rapidity plots based on the singles data. In order to assess the upper limit for fragment kinetic energies, cooling by H and He emission was blocked in this calculation. Even for this extreme assumption, the calculated total kinetic energies fail to approach the experimental distribution, suggesting that the fragment acceleration mechanism is more complex than simple Coulomb repulsion from a thermal source.

Comparison of the summed IMF charge and kinetic energy distributions for $M \geq 2$ in Fig. 12 reveals qualitatively

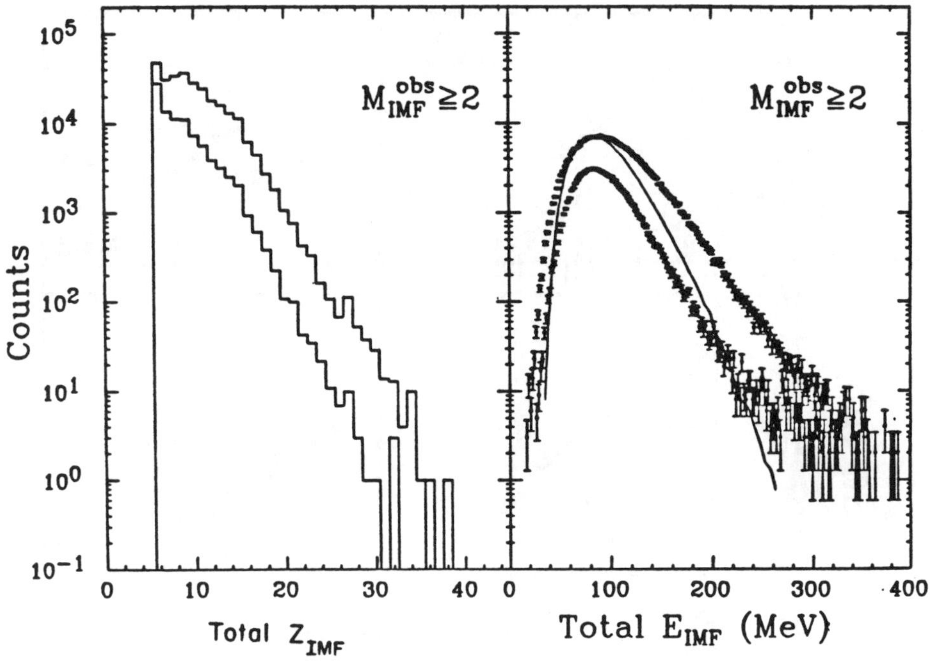

Figure 12. Distributions of summed IMF charge for IMFs with $M_{IMF}^{obs} \geq 2$ for the 3.6 (top) and 0.90 (bottom) GeV $^{3}He + {}^{nat}Ag$ reactions. Distributions of summed detected kinetic energy per event for IMFs for $M_{IMF}^{obs} \geq 2$ for 3.6 (top) and 0.90 (bottom) GeV. Solid line is result of simulation based on sequential statistical emission, normalized to the peak of the 3.6 GeV data.

similar shapes at 900 MeV and 3.6 GeV. This is in contrast to the IMF singles data at these two energies, where distinct changes in the charge distributions and energy spectra have been reported.[39] We interpret this result as due to a major contribution to the IMF singles yields from binary pre-equilibrium processes at 900 MeV, whereas at 3.6 GeV there is a much higher relative probability for multiple fragment emission. However, when multifragment events are selected out, as in Fig. 12, the ejectile behavior is similar at both energies.

One of the most distinctive features of previous inclusive studies (Figs. 9 and 10) was a transition in the shapes of the IMF energy spectra above 1 GeV bombarding energy. Refs. 37 and 56 report similar results for proton-induced reactions. This behavior is examined more quantitatively in Fig. 13, where the kinetic energy spectra

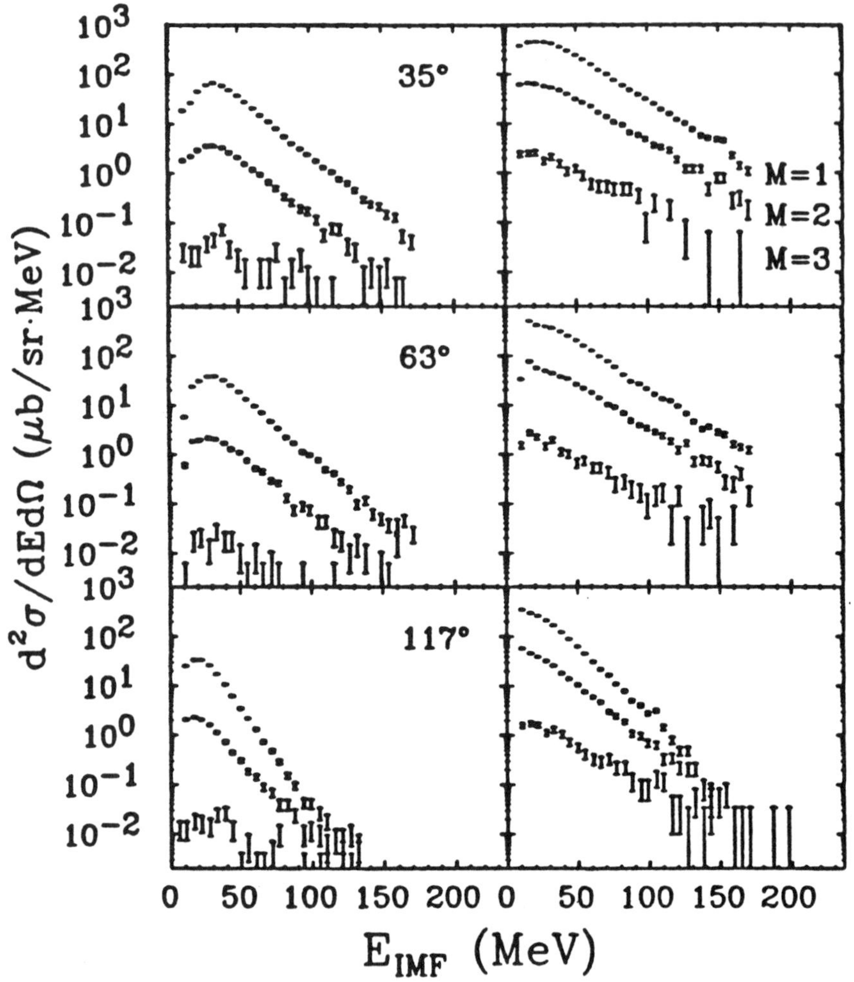

Figure 13. Energy spectra of carbon fragments for 900 MeV (left column) and 3.6 GeV (right column) observed at 35° (top row), 63° (middle row) and 117° (bottom row). In each frame top spectrum is for M_{IMF}^{obs}=1, middle is for M_{IMF}^{obs}=2 and bottom is for M_{IMF}^{obs}=3. Errors are statistical, based on total counts.

for carbon fragments measured at each of the three angles are shown as a function of M_{IMF}^{obs} for both bombarding energies. Similar results are observed for all IMF Z values. Each frame presents spectra gated on M_{IMF}^{obs} = 1, 2 and 3, respectively, from the entire detector array.

When the spectra are examined as a function of M_{IMF}^{obs}, one observes marked changes in which the spectral shapes become increasingly flat with increasing multiplicity. This is most apparent in the backward-angle data. A simple Maxwellian fit to the θ=117° spectra at 3.6 GeV gives slope parameters of T ≅ 13, 15 and 19 MeV for M_{IMF}^{obs} = 1, 2 and 3, respectively. Similar behavior is exhibited by all other complex fragments. In these fits the Coulomb barriers were much smaller than for touching spheres, suggesting expansion of the source[11] and/or significant charge loss prior to fragment emission. For the 900 MeV data, the spectra for high M_{IMF}^{obs} events are found to be much flatter and exhibit Coulomb peaks which extend to much lower energies than for the singles observations. This is consistent with our interpretation that at an energy of 0.90 MeV, the IMF yield is dominated by binary nonequilibrium emission or statistical decay. At 3.6 GeV the probability for multifragment emission is greatly enhanced, thus masking the spectra of simpler binary-like processes and producing a more isotropic emission pattern.

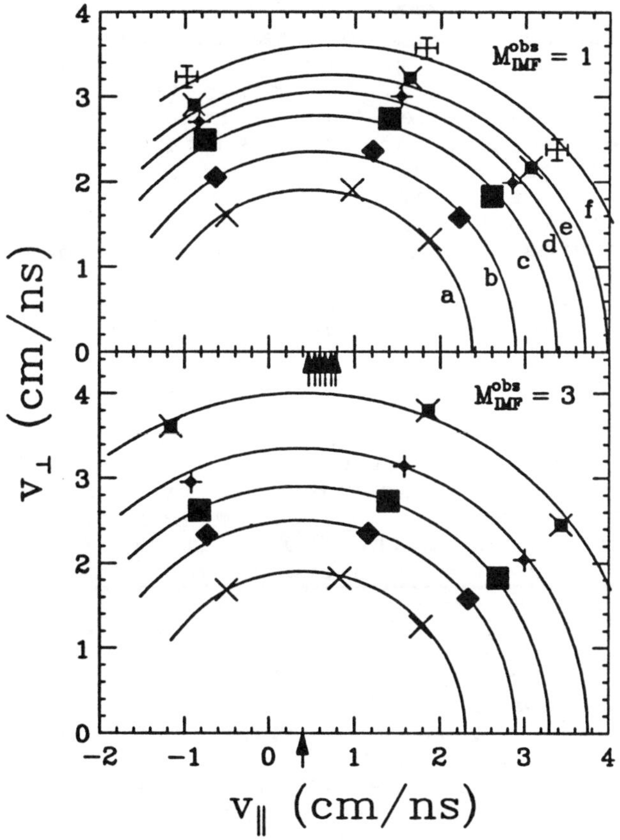

Figure 14. Invariant cross sections for carbon fragments produced in the 3.6 GeV + ^{nat}Ag reaction. Top: Fragments gated on M_{IMF}^{obs}=1; arrows are source velocity centroids of best-fit contour lines through data, sequentially labelled a, b, c, d, e, f. Bottom: Fragments gated on M_{IMF}^{obs}=3; arrow is source velocity centroid for all contour lines.

Another perspective on the influence of multiplicity on reaction observables is illustrated by the rapidity plots shown in Fig. 14. Here we compare invariant cross section contours as a function of longitudinal and transverse velocity components for carbon fragments emitted in the 3.6 GeV bombardment. For M_{IMF}^{obs} = 1 (top) the source velocity appears to increase approximately linearly from $v_s \approx 0.4 - 0.8$ cm/ns as a function of IMF velocity. Similar results have been noted in previous inclusive studies,[57] and are observed in the 0.90 GeV data. In contrast, for M_{IMF}^{obs} = 3 all the data are described by a single source velocity, $v_s \approx 0.4$ cm/ns (bottom). This result implies that the reaction mechanism associated with these high multiplicity events may be distinctly different than a simple sequential decay process. For example, one possible explanation for this result is that collisions in which large transverse momentum transfers occur may be responsible for these high excitation energy events. Earlier studies by Wilkins et al. are consistent with such a picture. Clearly, data of a 4π character and realistic model calculations are required to test this assertion quantitatively.

Finally, comparison of the spectra at the two bombarding energies in Fig. 13. shows that at 3.6 GeV the Coulomb peaks are systematically broadened toward lower fragment energies and the spectral tails are much harder. This behavior indicates that for IMF emission above a bombarding energy of ~1 GeV, the properties of the emitting source are significantly modified relative to the situation at lower energies. This is especially pronounced when IMFs are observed in the backward-angle detectors. It is also consistent with a picture in which the most violent events are selected when IMFs are observed at backward angles.

SUMMARY

From the preceding account, it is clear that the dynamical evolution of nuclear reactions leading to the formation of intermediate-mass fragments is a highly complex process. These events sample not only sources which approach the upper limits of thermal stability for nuclei, but also represent collective phenomena which occur on very short time scales. The theoretical task in creating a unified nuclear reaction model which quantitatively describes these processes is a major challenge -- demanding both mean-field and nucleon-nucleon collision approaches, as well as consideration of the nuclear equation-of-state and (perhaps) subnucleonic degrees of freedom.

In this report we have presented a schematic representation that identifies three well-characterized sources of IMF emission. These include: equilibrium emission from sources produced in central collisions, nonequilibrium emission -- presumably associated with intermediate-impact-parameter collisions, and projectile-like remnants for peripheral reactions with heavy-ion projectiles. A possible fourth source is instantaneous disintegration into multiple IMFs, or multifragmentation. While this phenomenon has not yet been verified experimentally, intensive effort is currently being directed toward its understanding.

The study of these mechanisms as a function of bombarding energy reveals a constant increase in emission probability up to total beam energies of several GeV. Near threshold ($E^* \approx$ 50-100 MeV), the equilibrium and peripheral components are most important. These components appear to persist in the spectra up to relatively high bombarding energies. Near the Fermi energy nonequilibrium emission also becomes important, producing fragments via a fast collective mechanism that appears to originate from a localized region of excitation within a nucleus. As the projectile energy exceeds the total binding energy of the colliding system, multifragment emission becomes observable. At least in part, some of these events must be due to sequential emission processes. However, some signatures in the inclusive data suggest that a new mechanism may also come into play at energies above 1 GeV; e.g. the saturation in the charge distribution parameter τ at $\tau \approx 2$ (Fig. 9); distinct changes in the spectral shapes (Fig. 10); changes in the angular distributions toward greater isotropy,[39] and strong increases in the production cross sections for heavier fragments.

Multifragment emission has been studied in numerous systems with complex detector arrays (see, for example, papers by Hagel, Kim and Trautman, these proceedings). In this review we have presented results for the ^{3}He + ^{nat}Ag system which confirm a strong increase in IMF multiplicity with bombarding energy. These data do not appear to be consistent with a sequential evaporation picture. The energy spectra for high multiplicity events differ markedly from the inclusive results -- especially when one gates on backward-angle events. High IMF multiplicities appear to have flatter spectra and Coulomb peaks that are strongly broadened toward lower energies. These characteristics correspond to those predicted by the expanding-evaporating source model of Friedman.[11] However, better experiments and more complete calculations are clearly required before it will be possible to establish whether these events are true multifragmentation or the compounding of statistical decays from very excited nuclei. Even in the latter event, multifragment emission promises to be a valuable tool for guiding our understanding of excited nuclei at the upper limits of temperature.

The authors wish to thank their many collaborators in the work discussed here. These include: at Indiana

University -- S.J. Yennello, J.L. Wile, M. Fatyga, D.E. Fields, E. Renshaw, K.B. Morley, R. N. Yoder and L.W. Woo; at Michigan State University -- G.K. Gelbke, W.G. Lynch, M.B. Tsang and Sam M. Austin; at CEN Saclay -- E.C. Pollacco, C. Volant, R. Dayras, R. Legrain, Y. Cassagnou, and S. Harrar. We also benefitted by many useful discussions with W. Friedman and W. Bauer. This list is only partial, as many others have been involved in this work; we appreciate their contributions.

REFERENCES

1. R. Wolfgang et al., Phys. Rev. 103, 394 (1956); A. Caretto et al., Phys Rev. 110, 1130 (1958).
2. N.A. Perfilov et al., Soviet Phys. Usp. 3, 1 (1960).
3. V.P. Crespo et al., Phys. Rev. 131, 1765 (1963); R.G. Korteling and E.K. Hyde, Phys. Rev. 136, B425 (1964).
4. J. Hudis, in Nuclear Chemistry (Academia Press, New York and London), 169 (1968).
5. W.G. Lynch, Ann. Rev. Nucl. Sci. 37, 493 (1987).
6. G.F. Bertsch and Das Gupta, Physics Reports 160, 189 (1988).
7. C. Gregoire et al., Nucl. Phys. A 471, 399c (1987).
8. J. Aichelin et al., Phys. Rev. C 37, 2451 (1988).
9. J. Bondorf et al., Nucl. Phys. A 443, 1245 (1985).
10. D.H.E. Gross et al., Phys. Rev. Lett. 56, 1544 (1986).
11. W.A. Friedman, these proceedings and Phys. Rev. C 42, 667 (1990).
12. J. Randrup and S.E. Koonin, Nucl. Phys. A 471, 355c (1987).
13. R.G. Stokstad, Comments on Nucl. and Part. Physics 13, 231 (1984).
14. M. Burgel et al., Phys. Rev. C 36, 90 (1987).
15. B.B. Back et al., Phys. Rev. C 22, 1927 (1980).
16. T.C. Awes et al., Phys. Rev. C 24, 89 (1981).
17. A.M. Poskanzer et al., Phys. Rev. C 3, 882 (1971).
18. D.J. Fields et al., Phys. Rev. C 30, 1912 (1984).
19. R.E.L. Green and R.G. Korteling, Phys. Rev. C 22, 1594 (1980).
20. K. Kwiatkowski et al., Phys. Lett. B 171, 41 (1986).
21. D.E. Fields et al., Phys. Lett. B 220, 356 (1989).
22. M. Fatyga et al., Phys. Rev. Lett. 58, 2527 (1987).
23. B. Borderie et al., J. de Physique 47, 251 (1986).
24. L.G. Sobotka et al., Phys. Rev. Lett. 51, 2187 (1983).
25. M. Fatyga et al., Phys. Lett. B 185, 321 (1987).
26. A.I. Warwick et al., Phys. Rev. C 27, 1083 (1983).
27. Y. Kim et al., Phys. Rev. Lett. 63, 494 (1989); also these proceedings.
28. K. Hagel, these proceedings.
29. J. Peter et al., Proc. 4th Int. Conf. on Nucleus-Nucleus Collisions (Karazawa, Japan, June, 1991), to be published in Nucl. Phys.

30. S.J. Yennello et al., Phys. Rev. Lett. (in press).
31. W. Trautman, these proceedings; C.A. Ogilvie, preprint GSI-91-18 (1991).
32. W.A. Friedman, Phys. Rev. C 40, 2055 (1989).
33. R. Trockel et al., Phys. Rev. C 39, 729 (1989).
34. N. Papadakis et al., Phys. Lett. B 240, 317 (1990).
35. D. Bowman et al., LBL preprint - 27691 (1989).
36. W. Trautman, these proceedings.
37. N. Porile et al., Phys. Rev. C 39, 1914 (1989); M. Mahi et al., Phys. Rev. Lett. 60, 1936 (1988).
38. J.L. Wile et al., Phys. Lett. B (in press).
39. S.J. Yennello et al., Phys. Lett. B 246, 26 (1990).
40. L.G. Moretto, Nucl. Phys. A 247, 211 (1975).
41. G.D. Westfall et al., Phys. Rev. C 17, 1368 (1978).
42. A. Kiss et al., Nucl. Phys. A 499, 131 (1989).
43. S. Kox et al., Nucl. Phys. A 420, 162 (1984).
44. R. Trockel et al., Prog. Part. Nucl. Phys. 15, 225 (1985).
45. A.D. Panagiatou et al., Phys. Rev. Lett. 52, 41 (1984).
46. M.B. Tsang et al., Phys. Rev. Lett. 52, 1967 (1984).
47. M. Fatyga et al., Phys. Rev. Lett. 55, 1376 (1985).
48. T. Batsch et al., Phys. Lett. B 189, 287 (1987).
49. V.E. Viola et al., Phys. Rev. C 31, 706 (1985).
50. T. Suzuki, Proc. 4th Int. Conf. on Nucleus-Nucleus Collisions (Karazawa, Japan, June, 1991), to be published in Nucl. Phys.
51. B.V. Jacak et al., Phys. Rev. C 35, 1751 (1987).
52. S.J. Yennello et al., Proc. 6th Winter Workshop on Nuclear Dynamics (Key West, FL, Jan. 1991), to be published.
53. F. Auger et al., Phys. Rev. C 35, 190 (1987).
54. R. Charity et al., Nucl. Phys. A 483, 371 (1988).
55. Y. Yariv and Z. Frankel, Phys. Rev. C 26, 2138 (1982).
56. A.M. Zebelman et al., Phys. Rev. C 11, 1280 (1975).
57. R.E.L. Green et al., Phys. Rev. C 29, 1806 (1984).
58. B.D. Wilkins et al., Phys. Rev. Lett. 43, 1080 (1979).

TWO-DIMENSIONAL LANGEVIN APPROACH TO NUCLEAR FISSION

N. CARJAN
Centre d'Etudes Nucléaires de Bordeaux, 33175 Gradignan Cedex, France

T. WADA*
RIKEN(Inst. of Physical and Chemical Res.,) Wako-Shi, Saitama 351-01, Japan

Y. ABE
Yukawa Institute for Theoretical Physics Kyoto University, Kyoto 606, Japan

ABSTRACT

The bidimensional Langevin equations for the symmetric fission of ^{213}At were solved without approximations using computer generated stochastic forces. Two dissipation mechanisms have been studied: a typical weak one (hydrodynamical viscosity) and a typical strong one (energy transferred to nucleonic motions from their changing self consistent potential) Calculated transient times (2.10^{-20} sec or less) are comparable with those extracted from measurements of prescission neutron multiplicity. The kinetic energy distribution of the fission fragments was calculated for nuclear temperatures T = 1, 2 and 3 MeV. A good agreement between theory and experiment was found for the variation with T of both its mean value and its variance. The calculated absolute values are however slightly lower than the empirical ones. At the same time it was also possible to test the validity of a deterministic (Lagrange) approach to the mean value of the prescission kinetic energy. The error involved was found to decrease with increasing T and to be less than 10% .

INTRODUCTION

Since the discovery of dissipative phenomena in large amplitude collective motions (like heavy-ion induced deep inelastic or fusion-fission reactions) methods of non-equilibrium statistical mechanics have been extensively applied to nuclear physics. In such processes the energy of a few slowly varying collective degrees of freedom (with large associated inertias) is dissipated among many randomly distributed, rapidly varying nucleonic degrees of freedom. The association with a 'Brownian System' was immediate and the collective and intrinsic variables were viewed as a Brownian particle and a heat bath respectively. In this context, the old Kramer's picture[1)] of nuclear fission as a diffusion over a potential barrier was

* Special Researcher, Basic Science Program

reactualized and the corresponding time-dependent Fokker-Planck equation was solved both analytically (for an inverted oscillator)[2] and numerically .[3] In this way it was possible to calculate new interesting properties of the fission process like transient times and distributions of kinetic-energy or mass.

It was only recently shown[4] that the modern computer technics enable us to solve also a more general transport equation: the Langevin equation and to apply it to the dynamics of nuclear dissipation. Its stochastic force does not allow an analytic treatment. Besides being more intuitive, the Langevin equation has two main advantages over the Fokker-Planck eq. which open numerous possibilities for future applications:

1. It can be generalized to non-Markovian processes by introducing a retarded friction and this makes it more suitable for nuclear physics problems.

2. It is an ordinary differential equation and not a partial differential one. Consequently, increasing the dimensionality from 1 to N increases the computing time only by a factor $= N$ and not by N orders of magnitude as in the case of Fokker-Planck eq. The Langevin approach is therefore more appropriate for the study of nuclear fission which is a typical multidimensional process. To describe it correctly one has to take into account at least three collective coordinates: the separation distance between the nascent fragments, the fragment deformation and the mass ratio. Because of their complexity, the multidimensional Fokker-Planck calculations performed so far have used unjustified Gaussian approximations for the distribution of these collective variables. The bundle propagation method[5] for instance, supposed Gaussian momentum distributions at each point in coordinate space and at every time step whereas the global momentum approximation[6] used Gaussian distributions for both coordinates and momenta but with time dependent mean values and widths. Moreover these calculations did not started at the ground state but at the saddle point where it is more difficult to guess the initial distribution. Such approximations are no more necessary if one chooses the Langevin approach.

The present paper represents a first attempt to solve the bidimensional Langevin equation for the fission process. The values obtained for the transient time and for the distribution of fragment kinetic energy are compared with existing measurements for the fission of ^{213}At. Sect. 2 is a reminder of the multidimensional Langevin equations with the method for their numerical integration explained in Appendix. The macroscopic fission model used (including shape parametrization, potential energy, inertia and dissipation tensors) is described in Sect. 3. Different dissipation tensors are studied in Sect. 4 through their influence on the time-dependent escape rate (transient time). The distribution of the fission-fragment kinetic energy is calculated in Sect. 5 where it is also compared with existing experimental results. Conclusions are drawn in Sect. 6.

MULTIDIMENSIONAL LANGEVIN EQUATIONS

The coupled Langevin equations for N collective coordinates (q_i) and their conjugate momenta (p_i) have the form:

$$\begin{cases} \dot{q}_i = (m^{-1})_{ij} p_j \qquad i = 1, \, .. \, N \, . \\ \dot{p}_i = -\partial V / \partial q_i - \frac{1}{2} \frac{\partial (m^{-1})_{jk}}{\partial q_i} p_j p_k - \eta_{ij} (m^{-1})_{jk} p_k + g_{ij} R_j(t) \end{cases} \tag{1}$$

with summation from 1 to N over repeated indices. V is the potential energy and m_{ij}, η_{ij} represent the shape dependent collective inertia and the dissipation tensors respectively. The coupling of the collective coordinates occurs both through the dependence of these quantities on coordinates and through the off-diagonal elements of m_{ij} and η_{ij}. In order that Eqs(1) provide a suitable description of a 'Brownian motion' m_{ij} has to be much larger than the nucleonic mass (not infinitely large however). In the right hand side of the second equation one recognizes the four forces governing the time evolution: conservative, inertial, continuous damping and stochastic (random) forces. The random force was defined by the following general (averaged over the ensemble) properties:

1. $\langle R_i(t) \rangle = 0$

since $\langle p_i(t) \rangle$should satisfy the deterministic equation.

2. $\langle R_i(t_1) R_j(t_2) \rangle = 2 \cdot \delta(t_1 - t_2) \cdot \delta_{ij}$

which implies a) independent nucleonic collisions with the nuclear surface, both along different axes $(i, \ j)$ and at different times $(t_1, \ t_2)$ b) duration of collision much smaller than the relaxation time for the collective modes. In the present calculations the fission process was therefore supposed to be Markovian. Since one usually assumes that the random force $R_i(t)$ has a Gaussian distribution it is not necessary to define higher order correlations. The strength of the random force is determined from a generalization of the fluctuation-dissipation theorem to the multidimensional case:

$$g_{ik} g_{jk} = \eta_{ij} T$$

where T is the nuclear temperature.

The numerical method used to solve the system of $2N$ coupled equations (1) is presented in the Appendix. One can apply it to any physical problem if the appropriate potential, inertia and dissipation are provided.

SHAPE-DEPENDENT POTENTIAL, INERTIA AND DISSIPATION

For realistic applications a flexible parametrization of the nuclear shape is necessary. The Legendre-polynomial expansion in cylindrical coordinates[7] :

$$\rho^2 = R_0^2 \sum_{n=0}^{N} \alpha_n P_n(\frac{z-\bar{z}}{z_0}) \tag{2}$$

was used since it is able to describe a large variety of fission modes: compact, elongated, extremely mass-asymmetric or even divisions into more than two fragments. The distance between the end points $2z_0$ is obtained from the requirement of volume conservation $\rightarrow z_0 = 2R_0/3\alpha_0$, where R_0 is the radius of the spherical nucleus. α_0 is fixed by the evident geometric condition $\rho^2(z_0) = 0 \rightarrow \alpha_0+\alpha_2+\alpha_4 = 0$. The simple case of the bidimensional symmetric fission was chosen here

$$q_1 = \alpha_2 \quad q_2 = \alpha_4 \quad \alpha_1 = \alpha_3 = 0$$

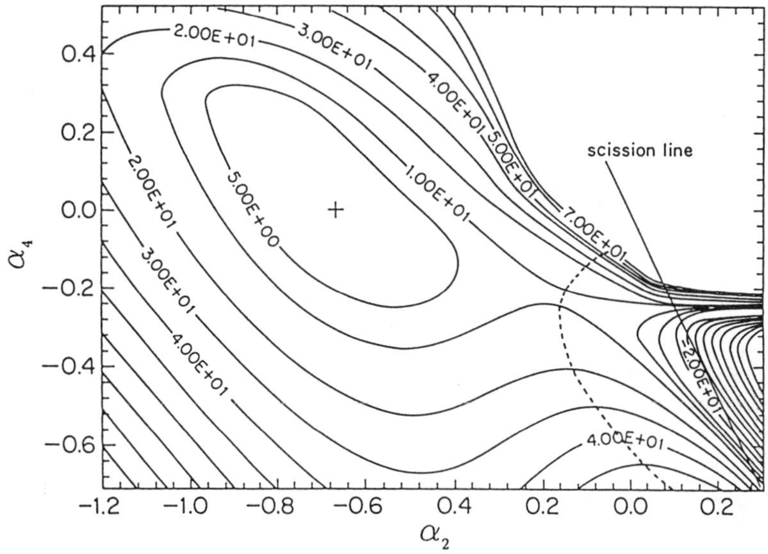

Figure 1. Potential energy as a function of two deformation parameters.

The resulting potential energy of deformation $V(\alpha_2, \alpha_4)$ is represented in Fig.1. It contains two terms[8] :

1. A generalized surface energy:

$$E_s \sim \int\int (\frac{\sigma}{a} - 2)\frac{e^{-\frac{\sigma}{a}}}{\sigma} d^3r d^3r' \tag{3}$$

with an exponential-minus-Yukawa folding function to account for the finite-range a of the nuclear forces. $\sigma = |r - r'|$.

2. A Coulomb energy for a diffuse charge distribution:

$$E_c \sim \int\int \frac{1}{\sigma} d^3r d^3r' + \Delta E_c \ , \tag{4}$$

i.e., that of a sharp charge distribution plus a first order correction due to finite diffuseness ($a_c \neq 0$). The inertia and dissipation tensors have been calculated for an incompressible and nearly irrotational nuclear flow using the Werner-Wheeler method.[9] They are defined as the coefficients of $\dot{q}_i \dot{q}_j$ in the expansions of the collective kinetic energy and of the Rayleigh function respectively:

$$T = \frac{1}{2} \sum m_{ij}(q) \dot{q}_i \dot{q}_j \tag{5}$$

$$F = \frac{1}{2} \sum \eta_{ij}(q) \dot{q}_i \dot{q}_j \tag{6}$$

$2F = dE/dt$ i.e. it gives the rate of dissipation of collective energy into internal excitation. For an incompressible fluid:

$$T = \frac{1}{2} \rho_m \int v^2 d^3r \tag{7}$$

$$F = \frac{1}{2} \mu \int (\nabla^2 v^2 + \omega^2 - 2\vec{\nabla}(\vec{v} \times \vec{\omega})) d^3r \tag{8}$$

where $\vec{v} = (\dot{z}, \dot{\rho})$ is the velocity field and $\vec{\omega} = \vec{\nabla} \times \vec{v}$ is the vorticity. The integrals are over the nuclear volume (where the shape dependence comes from) and their evaluation was performed under a simplifying assumption concerning the flow pattern (nearly irrotational circular layers of fluid). Expressions for m_{ij} and η_{ij} can thus be obtained by the comparison of their definitions (5) and (6) with the explicit results in (7) and (8). This Werner-Wheeler method is considered (at last for inertia) to be approximately valid at high excitation energies and large deformations.

The numerical calculations were performed with a viscosity coefficient

$$\mu = 0.02TP = 1.25 \times 10^{-23} \text{MeV} \cdot \text{s}/fm^3 \ .$$

This value best reproduces the experimental mean fragment kinetic energies for fission at high excitation energy throughout the Periodic Table when deterministic equations ($R_i(t) = 0$) and 5 shape parameters are used like in ref.10.

An alternative, much stronger dissipation mechanism (the so called 'wall and window'[11]) was also used to calculate $\eta_{ij}(q)$. In this case the dissipation arises from:

1. Collisions of the nucleons with the moving wall of the nuclear potential

$$F_{\text{wall}} = \frac{\rho \bar{v}}{2} \oint \dot{n}^2 dS \tag{9}$$

where ρ is the nuclear density, $\bar{v}$ and $\dot{n}$ are respectively, the average nucleonic and normal surface velocities with respect to the drift velocity of the system.

2. Transfer of nucleons through the window separating the nascent fragments:

$$F_{\text{window}} = \frac{1}{8} \rho \vec{v} \delta\sigma (2u_{\parallel}^2 + u_{\perp}^2) \tag{10}$$

where $u_{\parallel}$ and $u_{\perp}$ are the components of the relative velocity of the two fragments perpendicular and parallel to the plane of the window and $\delta\sigma$ is the area of the window.

The transition from mononuclear shapes (around the ground state) where the wall formula applies to dinuclear shapes (between saddle and scission lines) where the wall - and - window formula applies was done through a smooth interpolation[10]:

$$\eta = \sin^2(\frac{\pi}{2}\beta)\eta_{\text{wall}} + \cos^2(\frac{\pi}{2}\beta)\eta_{\text{wall-and-window}} \tag{11}$$

where $\beta = (r_{\text{neck}}/z_0)^2$.

Having defined all the ingredients of the Langevin equation one can now apply it to nuclear fission.

TIME-DEPENDENT ESCAPE RATE

An important result of the Brownian picture for nuclear fission is that the decay constant becomes time dependent; more precisely there is a transient time before the stationary (Kramer's) limit is reached. This affects the competition between neutron emission and fission since additional neutrons can be emitted both during this transient time (t_{tr}) and during the descent from saddle to scission (t_{ssc}). The measured number of these prescission neutrons represents therefore a valuable test for any dynamical model of nuclear fission.

These characteristic times were calculated by solving the system of Langevin equations (1) for a very large number $N_0(\sim 10^6)$ of initial conditions randomly chosen inside with the following distribution:

$$d_0(\alpha_2, \alpha_4, p_{\alpha_2}, p_{\alpha_4}) = \text{const} \cdot \delta(\alpha_2 + 2/3)\delta(\alpha_4)\exp[-\frac{1}{2T} \sum_{i,j=2,4} (m^{-1})_{\alpha_i\alpha_j} p_{\alpha_i} p_{\alpha_j}] \quad (12)$$

This means that, initially, the fissioning system was supposed to be in its ground state shape (sphere) and to have thermally distributed collective momenta.

At each time step, the number N of trajectories (examples) which did not escape (either beyond the saddle or the scission lines) was recorded and the fission rate was calculated

$$\lambda(t) = -\frac{1}{N}\frac{dN}{dt} \quad (13)$$

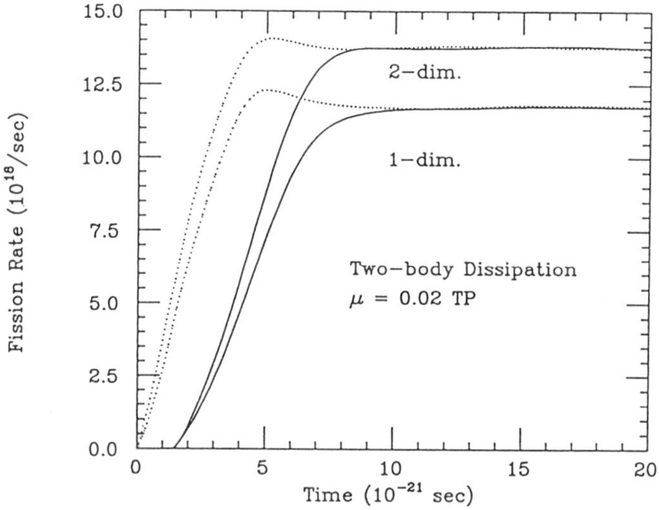

Figure 2. Time dependence of the fission rate for the symmetric fission of ^{213}At. Dashed and solid lines correspond to fission rates at saddle and scission respectively.

The resulting time dependence is plotted in Fig.2 for T = 3MeV and a two-body dissipation coefficient μ = 0.02TP. One notices that in this case t_{tr} and t_{ssc} (calculated at a fission rate level of 90% of its stationary value) are 3.4×10^{-21} sec and 2.8×10^{-21} sec respectively. Fig.3 shows that, as expected these values increase with increasing dissipation strength. The results obtained with the one-body dissipation and T = 3MeV (Fig.4) show indeed much larger transient and saddle to scission times (cca $2 \cdot 10^{-20}$ sec). They are in agreement with those deduced from a recent measurement of prescission neutrons.[12)] The precision for calculating these characteristic times is however not so good as in the case of two-body dissipation since the fission rate is 10 times lower and the statistical error is therefore larger. For the same reason we were not able to calculate the case $T = 1$ MeV (see Tables 1). For large dissipation ($\mu = 0.06$TP or one-body) the time dependence of the fission rate at scission cannot be obtained by simply displacing that at saddle by t_{ssc}. It looks like the trajectories which escape at the beginning of the transient period have shorter saddle to scission times.

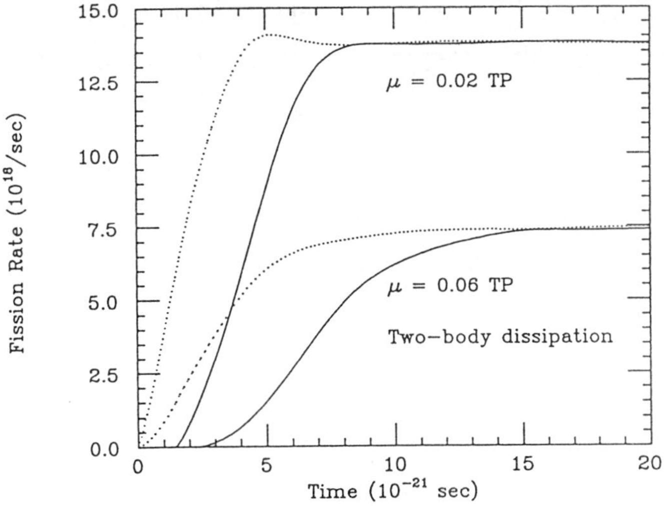

Figure 3. Time dependence of the fission rate of the symmetric fission of ^{213}At. Dashed lines denote the fission rate at saddle and solid lines denote those at scission. Two sets of lines correspond to the friction constant $\mu = 0.02$ and 0.06 TP.

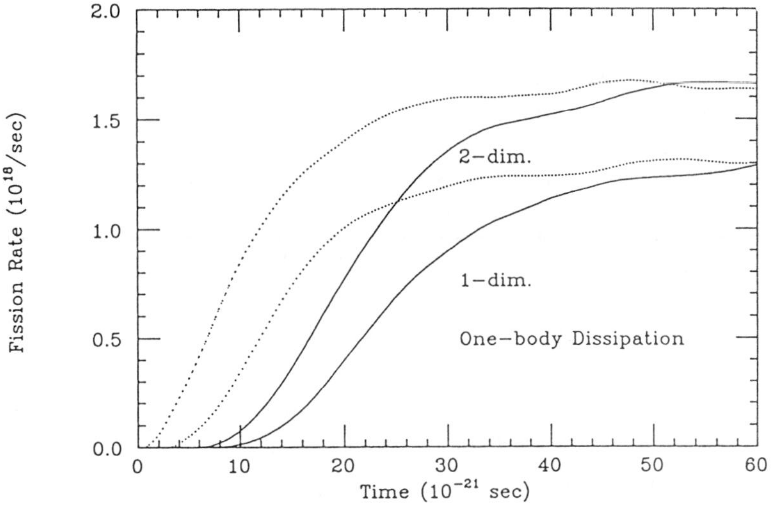

Figure 4. The same as in Fig.2 but for one-body dissipation.

Dissipation	Two-body ($\mu = 0.02$)			One-body	
T	1	2	3	2	3
$\langle E_c \rangle$	126.75	125.37	124.85	128.34	129.20
$\langle K_t \rangle_s$	19.85	21.78	23.12	13.63	14.33
$\langle K_t \rangle_d$	17.53	20.35	22.45	12.32	13.21
$\langle TKE \rangle_s$	146.60	147.15	147.96	141.97	143.53
$\langle TKE \rangle_d$	144.28	145.72	147.30	140.66	142.41
$\langle K_v \rangle_s$	2.65	3.54	4.39	2.02	2.85
$\langle K_v \rangle_d$	1.36	1.86	2.48	0.95	1.52
σ_c^2	48.83	94.15	142.18	101.60	154.67
σ_t^2	25.44	46.86	71.44	39.95	58.95
σ_{TKE}^2	45.89	92.72	144.80	107.72	168.83
σ_v^2	7.67	15.23	23.94	6.82	13.05

Table 1. Calculated mean values and variances of various energies at scission: Coulomb repulsion (c), translational (t) and vibrational (v). $TKE = E_c + K_t$ is total kinetic energy of the fission fragments. The suffices s and d outside of the brackets denote stochastic and derministic respectively.

One-dimensional results were also plotted in Figs.2 and 4 for comparison. The one-dimensional path was chosen to be the mean path of the two-dimensional calculation. (which by the way does not coincide with the bottom of $V(\alpha_2, \alpha_4)$) One should remember that in general it is not possible to find a one-dimensional problem completely equivalent with a two-dimensional one. Although the one-dimensional fission rate rises slower and attains lower stationary values, the characteristic times involved t_{tr} and t_{ssc} are about the same.

TOTAL KINETIC ENERGY DISTRIBUTION

Recording energy values for each trajectory which crosses the scission line and summing up their numbers in each energy interval, one can construct distributions of various energies at scission.

Their mean values are presented in the upper part of Table 1 for the two types of dissipation mechanisms studied. It is interesting to note that with increasing nuclear temperature a) the scission configurations become more elongated for two-body dissipation and more compact for the one body dissipation and b) prescission kinetic energies become larger for both cases. The stronger one body-dissipation leads, as expected, to lower mean values for all kinetic energies but to higher Coulomb repulsion at scission (more compact shape).

Since $\langle (m^{-1})_{ij} p_i p_j \rangle \neq \langle (m^{-1})_{ij} \rangle \langle pi \rangle \langle pj \rangle$ (or $\langle p^2 \rangle \neq \langle p \rangle^2$ in the one dimensional case), the results obtained with an equivalent deterministic approach ($\langle K_t \rangle_d$, $\langle TKE \rangle_d$ and $\langle K_v \rangle_d$) are different although we are talking about mean values. These differences were however found to be relatively small (less than 10%) and decreasing with T.

The lower part of table 1 shows the corresponding variances. They are all increasing almost linearly with T. Although the stronger one body-dissipation leads, to narrower distributions for the prescission kinetic energies E_t and E_v, the distributions of E_c and of the observable TKE are broader.

In the one dimensional case the scission point is fixed and the variance comes only from fluctuations of the pre-scission kinetic energy σ_t^2. On the other hand σ_c^2 is a pure multidimensional quantity and it is seen to be the main contribution to σ_{TKE}^2. From $\sigma_{TKE}^2 = \sigma_c^2 + \sigma_t^2 + 2\sigma_{ct}$ and noticing that $\sigma_{TKE}^2 \simeq \sigma_c^2$ one can infer $\sigma_{ct} < 0$ (anticorrelation).

The only directly observable distribution is that of TKE. There are two sources for its width: 1) The fissioning system crosses the saddle line with a distribution of positions and momenta which leads to different fission paths. This mainly affects E_c but also K_t (slightly) since the amount of dissipation is different along different paths. 2) The random force induces comparable fluctuations both in E_c and in K_t.

The calculated dependences of $\langle TKE \rangle$ and of σ_{TKE}^2 on nuclear temperature are compared in Fig.5 with experimental data from refs.13, 14 and ref.15 respectively. Both the absolute value of the variance and its variation with T are better reproduced by our two-dimensional Langevin calculations when the one-body dissipation tensor is used. The corresponding mean values are however slightly too small. This suggests that the consideration of an even stronger dissipation mechanism is not going to improve the overall agreement further. It is more probable that the remaining small discrepancies are due to the omission of higher order ($n > 4$) or of mass-asymmetry (n = odd) shape parameter (see Eq.(2)) in the present calculation.

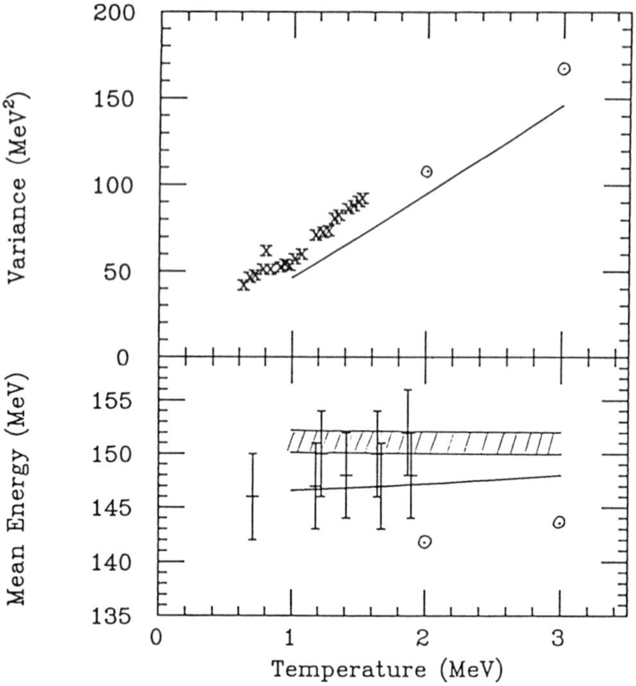

Figure 5. Comparison between calculated and experimental (ref.13 and ref.15) mean values and variances for the fragment kinetic energy as a function of nuclear temperature. The solid lines correspond to two-body dissipation and the circles to one-body dissipation. The shaded zone represent the result of an empirical systematics of $\langle TKE \rangle$.[14)]

CONCLUSIONS

It was possible to study the dynamical evolution of a fissioning nucleus from its ground state to the scission point using Langevin equations for two collective coordinates. Computer generated random forces can be easily and accurately handled in this context so that one can safely foresee the possibility (in the near future) of solving 3- and even 4-dimensional problems (like the asymmetric fission and the ternary fission) or of exploring non-Markovian systems. Calculated transient times (of the order of 10^{-20} sec) are in agreement with those extracted from very recent measurements[12)] of prescission neutrons. It should be pointed out that although a dissipation mechanism stronger than those studied here could explain transient times of the order of 10^{-19} sec (as it was previously suggested[16)]), this would disagree with the measurements of TKE distributions (see below). The two-dimensional transient time seems to be comparable with the one corresponding to the equivalent one-dimensional problem.

The calculated temperature dependence of the mean value and of the width of the TKE distribution is in good agreement with the existing experimental data. However the measured width is about 10% larger and the measured mean value is about 3% larger than the corresponding theoretical values. This is probably due to the mass-asymmetry degree of freedom which was not taken into account in the present calculation and not to a dissipation mechanism stronger than the one-body dissipation: this latter assumption would predict too lower values for $\langle TKE \rangle$ and probably too higher values for σ^2_{TKE} (if one or two more dimensions are added to these calculations.)

It was also possible to test the validity of a deterministic approach (Lagrange) to the mean value of the prescission kinetic energy. Such an approach underestimates this energy by 10% or less depending on the value of the nuclear temperature T.

Y.A. is supported by a Grant-in-Aid for scientific research of the Ministry of Education, Science and Culture in 1991-1993. (Grant No. 03640269)

REFERENCES

1. H.A. Kramers, Physica **7** (1940) 284.
2. P. Grangé, J.Q. Li and H.A. Weidenmüller, Phys. Rev. **C27** (1983) 2063. J.R. Nix, A.J. Sierk, H. Hofmann, F. Scheuter and D. Vautherin, Nucl. Phys. **A424** (1984) 239.
3. P. Grangé, H.C. Pauli and H.A. Weidenmüller, Phys. Lett. **88B** (1979) 9. F. Scheuter and H. Hofmann, Nucl. Phys. **A394** (1983) 477.
4. Y. Abe, C. Grégoire and H. Delagrange, J. Physique **47** (1986) C4-329.
5. F. Scheuter, C. Grégoire, H. Hofmann and J.R. Nix, Phys. Lett, **149B** (1984) 303.
6. G.D. Adeev and V.V. Pashkevich, Nucl. Phys. **A502** (1989) 405c.
7. S. Trentalange, S.E. Koonin and A.J. Sierk, Phys. Rev. **C22** (1980) 1159.
8. H.J. Krappe, J.R. Nix and A.J. Sierk, Phys. Rev. **C20** (1979) 992.
9. K.T.R. Davies, A.J. Sierk and J.R. Nix, Phys. Rev. **C13** (1976) 2385.
10. N. Carjan, A.J. Sierk and J.R. Nix, Nucl. Phys. **A452** (1986) 381.
11. J. Blocki, Y. Boneh, J.R. Nix, J. Randrup, M. Robel, A.J. Sierk and W. Swiatecki, Ann of. Phys. **125** (1980) 193.
12. E. Mordhorst, M. Strecker, H. Frobeen, M. Gasthuber, W. Scobel, B. Gebauer, D. Hilsher, M. Lehmann, H. Rossner and Th. Wilpert, Phys. Rev. **C43** (1991) 716.
13. F. Plasil, D.S. Burnett, H.C. Britt and S.G. Thompson, Phys. Rev. **142** (1966) 696.
14. V.E. Viola, K. Kwiatkowski and M. Walker, Phys. Rev. **C31** (1985) 1550.
15. E.N. Gruzintsev, M.G. Iktis, V.N. Okolovich and G.N. Smirenkin, Sov. J. Nucl. Phys. **39** (1984) 844.
16. D.J. Hinde, D. Hilscher and H. Rossner, Nucl. Phys. **A502** (1989) 497c.

APPENDIX

The numerical method to integrate the system of coupled Langevin equations is presented below for the case in which the inertia m_{ij} and the dissipation η_{ij} tensors depend on collective coordinates $\{q\}$. The system of differential equations can be transformed in a system of difference equations:

$$\begin{cases} p_i(t+\tau) - p_i(t) = \int_t^{t+\tau} h_i(p', q')dt' + g_{ij} \int_t^{t+\tau} \Gamma_j(t')dt' \ , \\ q_i(t+\tau) - q_i(t) = \int_t^{t+\tau} v_i(p', q')dt' \ , \end{cases}$$

with the notations: $p_i' = p_i(t')$, $q_i' = q_i(t')$

$$h_i(p,q) = -\frac{\partial V}{\partial q_i} - \frac{1}{2} p_j \frac{\partial}{\partial q_i} (\frac{1}{m})_{jk} p_k - \eta_{ij} v_j \ ,$$

$$v_i(p,q) = (\frac{1}{m})_{ij} p_j$$

and the following properties of the random force are assumed,

$$\langle \Gamma_i(t) \rangle = 0 \ , \tag{A}$$

$$\langle \Gamma_i(t_1) \Gamma_j(t_2) \rangle = 2\delta(t_1 - t_2)\delta_{ij} \ , \tag{B}$$

$$g_{ik} g_{jk} = \eta_{ij} T \ . \tag{C}$$

Expanding:

$$\begin{aligned} h_i(p', q') &= h_i(p,q) + \frac{\partial h_i}{\partial q_j}(q_j' - q_j) + \frac{\partial h_i}{\partial p_j}(p_j' - p_j) + \cdots \ , \\ v_i(p', q') &= v_i(p,q) + \frac{\partial v_i}{\partial q_j}(q_j' - q_j) + \frac{\partial v_i}{\partial p_j}(p_j' - p_j) + \cdots \ , \\ g_{ij}(q') &= g_{ij}(q) + \frac{\partial g_{ij}}{\partial q_k}(q_k' - q_k) + \cdots \ . \end{aligned}$$

and retaining only the first terms one obtains the solution to first order in τ:

$$\begin{cases} p_i(t+\tau) = p_i(t) + \tau h_i + g_{ij} \tilde{\Gamma}_j^{(1)}(t, \tau) \ , \\ q_i(t+z) = q_i(t) + \tau v_i \ , \end{cases}$$

where $\tilde{\Gamma}_i^{(1)}(t) = \int_t^{t+\tau} dt' \Gamma_i(t')$. We are left with the above integral over the random force (not being continuous and differentiable we couldn't use the same trick). We can however relate it to another gaussian random number:

$$\langle \tilde{\Gamma}_i^{(1)} \tilde{\Gamma}_j^{(1)} \rangle = \int_t^{t+\tau} dt_1 \int_t^{t+\tau} dt_2 \langle \Gamma_i(t_1) \Gamma_j(t_2) \rangle = 2\delta_{ij} \int_t^{t+\tau} dt = 2\tau\delta_{ij} \ ,$$

therefore

$$\tilde{\Gamma}_i^{(1)} = \tau^{1/2} \omega_i^{(1)}(t) \ ,$$

since any linear combination (like an integral) of gaussian random numbers is a gaussian random number. The new random numbers $\omega_i^{(1)}$ have the properties

$$\begin{cases} \langle \omega_j^{(1)}(t) \rangle = 0 \ , \\ \langle \omega_i^{(1)}(t_1) \omega_j^{(1)}(t_2) \rangle = 2\delta(t_1 - t_2)\delta_{ij} \ . \end{cases}$$

Retaining also the other terms and using the first order solution for $(q' - q)$ and $(p' - p)$ one obtains the solution to second order in τ:

$$\begin{cases} p_i(t+\tau) - p_i(t) = h_i\tau + \frac{1}{2}(\frac{\partial h_i}{\partial q_j} v_j + \frac{\partial h_i}{\partial p_j} h_j)\tau^2 + \\ \qquad\qquad g_{ij}\tilde{\Gamma}_j + \frac{\partial h_i}{\partial p_j} g_{jk}\tilde{\Gamma}_k + \frac{\partial g_{ij}}{\partial q_k} v_k \hat{\Gamma}_j \ , \\ q_i(t+\tau) - q_i(t) = v\tau + \frac{1}{2}(\frac{\partial v_i}{\partial q_j} v_j + \frac{\partial v_i}{\partial p_j} h_j)\tau^2 + \frac{\partial v_i}{\partial p_j} g_{jk}\tilde{\Gamma}_k \ , \end{cases}$$

where

$$\tilde{\Gamma}_i^{(2)} = \int_t^{t+\tau} dt'' \int_t^{t+\tau} \Gamma_i(t')dt' = \tau^{3/2}[\ \frac{1}{2}\omega_i^{(1)}(t) + \frac{1}{2\sqrt{3}}\omega_i^{(2)}(t)\ ] \ ,$$

$$\hat{\Gamma}_i^{(2)} = \int_t^{t+\tau} (t' - t)\Gamma_i(t')dt' = \tau^{3/2}[\ \frac{1}{2}\omega_i^{(1)}(t) - \frac{1}{2\sqrt{3}}\omega_i^{(2)}(t)\ ] \ ,$$

with

$$\begin{cases} \langle \omega_i^{(\alpha)}(t) \rangle = 0 \ , \\ \langle \omega_i^{(\alpha)}(t_1) \omega_j^{(\beta)}(t_2) \rangle = 2\delta_{ij}\delta_{\alpha\beta}\delta(t_1 - t_2) \ . \end{cases}$$

A DESCRIPTION OF HEAVY ION INDUCED FISSION BASED ON DIFFUSION MODEL ①

Wu Xizhen,Feng Renfa,Bao Jingdong,Sun Zemin,Zhuo Yizhong

(Institute of Atomic Energy,P.O. Box 275, Beijing, China)

ABSTRACT

In this talk a review on our recent work on fission theory from point of view of diffusion process is presented. The transient phenomena of fission rate are revealed and studied carefully. By comparison of the fission rate calculated both at the saddle and seission points, the latter is emphasized for actual calculation. Our investigation related to the enhancement of the neutron emission prior to fission induced by heavy ion reactions is reported. The quantum aspect of the deffusion model is studied, which is very important for low energy fission. Finally, the extension of one dimensional diffusion model to multi–dimensional cases is also discussed.

I INTRODUCTION

Considerable progress has been made recently in the heavy ion induced fission. Smooth systematical behavior with fissility or excitation energy is observed experimently for total kinetic energy release, fission fragment mass and angular distributions as well as for the multiplicities of the particles emitted prior to fission. Measurements of associate particles demonstrate large prescission probabilities, compared to expectations based on statistical model calculations. The implication of this enhanced emission is that the compound nucleus formed by the colliding heavy ions, moves relatively slowly towards scission, compared to the time scale of particle emission. The particles can be emitted as the collective deformation coordinate starts moving towards the saddle and scission point.

Since the excitation energy in heavy ion induced fission can reach and exceed the fission barrier height, the transient behavior of fission becomes more

---

① * This work is supported by the National Natural Science foundation of China and more important. From the apparently anomalous angular distribution observed in a number of heavy ion induced fissions, it has been shown that the fission width becomes sufficiently large for the composite systems with fission barrier heights comparable their temperature. It seems to come from a fraction of the fission events taking place during the time (order of 8×10^{-21}s comparable to the characteristic relaxation time of the K degrees of freedom. Therfore, in such cases it will be necessary to consider the process to reach the saddle or scission points. However, the standard way to calaulate the fission rate (width), i.e. the Bohr Wheeler formula, is based on the assumption of attaining instantaneous thermodynamic equilibrium between the compound nucleus configuration and the saddle point. It can not fully account for the entire fission process. espeatially the transient process. Thus, as early as in 1940 Kramers(1) suggested to describe the nuclear fission as a diffusion process by solving Fokker–Planck equation (FPE). Nevertheless, he solved FPE only in a quasi–stationary approximation for limiting cases of small and large viscosity coefficients. In 1978 we restudied the old suggestion of Kramers on fission problems(2). Since then the earlier Kramers work has been re–examined and greatly extended by us and various authors(3). The task of this talk is to give a brief report on our recent work in nuclear fission.

II THE CLASSICAL DIFFUSION MODEL

Basic consideration and definition

In the diffusion model of nuclear fission the motion of shape deformation of fissioning nucleus is treated as diffusion of Brownian Particle, the motion of nucleons inside the nucleus as a heat bath and the collision of nucleons with the nuclear wall as a classical random force. Their strength is measured by the viscosity coefficient β. In this model, the shape deformation coordinate x and its canonical conjugate momentum p are treated as classical variables and the distribution function W(x,p) is assumed to obey a Fokker– Planck equation (FPE). In two–dimensional phase space this equation is written as

$$\frac{\partial W(x,p,t)}{\partial t} = -\frac{p}{m}\frac{\partial W(x,p,t)}{\partial x} + \frac{\partial}{\partial p}[(\beta p - F(x))W(x,p,t)] + D\frac{\partial^2 W(x,p,t)}{\partial p^2} \quad (1)$$

Here, $$F(x) = -\frac{\partial}{\partial x}U(x) \quad (2)$$

relates to the potential of the fissioning system U(x), and β is the viscosity coefficient mentioned above. The diffusion coefficient D is given by

$$D = \beta T m \tag{3}$$

where m is the inertia mass of the fissioning system and T is the nuclear temperature which is related to the excitation energy E and the level density parameter α gived by the empirical formula as

$$E = \alpha T^2 \tag{4}$$

Supposed the system is initially given by a narrow distribution, which evolves with time and widens due to random collisions. In this way, the diffusion current across the saddle point x^s to reach the scission point x^{sc} is finally obtained and denoted as:

$$J(x^i,t) = \int_{-\infty}^{\infty} \frac{p}{m} W(x^i,p,t)dp \tag{5}$$

with $x^t = x^s$ or x^{sc}.

Let

$$\pi(x^i,t) = \int_{-\infty}^{x^t} dx \int_{-\infty}^{\infty} W(x,p,t)dt \tag{6}$$

be the probability that the system is to the left of x^i, then the time–dependent fission rate or fission width $\Gamma_f(t)$ can be defined by

$$\Gamma_f(t) = h\lambda_f(t) = hJ(x^i,t) / \pi(x^i,t) \tag{7}$$

For sufficiently large t, and for values of β which are not unreasonably small Γ_f is expected to attain the quasistationary values given by Kramers

$$\Gamma_f^k = \Gamma_f^{BW} \{[1 + (\beta / 2W_0)^2]^{\frac{1}{2}} - \frac{\beta}{2W_0}\} \tag{8}$$

where W_0 is the frequency of the inverse harmonic oscillator potential that corresponds to the fission barrier at the saddle point. Γ_f is the one which often refers to as being given by the transition state method and can be identified with the Bohr–Wheeler expression. It is noted that the modification of the Bohr–Wheeler formula due to Kramers by including a dissipation dependent factor reduces the value of the fission width for finite β. Another fundamental problem in diffusion model is to study the transient behaviour in fission process.

For this we firstly developed a new numerical method [20] for solving exactly one demensional FPE in 1981 and then it was improved by another authors[21] . Although this method is very computation time consuming it is necessery to be developed for examining and checking various approximate methods to study diffusion model of nuclear fission. At the same time several approximations were studied. One of the useful approximation is Locally Harmonic Approximation (LHA)

<u>Time depedent fission rate and transient behaviour of fission</u>

For illustration purpose, we just consider a schematic model with a potential of the fissioning system consisting of two smoothly joined parabolas. The examples given below with parameters specified as: the inertia mass $m=(A/4)m_0$ with m_0 the nucleon mass, and $A=240$. Assuming that m, T and β are all independent of x, we solve the FPE to obtain the fission rates $\lambda_f(t)$ by Eq. (7) shown in Fig.1, for both at saddle point and scission point.

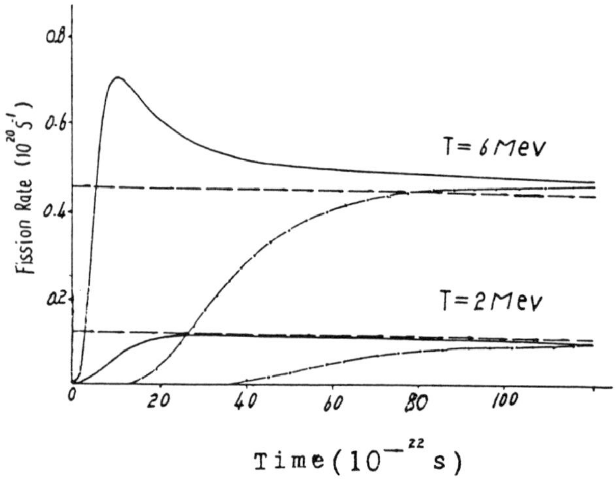

Fig.1. The fission rate (in units of $10^{20}sec^{-1}$) for $E_f=4$Mev, $\beta = 7.5 \times 10^{21}sec^{-1}$ and for two values of T as indicated versus time t (in units of 10^{-22}sec). The full curves are the results at saddle point (λ_f^s)and the dashed–dotted curves are the results at scission point (λ_f^{sc}). The dashed straight lines correspond to the quasi–stationary values of Kramers.

As the solution of the FPE describes the gradual spreading of the initial distribution with the probability current over the fission barrier and finally arriving at scission point,one would expect the trasient process to occur. It is shown in Fig.1., however, the transient behavior at the saddle point is quite different from that at the scission point for different ratio of T/E_f. At the saddle point, for $T<E_f$, $\lambda_f^s(t)$ rises gradually from 0 at $t=0$ to the quasi–stationary value after a time interval τ (which is often denoted as transient time), while for

$T > E_f$, $\lambda_f(t)$ increases rapidly first and decreases later to the quasistationary value. Thus, according to the differences in the transient behavior at the saddle point, one would expect that there exist two regimes of T/E_f. We call $T < E_f$ the Kramers regime, in which the enhancement of the neutron emission prior to fission is anticipated and call $T > E_f$ the overshooting regime, in which the opposite effects are expected. For a realistic description, however, it is also necessary to evaluate the fission rate at the scission point. It is very interesting to see in Fig.1, that in addition to a time delay due to the system to transverse the distance from the saddle point to the scission point, the fission rate λ_f^{sc} (t) always gradually rises to the quasistationary value even in case of $T > E_f$. Thus it seems more reasonable to introduce the transient time τ, the time required for the current to reach, say, 90% of the quasi–stationary value at the scission point. If τ is comparable to the average life time Γ_n / h for neutron emission, where Γ_n is the neutron width, one expects to find the order of one more neutron emitted prior to fission than that calculated with a statistical model.

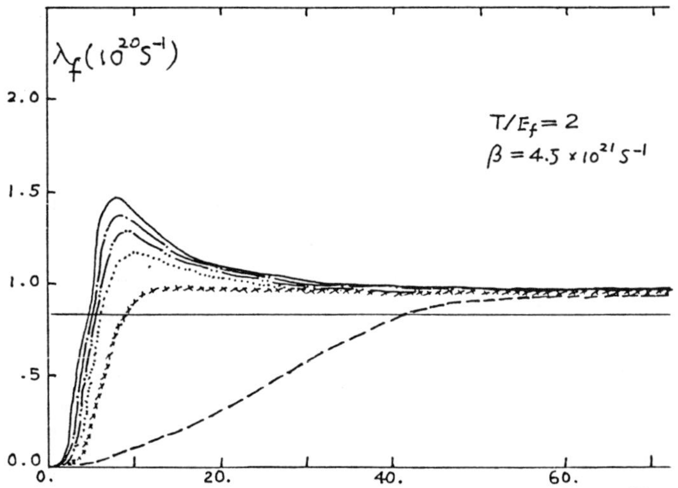

Fig. 2

Recently, we have paid a special attention to the transient behavior of fission rate at the high temperature case ($T > E_f$). Fig.2 shows the time evolution of the fission rate calculated at varous points between the saddle and scission point. The full and dashed curves are calculated at the saddle and scission points, respectively. The others are in between. From this figure we can find that the overshooting gradually disappears as the system approaches to the scission point. But for the system in which the saddle and scission points are close enough, the overshooting might still exist at the scission point. In this case the overshooting will depress the enhancement of the neutron emission. Therefore we should point out that the current opinion that the transient feature of fission rate can explain the enhancement of neutrom miltiplicity may not be appropriate for the regime of $T > E_f$, where the depression of the neutron emission enhancement may become dominant, especially for the system in which the saddle and scission points are close enough. Till now our theoretical prediction is only qualitative. We expect the futher study about this problem both theoretically and experimently.

Ⅲ QUANTUM ASPECT IN DIFFUSION MODEL

The diffusion model discussed in previous section is purely classical one which is applicable only for describing the fission process at high temperature. In order to extend the classical diffusion model to low temperature regime the quantum effects should be included. For this purpose a simple model has been established[17] to reproduce the classical FPE at high temperature and to retain the quantum feature at low temperature case. The model Hamiltonian has three parts:

$$H = H_A + H_B + H_I \tag{9}$$

Where H_A is the Hamiltonian of the subsystem A considered in the problem, that is the quantum Brownian motion in the fission potential. H_B is the Hamiltonian of "heat bath" consisting of N harmonic oscillators, and H_I is a linear coupling term between the subsystems A and B with a coupling strength C_K (k runs from 1 to N). In order to make a connection between this microscopic coupling strength with the macroscopic viscosity and to reproduce the FPE at high temperature, the distribution density of oscillator frequencies for subsystem B is introduced as done, for instance, in solid state physics[21]

$$\rho_D(\omega)C^2(\omega) = \begin{cases} \dfrac{2m_b \eta \omega^2}{\pi} & \omega < \Omega \\ 0 & \omega > \Omega \end{cases} \tag{10}$$

Here, η is the viscosity coefficient, ρ_D the distribution density of oscillators in subsystem B with a high frequency cut off Ω. According to formula (10), a discrete distribution of oscillator frequencies is replaced by a continuous one. This replacement is very crucial in a sense that leads the microscopically reversible process to the macroscopically irreversible one. Hence, the subsystem B has become a real heat bath now.

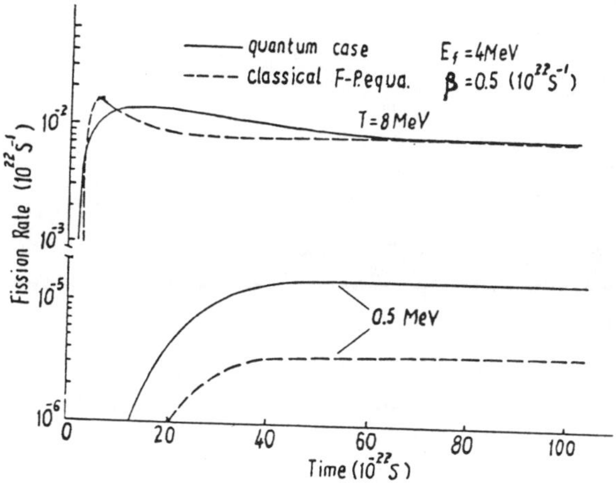

Fig.3. The fission rates for both quantum and classical cases are versus time t. All results are calculated for a fission barrier E_f= 4 Mev and a reduced viscosity coefficient $\beta = 5 \times 10^{21}sec^{-1}$.

So far the potential in the subsystem A is general. In the investigation of the diffusion process it has already been proved that the locally harmonic approximation can be used to solve the FPE for real fission potential quite well except for very small viscosity[16]. It seems reasonable to extend this approach to the quantum case. In this approximation, the fission potential can be divided into many small regions, each of them can be approximated by a harmonic oscillator with the frequencies depending on collective coordinate x. Based on this simplified model and by means of Feynman–Vernon theory an analytical expression for the reduced density operator can be obtained for each small region of the fission potential. The final reduced density operator for the whole fission potential can easily be carried out numerically as done in the classical case within the locally harmonic approximation.

The calculated results of the fission rates at the saddle point are illustrated in Fig.3 for both quantum (solid curves) and classical (dashed curves) cases at temperature of T = 0.5 and 8 Mev. It is seen that at T = 8Mev, the results of both quantum and classical approaches are very similar. This implies that the classical diffusion is the limiting case of the quantum diffusion at high temperature. In the lower part of Fig.3, the results for low temperature of T = 0.5 Mev are shown. Because of the enhancement of the diffusion due to quantum effects the fission rate in quantum case is considerably larger than that in classical case in low temperature region as expected. Thus, this model has improved the behavior of the classical diffusion at low temperature by including the quantum effects and kept its reasonable features at high temperature. Further work concerning its link with the microscopic foundation and comparison with experiments is very much encouraged.

IV MULTI–DIMENSIONAL CASE

Generally speaking, for a description of the shape deformations occuring during the fission process several degrees of freedom are necessary. In this respect, the induced fission is described in terms of a FPE for N degress of freedom. Owing to the difficulty in solving multi–dimensional FPE in a general way, progress has only been made in some specific cases with various approximations. Solving directly a multi–dimensional Langevin equations may be also a effective way for studies of multi–dimensional diffusion process.

Stationary diffusion over a multi–dimensional fission barrier

The Kramers′ formula has been generalized to stationary diffusion over a potential barrier for two or N degrees of freedom[13,9,14]. We have shown that the two–dimensional correction to the fission width Γ_f is mainly due to the geometrical behavior of the potential energy surface based on a rather realistic parametrization. The multidimensional effect is very important in the calculation for fission width. That is to say, within the same parameter space the number of degree of freedom can not be simply frozen and reduced to a fewer number of degree of freedom for the calculation of Γ_f. As demonstrated by Baran that there certainly exists a set of parameters for a restricted number of degree of freedom giving the same value for fission width, those parameters would be effective ones. Thus, in order to calculate Γ_f in one–dimensional approximation, it is necessary first to obtain the effective parameters with same physical relevance in accordance with the N–dimensional case.

Fission–fragment kinetic energy and mass distributions

In order to describe the kinetic energy or mass distribution of the fission fragments more than two degrees of freedom for shape deformations are necessary. Recently we have performed this kind of calculation based on (c,h,α) parameterization[19]. Two deformation parameters of c and h describing the elongation and necking. The parameter α characterizes the left right asymmetry of the mass.

The variances of mass distribution of fission fragments from a number of excited nuclei (the temperature of these nuclei at the saddlc point $T_s \approx$ 1.5Mev) have been calculated based on FPE which are shown in Fig.4. (dashed curves). Some preliminary results calculated with Langevin equation are also shown there (full triangles).

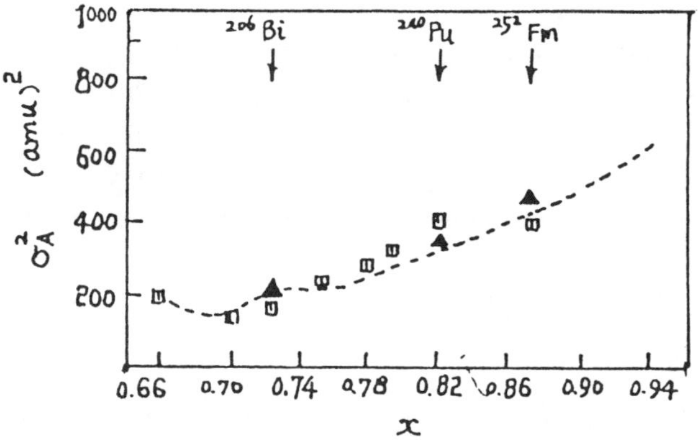

Fig.4. The variances of mass distribution versus fussibility parameter x. The squares correspond to experimental values[22,23]. The dashed curves represent the calculated results taken from Ref.18, while the full triangles are also the theoretical values taken from Ref.19,.

It is seen that the calculations can reproduce the experimentally observed increase of the mass variance with the growth of the fissibility parameter x. This increase of the calculated variances is mainly due to a considerable decrease in the stiffness coefficients corresponding to the saddle points for heavy fissioning nuclei and to the non–equilibrium character of the decent from saddle to scission.

The variances of the kinetic energy distribution calculated both with FPE and Langevin equation [19]are shown in Fig.5.

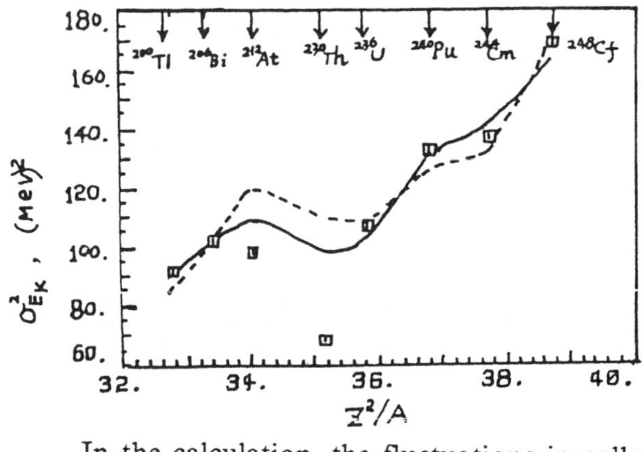

Fig.5. The variance of the kinetic energy distribution versus parameter Z^2/A. The squares represent the experimental data[24] while the dashed and solid curves correspond to the calculations based on FPE[18] and on Langevin equation[19], resppectively.

In the calculation, the fluctuations in collective coordinates near scission and prescission kinetic energy have been taken into account simultaneously. Owing to the fact that heavy fissioning nucleus can attain greater elongation without neck rupture than the light one the fluctuation in collective coordinate rapidly increases with z^2/A. As a result the variances increase with Z^2/A are understandable, since the fluctuations in collective coordinates contribute the main part to the variances. The calculations again are in agreement with the experimental data.

Regarding the extension to multi–dimensional calculation based on diffusion model, it seems that the Langevin equation which is equivalent to FPE has the advantage in saving the computer time and is more convenient and feasible to compute for cases with coordinate dependent parameters by Monte–Carlo simulation method.

V CONCLUSION

The recent progress in nuclear fission theory based on diffusion model has been briefly reported. The basic assumption of this model lies in the fact that the single–particle degrees of freedom are equilibrated (10^{-22}sec) much faster than the relaxation time of deformation shape (10^{-21}sec).

The standard statistical and dynamecal models are two limiting cases of difusion model.

The results obtained so far with this model are encouraging. This provides us with a good basis for further studies.

Up to now only the classical one dimensional problems have been investigated rather thoroughly, further studies on quantum effects and extension to multi–dimensional cases are needed.

It is hopeful that within a few years we can eventually establish a self– consistent theory to treat all fission problems in an unified way, which might also be useful in nuclear data evaluation.

References

1. Kramers. A, Physica 7,284 (1940)
2. Wu Xizhen, Zhuo Yizhong, Studying Nuclear Fission Rate from Point of View of Brownian Motion', Report on the Annual National Confereuce of Chinese Physical Society, 1978, Lushan, China, Chin.Phys.(New York), 1, 671(1981)

3. Wu. X. Z, Zhou. Y. Z, Zhang. X. Z, Yang. Y. H, Ma. Z. Y and Fang. R. F , Comun. Theor. Phys. (Beijing), 1, 769(1982)
4. Grange. P, Li. J. Q and Weidenmuller. H. A, Phys. Rev., C 27,2063 (1983)
5. Hofmann. H and Nix. J. R, Phys. Lett., 122B 117(1983)
6. Scheuter. F, Hofmann. H Nucl. Phys., A 394,477(1983)
7. Nix. J. R,Sierk. A. J, Hofmann. H, Scheuter. F, and Vautherin. D , Nucl. Phys., A 424, 239(1984)
8. Feng. R. F, Zhou. Y. Z and Li. J. Q, Chin. J. Nucl. Phys., 6,113(1984)
9. Hu Jimin and Zhong Yunxiao, Chin. Phys., 2,157(1982)
10. Adeev. G. D and Gonchar. I. I , Z. Phys., 320,451, 322,479 (1985)
11. Lu. Z. D , Feng. R. F, Zhang. J. S, and Zhuo. Y. Z, Z. Phys., A 323,477(1986)
12. Wu Xizhen, Feng Renfa Zhuo Yizhong,Chinese Phys. Lett., 5,1(1988)
13. Wu Xizhen, Zhuo Yizhong Chin. Phys., (New York), 1, 693(1981)
14. Weidenmuller H.A. and Zhang, J.S. Phys. Rev., C29, 879(1984)
15. Feng Renfa, Wu Xizhen Zhuo Yizhong, Chin. Phys. (New York) , 9,850,9,774(1989)
16. Sun. Z. M , Wu. X. Z , Feng. R. F , Zhuo. Y. Z , Xin. J. R and Zhang. S. F , (to be published)
17.Wu. X. Z , Li. Z. X, Maruhn. J. A , Greiner. W and Zhuo. Y. Z , J. Phys. G. Nucl. Phys., 14, 1049(1988)
18.Ageev. G. D, Gonchar. I. I, Marchenko. L. A , Pishasov. N .I ,Yad. Fiz. (USSR), 43,1137(1986) 46,710(1987)
19.Bao. J .D , Zhuo. Y. Z , (to be published)
20. Yang Yinghui, Wu. Xizhen Peng Qingquan, Commun in theor. Phys., 1, 597(1982)
21. Xing Jingru, Zhang Shufa, (to be published).
22.Gruzintsev. Y. N , Itkis. M.G, Okolovich.V .N , Rusanov. A. Ya , Tolstikov. V. N , Smirenkin. G. N , Z. Phys., A 316,61(1984)
23.Sahm. C. C, Schulte. H, Vermeulen. D , Clerc. H. G , Schmidt. K. H , Heß berger. F , Munzenberg. G, Z. Phys., A 297, 2(1980)
24.Lazarev. Yu. A , At. Evergy Rev., 15,75(1977)
25.Baran. A et. al. Nucl. Phys., A361, 83(1981)

DISAPPEARANCE OF THE GIANT DIPOLE RESONANCE IN HOT NUCLEI*

J. Kasagi
Department of Physics, Tokyo institute of Technology, Oh-okayama,
Meguro, Tokyo 152, Japan

ABSTRACT

The properties of the GDR in hot nuclei were investigated by measuring high energy γ-rays in heavy ion fusion reactions. It has been confirmed that the width of the GDR increases rapidly for $E_x \geq 100$ MeV and our previously introduced $\Gamma(E_x)$ can explain the data well for $80 \leq E_x \leq 120$ MeV. The limiting temperature of the nucleus for existing of the GDR has been deduced for A ~ 70, 105 and 130, and it is concluded that the GDR disappears above the temperature of about a quarter of the resonance energy.

INTRODUCTION

It is interesting to study the properties of the GDR at very high temperature where, in a naive picture, the thermal random motion of nucleons may destroy a coherent collective motion such as the GDR. So far, the width of the GDR has been reported by several authors in various excitation energy region. In the low excitation energy region ($E_x \leq 150$ MeV), the spreading of the GDR strength is considered to be mainly due to the effect of the angular momentum [1-3]. Bracco et al.[4] analyzed the γ-ray spectra from ^{110}Sn at $E_x \sim 240$ MeV and concluded that the GDR width stays almost constant for $E_x \geq 150$ MeV. It was claimed that the saturation in width is due to the limitation of the angular momentum brought into the compound nuclei. Gaardhoje et al.[5] measured γ-rays for the ^{70}Ge+^{40}Ar reaction at E/A = 15 and 24 MeV and suggested that the GDR γ-ray decay is suddenly quenched above $E_x \sim 350$ MeV.

These results seem to indicate that the width of the GDR depends weakly on the thermal excitation energy.

In our previous work[6], we have measured yields of decay γ-rays from hot nuclei produced in the $^{92}Mo+^{40}Ar$ reaction at E/A = 21 and 26 MeV. In order to explain observed saturation of the excitation function of the yield for γ-rays with energies between 12 and 20 MeV, we have introduced rather strong energy dependence of the width described as $\Gamma(E_x) = 4.8 + 0.035E_x + 1.6\text{x}10^{-8}E_x^4$. This strong energy dependence leads to the conclusion that the GDR disappears at $E_x \sim 200$ MeV due to excessive broadening of the resonance.

Therefore, there are serious discrepancies on the GDR for $E_x \geq 100$ MeV. As mentioned, Bracco et al. claimed that the width of the GDR stays constant above $E_x \geq$ 150 MeV due to the saturation of the angular momentum, while we proposed that the width increases as the excitation energy increases. Although it is agreed qualitatively that the GDR disappears at very high temperature, Gaardhoje et al. suggested the critical excitation energy, above which the GDR decay γ-rays cannot be seen, to be much higher than the one we obtained. Thus the questions to be solved are those; Does the width of the GDR increase strongly as a function of the excitation energy?; Above what energy (temperature) does the GDR disappear?

In order to answer the questions, we have done a series of experiments at JAERI and at RIKEN. At the Tandem Accelerator Facility at JAERI, we have measured the GDR decay γ-rays as a function of the excitation energy between 80 and 120 MeV, in order to test the validity of the excitation energy dependence of the width introduced empirically in ref. 6. For very high temperature, we have carried out the measurements of high energy γ-rays at the RIKEN Ring Cyclotron (RRC), in order to obtain mass dependence of a limiting excitation energy for the existence of the GDR.

APPARENT WIDTH OF THE GDR

The difficulty of the analysis of the decay of highly excited nuclei may cause the disagreement mentioned above. Since the decay of fusion nuclei is considered as a statistical process, the nuclei emit γ-rays according to the decay probability from the initial decay step. Although the branching ratio of the γ-ray decay is very small ($\sim 10^{-3}$), the γ-rays are emitted in competition with particle evaporations at every decay step.

Therefore, experimentally obtained γ-ray spectra include information not only on the initial fused nuclei but also on all nuclei in a decay chain starting from the fused nuclei. If the resonance energy (E_G) and the width (Γ_G) of the GDR do not depend on the excitation energy and the mass number, one can do good job to determine the resonance energy and the width by comparing the spectra with the statistical model calculations. In the statistical calculation, the γ-ray decay probability is given as the product of the photo absorption cross section and phase space factors including the level density. Since the photo absorption cross section is usually approximated to be a Lorentzian function, one only has to fix the values of parameters E_G and Γ_G in the calculation. The conclusion of ref. 1,4 and 5 is deduced based on this kind of analysis. As pointed out in the previous section, the resonance width does depend on the excitation energy. Thus, the obtained width with the calculation with the fixed Γ_G is not the width at the initial excitation energy but the averaged one, and is better to be called apparent width.

In ref. 6, we calculated the γ-ray spectra with changing the width as a function of the excitation energy. Comparing the results of both calculations, we found the following. The apparent width is always narrower than the width given at the initial excitation energy when the width increases as the excitation energy increases. The apparent width does not increase and seems to be saturated above a certain excitation energy where the real width becomes so large that the strength function becomes almost structureless. Therefore, the saturation of the apparent width does not necessarily correspond to the saturation of the real width. The saturation is expected not only for the apparent width but also for the γ-ray yield of the GDR, as a result of the increase of the width with increasing of the excitation energy. Thus, it is very important to measure the excitation function of the GDR yield for the excitation energy region where the apparent width is being saturated.

THE GDR AT $E_x \sim 100$ MEV

In order to see the evolution of the apparent width and the GDR yield for mass number around 130, γ-rays were measured in the $^{32}S+^{100}Mo$ reactions at incident energies (E_i) of 150, 180 and 210 MeV, at the Tandem Accelerator Facility at JAERI. These incident energies correspond to excitation energies of 80, 103 and 125 MeV, respectively, in ^{132}Ce. High energy γ-rays were detected with a large BaF_2 detector consisting of 7 hexagonal crystals, placed at 90°. Background events due to cosmic-rays were rejected by surrounding plastic scintillators serving as veto counters. The γ-ray multiplicity associated with the high energy γ-rays was measured with a multiplicity-filter array consisting of 25 BaF_2 detectors. Angular momenta brought into compound nuclei can be deduced from the γ-ray multiplicity data. The γ-ray multiplicity is related to the number of fired detectors (N_γ) in the multiplicity detector array, and hence, observed γ-ray spectra were gated on N_γ in the later off-line analysis. Low multiplicity events were found to include non-negligible contributions of the peripheral reaction process, so that the events with $N_\gamma \geq 9$ were analyzed and discussed.

In Fig. 1, the γ-ray spectra for $N_\gamma \geq 9$, which correspond roughly to the inclusive spectra, are shown for each incident energy. As seen, the spectra deviate from an exponentially decreasing function above $E_\gamma \sim 10$ MeV, because of the GDR decay. The statistical model calculations were performed using the code CASCADE[7]. In the , a single Lorentzian function was used as the strength function of the GDR. The solid lines in Fig.1 are the results of the calculation with fixed values of E_G and Γ, which are determined so as to reproduce the spectra. With only two adjustable parameters, the calculations explain the spectra very well. The obtained E_G and Γ for each incident energy are plotted at the right side of Fig. 2. The fact that the values of E_G are in good agreement with each other indicates that the resonance energy does not change for different excitation energies, and is consistent with the results so far reported[2,8]. On the other hand, Γ does depend on the incident energy; it increases as the incident energy increases. It should be noted that the real width of the GDR increases more rapidly than the obtained apparent width Γ.

Results of the similar analysis for the spectra gated on N_γ are also shown in Fig. 2. As shown, E_G seems to decrease slightly as N_γ increases. Γ seems to stay constant for $E_i = 150$ and 180 MeV, but it decreases as N_γ increases for $E_i = 210$ MeV. This is just the opposite of the conclusion of ref. 2 and 3, which claims that the increase of the angular momentum causes the increase of the width. Actually, the angular momentum distribution of the initial compound nucleus is calculated to be very similar for $E_i = 180$ and 210 MeV, because of the limitation set by the fission decay. Therefore, the increase of the apparent width observed in the present work is mainly not due the angular momentum but due to the thermal excitation energy.

In Fig.3, the apparent width (open circles) obtained from the spectra for $N_\gamma \geq 9$ is plotted as a function of the excitation energy, together with the apparent width (cross) obtained in the $^{92}Mo + ^{40}Ar$ reaction at E/A = 21 MeV[6]. As the value of Γ for higher excitation energies is the same as the one plotted at Ex ~ 240 MeV, the apparent width is saturated at the excitation energy between 150 and 230 MeV. The dashed-dotted line in Fig. 3 is the energy dependence of the apparent width given by $\Gamma = 4.8 + 0.0026E_x^{1.6}$ for $E_x < 130$ MeV reported in ref. 2. It explains well the present result of A ~ 130, although the dependence was deduced for the data of A ~ 110. As described in ref. 6, we have introduced empirically the excitation energy dependence of the width so as to explain this energy dependence of the apparent width together with its saturation for $E_x \geq 240$ MeV. Therefore, the excitation energy dependence of the width $\Gamma(E_x) = 4.8 + 0.036E_x + 1.6x10^{-8}E_x^4$, which is shown also in Fig. 3 with a solid line, can well explain the evolution of the observed apparent width from $E_x \sim 80$ to $E_x \geq 240$ MeV for A ~ 120. As can be seen in Fig. 3, difference between the real width given by $\Gamma(E_x)$ and the obtained apparent width is getting larger and larger as the excitation energy increases. Even at $E_x \sim 130$ MeV, the difference is more than 2 MeV.

In Fig. 4, γ-ray yields of the GDR region ($12 \leq E_\gamma \leq 20$ MeV) are plotted against the excitation energy. The data below $E_x \sim 150$ MeV (squares) are from the reaction $^{100}Mo + ^{32}S$, while those for $E_x \geq 240$ MeV (circles and diamonds) are from ref. 6. The nuclei emitting the GDR decay γ-rays cannot be exactly same for whole excitation energy region because of the difference not only of the reactions but also of the decay steps. Nevertheless, we can discuss the general trend of the excitation function of the GDR decay γ-rays for A ~ 120. As expected, the γ-ray yield is increasing with the increase of the excitation energy for $E_x \leq 150$ MeV. However, the increment of the yield between $E_x \sim 120$ and $E_x \sim 100$ MeV is about a half of that between $E_x \sim 100$ and $E_x \sim 80$ MeV. This indicates that the yield is being saturated above $E_x \sim 100$ MeV and finally stays constant above $E_x \sim 240$ MeV. Considering the fact that the apparent width strongly increases for E_x between 80 and 130 MeV and is saturated for E_x above 240 MeV, we can naturally interpret the evolution of the GDR yield as a result of the rapid increase of the real width of the GDR with the increase of the excitation energy. The solid line in Fig. 4 is the result of the calculation for the $^{92}Mo + ^{40}Ar$ reaction with the $\Gamma(E_x)$ described above. The calculation reproduces the trend of the γ-ray yield for lower excitation energies as well.

In Fig. 5, we compare the γ-ray energy spectra for $N_\gamma \geq 9$ with the calculation with the $\Gamma(E_x)$ as shown with solid lines. The calculation can explain the experimental spectra very well for all the incident energies as already expected from the discussion above. Therefore, summarizing the results for $E_x \sim 100$ MeV, we can say that the width of the GDR depends strongly on the thermal excitation energy rather than the angular momentum, and the excitation energy dependence of the width introduced in the previous work[6] can explain the evolution of the GDR for A ~ 120. The present conclusion supports our previous interpretation that the GDR gradually disappears due to excessive broadening at higher temperature.

LIMITING TEMPERATURE FOR EXISTING OF THE GDR

In order to obtain systematic information on the disappearance of the GDR at very high temperature, high energy γ-rays were measured in coincidence with fusion-like residues in Ni, ^{92}Mo, ^{122}Sn + ^{40}Ar reactions at E/A = 26 MeV. The experiments were carried out at the RIKEN Ring Cyclotron. The fusion-like residues were detected in an annular parallel plate avalanche counter (PPAC) placed at 40 cm from the target in forward direction as described in ref. 6, which provided both a measurement of the time-of-flight of residues and their energy losses. The high energy γ-rays were detected with a BaF_2 2π detector system consisting of 80 BaF_2 scintillators, which covered the hemisphere from 30° to 210° with respect to the beam axis. In order to verify the excitation energies of the fused nuclei, neutrons were also measured in coincidence with the fusion residues in a separate run. Seven NE213 scintillators were placed at angles between 30° and 160° and at a distance between 100 and 180 cm from the target.

Fusion events were selected by setting a gate on the two-dimensional spectrum (time versus energy) of residues detected with the PPAC. A good separation between peripheral collisions and the fusion reactions was seen in the two-dimmensional spectrum. In order to test the initial excitation energy, the neutron spectra obtained by setting the velocity gates on the fusion residues were analyzed. The data were fitted with two thermal moving sources, which correspond to the fusion nuclei and to the pre-equilibrium neutron emission[9]. The initial excitation energies of the thermally equilibrated nuclei were estimated with the amount of the transfered linear momentum and the multiplicity and the energy of emitted particles at the pre-equilibrium stage. In Fig. 6, the deduced temperature of the fusion nuclei are plotted against the excitation energy. The data plotted with squares, circles and diamonds are obtained with Ni, ^{92}Mo and ^{122}Sn targets, respectively. The deduced temperature, as well as the multiplicity of neutron, increases smoothly with increasing excitation energy, as expected for a statistical decay from an equilibrated system. The obtained temperature-excitation energy relation is well described with a level density parameter of a = 9/A as shown with a dashed line in Fig. 6.

In Fig. 7 are shown the γ-ray spectra which were measured with the BaF_2 2π detector system and gated on all fusion events in the $^{92}Mo + ^{40}Ar$ reaction. The spectra being multiplied by 10 for each angle are shown for eight different angles between 50° and 160°. As seen, the spectrum consists roughly of three exponentially falling components; γ-ray emission after particle decay ($E_\gamma \leq 10$ MeV), the GDR γ-ray decay competing with particle evaporation ($10 < E_\gamma < 20$ MeV), the bremsstrahlung production during the collision ($E_\gamma > 30$ MeV). The bremsstrahlung is considered to be originated in the proton-neutron collisions in the early stage of the reaction, and hence the velocity of the emission source is expected to be about a half of the velocity of the projectile. The other two components correspond to the decay of the fused nucleus so that the velocity of the source is just the observed one with the PPAC. The solid lines are the result of the moving source fit with the three sources. The simple exponential functions with different slope parameters were assumed for the shape of spectra from each emitting source. As shown in Fig. 7, the spectra can be well fitted with this simple assumption. Contributions of the bremsstrahlung were determined in this way.

The emitted γ-rays from the hot nuclei were calculated, firstly, by using the code CASCADE with fixed Γ, and were compared with the spectra. Such comparisons are shown for the spectra at 90° obtained in the Ni + ^{40}Ar, $^{92}Mo + ^{40}Ar$ and ^{122}Sn + ^{40}Ar reactions, respectively, in Figs. 8, 9 and 10. Dashed curves in Figs. 8-10 show the result of the calculation with Γ = 15 MeV and solid curves with Γ = 20 MeV. Dotted lines show the contributions of the bremsstrahlung. The initial excitation energies of the fused systems used in the calculation are varied so as to reproduce the spectra, and the values are indicated in the figures. If the γ-ray decay is suddenly quenched above these energies, we can call them critical excitation energy, above which the nucleus does not emit γ-rays. The critical excitation energies determined are 230±30, 230±30 and 190±20 MeV for the $^{122}Sn+^{40}Ar$, $^{92}Mo+^{40}Ar$ and $Ni+^{40}Ar$ reactions, respectively. Since the width increases with the increase of the excitation energy, the critical excitation energy deduced by this method should be considered to be an upper limit of the critical excitation energy for the existence of the GDR.

As discussed in the previous section, the width of the GDR strongly depends on the excitation energy and the neglect of the excitation energy dependence cannot be justified. However, the excitation energy dependence is not known precisely, except for mass number about 120 where the calculation with the empirical $\Gamma(E_x)$ obtained in ref. 6 can well reproduce the spectrum at $E_x \leq 120$ MeV, as discussed in the previous section. Therefore, we applied the same function to the calculations for the $^{122}Sn+^{40}Ar$ and $Ni+^{40}Ar$ reactions, also. In Fig. 11, γ-ray spectra gated on all fusion events are shown for each target. Solid curves in Fig. 11 are the results of the calculations with $\Gamma(E_x) = \Gamma_0 + \alpha E_x + \beta E_x^4$, where Γ_0 is the width of the GDR of the ground state. As seen, a satisfactory agreement is obtained for the $^{122}Sn + ^{40}Ar$ reaction with the same values of the coefficients α and β as those for the $^{92}Mo + ^{40}Ar$ reaction. For the Ni + ^{40}Ar reaction, however, slightly larger values (~ 3 %) than those for the $^{92}Mo + ^{40}Ar$ reaction were required. The critical excitation energy for the existence of the GDR cannot be defined in this case, since the GDR disappears not suddenly but continuously. Therefore, in stead of the critical excitation energy, we had better define the limiting excitation energy or temperature above which the width of the GDR becomes larger than 30 MeV and thus the GDR does not exist any more.

Table 1 Deduced limiting excitation energy (E_l) and temperature (T_l) for GDR

Reaction	A_f	E_l(MeV)		T_l(MeV)	
		fixed Γ^a	$\Gamma(E_x)^b$	fixed Γ^a	$\Gamma(E_x)^b$
$^{122}Sn+^{40}Ar$	130	230±30	180	3.8±0.2	3.3
$^{92}Mo+^{40}Ar$	105	230±30	180	4.1±0.2	3.6
$Ni+^{40}Ar$	72	190±20	140	4.6±0.2	3.9

$^a\Gamma = 15 \sim 20$ MeV. $^b\Gamma = \Gamma(g.s.) + \alpha E_x + \beta E_x^4$; $\Gamma > 30$ MeV.

DISCUSSION

The limiting excitation energies and temperatures deduced in this way are listed in Table 1 together with the values obtained by the analysis with the fixed Γ. The corresponding temperatures are calculated by using a simple relation $E_x = A/8\ T^2$. The analysis with the fixed Γ always gives larger values than that with the energy dependent

Γ, since the former just gives the upper limit as already pointed out. The deduced limiting temperature (T_l) decreases as the mass number increases as shown in Fig. 12. The mass dependence seems to follow $A^{-1/3}$ as presented with the solid line in the figure. Since the resonance energy is approximated to be $80A^{-1/3}$ MeV, it can be said that the GDR disappears at the temperature of about a quarter of the resonance energy.

The γ-ray spectra for $E_x > 80$ MeV can be explained well with the statistical model calculation which requires the dependence of the GDR width only on the excitation energy. The disappearance of the GDR is realized quite naturally as a consequence of the rapid increase of Γ above $E_x \sim 100$ MeV. However, the fact that the GDR is described by a single Lorentzian function in the present analysis means that a change of Γ includes all effects; increases of the direct escape and the spreading width, and a separation and fluctuations of the resonance energy. The separation and fluctuations of the resonance energy are related to a change and fluctuations of the nuclear shape, which are predicted to be associated with a change of the angular momentum[10]. Our observation clearly shows that Γ slightly depends on the angular momentum at $E_x \sim 100$ MeV. Therefore, a strong increase of the spreading width mainly contributes to the increase of Γ, since the direct escape width is expected to be about 1 MeV.

The present result shows that the limiting temperature for the existence of the GDR decreases with increasing the mass number. It is interesting to compare this tendency with the lifetime of the hot nuclei. In Fig. 13, equi-lifetime lines are shown on the temperature vs. mass number plot, together with the deduced limiting temperatures. As seen, the limiting temperature of the nucleus for the GDR existing corresponds to the lifetime of about $5 - 8\text{x}10^{-22}$ sec. Thus, we can say, in other words, that the nucleus with the lifetime less than $5\text{-}8\text{x}10^{-22}$ sec, which is accidentally almost equal to the time of a period of the GDR vibration, cannot decay through the GDR vibration. One can also argue that the lifetime obtained is a characteristic time for the hot nuclei to have the coherent collective motion. This characteristic time should be related to the spreading width of the GDR, which corresponds to the decay time of the coherent collective motion to complicated thermally excited states. However, the time obtained from the spreading width of about 30 MeV is much shorter than the lifetime of the hot nuclei.

So far, many theoretical works have shown that the GDR is very stable for being heated up. None of the calculations predicted the strong increase of the width for the temperature up to ~ 6 MeV. Very recently, however, a Vlasov type calculation by Smerzi et al.[11] has shown that the width increases dramatically due to the enhancement of nucleon-nucleon collisions.

This work was supported in part by the Grant-in-Aid for General Scientific Research (No. 02640219) by the Ministry of Education, Science and Culture.

*This work has been done with the following collaborators: The experiment at RRC were performed with K.Yoshida[1], H. Hama[2], K. Furutaka[3], K. Ieki[4], W. Galster[5], T. Kubo[1], M. Ishihara[1,6], and A. Galonsky[7]. The experiment at JAERI were performed with K. Furutaka[3], T. Murakami[3], A. Yajima[3], M. Ohshima[8], S. Niiya[9], and H. Tominaga[10]. The institutions are follows:
[1]RIKEN, Wako-shi, Saitam 351-01, Japan
[2]UVSOR, Institute for Molecular Science, Myodaiji, Okazaki 444, Japan
[3]Department of Physics, Tokyo Inst. of Tech., Oh-okayama, Meguro, Tokyo, Japan
[4]Department of Physics, Rikkyo University, Toshima, Tokyo 171, Japan
[5]Chemin du Cyclotron, Univ. Catholique de Louvain, Louvain-La-Neuve, Belgium
[6]Department of Physics, Univ. of Tokyo, Hongo, Tokyo 113, Japan
[7]NSCL, Michigan State Univ., East Lansing, Michigan 48824, USA
[8]Department of Physics, JAERI, Tokai-mura, Ibaraki, Japan
[9]Department of Physics, Kyushu Univ., Hakozaki, Fukuoka, Japan
[10]Department of Physics, Hiroshima Univ., Hiroshima, Japan

REFERENCES

1. J.J. Gaardhoje et al., Phys. Rev. Lett. 56 (1986) 1783.
2. D.R. Chakrabarty et al., Phys. Rev. C36 (1987) 1886.
3. M. Gallardo et al., Nucl.Phys. A443 (1985) 415.
4. A. Bracco et al., Phys. Rev. Lett. 62 (1989) 2080.
5. J.J. Gaardhoje, et al., Phys. Rev. Lett. 56 (1986) 1783.
6. K. Yoshida, et al., Phys. Lett. 245B (1990) 7.
7. F. Puhlhofer, Nucl. Phys. A280 (1977) 267.

8. K. Snover, Ann. Rev. Nucl. Part. Sci. 36 (1986) 545.
9. D. Hilscher et al., Phys. Rev. C36 (1987) 208.
10. P.F. Bortignon et al., Nucl. Phys. A495 (1989) 155c.
11. A. Smerzi, A. Bonansera and M. Di Toro, private communication.

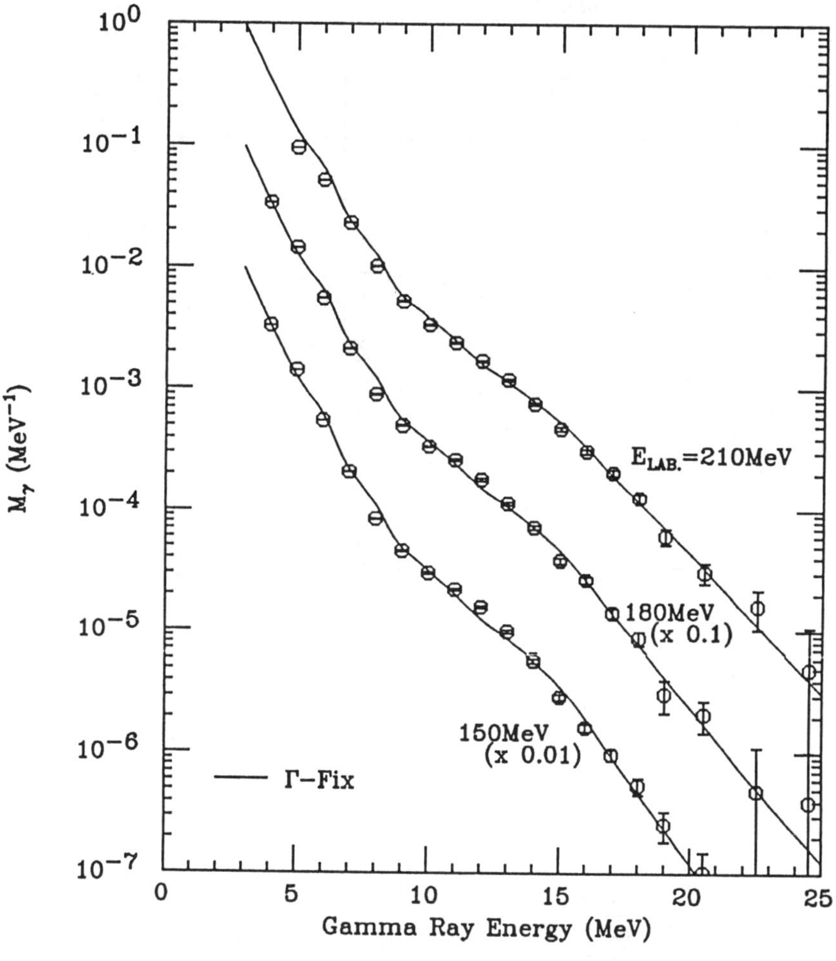

Fig. 1. Gamma-ray spectra for $N_\gamma \geq 9$ obtained in the $^{32}S + ^{100}Mo$ reactions. The solid curves are the results of the statistical model calculations with fixed E_G and Γ.

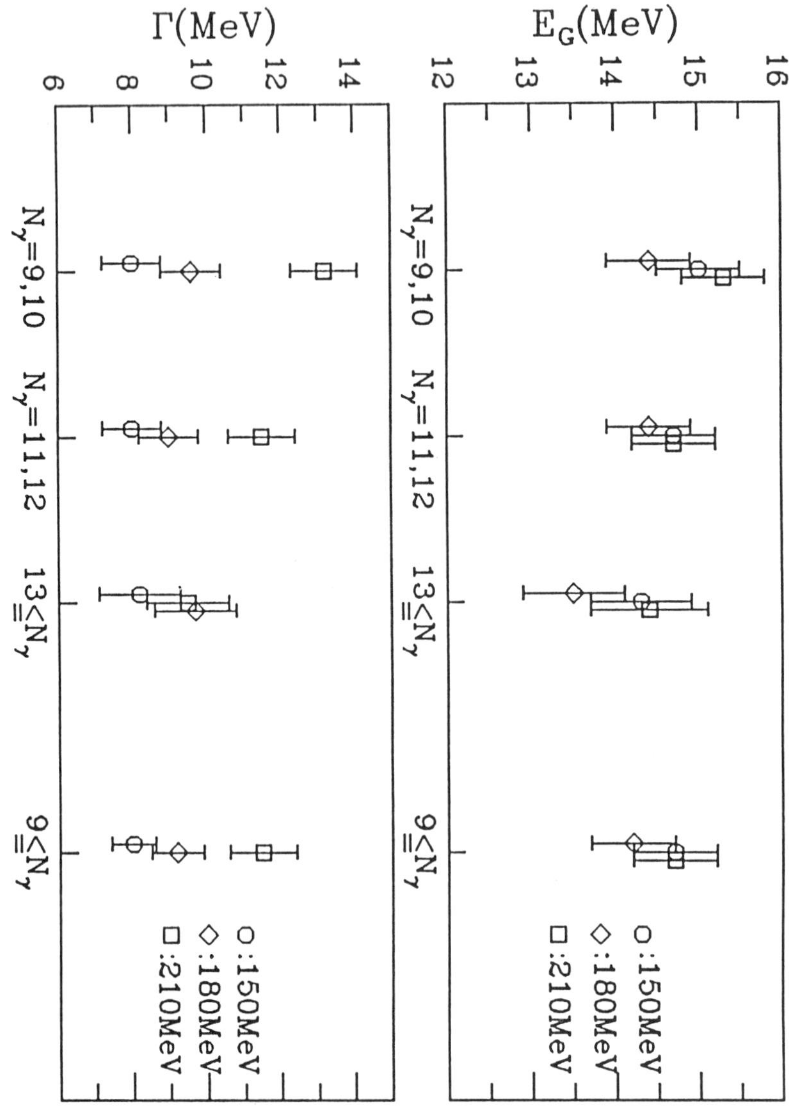

Fig. 2. Deduced resonance energy (E_G) and apparent width (Γ) against N_γ.

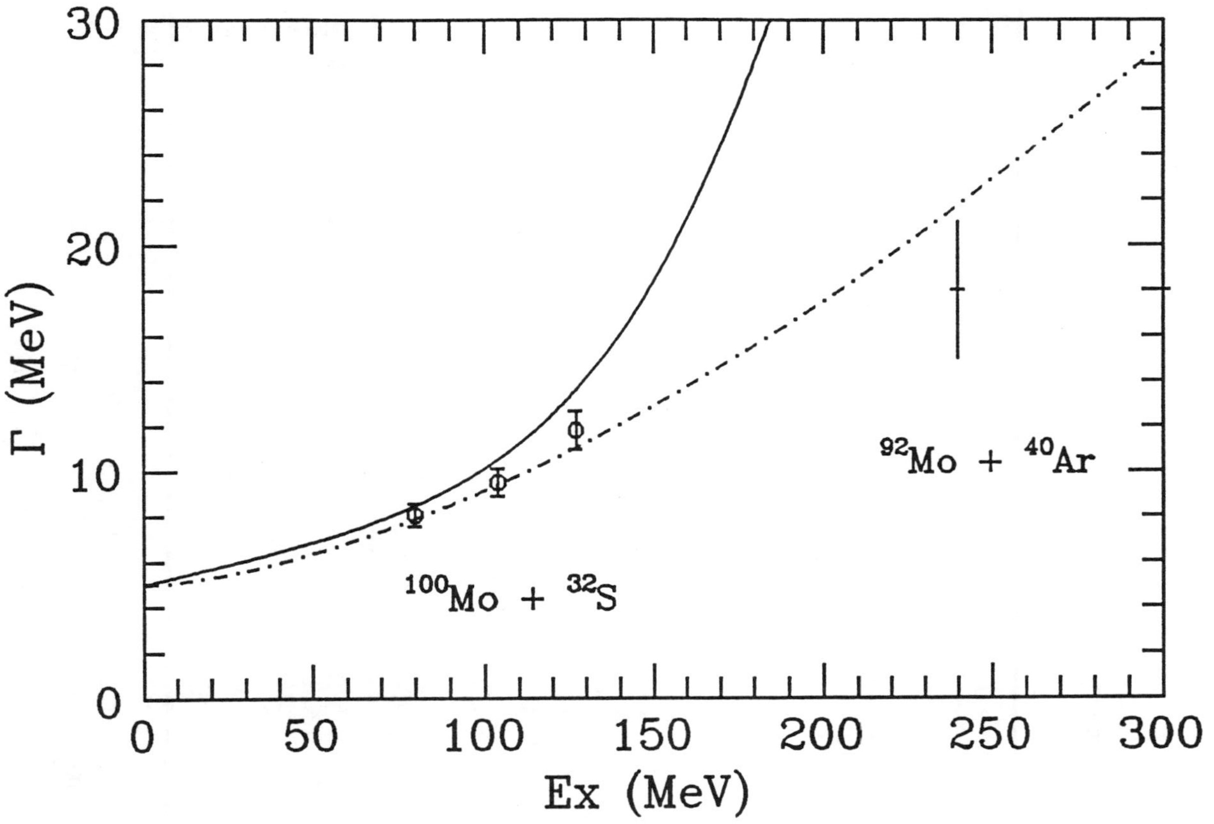

Fig. 3. Apparent width obtained in the ^{100}Mo + ^{32}S reaction (circles) and in the ^{92}Mo + ^{40}Ar reaction (cross) against excitation energy.

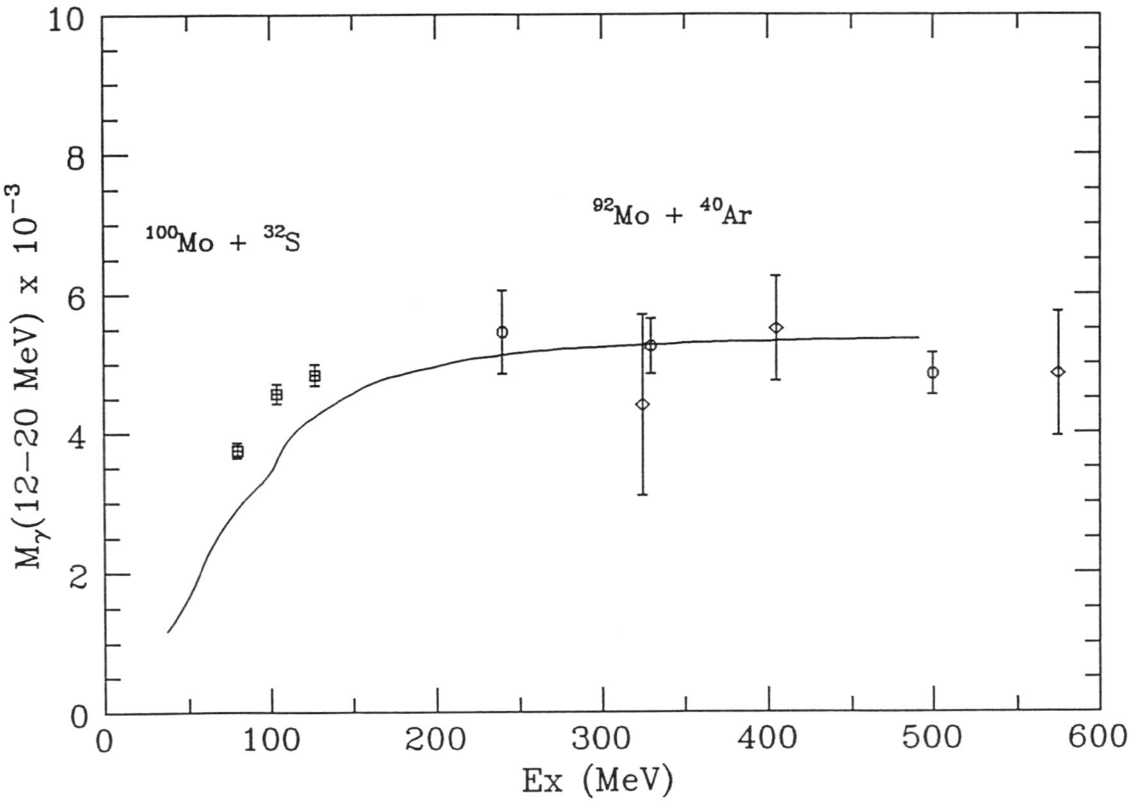

Fig. 4. Gamma-ray yield between 12 and 20 MeV obtained in the $^{100}Mo + ^{32}S$ reaction (squares) and in the $^{92}Mo + ^{40}Ar$ reaction (circles and diamonds) versus excitation energy.

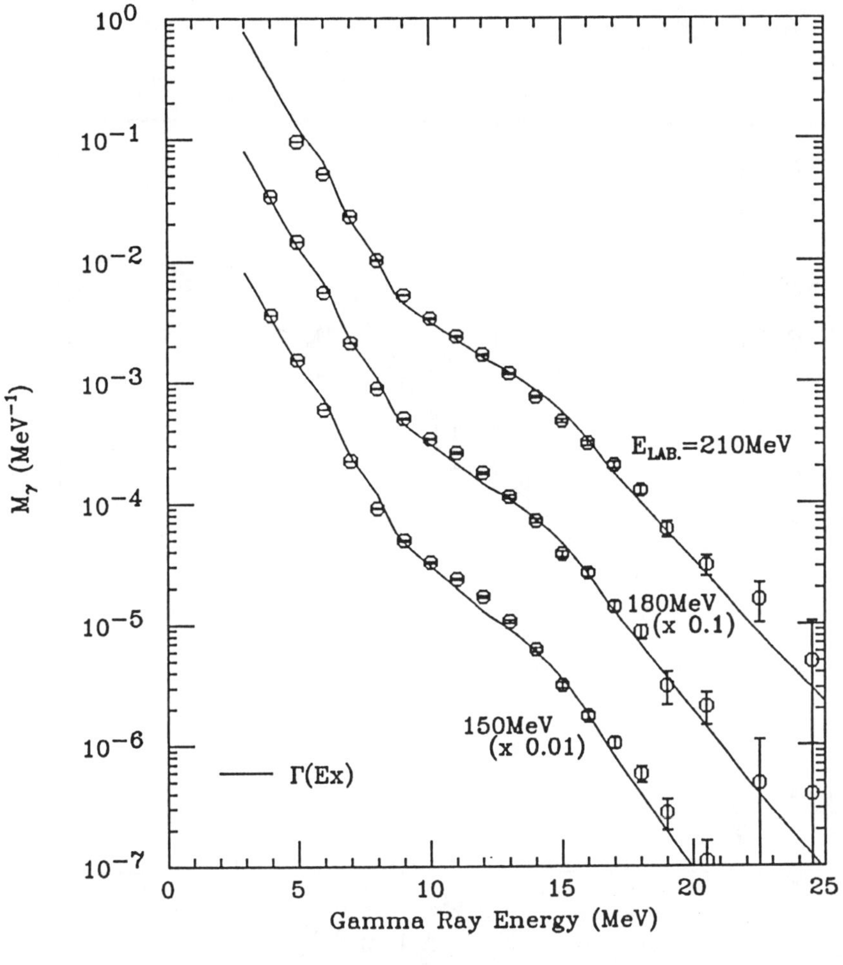

Fig. 5. Gamma-ray spectra for $N_\gamma \geq 9$ obtained in the ^{32}S + ^{100}Mo reactions. The solid curves are the results of the statistical model calculations with $\Gamma(E_x)$.

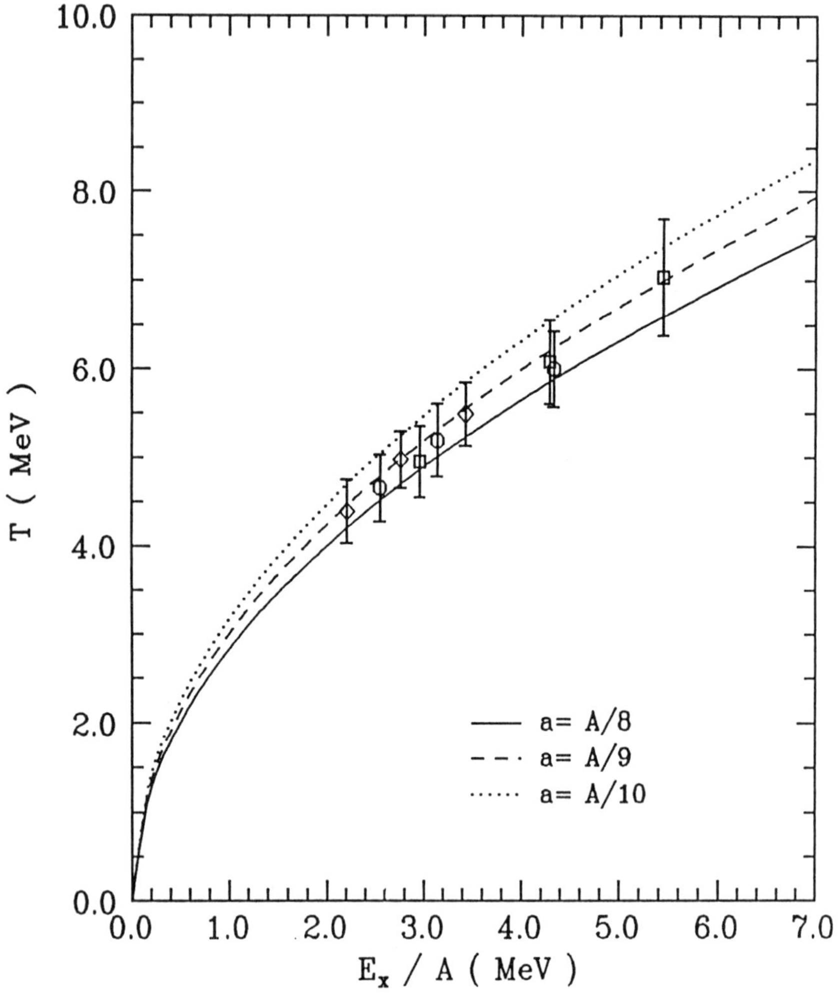

Fig. 6. Deduced initial temperatureof the fusion nuclei against excitation energy. The data plotted with squares, circles and diamonds are obtained in the ^{40}Ar + Ni, ^{92}Mo and ^{122}Sn reactions, respectively.

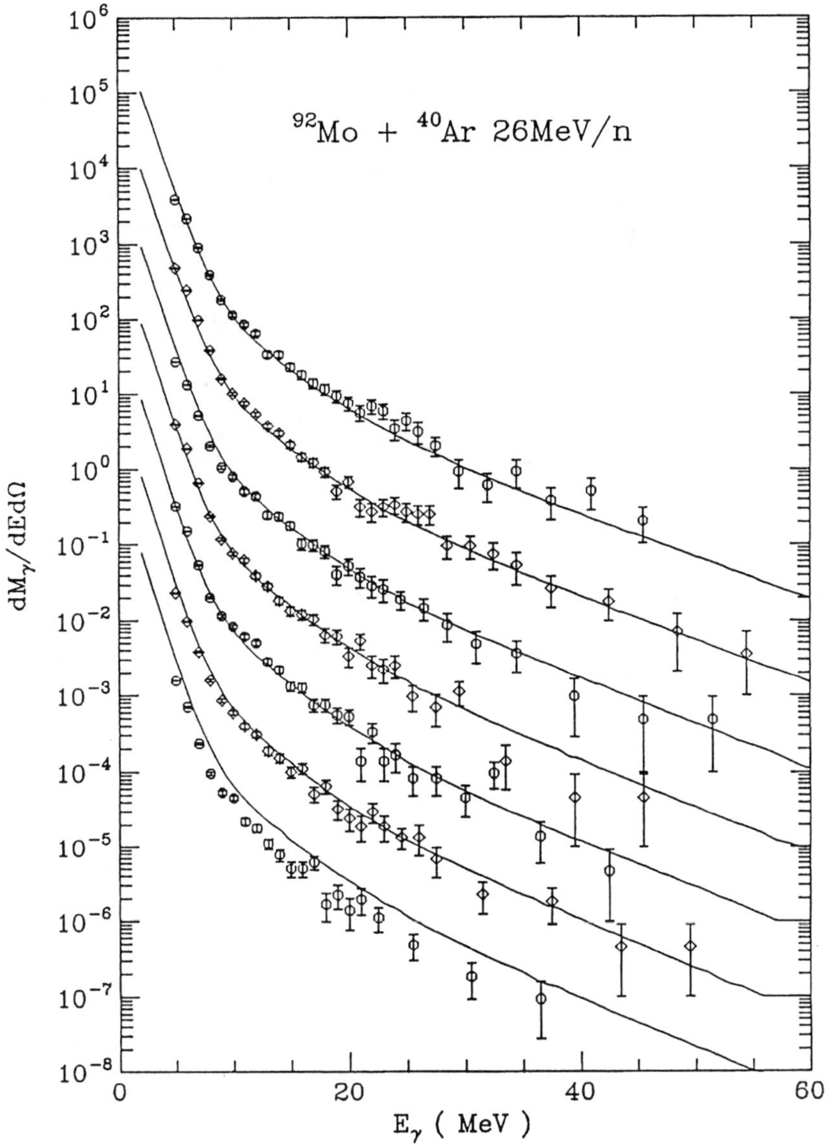

Fig. 7. Gamma-ray spectra gated on fusion events obtained in the ^{92}Mo + ^{40}Ar reaction with the BaF_2 2π detector system. The specta shown from top to bottom are measured at angles of 53°, 68°, 83°,98°, 113°, 128°, 143° and 158°.

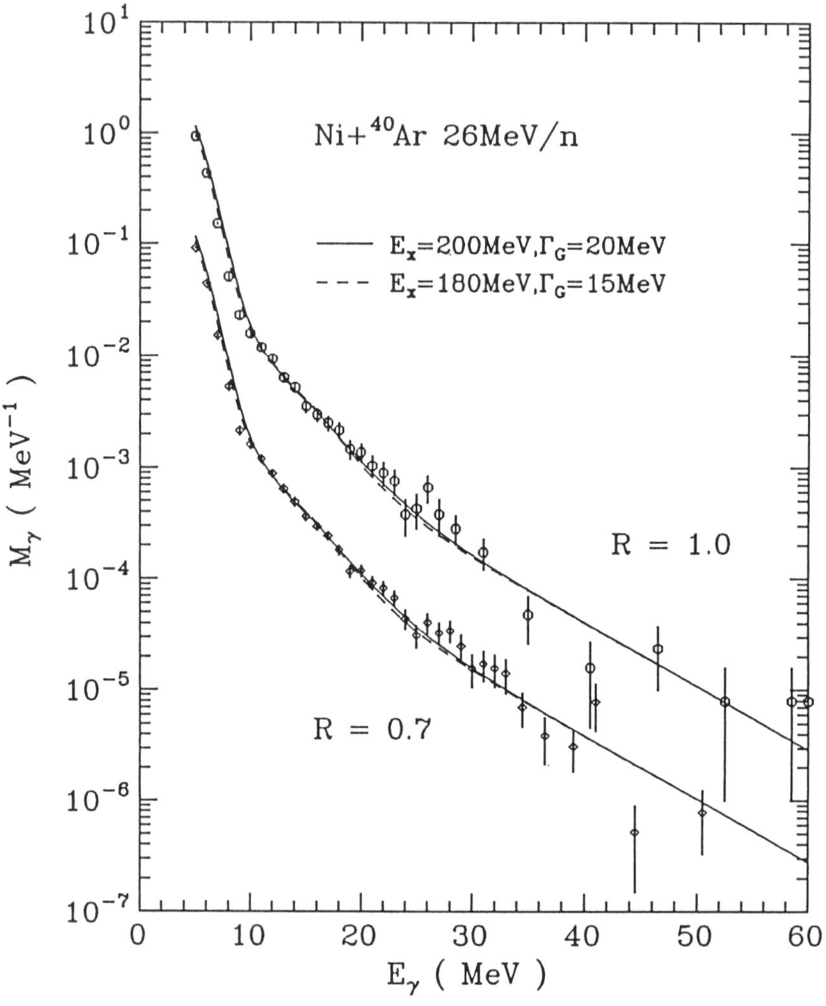

Fig. 8. Gamma-ray spectra obtained in the Ni +^{40}Ar reaction at E/A = 26 MeV compared with the calculation with fixed Γ.

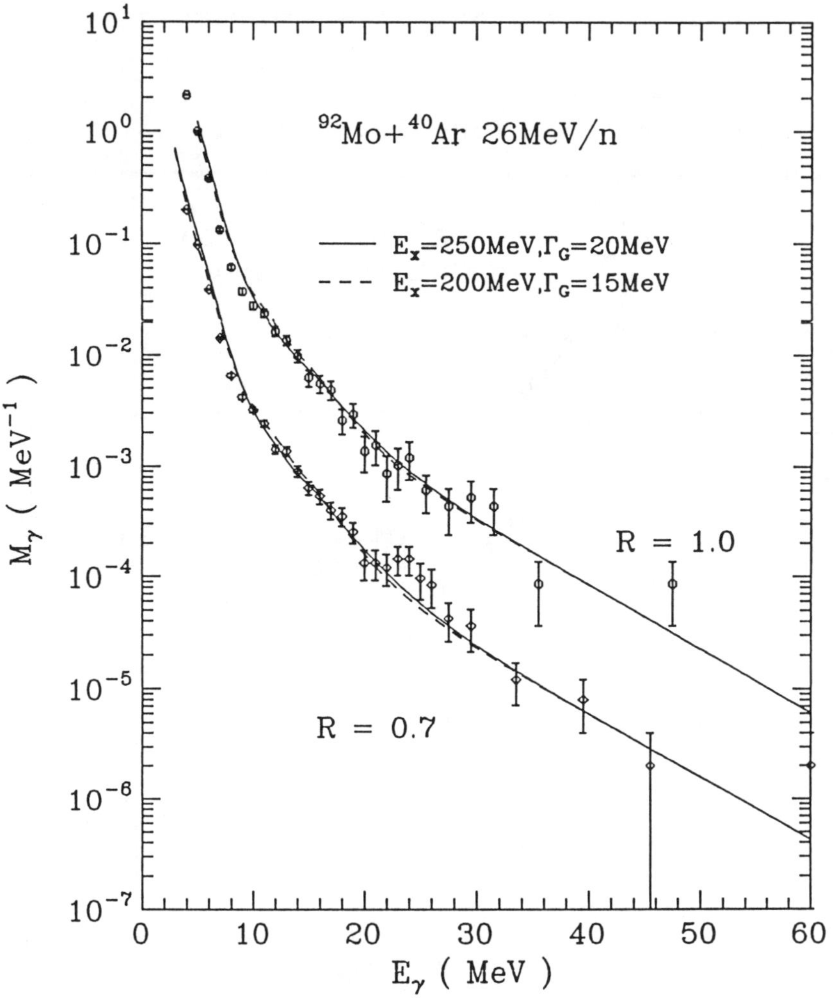

Fig. 9. Gamma-ray spectra obtained in the ^{92}Mo +^{40}Ar reaction at E/A = 26 MeV compared with the calculation with fixed Γ.

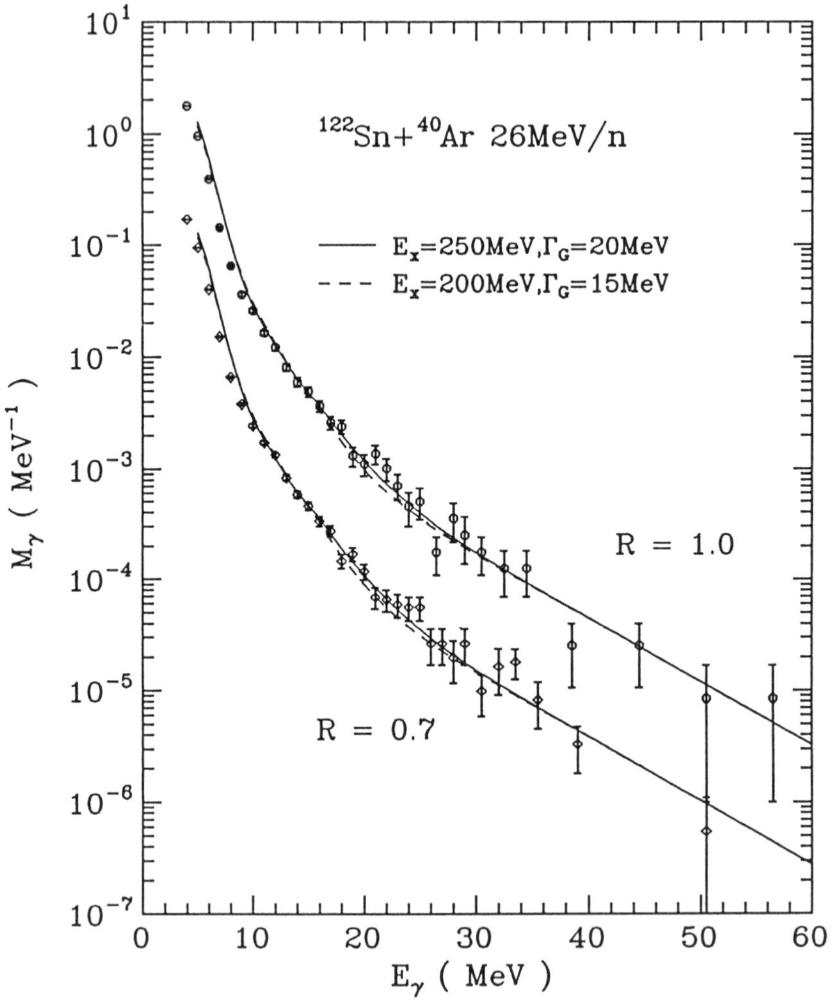

Fig. 10. Gamma-ray spectra obtained in the ^{122}Sn +^{40}Ar reaction at E/A = 26 MeV compared with the calculation with fixed Γ.

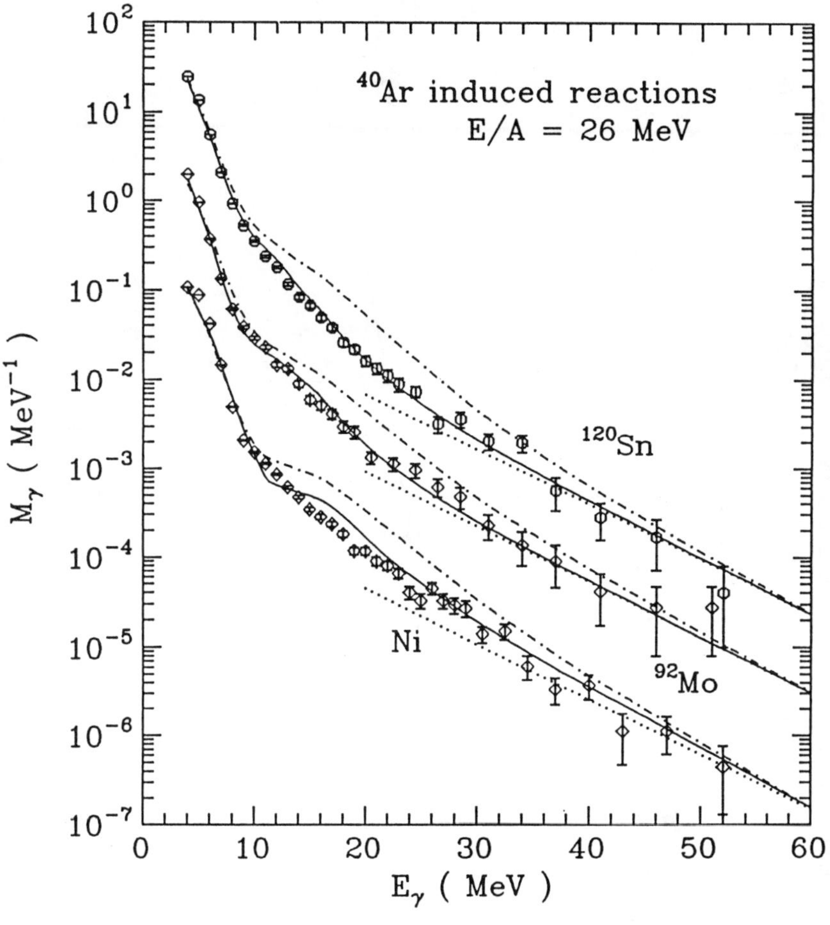

Fig. 11. Gamma-ray spectra obtained in the ^{40}Ar + ^{122}Sn, ^{92}Mo, Ni reactions at E/A = 26 MeV gated on fusion events. The calculations with $\Gamma(E_x)$ are shown with the solid lines.

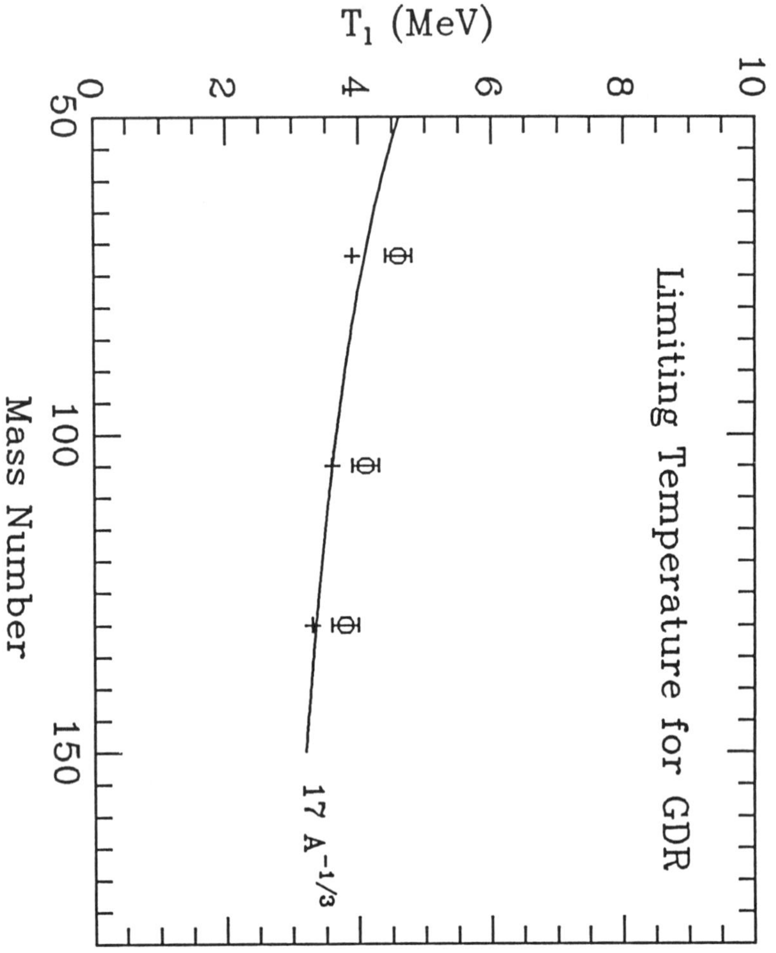

Fig. 12. Deduced limiting temperature (T_l) of the nucleus for existing of the GDR against mass number. Open circles are obtained by the calculations with fixed Γ and crosses with $\Gamma(E_x)$.

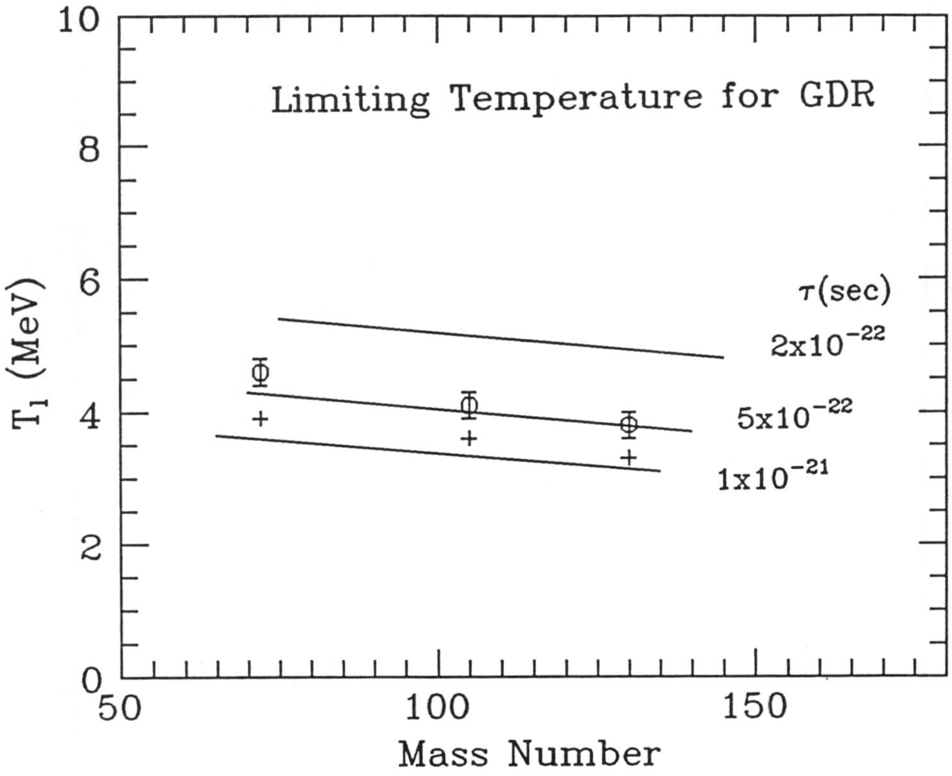

Fig. 13. Same as Fig. 12. The solid lines present the lifetimes of the nuclei at the given temperature.

MICROCLUSTER IMPACT FUSION

EXPERIMENTAL DEMONSTRATION OF DD FUSION BY CLUSTER ION IMPACT

R.J.Beuhler, Y.Y.Chu, G.Friedlander & L.Friedman

Chemistry Department, Brookhaven National Laboratory, Upton, NY 11973

ABSTRACT

The recently discovered phenomenon of DD fusion induced by cluster ion impact on deuterated surfaces is reviewed, along with experimental investigations designed to prove that the phenomenon is not caused by artifacts. Particular emphasis is given to time-of-flight experiments which unambiguously establish that the observed fusion events are caused by the slow-moving cluster ions rather than by faster small-ion impurities.

In this paper we review our motivation for attempting to produce DD fusion by cluster ion impact, the experimental observation of the phenomenon, and the demonstration, particularly by time-of-flight (TOF) studies, that the observed fusion events are indeed produced by cluster impact rather than by light-ion beam impurities or other artifacts.

The generation, acceleration, analysis, and detection of singly charged water cluster ions have been studied at Brookhaven National Laboratory for many years.[1] These cluster ions have the general formula $H^+(H_2O)_n$; they have the approximate density of ice and are held together by hydrogen bonds (0.4 eV/molecule). For n=100 the cluster radius is ~1 nm.

Two important properties of cluster ion beams accelerated to a few hundred kiloelectronvolts are: 1) Because of their relatively low velocities ($<10^8$ cm s^{-1}), very little of their kinetic energy is, on impact, lost to electronic degrees of freedom; most of it goes to kinetic energy of struck atoms. 2) Because of their compact sizes and short ranges they can deliver very large particle currents to the struck surface. Consider, for example, a cluster of 100 water molecules at 300 keV. It moves with a velocity of about 1.7×10^7 cm/s and will impact a surface area of about 10^{-14} cm^2. Penetrating a layer of 10^{-7} cm in 6×10^{-15} s, the 300 atom cluster delivers a particle current density of $300/(6 \times 10^{-15} \times 10^{-14}) = 5 \times 10^{30}$ cm^{-2} s^{-1} and thus deposits more than 2×10^{17} watts/cm^2.

These considerations gave rise to the notion that the particle and energy densities achieved when D_2O clusters impact a deuterated surface might be sufficiently large for observation of DD fusion reactions.

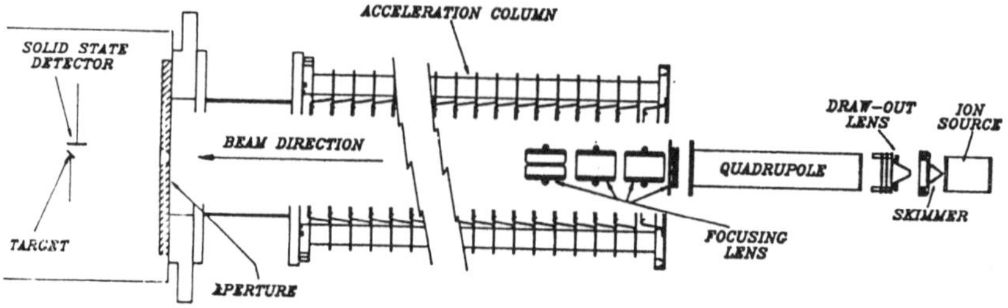

Fig. 1. Schematic diagram of apparatus used. Helium gas saturated with D_2O enters the ion source at pressures of 300 to 750 Torr. Water is ionized in a corona discharge and the weakly ionized plasma is expanded through a supersonic nozzle into a differentially pumped region between nozzle and skimmer. Cluster ions are extracted from the skimmer with the draw-out lens and mass-analyzed in the quadrupole mass spectrometer. The ions are then focussed and accelerated in a Cockcroft-Walton column and pass through an aperture into the target chamber. A monitor grid on the aperture is used to measure the cluster ion current. A silicon solid-state detector about 1 cm from the deuterated target is used to detect the charged-particle products of fusion reactions.

The apparatus used is shown in Figure 1. In the initial experiments a 300 kV Cockcroft-Walton accelerator was used. More recently a 750 kV Cockcroft-Walton accelerator became available and this machine has been used in all the TOF experiments.

The two branches of the DD reaction are:

1. $D + D = {}^1H + {}^3H + 4$ MeV;

2. $D + D = n + {}^3He + 3.3$ MeV.

Our experiments were designed for detection of the three charged reaction products in a surface barrier silicon detector. A pulse height spectrum showing the three peaks is reproduced in Figure 2. The area under the 3 MeV proton peak was generally used as a measure of the fusion rate.

The initial experiments with $D^+(D_2O)_n$ clusters impinging on targets of TiD, $(C_2D_4)_m$, and $ZrD_{1.65}$ have been described in previous publications.[2,3] Here we will only briefly recapitulate the results.

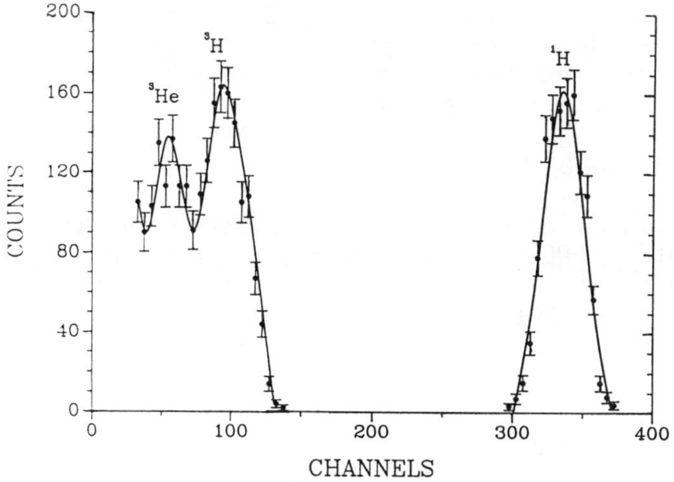

Fig. 2. Pulse height spectrum showing ^1H, ^3H, and ^3He peaks produced by impact of 275 keV $(D_2O)_{115}$ ions on $(C_2D_4)_m$.

The key findings were:

a) For clusters with n=100 the DD fusion yields increased by nearly an order of magnitude with an increase in beam energy from 200 to 300 keV.

b) The absolute rate at 300 keV was of the order of 10^{-10} fusions per cluster impact.[3] Typical excitation functions are shown in Figure 3. Over the range 100<n<400, rather weak cluster size dependence was observed at 300 keV, with the peak yield in the neighborhood of n=200.

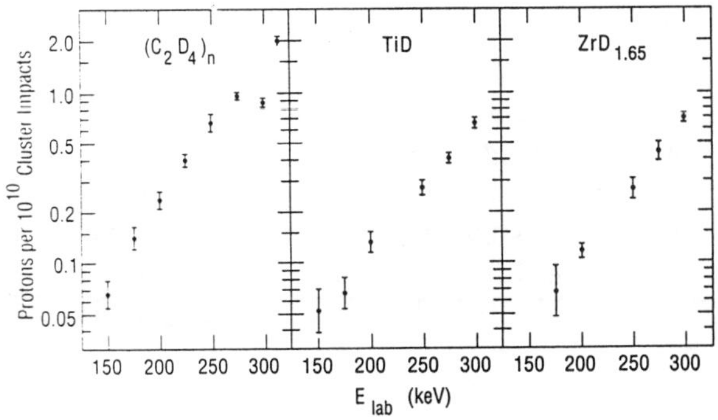

Fig. 3. Excitation functions for DD fusion protons produced by impact of $(D_2O)_{100}$ clusters on three different targets. The proton yields have been corrected for detector geometry.

The energy dependence of the cross section for DD reaction 1 in the range from 3 to 1,000 keV (laboratory energy) is shown in Figure 4. The cross section can be extrapolated to lower energies (where there are no experimental data) by the approximate formula

$$\sigma = (55x10^{-24}/E)\exp(-31.28/E^{1/2})\ cm^2,$$

where E is the center-of-mass energy in keV. On the assumption that a n=100 cluster accelerated to 300 keV behaves as a collection of 200 deuterons, each with 0.3 keV laboratory energy (E_{cm} = 0.15 keV), the extrapolated cross section would be about 10^{-54} cm^2 per cluster. The observed fusion rates correspond to cross sections some 25 orders of magnitude larger.

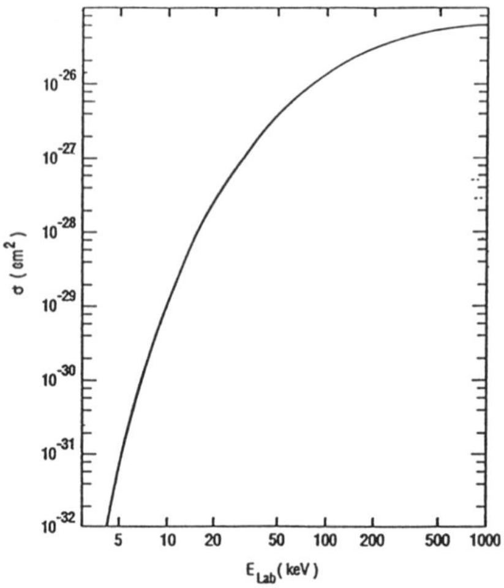

Fig. 4. Excitation function for the reaction $^2H + ^2H = ^1H + ^3H$

This enormous discrepancy between observed rates and expectations based on a naive model made it imperative to search for artifacts that could be responsible for the observations. In particular, the possible presence of deuterons or other small ions such as D_2O^+, D_3O^+, etc. in the beam had to be carefully investigated. A number of experiments designed to this end wll be briefly reviewed; they are more fully described in ref. 3.

When the D_2O clusters in a 300 keV beam were replaced with H_2O clusters, DD fusion was still observed (even after exhaustive flushing of the system with H_2O), but at a rate about 5% of that found with D_2O clusters. This result is not compatible with any explanation based on the presence of deuterium-containing small-ion impurities (such as D^+, D_2O^+, D_3O^+) in the beam.

The steep increase in fusion yield with energy in the 200 to 300 keV range (Fig. 3) rules out fusion by fully accelerated D^+ or D_2^+, as can readily be seen by reference to the DD excitation function in Fig. 4. Subsequent measurements have shown that cluster fusion yields continue to rise steeply between 300 and 600 keV, which very much strengthens this conclusion and also eliminates fully accelerated D_3^+ as the responsible projectile.

Experiments were performed with perdeuteriopolyethylene targets covered with 100 to 500 μg/cm^2 of gold. These gold films were thick enough to stop clusters but thin enough to transmit D^+ and D_2^+ beams with only small energy losses. In these experiments no fusions were observed initially, but the fusion rate was found to increase gradually with time, as the gold coating was sputtered off by the cluster beam. Examination of the target after such a run showed the gold coating missing in the area struck by the beam. These experiments again rule out fully or nearly fully accelerated D^+ and D_2^+ as artifacts responsible for the observed fusions.

Another set of experiments was devised to exclude the possibility that the observed fusions were due to partially accelerated small ions produced somewhere in the accelerating column by collisional dissociation or electron impact. For this purpose a parallel plate electrostatic analyzer was installed ahead of the deuterated target, which was now placed off axis. With appropriate deflection voltages on the analyzer plates neither ions with less than about 75% of the full energy nor neutrals could strike the target. Yet the fusion yields were, within experimental error, the same as in the normal, on-axis configuration.

Although the aggregate of these various tests should leave little room to doubt that we had indeed observed fusion induced by cluster impact, it was suggested[4] that even more conclusive proof could come from time-of-flight (TOF) experiments.

As soon as we had access to the larger (750 kV) Cockcroft-Walton accelerator, we therefore undertook TOF studies.[5] Figure 5 is a schematic sketch of the apparatus used. The beam was pulsed electrically. Sufficient positive voltage (a few hundred electronvolts) was applied to the timing grids at the exit of the quadrupole mass analyzer to retard the positive beam; this voltage was removed for 2 to 10 μs every 100 to 150 μs by means of a pulsing circuit. At the target end the beam could either travel on axis or be deflected electrostatically to an off-axis position. The $(C_2D_4)_m$ target could be placed on axis, with the electron multiplier assembly off axis, or vice versa.

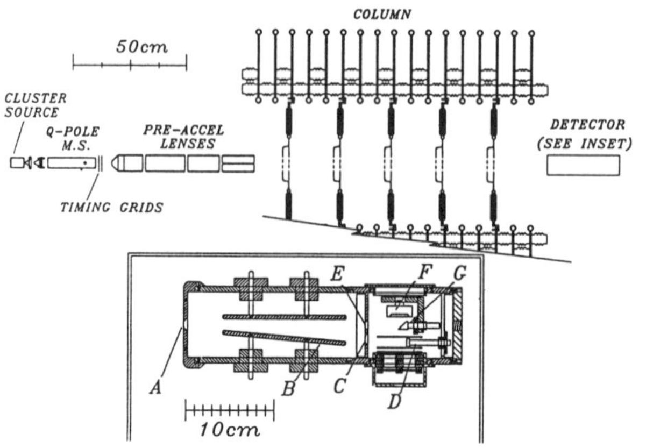

Fig. 5. Sketch of TOF apparatus. The beam is pulsed by the pair of grids at the exit of the quadrupole mass analyzer. The inset shows the target assembly. A - entrance aperture, B - deflection plates, C - off-axis aperture, D - copper probe and electron multiplier, E - on-axis aperture, F - proton detector, G - perdeuteriopolyethylene target.

The electron multiplier assembly was used to measure the electron distribution produced by ion impact on the copper probe. Cluster ions with energies of hundreds of kiloelectronvolts produce of the order of 100 electrons per impact[6], whereas small water ions such as D_2O^+, D_3O^+, $D_5O_2^+$ produce less than 10 electrons per impact. Thus the secondary electron distributions provide a very useful, in fact crucial diagnostic for clusters. Figure 6 shows an electron distribution generated by impact of 250 keV $D^+(D_2O)_{133}$ clusters on copper.

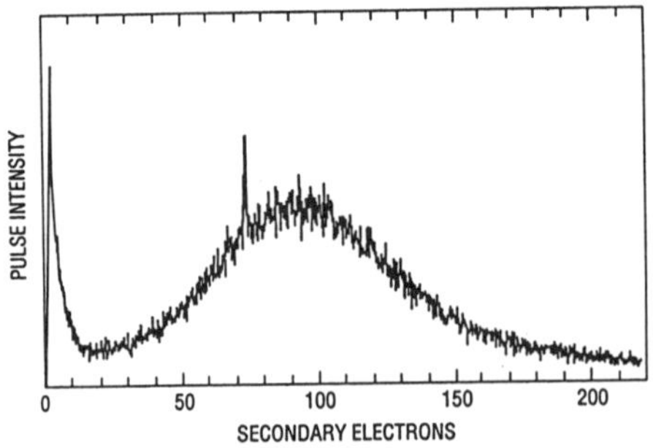

Fig. 6. Secondary electron distribution produced by impact of 250 keV $(D_2O)_{133}$ ions on copper. The spike near 70 electrons is generated by a pulser.

A significant problem with cluster fusion TOF experiments arises from the sensitivity of fusion yields to buildup of surface films on targets. In our oil-pumped vacuum system with pressures of $(5\text{-}10)\times10^{-7}$ Torr in the target chamber we found that time-averaged beam currents of the order of 1 nA were required to keep the target surface clean. At lower currents the yield would gradually drop until the target was "cleaned" by sputtering off the surface film with a more intense beam. With the duty cycle of the TOF experiments the average beam currents were much too small to keep the target clean. Consequently the average fusion rates in the TOF studies were even much smaller than what would correspond to the duty cycle, and the target surface had to be periodically cleaned by bombardment with steady beams.

Another important aspect of the TOF experiments was absolute calibration of the system's time scale. This was accomplished by means of completely resolved TOF spectra for low-mass clusters such as the example shown in Figure 7. In this way any delays in the electronics are automatically taken into account.

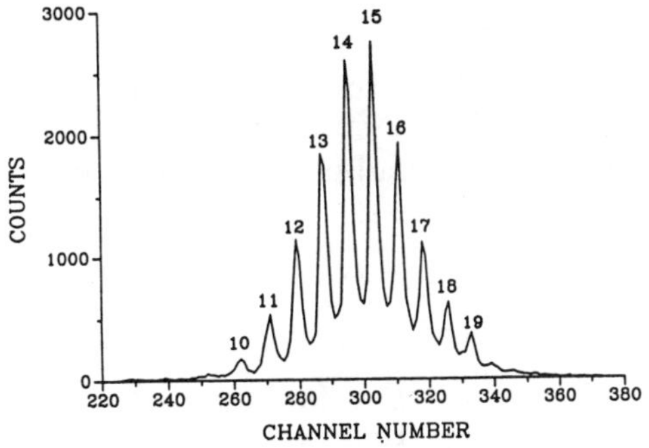

Fig. 7. Resolved TOF spectrum for small D_2O clusters.

Two types of TOF experiments were performed. In the first, the perdeuteriopolyethylene target was placed on axis and the copper probe and electron multiplier assembly off axis as shown in Figure 5. Cluster arrival times were measured with the multiplier assembly and compared directly with the times at which fusion protons from the target were detected in the silicon detector. An example of the comparison between the two TOF spectra for $(D_2O)_{115}$ clusters accelerated to 520 keV is shown in Figure 8. The two peaks coincide within 1 μs, and about 90% of all fusion protons are contained in the peak that corresponds to the TOF of the clusters. Since these clusters spend about 11 μs in the acceleration column, the observed time coincidence rules out low molecular weight species formed prior to or during early stages of acceleration as major contributors to the observed fusions.

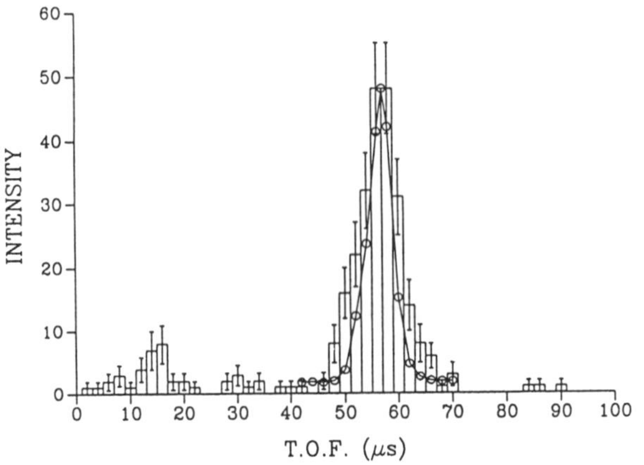

Fig. 8. TOF spectra for $(D_2O)_{115}$ clusters accelerated to 520 keV. Smooth curve: arrival time of clusters as measred with the electron multiplier. Histogram: times at which 3 MeV fusion protons were detected. The ordinates for the two data sets were normalized to the same peak height at 57 μs.

Similar TOF spectra were taken for fusion protons produced by clusters containing between 35 and 200 D_2O molecules. Figure 9 shows that, as expected, the time between start pulse and proton detection is proportional to the square root of the cluster mass.

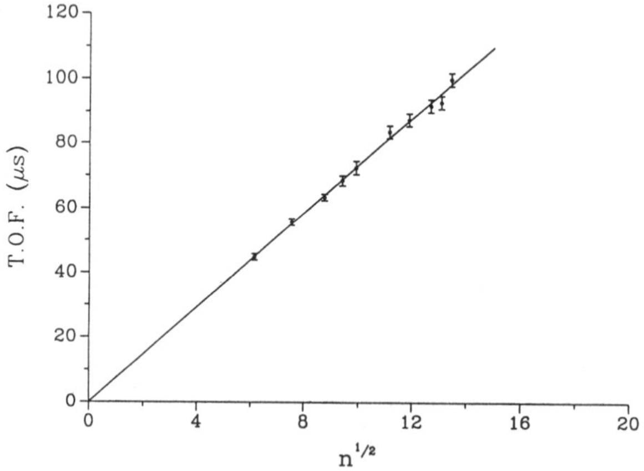

Fig. 9. Arrival times of fusion protons for $(D_2O)_n$ clusters accelerated to 525 keV. (The arrival time for n=115 differs from that shown in Fig. 8 because different preacceleration voltages were used in the two sets of experiments.)

The second type of TOF experiment was a deliberate search for low-mass contaminants in the cluster beam, aimed at setting limits on the possible fusion contributions of such contaminants formed in the later stages of beam acceleration.

By using a single channel analyzer to set a "window" on the pulse amplitudes of the electron multiplier it was possible to select pulses containing either large or small numbers of secondary electrons per impact, depending on the multiplier voltage.

With the multiplier assembly in the off-axis position to minimize detector noise, the TOF spectra shown in Figure 10 were obtained.

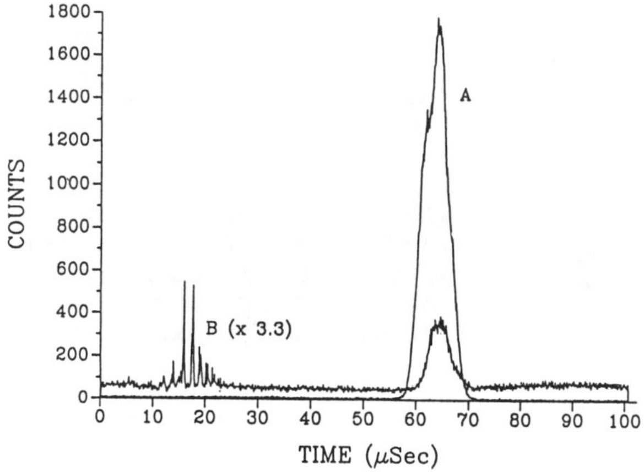

Fig. 10. TOF spectra taken off-axis with electron multiplier. For spectrum A only large pulses (40-120 electrons/impact) were accepted, for spectrum B only small pulses (4-12 electrons/impact).

Spectrum A was acquired with the window set to accept pulses of 40 to 120 electrons per impact, spectrum B with the window set for 4 - 12 electrons per impact. Thus A represents the TOF spectrum of cluster ions, with arrival times around 63 μs, whereas B is the TOF spectrum of small ions. The resolved peaks in spectrum B at 12 to 22 μs are due to ions containing from 1 to 7 D_2O molecules. Their flight times indicate that these ions originated at or near the timing grids. It is interesting to note that unsolvated D_3O^+ represents no more than a few percent of the small-ion fragments produced by collisional decomposition of cluster ions.

Since, in the off-axis experiment just described, only ions with nearly the energy of the fully accelerated beam are detected, a similar experiment without electrostatic deflection was performed to search for small-ion impurities of reduced energy formed somewhere in the acceleration column. Clusters of 130 D_2O molecules, accelerated to 300 keV were used in this experiment. Again the detector window could be set to accept either large pulses characteristic of clusters or small pulses due to water ions of low molecular weight. The two spectra so obtained are shown in the upper and lower parts of Figure 11. The small peak in the lower spectrum coincident with the cluster arrival time was shown in separate experiments to be almost entirely due to x rays produced by cluster impacts in the vicinity of the detector, the x rays in turn generating small numbers of electrons.

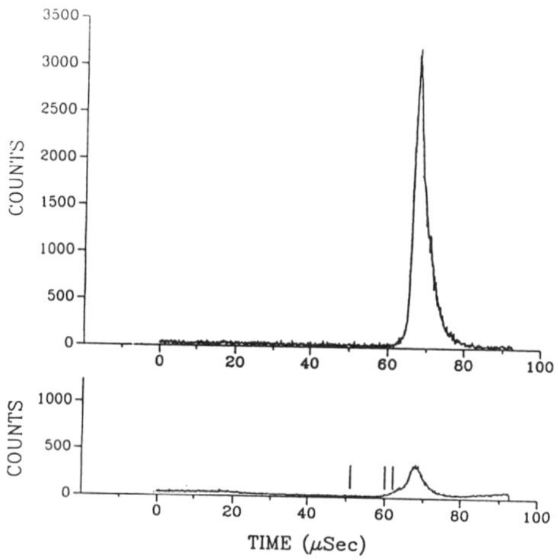

Fig. 11. TOF spectra, taken with electron multiplier detector on axis, when $(D_2O)_{130}$ clusters were accelerated to 300 keV. Top and bottom curves were taken with detector set to accept large and small pulses, respectively. See text.

After correction for this x-ray contribution, upper limits could be set on the amounts of water ions that could be present in the small-ion TOF spectrum. The three vertical lines in that spectrum mark the flight times for D_3O^+ at the entrance to the column, halfway down the column, and two thirds down the column. D_3O^+ ions formed at these three locations would have acquired energies of approximately 300, 150, and 100 keV, respectively; the corresponding energies per deuteron are ~ 27, 14, and 9 keV. Folding the ranges and DD cross sections for deuterons of these energies with the upper limits for D_3O^+ abundances from Fig. 11, one arrives at upper limits for fusion rates due to monomer ions formed at the three positions in the column. These limits are two to four orders of magnitude below the observed cluster fusion rates. For dimers etc. the limits are even much lower.

The sum total of experiments described in this paper confirms that impact of heavy-water cluster ions on deuterated surfaces produces DD fusion reactions. The fusion rates are so large as to suggest that some novel collective mechanism is at work that involves concentration of energy on a fraction of the deuterons in the hot assembly of atoms created by the cluster impact.

The authors are indebted to the Brookhaven AGS Department for making the 750 kV Cockcroft-Walton accelerator available for this work. Special thanks go to J. G. Alessi and V. LoDestro for valuable help and advice in the TOF experiments.

This research was carried out at Brookhaven National Laboratory under contract DE-AC02-76CH00016 with the U.S. Department of Energy and supported by its Division of Chemical Sciences, Office of Basic Energy Sciences.

REFERENCES

1. R. Beuhler and L. Friedman, Chem. Revs. 86, 521 (1986).

2. R. J. Beuhler, G. Friedlander, L. Friedman, Phys. Rev. Lett. 63, 1292 (1989).

3. R. J. Beuhler, Y. Y. Chu, G. Friedlander, L. Friedman, W. Kunnmann, J. Phys. Chem. 94, 7665 (1990).

4. D. Alburger, J. Phys. Chem. 94, 8494 (1990).

5. R. J. Beuhler, Y. Y. Chu, G. Friedlander, L. Friedman, J. G. Alessi, V. LoDestro, J. P. Thomas, Phys. Rev. Lett. 67, in press (1991).

6. R. J. Beuhler and L. Friedman, Nucl. Instr. and Meth. 170, 309 (1980).

SEARCH FOR NUCLEAR FUSION INDUCED BY D_n^+ AND N_n^+ CLUSTER IONS IN DEUTERATED TARGETS

M. Fallavier , J. Kemmler , R. Kirsch , J.-C. Poizat , J. Remillieux , and J.-P. Thomas
Institut de Physique Nucléaire de Lyon , IN2P3-CNRS/Université Claude Bernard
Lyon I , 43 Bd. du 11 Novembre 1918 , 69622 Villeurbanne Cedex , France

ABSTRACT

According to the results of the Brookhaven group , d,d fusion can be achieved in deuterated targets when using water clusters , deuterated or not , for energies per unit mass even much below 1 keV . We have investigated the extreme situation of pure deuterium cluster bombardment , choosing a similar range of energy per deuteron in order to make a significant comparison .

With such projectiles , magnetically mass selected after acceleration , no fusion event has been detected and our experimental upper limit for the fusion rate was more than two orders of magnitude below the typical values reported with deuterated water clusters . This result supports the picture reported in most of the theoretical models that any favourable energy transfer to the deuterons requires the presence of a heavier collision partner in the projectile .

In the course of these experiments , the only possibility for assessing the effect was to use nitrogen instead of deuterium . Unfortunately , only the deflection of the low mass projectiles off the target direction was experimentally feasible . Since fusion was observed both with the pulsed cluster beam and with the continuous leak beam of atomic and molecular species , a contamination by residual deuterons is most likely responsible for the effect .

INTRODUCTION

Since the first report of Beuhler et al. [1] at Brookhaven National Laboratory (BNL) on d,d fusion induced by deuterated water clusters on deuterated targets , experimental confirmations have been brought not only by the same group on another machine [2] but also by Bae et al. [3] at Stanford Research Institute (SRI) . The observed phenomenon can be summarized in three points : 1) From the cross-sections expected in the energy per deuteron range (below 1 keV) , an enhancement of about thirty orders of magnitude can be reported . 2) Between 150 and 600 keV the excitation curve exhibits a steeper slope than the well known one for single deuterons . 3) At a given energy , the fusion yield reaches a maximum as a function of the number n of molecules in the cluster (around n = 200 at 300 keV) . However , while many experimental parameters were modified , the same projectiles have been used .

One of our motivations in such a context was to investigate , within our experimental capabilities , the two extreme situations where the incident clusters are formed from pure deuterium or from heavier atoms . Since fusion was also observed under ligth water cluster bombardment , although at a rate 20 times smaller [4] , it was reasonable to expect a similar behaviour with nitrogen clusters . Since the projectile mass selection of our machine is achieved after acceleration with a magnet of a given maximum mass-energy product , our experiments have been performed with 100 - 150 keV D_{200}^+ - D_{300}^+ clusters , in order to be in the energy - per - deuteron range reported in the BNL paper .The net result reported in the following is that no fusion event was detected [5] . In the second set of experiments , nitrogen clusters of sufficient number of atoms could not be magnetically analyzed because of their larger masses . Then , we chose to work on a direct beam line , but deflecting the lightest masses off the target direction (low mass threshold experiment) . Fusion was indeed observed , but has to be attributed to deuterons formed in the residual gas contaminated by the previous use of deuterium in the source .

EXPERIMENTAL

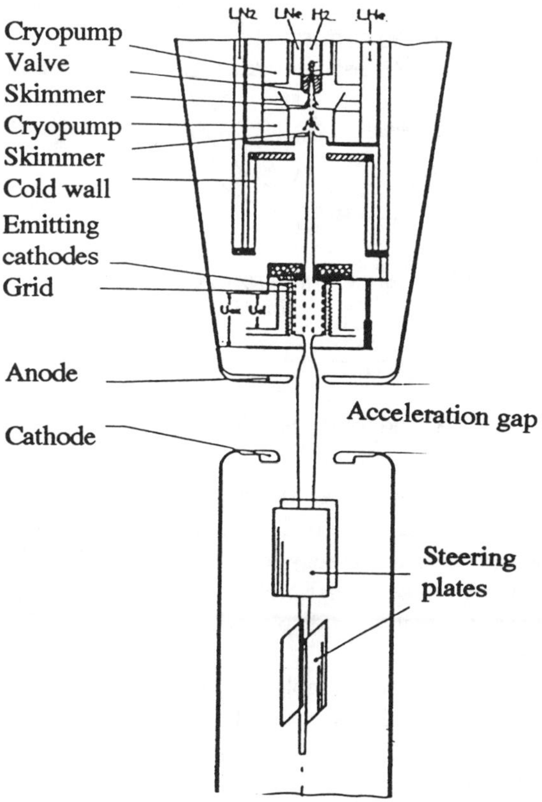

Fig. 1. Sketch of the cryogenic source of the Cockroft-Walton accelerator.

The cryogenic source of the Lyon accelerator was originally designed for hydrogen cluster production , from the expansion of hydrogen gas at a given pressure and temperature . The various cooling enveloppes used to set this last parameter are indicated on the sketch of the source presented in Fig. 1 . The external reservoir is filled with liquid helium which is consumed during the experiment and this cold wall is also aimed to maintain a vacuum compatible to the survival of the clusters leaving the nozzle . However , such a pumping system cannot sustain a continuous expansion and an electromagnetic valve has to be repeatedly operated . Then , clusters are formed during pulses (typically 100 ms every 1s , i.e. 10% duty cycle) while a "leak" beam resulting from ionisation of the residual gas (monoatomic species)can be observed between pulses at a much lower intensity . Such a setup cannot be easily adapted to the production of water clusters formed at room temperature and requiring powerful pumping for a continuous beam (the source is grounded in the BNL experiment) . On the other hand , only minor adjustments are needed for deuterium and nitrogen clusters production .

After acceleration , the projectiles of a given energy, but with various masses, are deflected by a 74° electrostatic analyzer and enter a 16° bending magnet with a mass-energy product of 60 MeV.u . Then , as sketched in Fig. 2 , they can be sent on beam line A or B , depending of their nature . For D_n^+ (line A) , the magnet allows the mass selection with a mass resolution of about 3000 at the target site , defined by the exit slit of the magnet and an opaque collimator located 44 cm upstream from the target . For N_n^+ (line B) , the applied magnetic field only prevents the low-mass projectiles from reaching the target: for example, at a total energy of 500 keV, 0.05 T is sufficient to deflect N_{16}^+ and lighter clusters from the target area .

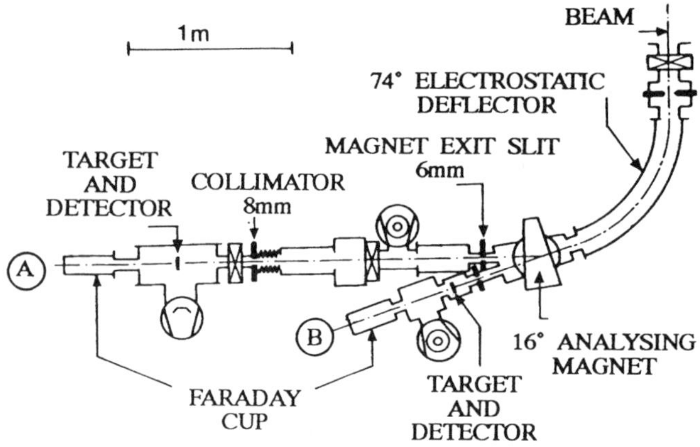

Fig. 2. Experimental setup for the D_n^+ (line A) and for the N_n^+ (line B) experiments.

In these experiments , the beam intensity was determined , for a given projectile , from a normalisation between the value obtained on the target and the one obtained onto a specially shielded and biased Faraday cup placed behind the removable target . This has been possible because of the stability of the pulsed beam (1 Hz - duty cycle of 10%) and the beam intensity was displayed on-line on a digital oscilloscope and integrated for dose measurement .

The target-detector assembly was quite comparable to the one used in the BNL experiment .The target was tilted at 45° with respect to the beam axis and the 900 mm^2 Si solid state detector was placed at 90° with a solid angle of about 2 sr . A thin absorber was used to protect the detector against low-energy scattered fragments and from light . A calibration of the detection system as well as a check of the integrity of the target was easily performed between experiments by simply tuning the magnet on low-mass deuterium clusters . As shown in Fig. 3a , from the use of 100 keV D_3^+ the fusion products (^3He , t , p) are clearly identified .

Two types of deuterated targets have been used . The TiD target was elaborated [6] by deposition of 320 μg.cm^{-2} Ti on a copper backing followed by high pressure D loading up to $TiD_{1.7}$ stoichiometry . Because of possible D depletion at the very surface of the TiD target we also used deuterated polyethylene disks , 0.7 mm thick . Moreover , huge sputtering effects arising with nitrogen clusters prevent the use of thin TiD targets .

Conventional electronics was used to inhibit transient signals generated in the environment of our pulsed accelerator . Moreover , a gate signal was applied to trigger the counting not only during the beam pulses , but also between the pulses (leak beam of atomic or molecular species) .

RESULTS

1) D_n^+ experiments :

Considering the limitations of our analyzing magnet , the choice of the mass and energy range of the deuterium clusters has been decided on the following bases : if the BNL data of Ref 1 are represented by the proton yield per 10^{15} incident deuterons as a function of the cluster size , it appears that , with the extreme values of D_{200}^+ -150 keV and D_{300}^+-120 keV , we investigate a range of energy per deuteron where a maximum yield had been obtained . According to the shape of the background spectrum taken at zero voltage on the terminal (Fig. 3b) , it has been decided to valid the fusion events from a proton energy window ranging from 2.8 to 3.2 MeV , and not to consider t and ^3He particles . For a net acquisition time of 2 x 10^4 s , there are fifteen background counts from 1.4 to 7.4 MeV , but *none* in the proton window . Even if we assume a uniform distribution of background events within this energy range ,

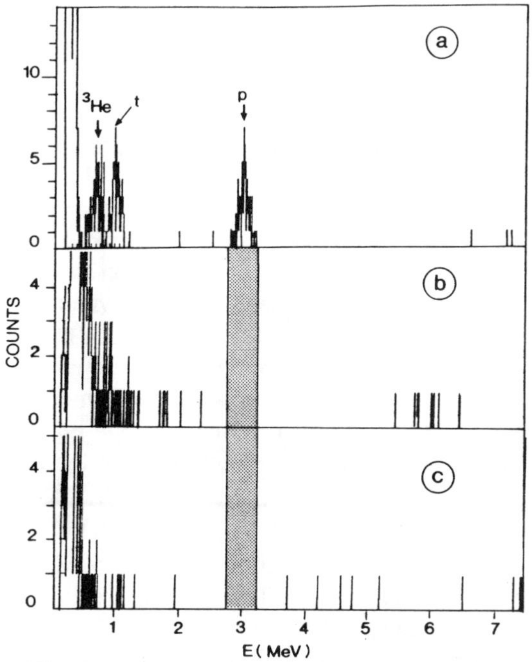

Fig. 3. Energy spectra obtained for the following conditions : (a) 100 keV D_3^+ , (b) No voltage (2 x 10^4 s) , (c) summation of spectra corresponding to the experimental conditions described in the text (total time 8400 s).

the background count rate is expected to be less than 5 x10^{-2} per 1000 s and we should be able to measure proton yields 2 orders of magnitude below those reported in the BNL experiments .

Five measurements have been performed with 100-150 keV D_n^+ clusters (n=200 , 250 and 300) bombarding deuterated titanium and polyethylene targets . Net acquisition times were ranging between 950 and 4200 s . We never obtained any count in the proton window , even at the highest energy of 750 eV per deuteron and for the longest acquisition time . From this last case , one can at least determine an upper limit of the count rate which , for a confidence level of 95% , appears to be more than 10 times below the corresponding result of BNL (300 keV - $(D_2O)_{40}^+$) . Moreover , as seen in Fig. 3c , there is no significant difference between the spectrum resulting from the summation of the spectra corresponding to the various measurements (8400 s) and the background spectrum : the resulting count rate in the whole energy range 1.4 - 7.4 MeV is within the error bar of the background of the detection system (measured over 2 x 10^4 s) . From this argument it is reasonable to conclude that with D_n^+ projectiles the proton yields per incident deuteron are 2 orders of magnitude below the BNL values , i.e. , very difficult to measure under our detection conditions .

2) N_n^+ experiments :

As previously indicated , on the "direct" beam line B (see Fig. 2) fully accelerated projectiles of all masses reach the deuterated polyethylene target if the magnet is turned off , and intensities in the μA range are easily obtained . Under these conditions , we did observe fusion with a rate of several protons per second . However , since beam contamination from previous experiments with deuterium clusters is a well known phenomenon , the proton counting was also performed between the beam pulses where the remaining "leak" beam must be ascribed to the ionisation of the residual gas . As a reasonable assumption this contribution should be constant during the pulse and between two consecutive pulses . The first evidence for the main contribution of the "leak" beam to the fusion rate can be seen from the data obtained with the magnet off , reported in Table 1 . If the proton rate is related to the number of impacts , it increases by more than a factor of ten when performed with the "leak" beam , but it is almost constant when related to the effective counting time. Our previous assumption is also confirmed by the expected variations with the beam intensity .

With a magnetic field of typically 0.05 T (deflecting N_{16}^+ and lighter clusters off the target at 500 keV) , the proton rate is reduced by 3 orders of magnitude while the intensity drops only by about one order of magnitude . Then , as shown in Table 1 it is confirmed that the fusion process is mainly induced by the "leak" beam . To answer the remaining question of what particles , atomic and molecular deuterons or atomic and molecular nitrogen ions through capture or dissociation leading to neutrals , are responsible for the effect , it must be noted that N^+ , N_2^+ , and N_3^+ are the predominant species in the beam , with intensities more than two orders of magnitude larger than those of the clusters of higher masses . Then , there are no reasons for the two - orders - of - magnitude increase of the proton yield per impact from the pulsed beam to the leak beam . Moreover , the almost constant values per second , still found in both cases , are a good indication of a *deuteron contamination* .

Exp. Conditions	I pulse target (nA)	p / 10^{10} impacts		p / s	
		pulse	leak	pulse	leak
500 kV 0 tesla	750	1.8 10^{-2}	2.5 10^{-1}	1.5	1.8
	6400	1.9 10^{-2}	1.3 10^{-1}	6.3	5.6
	500	2. 10^{-2}	1.4 10^{-1}	0.8	1.0
	2700	1. 10^{-2}	2.3 10^{-1}	2.7	2.5
500 kV 0.03 tesla	1200	3.5 +-0.9 10^{-4}	4.9 10^{-1}	4.3 +- 1 10^{-2}	4.3 10^{-2}
500 kV 0.035 tesla		2.6 +- 1.2 10^{-4}	3.4 +- 1.7 10^{-2}	1.0 +- 0.4 10^{-2}	2.3 +- 1.2 10^{-3}
500 kV 1.2 tesla	28	7.7 +- 7.7 10^{-4}	2.4 +- 1 10^{-1}	2.3 +- 2.3 10^{-3}	1.8 +- 0.8 10^{-3}
500 kV 0 tesla	2560	3.8 10^{-3}	1.2 10^{-1}	1.6	0.5
300 kV 0.05 tesla	160	3.6 +- 1.9 10^{-4}	1.1 +- 0.5 10^{-1}	9 +- 0.5 10^{-3}	3.4 +- 1.7 10^{-3}

Table 1. Numbers of detected protons in various experimental conditions of N_n+ bombardment , for a given number of impacts and for a net acquisition time , respectively.

The unanswered question concerns the value of the reduction factor F between the count rate per second with and without applying the magnetic field . Such a value does not exceed 10^3 while any estimate based on a D_2^+ dissociation leading to D^0 production predicts a value larger than 10^4 . Additional effects related to scattering could also play a role in such a short beam line : any attempts to mass - analyze this contribution with the 16° magnet have been unsuccessful mainly because of the strongly divergent beam .

DISCUSSION

Among the theoretical investigations on the possible processes leading to fusion under cluster impact , most have failed to reproduce the BNL data . The many orders of magnitude enhancement of the fusion cross-section (around 25) has been mainly attributed to particular collisions leading to acceleration of deuterons . The first attempts of Carraro et al. [7] to introduce molecular dynamics calculations were unsuccessful mainly because of insufficient statistics in simulating Fermi shuttle processes [8] . Such processes in which light particles are accelerated by repeated scattering back and forth between the heavy components of the beam and the target have been since more carefully considered , leading to significant evidence that a fraction of deuterons is accelerated far beyond the prediction of a single collision . Consequently , there is a general agreement for a consistent increase of the fusion cross-section (see for example ref 9) even if the predictions are still below the experimental values . Among the positive predictions, and if one excepts the early qualitative agreement of Echenique et al. [10] , based on the assumption of a high energy tail of the deuteron energy spectrum postulated to be exponential , the only quantitative agreement , still based on the necessity of an heavy collision partner in the cluster , is the one of Kim et al. [11] . In their model , the local Maxwell-Boltzmann distribution , resulting from the multiple collisions previously described , can account for a non-equilibrium situation , in which the average deuteron transport velocity v_d does not go to zero (partial thermalization compared to $v_d = 0$) . One interesting feature in such a model is that it can quantitatively predict both BNL results and our null results , as it can be seen in Table 2 .

We have normalized the experimental data to the same number of incident deuterons at the same energy per deuteron and the calculations have been done for intermediate degrees of thermalization . As one can see , there is not only a good agreement with the BNL data but the right order of magnitude difference between their data and ours is also found .

The main problem with these models is still their unability to reproduce the experimental variation of the fusion rate with the cluster size . Another approach has been reported in this conference , based on the concept of shock wave [12] . Details on the validity of such a concept can be found in the related proceedings but it is worthwhile to note that such a description is also consistent with the unability for deuterium clusters to induce the fusion . This model is also able to predict the fusion probability of $(H_2O)_n^+$ impacting on deuterated targets : as demonstrated by another paper [13] in this Conference and by our experiments with nitrogen clusters , the effect cannot be simply ascribed to high energy atomic or molecular heavy ions resulting from unsufficient mass selection .

n	Energy /D (eV)	Incident D's(10^{15})	$\tilde{R}_{calc}$ ($v_d = 0$)	$\tilde{R}_{calc}$ ($v_d = v_0/\sqrt{2}$)	$\tilde{R}_{calc}$ ($v_d = v_0$)	$\tilde{R}_{exp}$ (Ref. 1)
40	750	6.15	153	137	102	30
60	500	3.3	5.9	5.1	3.4	16
75	400	2.2	0.8	0.7	0.2	8

n	Energy /D (eV)	Incident D's(10^{15})	Target	$\tilde{R}_{calc}$ ($v_d = 0$)	$\tilde{R}_{calc}$ ($v_d = v_0/\sqrt{2}$)	$\tilde{R}_{calc}$(mono) (δ-function)
200	750	6.15	$(CD_2)_n$	1.71×10^{-2}	1.64×10^{-3}	1.00×10^{-13}
200	500	3.68	$(CD_2)_n$	1.78×10^{-4}	9.40×10^{-6}	4.32×10^{-19}
200	500	2.96	$TiD_{1.7}$	1.43×10^{-4}	7.56×10^{-6}	5.02×10^{-19}
250	400	2.27	$TiD_{1.7}$	9.05×10^{-6}	3.30×10^{-7}	2.50×10^{-22}
300	400	2.15	$TiD_{1.7}$	8.57×10^{-6}	3.13×10^{-7}	1.68×10^{-22}

Table 2. Comparison of the calculated number of protons R_{calc} with the experimental data of BNL normalized to our experimental conditions as indicated in the text (upper part) . Similar calculations have been made for interpreting our null results (lower part) . Computations are made for the cases where half the kinetic energy becomes thermal ($v_d = v_0 / 2^{1/2}$) and none of the kinetic energy becomes thermal (δ-function velocity distribution) (taken from Ref. 11).

CONCLUSION

For identical energy per deuteron (several hundred of eV range) and comparable number of deuterium atoms per cluster (around 100) , there is at least a two - orders - of - magnitude difference in the fusion rate induced by D_n^+ and $(D_2O)_n^+$ projectiles on similar deuterated targets . Our negative results with D_n^+ appear as a strong argument for the necessity of the presence of a heavy partner in the collision process such as the oxygen atom in the water cluster used at BNL , in agreement with the predictions cf most of the theoretical models presented so far . The unconclusive results obtained with the particular setup for N_n^+ bombardment demonstrate how careful one must be in order to avoid any beam contamination by fast particles . This justifies the many efforts of the BNL group to rule out such possibilities , as emphasized by the time-of-flight experiment reported in this Conference [14] .

One positive feature in our experiment is the effectiveness of the mass selection by the analyzing magnet : what is now needed is to investigate the effect in other experimental conditions . This will be done in the near future on our machine which has recently delivered analyzed beams of $(CH_4)_n^+$ clusters with intensities of a few nA for n around 50 .

REFERENCES

1. R.J. Beuhler , G. Friedlander and L. Friedman , Phys. Rev. Lett. , 63 ,1292 (1989).
2. R.J. Beuhler , Y.Y. Chu , G. Friedlander , L. Friedman , J.G. Alessi , V. LoDestro and J.-P. Thomas , to be published in Phys. Rev. Lett. 1991.
3. Y.K. Bae , D.C. Lorents and S.E. Young , to be published and these proceedings.
4. R.J. Beuhler , Y.Y. Chu , G. Friedlander , L. Friedman and W. Kunmann , J. Phys. Chem. , 94 , 7665 (1990).
5. M. Fallavier , J. Kemmler , R. Kirsch , J.-C. Poizat , J. Remillieux , J.-P. Thomas, Phys. Rev. Lett. , 65 , 621 (1990).
6. We thank P. Trocellier and C. Bonetti for providing us with the TiD target.
7. C. Carraro , B.Q. Chen , S. Schramm and S.E. Koonin , Phys. Rev. A 42 , 1379 (1990).
8. E. Fermi , Phys. Rev. 75 , 1169 (1949).
9. M. Hautala , Z. Pan and P. Sigmund , submitted to Phys. Rev. A (1991).
10. P.M. Echenique , J.R. Manson and R.H. Ritchie , Phys. Rev. Lett. , 64 , 1413 (1990).
11. Y.E. Kim , M. Rabinowitz , G.S. Chulick and R.A. Rice , Mod. Phys. Lett. B 5 , 427 (1991).
12. Y.E. Kim , M. Rabinowitz , Y.K. Bae , G.S. Chulick and R.A. Rice , to be published in Mod. Phys. Lett. B and these proceedings.
13. T. Mizota , K. Ohkushi , Y. Honjo , Y. Futami , Y.H. Pu , H. Toyokawa , J. Kasagi , B. Heusch , M. Hage-Ali , J. Stoquert and S.M. Lee , these proceedings.
14. G. Friedlander , these proceedings.

Search for Elementary Process of Cluster Impact Fusion Using Atomic Heavy Ions

T. Mizota, K. Ohkushi, Y. Honjo, Y. Futami, Y. H. Pu, H. Toyokawa and S. M. Lee
Tsukuba University, Ibaraki 305 JAPAN

J. Kasagi
Tokyo Institute of Technology, Ohokayama, Meguro-ku 152 JAPAN

B. Heusch, M. Hadge-Ali and J. Stoquert
CRN-Strasbourg, FRANCE

ABSTRACT

In order to investigate the role of heavy nuclei in the heavy water cluster impact d+d fusion, we have undertaken a series of d+d fusion experiments using atomic and molecular heavy ions (C, O, Ne, Ar, Kr and Xe) on the deuterated solid targets in the energy range of 0.4−4 MeV. The observed excitation functions of 3 MeV proton yields are well accounted for in terms of the two step knock-on cascade model. No distinct differences were observed in d+d fusion rate even when the number of atoms in the oxygen molecule (O^+, O_2^+ and O_3^+) were changed.

Extending the calculation of the excitation function of the atomic oxygen impact fusion rate up to the extremely low incident energies (~ keV), it has been found that the estimated proton yields are much smaller than the heavy water cluster impact fusion by 10 orders of magnitude.

This indicates that the recent observation of strong enhancement of cluster impact d+d fusion from Brookhaven cannot be explained by the sum of usual elementary processes but must be due to some kind of non-linear effect.

The present study was motivated by the recent Brookhaven experiment of heavy water cluster impact fusion[1] in which the unexpectedly large enhancement of d+d fusion rate was observed. When $(D_2O)_n^+$ clusters with total energy of $\sim$ 300 keV and n = 20−1000 were bombarded a TiD solid target, the measured maximum fusion rate is nearly 10^{10} times larger than the expected from free deuterons on the same energy ($\sim$ 300 eV) assuming that only the ten top layers of target atoms are participated in the fusion reaction.

This surprising results have been confirmed recently by an independent experiment of Y. K. Bae et al.[2]. Furthermore carrying the time of flight impact fusion measurements, Brookhaven group has also shown that the low molecular weight ion contaminants in the cluster beams cannot be responsible for the observed fusion events[3]. Therefore it appears that the strong enhancement of cluster impact d+d fusion is a real phenomenon and cannot be explained on the basis of a certain kind of artifacts.

Carraro et al.[4] tried to understand a number of mechanisms using usual linear concept of physics and are failed to account for the magnitude and cluster size dependence of the yields. Y. E. Kim et al.[5] developed a phenomenonological shock-wave theory and succeeded to reproduce the excitation function of cluster impact fusion rate.

The Lyon group has looked for d+d fusion induced by the pure deuterium clusters and did not observe any enhancement[6]. Furthermore performing d+d fusion experiments for 225 keV water cluster anions the Seattle group[7] has also shown that any fusion yield enhancement for clusters must occur for $n > 4$. On the other hand Beuhler et al.[1] reported that such enhancement was observed even by light water cluster $(H_2O)_n^+$ projectiles but with a rate 20 to 50 times smaller than $(D_2O)_n^+$.

These several experiments seem to show that the d(target)+d(target) collisions can be responsible in the observed d+d fusion rate and the role of oxygen in the projectile clusters must be investigated.

For investigating the role of heavy ions in projectile clusters we have performed a series of d+d fusion experiments using atomic and molecular heavy ion beams such as C, O, Ne, Ar, Kr and Xe in the energy range of 0.4−4 MeV.

The atomic and molecular beams of C, O and Ne were accelerated by 4 MV Van de Graaff of Tokyo Institute of Technology from 1.0 to 4.0 MeV and Ar and Xe atomic beams were obtained by 4 MV Van de Graaff of CRN Strasbourg from 0.4 to 2.0 MeV where the upper limit of energy was determined only by the power limit of the analysing magnet.

In fig. 1 is shown a typical experimental set-up. We used two silicon surface barrier detectors at $\pm155°$ to detect the 3 MeV protons and 1 MeV tritons coming from d+d fusion. Two detectors were located with the solid angle of 0.25 steradian and covered with thin mylar to prevent the scattered heavy ions. Magnets were mounted on upper and lower part of the target to reject the secondary electrons. The targets of TiD were made by the use of gas absorption method. The beam spot was always kept as $\sim$ 3 mm during the experiment.

Fig. 2 shows a typical energy spectrum observed using 800 keV Xe beam. The higher energy peak must come from 3 MeV protons and lower ones from 1 MeV tritons. The observed energies are a little bit smaller than the expected values : this must be due to the kinematical and energy loss effects. Proton and triton yields are nearly the same as is expected.

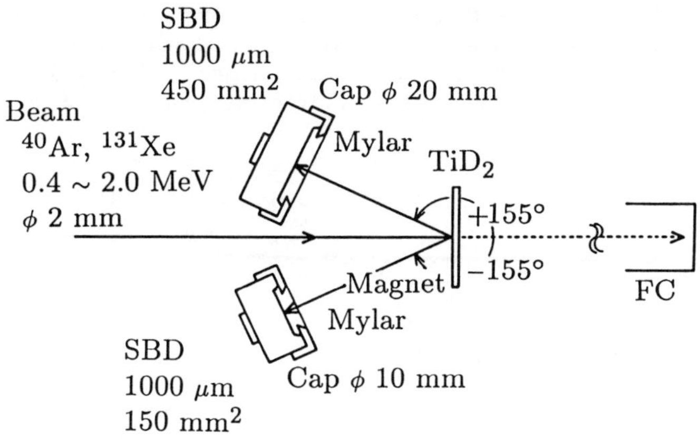

Fig. 1 The typical experimental set-up.

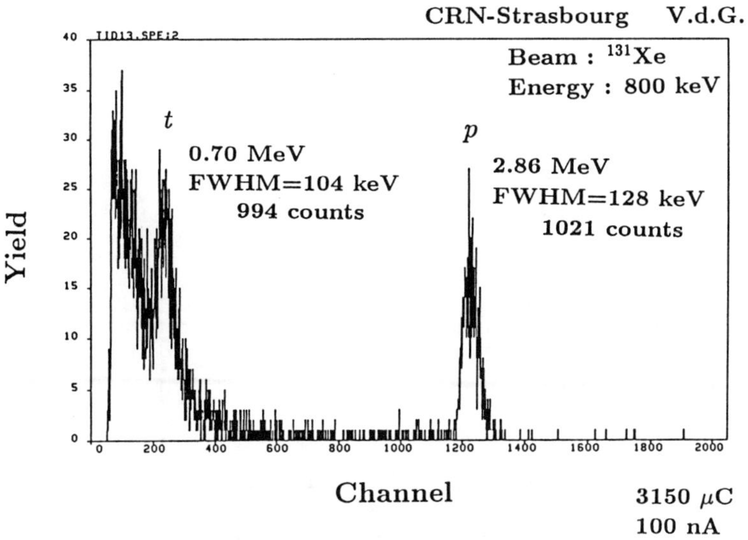

Fig. 2 The energy spectrum of d+d fusion induced by 800 keV Xe beam on the TiD target.

In fig. 3 and 4 are shown the excitation functions of d+d fusion rate per steradian using Xe, Ar, O and C beams. These excitation functions are tried to interpret in terms of two step process : at first the deuteron in target is accelerated by the Rutherford scattering of a projectile heavy ion and then the recoiled deuteron meets with the other deuteron to fuse each other.

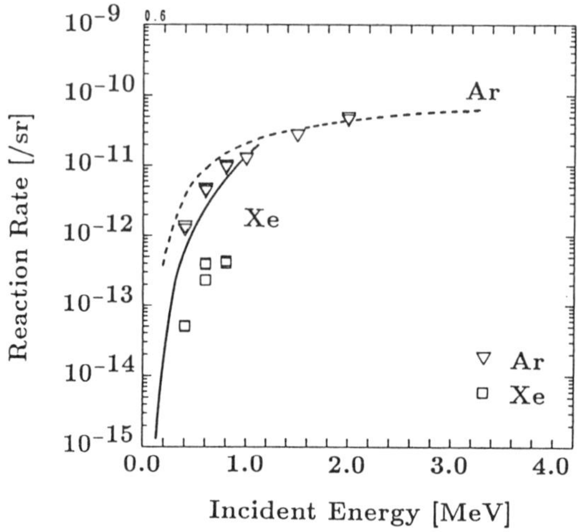

Fig. 3 The excitation functions of d+d fusion rate per steradian using Xe and Ar beams. ▽ and □ are the experimental values and the solid and dashed lines are the calculated ones.

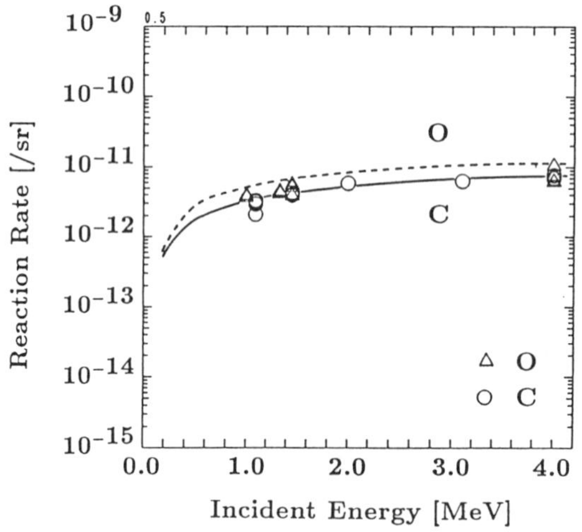

Fig. 4 The excitation functions of d+d fusion rate per steradian using O and C beams. △ and ○ are the experimental values and the solid and dashed lines are the calculated ones.

The cascade knock-on computer program is written to calculate such two step processes for the small segment of the incident energy ($E_0 \leq E \leq E_0 + \Delta E$) ranging from the induced energy to E = 0. Energy loss of heavy ion as well as that of deuteron was calculated by Ziegler's formalism[8]. For energy correction term electron screening effect[9] is taken into account. We made a normalization between the data points and calculated values at one data point of the highest incident energy of each heavy ion beam because of the uncertainty of effective deuterium target thickness. The experimental results are generally well reproduced by the present calculations, indicating that our assumption of two step processes must be an elementary process of d(target)+d(target) fusion at the relatively higher energies. Only in the case of Xe beam the discrepancies between the experimental results and the calculated ones are not negligibly small though the experimental trend was well reproduced. This might be due to two effects : one is due to the uncertainty of deuteron distribution on the surface region of titanium since the range of Xe is extremely small, and the other might be an application limit of Ziegler's formalism for high Z value of atom.

D+D fusion reaction rate are almost constant even if we change the number of oxygen atom and molecule such as O^+, O_2^+ and O_3^+ (fig. 5 (a) and (b)). This suggests that to see a considerable enhancement of fusion rate we need much large number of atoms in a cluster.

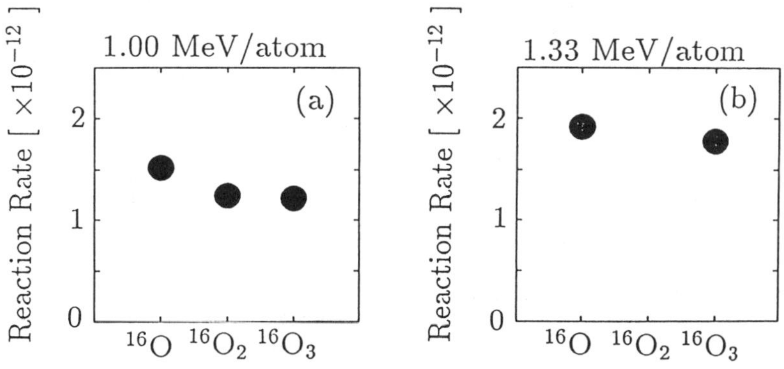

Fig. 5 The comparisons of d+d fusion reaction rate among O^+, O_2^+ and O_3^+ at 1.00 MeV/atom (a) and 1.33 MeV/atom (b).

To see whether BNL $(D_2O)_n^+$ impact fusion data can be reproduced by the sum of oxygen induced two step d+d processes or not, we have extended this calculation to extremely low energy. In fig. 6 are shown the solid line of such calculations together with the data points of $(H_2O)_n^+$ impact fusion[1]. The calculation is much smaller than their results by 17 orders of magnitude. Furthermore the shape of cluster size dependence are completely different from each other as shown in fig. 7 : for 100 atoms the calculated results are smaller than BNL data by over 13 orders of magnitude and for 200 atoms over 21 orders.

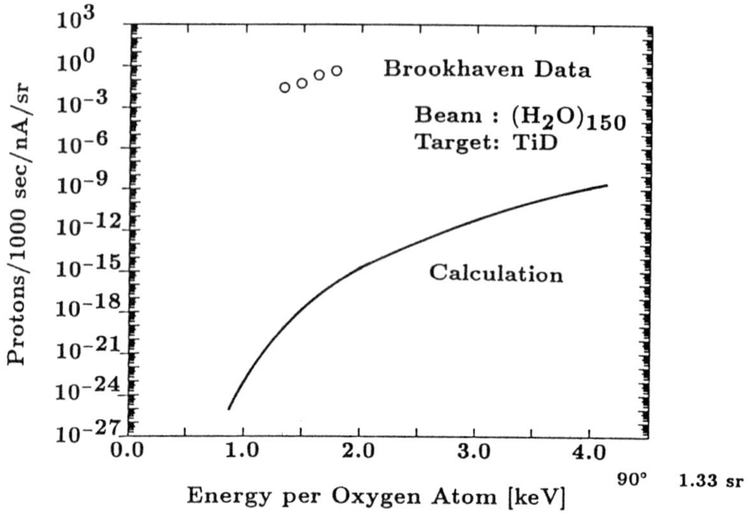

Fig. 6 The excitation function of d+d fusion rate. The open circles are the experimental results from $(H_2O)^{+}_{150}$ beam[1]. The solid line is extended calculation to the low energy (see the text).

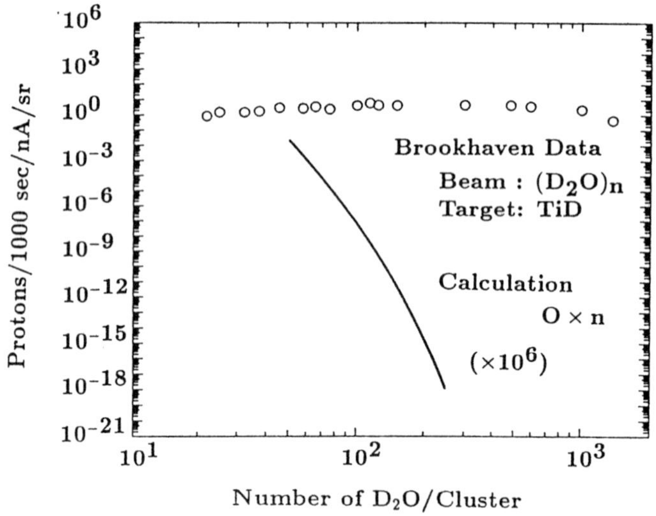

Fig. 7 The d+d fusion rate vs number of (D_2O) per cluster. The open cirles show Brookhaven data and the solid line is the calculated result (see the text).

To summarize the atomic heavy ion impact fusion for 0.1−4 MeV energy region can be well explained by the knock-on cascade model assuming two step elementary processes. No experimental discrepancies were observed among O^+, O_2^+ and O_3^+ for the same energy per single atom.

Therefore the recent observation of the unexpectedly large enhancement of d+d fusion using heavy water cluster impact cannot be accounted for in terms of the sum of the usual reaction process but rather this must be due to some kind of non-linear effect such as shock wave effect[5)] in short lived high temperature and condensed plasma in the solid.

REFERENCES

1) R. J. Beuhler, G. Friedlander and L. Friedman, Phys. Rev. Lett. 63(1989) 1292
R. J. Beuhler, Y. Y. Chu, G. Friedlander, L. Friedman and W. Kunnmann, J. Phys. Chem. 94(1990)7665

2) Y. K. Bae, D. C. Lorents and S. Young, contribution to this symposium.
Y. K. Bae, D. C. Lorents, S. Young and K. R. Stalder, Phys. Rev. Lett., submitted

3) R. J. Beuhler, G. Friedlander and L. Friedman, contribution to this symposium.
R. J. Beuhler, Y. Y. Chu, G. Friedlander, L. Friedman, J. G. Alessi, V. LoDestro and J. P. Thomas, Phys. Rev. Lett. 67(1991)473
R. J. Beuhler, G. Friedlander and L. Friedman, Acc. Chem. Res. 24(1991)198

4) C. Carraro, B. Q. Chen, S. Schramm and S. E. Koonin, Phys. Rev. A42(1990) 1379

5) Y. E. Kim, M. Rabinowitz, G. S. Chulick, R. A. Rice and Y. K. Bae, contribution to this symposium.
Y. E. Kim, M. Rabinowiz, Y. K. Bae, G. S. Chulick and R. A. Rice, Mod. Phys. Lett., to be published.

6) J. Fallavier, R. Kirsch, J. C. Poizat, J. Remillieux and J. P. Tomas, Phys. Rev. Lett. 65(1990)621

7) R. Vandenbosch, T. A. Trainor, D. I. Will, J. Neubauer and I. Brown, to be published.

8) H. H. Anderson and J. F. Ziegler, a series of *Stopping Powers and Ranges*, Pergamon Press, New York (1977)

9) L. C. Feldman and J. W. Mayer, *Fundamentals of Surface and Thin Film Analysis*,(1986)

EXPERIMENTAL CONFIRMATION OF CLUSTER-IMPACT FUSION

Y. K. Bae, D. C. Lorents, and S. E. Young
Molecular Physics Laboratory
SRI International
Menlo Park, CA 94025

ABSTRACT

We have investigated deuteron-deuteron fusion induced by impact of $(D_2O)_n^+$ and $(H_2O)_n^+$ clusters with n = 1-150 on deuterated polyethylene targets at energies of 135-225 keV. Both the energy dependence and magnitude of the fusion yield measured with $(D_2O)_{115}^+$ clusters confirm the results of Beuhler et al. The size dependence of the $(D_2O)_n^+$ fusion yields measured at the impact energy of 225 keV for small (n<10) clusters follow the "thermonuclear" model proposed by Carraro et al. However, for n>10 the size dependence rapidly deviates to higher yields from the model as n increases. For H_2O^+ and $(H_2O)_2^+$ at 225 keV, the yields roughly agree with the "knock-on" model by Carraro et al. No fusion event was observed for $(H_2O)_n^+$ clusters with n = 4-50, however, n = 115 clusters produced an observable fusion rate. The ratio between fusion rates of $(H_2O)_{115}^+$ and $(D_2O)_{115}^+$ is 5 ± 2%, in close agreement with the result, ~5%, measured by Beuhler et al. at 300 keV.

INTRODUCTION

Understanding the physics of materials in extreme conditions of pressure, density, and temperature is important to many scientific issues such as physics of interiors of stars, and to practical applications such as generation of nuclear fusion energy. One way of attaining such extreme conditions in laboratory scale experiments is to impact small particles with surfaces at high velocities. For example, in 1964 Winterberg proposed that shock waves triggered by impacting small particles at velocities greater than 100 km/s can achieve sufficiently high temperatures and pressures to ignite nuclear fusion.[1]

Recently Beuhler et al. have performed experiments based on such an idea.[2,3] In their experiment, singly charged D_2O cluster ions with diameters of ~10 Å were accelerated to ~100 km/s and impacted on surfaces of deuterated materials. Their observed DD fusion yield with $(D_2O)_{100}^+$ clusters was strikingly more than 25 orders of magnitude larger than the yields calculated with the conventional two-body nuclear fusion cross sections.[2,3] Furthermore, they observed DD fusion yield with $(H_2O)_{115}^+$ clusters; the rate was ~5% of that measured with $(D_2O)_{115}^+$ clusters. These results indicate the existence of a novel mechanism, perhaps an atomic scale analogue of macroscopic shock wave compression that initiates nuclear fusion reactions in cluster impact.

The results of Beuhler et al. have been challenged by an experiment performed by Fallavier et al.,[4] using pure deuterium clusters (D_{200}^+ - D_{300}^+) in the same range of incident energy per deuteron with that of Beuhler et al.[2,3] No fusion signal was observed. Several theoretical attempts to explain for cluster-impact fusion have had little success. It seems that further experimental confirmation and extension covering a wider range of cluster sizes and energies are needed for more complete understanding of this phenomenon. We report here the first experimental confirmation of the results by Beuhler et al. Moreover, we have extended the size range of the fusion measurement to investigate how the fusion yield evolves from that of isolated atom collisions to that of large cluster collisions.

THE ION SOURCE VACUUM SYSTEM ANDION OPTICS

Figure 1 shows the schematic diagram of the SRI cluster impact fusion facility. The ion source vacuum chamber consists of two sections vacuum-isolated by a skimmer. The first vacuum section contains the ion source and is made of a Lucite tube for electrical insulation. This section is pumped by a 300 L/s Roots Blower pump backed by a 40 cfm mechanical pump. The pressure of the first section is typically 0.1 Torr when the source is operating. The skimmer has a 1.6-mm-ID entrance aperature and is located 1 cm away from the exit of the nozzle. The cluster ions are focused by a potential difference applied between the skimmer and the nozzle. The requisite potential difference varies as a function of the mean cluster size. For example, focusing of cluster ions with a mean size of 100 molecules/cluster requires a skimmer potential of +30 V relative to the nozzle, whereas focusing of monomer ions requires -100 V.

After passing the first skimmer, the ions are accelerated through the second skimmer, which has a 6-mm-ID entrance aperture to ground potential. The vacuum section defined between the first and second skimmers is pumped by a 10-inch diffusion pump backed by a 60-cfm mechanical pump. The pressure of the first section is typically 10^{-4} Torr when the source is operating. After being accelerated to 500 eV, the cluster ions are focused by two sets of symmetric einzel lenses through 3-mm image apertures. Both lenses are operated at an electrical potential on their central elements equal to roughly two-thirds of the ion beam energy. Vertical and horizontal deflection plates follow each einzel lens to allow small corrections to the ion beam direction. Voltages applied to the deflection plates are usually less than one-tenth of the beam acceleration voltage.

The vacuum chamber that contains the ion optics is pumped by a 6-inch diffusion pump; with the source in operation, its pressure rose from 4×10^{-7} Torr to 2×10^{-6} Torr. The accelerated and collimated cluster ion beam is mass selected by a 40-cm-radius, 60° electromagnetic sector capable of bending 5500-keV-amu ions, passed through a 2-cm-diameter aperture, and further focused and collimated by the third set of einzel lenses and deflection plates. The beam is then sent to a movable Faraday cup located 65 cm away from the exit of the magnetic sector. The 2-cm aperture yields an effective mass resolution $\Delta M/M$ of ~0.05, which was measured with small cluster ions on the Faraday cup. Moving the Faraday cup out of the beam path allows the beam to pass through another 2-cm-diameter aperture and enter the acceleration column of the 300-keV accelerator. The effective mass resolution of the beam measured in front of the target after acceleration was also 0.05.

HIGH INTENSITY WATER CLUSTER ION SOURCE

The detailed cross sectional drawing of the ion source together with skimmers is shown in Figure 2. The ion source was built by modifying the body of the commercial pulsed valve (Lasertechnics). The source assembly is mounted on a

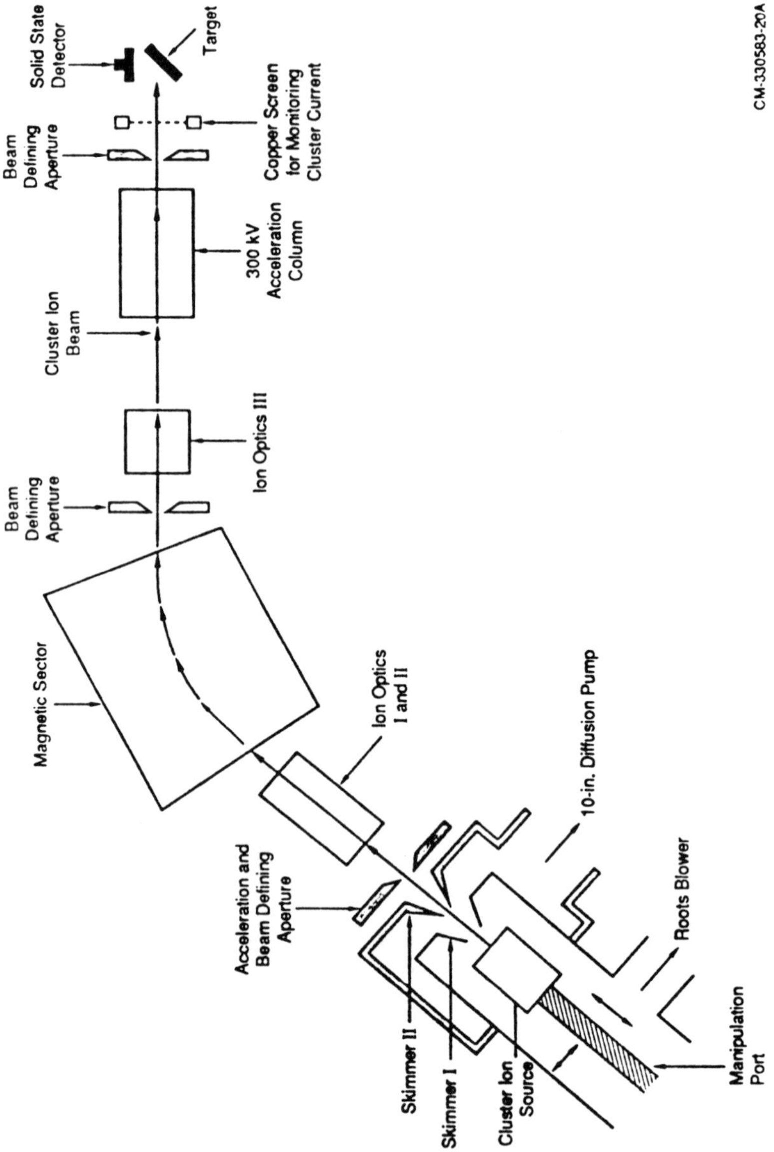

Figure 1. SRI Cluster Impact Facility.

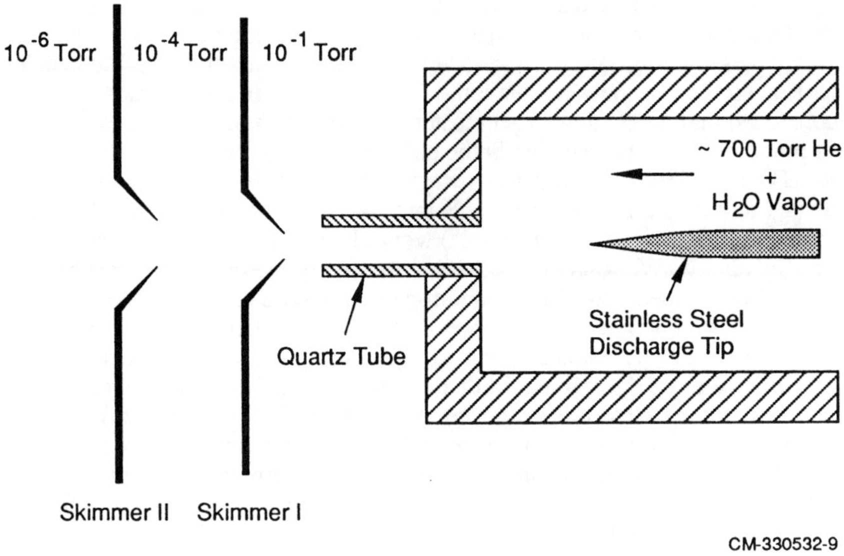

Figure 2. Schematic of the water cluster ion source.

manipulation port that allows tilt adjustments on two axes orthogonal to the ion beam plus translation of the assembly along the ion beam direction. The purpose of these adjustments is to allow precise positioning of the assembly relative to the skimmer and ion optics, which are fixed to the apparatus. The optimum distance from the nozzle to the skimmer depends on the mean size of the clusters produced. The typical distance for generating clusters with a mean size of 100 molecules/cluster is 1 cm. The generation of larger clusters requires a larger distance, and shorter ones require a smaller distance. The source assembly and the first skimmer are maintained at 500 V to establish the final kinetic energy of the cluster ions after they are accelerated to ground potential on the second skimmer.

The principle of the new cluster ion source is very similar to that described by Beuhler et al.[2,3] The ion source produces high intensity cluster ion beams by generating abundant seed ions in the cluster source and expanding the ions that nucleate the condensation of neutrals for the growth of the cluster ion in a He and water vapor mixture. The seed ions are produced in a discharge of He and water vapor mixture between a stainless steel tip and the wall of the source near the nozzle. The mixture is produced by bubbling He through water in a gas washing bottle at room temperature. Production of water clusters of 100 molecules requires a He pressure of ~700 Torr. The nozzle-discharge-tip assembly is made of a commercial Lasertechnics nozzle. The tip is held to the nozzle by an adapter made of Macor. The tip holder has holes that allow conductance of the He water vapor mixture. The typical distance between the tip and the nozzle is 4 mm, and the typical current and impedance of the discharge are 2-3 mA and 100-200 kΩ, respectively.

Initially we had tried a Lasertechnics nozzle with a 0.5-mm-diameter hole and observed cluster ions with a mean size only up to 30 molecules/cluster. Increasing the source pressure to generate clusters with a mean size larger than 30 molecules/ cluster triggered an arc between the ion source assembly and the vacuum chamber. We concluded that generating larger cluster ions with the nozzle requires a much higher pumping capacity to maintain the chamber pressure below the threshold of arcing. Thus, we investigated various ways of reducing the conductance of the nozzle and found that nozzles made of 8-mm-long, 0.5-mm-ID quartz tubes (which have much less conductance than a hole) worked best. The tubes were glued to the nozzle with epoxy. We found that covering the end of the quartz tube facing the discharge tip with metal prevents generation of an intense cluster ion beam. The reason for this finding is not well understood now. We also tried tubes made of other materials, such as stainless steel, alumina, and glass. The stainless steel tubes did not produce any cluster ions. Both alumina and glass tubes worked as well as quartz tubes, but the glass tubes cracked in only a few hours of operation.

A mass spectrum of the produced $(D_2O)_n^+$ cluster ions measured with the movable Faraday cup is shown in Figure 3. The spectrum has a full width at half maximum (FWHM) of 30% and should represent the mass distribution of the ion source convoluted with the resolution of the magnetic sector. The mean size of the clusters can be readily controlled by tuning the pressure of the source and the discharge current. Lower discharge currents and the higher pressures result in larger cluster ions. For generating monomer ions, no He pressure is necessary. With the source systems tuned for clusters of 100 molecules, the maximum observed current of cluster ions on the Faraday cup is 7 nA. Higher currents are generated with the source tuned for smaller clusters, and vice versa.

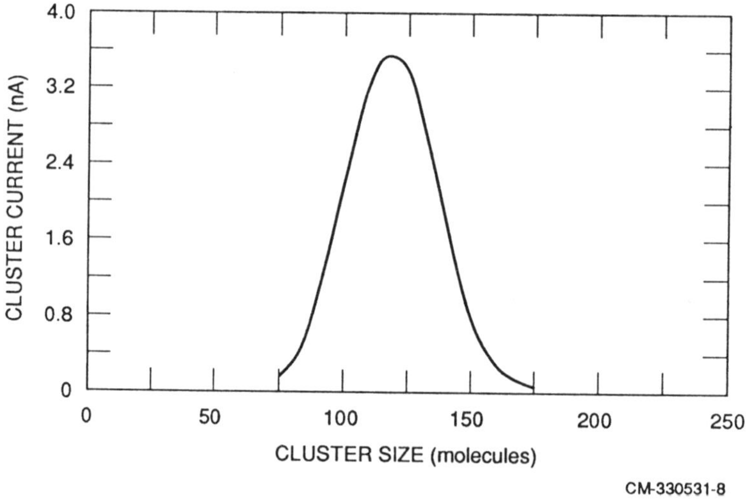

Figure 3. Size distribution of the generation $(D_2O)_n^+$ clusters from the ion source.

FUSION DIAGNOSTICS SYSTEMS

The mass selected and collimated cluster ions are accelerated through a 45-cm-long acceleration column of the 300-keV accelerator and, then enter a chamber that contains the target and the detector which is maintained at the full terminal voltage. The target is a 2-cm-diameter disk made of perdeuteropolyethylene [$(C_2D_4)_n$] and is tilted 45 degrees with respect to the beam axis. The current of the accelerated cluster ions is measured by monitoring secondary electrons from a copper grid upstream of the target. The grid is maintained at 50 V relative to the high voltage terminal, and the loss of the electrons from it is monitored with a DC current amplifier connected through a fiber optic link and monitored by an ADC in a CAMAC crate interfaced with a computer. Fusion product particles are detected by a particle detector, which is a 300-mm^2 Ortec "ruggedized" silicon solid-state detector.

The active surface of the detector is covered with a 50 μg/cm^2 layer of aluminum. The detector is located 2.5 cm from the center of the target. The amplified detector pulses are sent through a fiber optic link to a pulse height analyzer interfaced with a computer. The detector system, including the fiber optic link, is calibrated with 5.48-MeV α particles from ^{241}Am and with DD fusion particles, protons, tritons, and ^{3}He produced by D^+ impact on the $(C_2D_4)_n$ target. Typical fusion signals observed in the collisions of $(D_2O)_n^+$ clusters with a $(C_2D_4)_n$ target in this facility are shown in Figure 4.

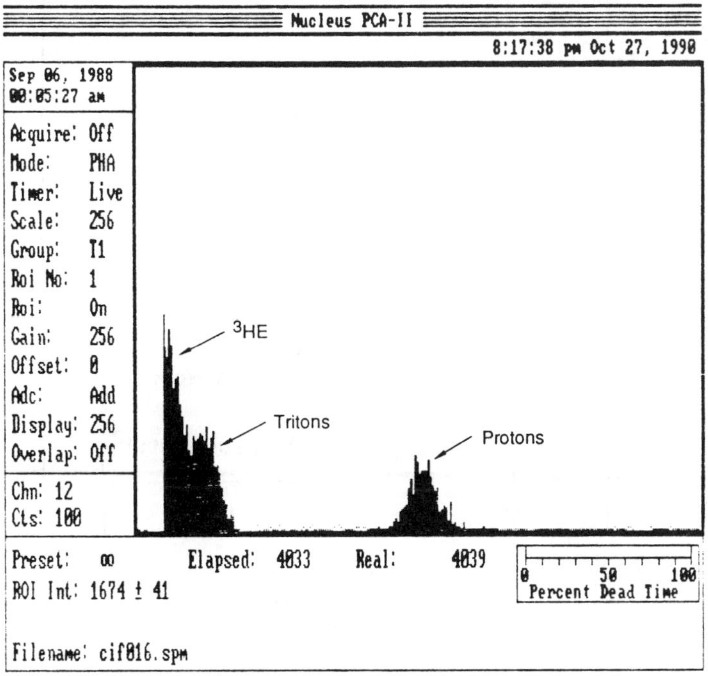

Figure 4. Typical DD fusion signals observed with the impact of $(D_2O)_n^+$ clusters on a $(C_2D_4)_n$ target.

RESULTS AND DISCUSSIONS

We measured DD fusion yields with $(C_2D_4)_n$ targets. The targets were made by compacting $(C_2D_4)_n$ powder mixed with 1 wt% colloidal carbon under ~1 kbar. Figure 5 shows the energy dependence of the 3-MeV fusion proton yields measured when D_2O clusters impacted on the targets. The incident clusters had a mass distribution with a peak at 115 molecules/cluster and 5% FWHM. In Figure 5, our data are shown as open squares and those of Beuhler et al. are shown as filled-in circles.[3] Both the magnitude and energy dependence of our data are in excellent agreement with those of Beuhler et al. The agreement is striking considering that our experimental arrangement is very different from theirs.

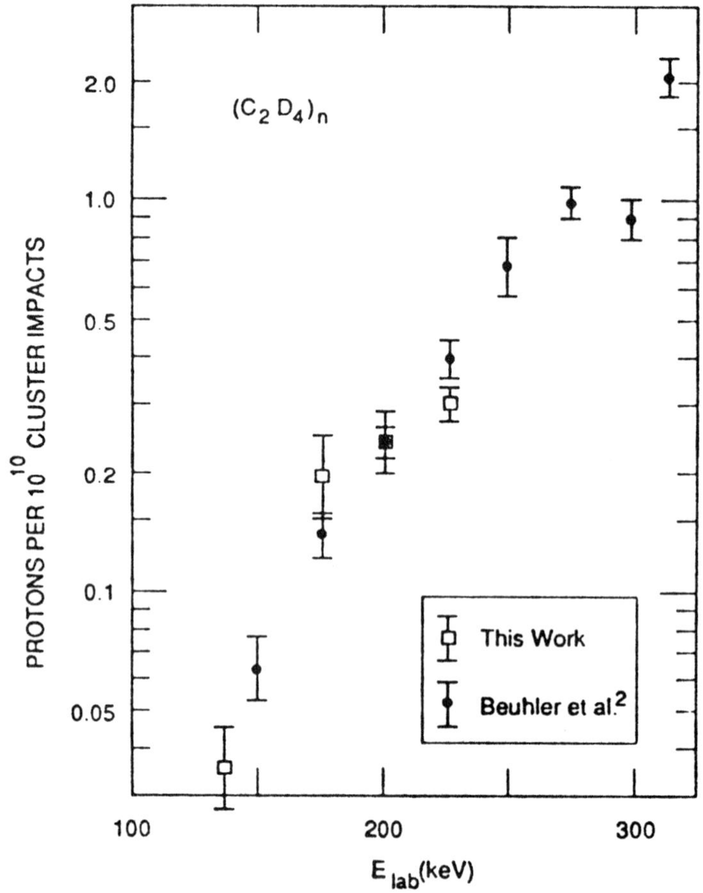

Figure 5. Energy dependence of the observed 3-MeV protons from DD reactions with the impact of $(D_2O)^+_{115}$ clusters on a $(C_2\,D_4)_n$ target.

Figure 6 shows the cluster size dependence of the total DD fusion yield measured at a cluster beam energy of 225 keV. Note that a log-log scale is used for this plot. Beuhler et al. obtained the cluster size dependence at 300 keV using clusters with a mass distribution of 38% FWHM compared to our 5% FWHM. Their yield has a peak at n = 200 and decreases by a factor of three at n = 100 and 350. Our measured yield at 225 keV has a less pronounced size dependence and seems to peak near n = 100. We believe that our size-dependence differs from that of Beuhler et al. because of our lower cluster beam energy.

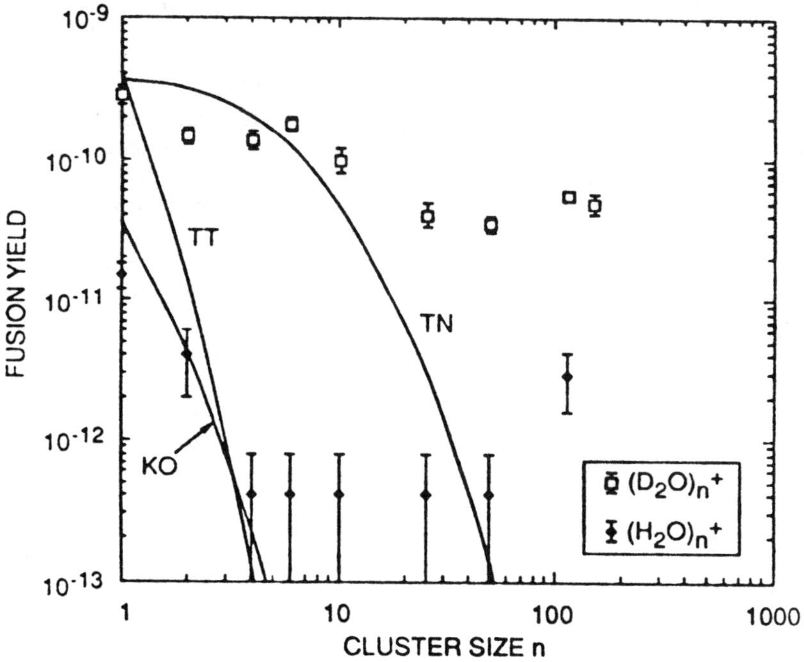

Figure 6. The total DD fusion yield at the impact energy of 225 keV as a function of the cluster size n. Note that a log-log scale is used. Open squares: fusion yields of $(D_2O)_n^+$ clusters. Filled-in diamonds: fusion yields of $(H_2O)_n^+$ clusters. Note that for $(H_2O)_n^+$ clusters with n = 4-50 no fusion event was observed. The data for n = 4-50 represent the estimated upper bounds of the fusion yields. TT curve: the calculated values using the thick-target model by Carraro et al.[8] KO curve: the calculated values using the knock-on model by Carraro et al.[8] TN curve: the calculated values using the thermonuclear model by Carraro et al.[8]

DD fusion reactions were also observed for $(H_2O)_n^+$ clusters impacting on the $(C_2D_4)_n$ targets. The results are also shown in Figure 3. Both the magnitude and cluster size dependence of the fusion yield of H_2O clusters are very different from those of D_2O clusters. For n = 1 and 2 significant fusion signals were observed.

No fusion events were observed for n = 4-50 clusters during an integration time of 3 hours for each datum. To estimate upper bounds of these rates, we assumed the rates of 0.5 cts/3 hours, an order of magnitude below the observed rate for $(H_2O)_2^+$ clusters. However, $(H_2O)_{115}^+$ clusters produced an observable fusion yield which is $5 \pm 2\%$ of the yield of $(D_2O)_{115}^+$ clusters. This ratio is remarkably similar to that of ~5% observed by Beuhler et al. at 300 keV.

Several researchers have suggested that the observed fusion yields are produced by contamination of the cluster ion beam with small ions, such as D^+, D_2O^+, or D_3O^+ that get accelerated to high energies. In our setup the magnetic sector shields the source from the accelerator column so that neutrals of molecules and small clusters, which can be ionized by the high energy electrons from the negative high-voltage terminal, cannot enter the accelerator column. This differs from the Brookhaven experiment which used a direct line of sight geometry. Furthermore, the mass distribution from the magnetic sector is triangular with 5% FWHM without a long tail towards small clusters. This also significantly differs from the Brookhaven experiment which used a quadrupole mass filter. Therefore, in our setup, three possible mechanisms that contribute to beam contamination are: 1) collisional dissociation of clusters by residual gas molecules in the vacuum chambers including the acceleration column, 2) ionization of the residual D_2O vapor in the accelerator column (migrating from the ion source) by high energy electrons from the negative high-voltage terminal of the electrostatic accelerator, 3) ionization by the cluster ion beam of sputtered and "splashed" D atoms from the target.[5-8]

The contribution of the first and second contamination mechanisms should depend on the pressure in the flight chamber and accelerator column. We have measured fusion yields at several different pressures between $2\text{-}8 \times 10^{-7}$ Torr in the flight chamber connected to the acceleration column. No dependence of the yield on pressure was detected over this pressure range. Therefore, we conclude that the contribution of the first and second mechanisms is negligible in the observed fusion signals.

The possible contribution of the third mechanism was checked by measuring fusion yields of $(H_2O)_n^+$ clusters which are chemically identical to $(D_2O)_n^+$, clusters. $(H_2O)^+_{100}$ and $(D_2O)^+_{100}$ clusters impacting at 200 keV to 300 keV sputter many more atoms and molecules from the target than from the projectile clusters. For example, Matthew et al.[5] observed that at 300 keV $(H_2O)_{100}^+$ can sputter as much as a factor 100 more target (carbon) material than the projectile material. Therefore, the "splash" effect should contribute much less than the target sputtering effect, and if the third mechanism was responsible for the observed fusion yields, the fusion yields for $(H_2O)_n^+$ and $(D_2O)_n^+$ clusters would be nearly identical. However, as we note earlier, the fusion yields of $(H_2O)_n^+$ clusters are not detectable for n = 4-50, and dramatically differ from those of $(D_2O)_n^+$ cluster for other values of n. In any case, the following estimates the upper limit for the fusion yields that the "splash" effect can produce. The "splashed" atoms and molecules are much more likely to be neutral. Then, the ejected neutrals from the target can penetrate into the accelerator column and accelerate back once they are charged by collision. Let us assume that all the projectile deuterium atoms are ejected and some of them reach the entrance of the accelerator to gain full acceleration energy. The impact of $\sim 10^{10}$ clusters/s (1.5 nA) produces 2×10^{12} deuterium atoms/s. (If molecular species are produced the fusion yields are less.) The upper limit of the fusion yield, Y_{sp}, contributed by such "splashed" deuterium atoms per cluster impact can be estimated from

$$Y_{sp} \sim (I_D/v)\, \sigma\, L\, Y^D{}_{TT} \tag{1}$$

where $I_D \sim 4 \times 10^7$ /s/cm^2 is the deuterium atom flux at the accelerator entrance (~65 cm away from the target), $\sigma \sim 10^{-14}$ cm^2 is the charge exchange cross section, $v \sim 10^5$ cm/s is the velocity of the deuterium atoms, L~10 cm is the length of the interaction region for allowing full energy near the accelerator entrance, and $Y^D{}_{TT} \sim 5 \times 10^{-7}$ is the thick target fusion yield of D^+ ion impacting on a $(C_2D_4)_n$ target at 225 keV. These values result in the upper bound of $Y_{sp} \sim 2 \times 10^{-17}$ which is much smaller than 5×10^{-11} for $(D_2O)_{100}^+$ cluster impact. Therefore, we conclude that the contribution of the third mechanism is negligible, and that the observed fusion signals are produced by clusters rather than small deuterated ions.

The first attempt to explain the experimental results was reported by Echenique et al.[9] Using a model based on the large energy and density fluctuations in swift-cluster impact, they have obtained qualitatively similar magnitudes and energy dependence of the fusion yield to those of the experimental data. In their model the major contribution to the enhanced fusion yield results from the small number of D atoms in the high-energy tail of the Maxwell-Boltzmann velocity distribution function. However, the existence of such a high-degree thermalization mechanism was questioned by Carraro et al.[8]

Carraro et al.[8] have developed analytical formulae for fusion yields using several different models. For the purpose of comparing the general trends and orders of magnitudes, Figure 3 also shows the results of theoretical calculations of fusion yields for $(D_2O)_n^+$ clusters impacting on a TiD target at 225 keV using various models. The TT curve represents the thick target model which assumes that deuterons in the projectile clusters undergo fusion reactions with the target deuterons while slowing down in the target due to the stopping power. The KO curve represents the knock-on model, which considers substrate deuterons knocked-on by oxygen atoms in incident clusters. This curve was obtained by dividing the Eq. (8) of Ref. 8 by a factor of 10 to compensate for the deficiencies of the parameterization in their modeling as suggested by the authors.[8]

The TN curve represents a thermonuclear model by Carraro et al.[8] The model assumes that the incident cluster thermalizes in the target and produces a hot gas of atoms. The model also assumes that the cluster thermalizes with an equal number of the target atoms. It is interesting to note that our measured data for $(D_2O)_n^+$ clusters roughly follow the TN model curve for $n<10$. For $n>10$, however, our data rapidly deviate from the model as n increases, indicating existence of some unknown effects. Furthermore, the equal thermalization assumed by Carraro et al. of the deuterons in the incident clusters and target cannot explain the dramatic difference of fusion yields between $(D_2O)_n^+$ and $(H_2O)_n^+$ clusters for $n<10$. The fusion yields of $(H_2O)_n^+$ clusters with $n = 1$ and 2 seem to agree with the value based on the steeply decreasing knock-on model which does not involve highly correlated collisions.

The large difference between the fusion yields for $(H_2O)_n^+$ and $(D_2O)_n^+$ clusters suggests that the energy enhancing mechanism of projectile D atoms is different from that of target D atoms. This point was suggested independently by both Rabinowitz et al.[10] and Bitensky and Parilis[11] very recently. They proposed that the energy of deuterons in the incident cluster is highly enhanced because of multiple backscattering between incident cluster deuterons, heavy target atoms, and heavy cluster atoms. As a result, the energy of deuterons in the clusters is enhanced much more than the energy of deuterons in the target explaining the experimental observation. Furthermore, the proposed energy enhancing mechanism requires presence of heavy atoms in the incident cluster, apparently explaining the null results by Fallavier et al.[4] with the pure D clusters.

In summary, we have measured deuteron-deuteron fusion yields by colliding $(D_2O)_n^+$ clusters with n = 1-150 on $(C_2D_4)_n$ targets at the impact energies of 135-225 keV. Although the geometry of our experimental apparatus and the method of mass-selection are very different from those by Beuhler et al., our results are in excellent agreement with those by Beuhler et al. strongly supporting the credibility of the newly discovered phenomenon, cluster-impact fusion. Further experimental and theoretical effort on this important fusion reaction process is clearly needed. We are in the process of extending both the impact energy and size ranges of the clusters with the use of multiply charged large clusters.

This work has been supported by the AFOSR, Contract F49620-90-C-0048, and by the Independent Research and Development Program of SRI International. We acknowledge the very helpful and stimulating discussions with Drs. K. R. Stalder, D. J. Eckstrom, R. J. Beuhler, G. Friedlander, L. Friedman, S. E. Koonin, M. Rabinowitz, Y. E. Kim, and J. R. Peterson. We are grateful to Prof. W. Kunkel for the loan of the TNC accelerator. We also appreciate technical contributions of M. Nicholson, W. Olson, and J. W. Stearns.

REFERENCES

1. F. Winterberg, Z. Natur. **A190**, 231 (1964).
2. R. J. Beuhler, G. Friedlander, and L. Friedman, Phys. Rev. Lett. **63**, 1292 (1989).
3. R. J. Beuhler, Y. Y. Chu, G. Friedlander, L. Friedman, and W. Kunnmann, J. Phys. Chem. **94**, 7665 (1990).
4. M. Fallavier, J. Kemmiler, R. Kirsch, J. C. Poizat, J. Remillieux, and J. P. Thomas, Phys. Rev. Lett. **65**, 621 (1990).
5. M. W. Matthew, R. J. Beuhler, M. Ledbetter, and L. Friedman, Nucl. Inst. Meth. Phys. Res. **B14**, 448 (1986).
6. M. H. Shapiro and T. A. Tombrello, Phys. Rev. Lett. **65**, 92 (1990).
7. F. E. Cecil and J. A. McNeil, Phys. Rev. Lett. **64**, 2210 (1990).
8. C. Carraro, B. Q. Chen, S. Schramm, and S. E. Koonin, Phys. Rev. **A42**, 1379 (1990).
9. P. M. Echenique, J. R. Manson, and R. H. Ritchie, Phys. Rev. Lett. **64**, 1413 (1990).
10. M. Rabinowitz, Y. E. Kim, R. A. Rice, and G. S. Chulick, Paper presented at Int. Work. Anol. Nucl. Eff. in Deuterium/Solid Syst., 1990 (to be published in AIP proceedings).
11. I. S. Bitensky and E. S. Parilis, submitted to Phys. Rev. Lett.

HOT PLASMA SHOCK–WAVE THEORY OF CLUSTER–IMPACT FUSION

Y. E. Kim,* M. Rabinowitz, Y. K. Bae,*** G. S. Chulick,* and R. A. Rice***

*Department of Physics, Purdue University, West Lafayette, IN 47907, USA
**Electric Power Research Institute, Palo Alto, CA 94303, USA
***Molecular Physics Laboratory, SRI International, Menlo Park, CA 94025, USA

ABSTRACT

This paper demonstrates that cluster–impact fusion can be understood as "**hot fusion**" on a microscopically small atomic scale. We show that it is improbable to account for the data as an artifact of contamination. By means of theoretical analysis based on a universal scaling equation, our high temperature model is shown to be consistent with the known data of recent deuteron–deuteron fusion experiments in which clusters of D_2O, H_2O, and D are accelerated onto deuterated targets. Furthermore, we show that the line broadening of the experimentally obtained proton spectrum supports our prediction of the high temperature in the impact region. Although we have focused on a temperature enhancing shock–wave model, our methodology and scaling equation are sufficiently general to encompass additional processes such as pinch instability heating. We estimate the present efficiency of this process, and predict higher efficiencies for various experimental conditions.

I. INTRODUCTION

The plausibility of generating nuclear fusion ignition temperatures ($\sim 10^8$ K) and densities ($\sim 1g/\text{cm}^3$) from shock waves created by micron–size particles impacting at $\sim 10^8$ cm/sec on a liquid tritium–deuterium target was proposed by Winterberg in 1963.[1] At about the same time, Harrison[2] independently suggested a similar possibility. Serious experimental pursuit of such ideas, in 1989 by Beuhler et al. (BFF[3] and BCFFK[4]) for much smaller molecular–size "cluster" particles, was successful in achieving unexpectedly high rates of $\sim 1-10$ sec^{-1} $/D-D$ for deuterium–deuterium ($D-D$) fusion. They accelerated singly charged clusters of D_2O molecules ($(D_2O)_n^+$, $n = 25-1300$) to 150 to 325 keV (with a beam current of $\sim 1nA$) onto TiD, C_2D_4 and $ZrD_{1.65}$ targets.[3,4] In addition, $D-D$ fusion was observed with $(H_2O)_{115}^+$ clusters impacting upon a C_2D_4 target.[4] A virtual invalidation of the former results was reported by Fallavier et al.[5] (FKPRT) who carried out similar experiments using pure deuterium clusters $D_{200}^+ - D_{300}^+$ with kinetic energies of 100 – 150 keV, which correspond to center of mass (CM) energies per incident D of 200 – 375 eV. They observed no fusion events upon bombarding TiD and C_2D_4 targets. However, the results of Beuhler et al.[3,4] have recently received confirmation by Bae et al.[6] (BLY) for $(D_2O)_{115}^+$ and $(H_2O)_{115}^+$ cluster beams impacting on a C_2D_4 target.

The experimental yields for $(D_2O)_{100}^+$ clusters are $\sim 10^{25}$ times higher than that expected[7,8] from single D^+ clusters at $D-D$ CM energies of 150 eV$/D$, and $\sim 10^{100}$ times higher than at 15 eV$/D$ for $(D_2O)_{1000}^+$. A number of theoretical models[9–13] have been proposed but they underestimate the observed fusion yields[3,4,6] by many orders of magnitude for large clusters with $n > 100$. Of these models, none but Refs. 12 and 13 explain the positive results[4,6] for $(H_2O)^+_{115}$ clusters and the negative results[5] for $(D)^+_n$ clusters.

In this paper, we present a cluster–impact shock wave model[14] which explains and reproduces the observed $D - D$ fusion rates[3–6] for $(D_2O)^+_n$, $(H_2O)^+_n$, and $(D)^+_n$ clusters striking TiD, C_2D_4, and $ZrD_{1.65}$ targets. The model is expressed in the form of a universal scaling equation with two physically reasonable parameters: (1) the temperature enhancement factor $\tau(n)$, and (2) the fraction $\nu(n)$ of projectile cluster and target D atoms reaching high temperature. These parameters are determined by fitting them to the experimental data for $(D_2O)^+_{100}$ clusters impacting on a C_2D_4 target. The same values for ν and τ also give good fits to the results of recent $(D_2O)^+_{100} - TiD$ and $(D_2O)^+_{100} - ZrD_{1.65}$ experiments.[4] We present predictions[15,16] for fusion yields for $(D_2O)^+_{100}$ clusters impacting on other targets (deuterated and tritated) for wider energy ranges than have been hitherto used experimentally, from our universal scaling equation using the same values for ν and τ as determined for C_2D_4. For the cases of $(D_2O)^+_{100} - C_2T_4$ and $(D_2O)^+_{100} - C_2D_4$, we give predictions for the energy production efficiency as a function of the cluster energy.

Although some suspect that these anomalously high yields are due to light ionic contaminants in the beam, we show[15,16] that this is highly unlikely because the magnitude and/or slope of the expected contaminant fusion yield as a function of energy differs significantly with the experimental data.[3,4] Light–fragment contaminants (e.g., D^+, D_2O^+, D_3O^+) produce proton yields that are shown to be insufficient to account for the experimental results given the determined upper limits[4] for these contaminants in the cluster–impact fusion experiments.

II. SHOCK–WAVE MECHANISM FOR CLUSTER–IMPACT FUSION

In the cluster shock–wave impact fusion mechanism, $(D_2O)^+_n$, $(H_2O)^+_n$ or $(D_2)^+_n$ molecular clusters impact on the deuterated target with a cluster velocity of $\sim 10^7$ cm/s in an impact period of $\Delta t \approx 10^{-14}s$. It has been observed in cluster impact experiments[17] that energetic molecular clusters can produce craters with a diameter comparable to the size of the cluster. Initially, the leading edge of the projectile cluster creates upon impact a plasma consisting of target and cluster atoms (ions) and electrons which is partially trapped in the micro–crater of several Å size between the target cavity surface and the trailing cluster atoms. The trailing cluster atoms will then move through this plasma before impacting on the cavity surface, thus creating a shock wave thorough the plasma as shown in Fig. 1. Although the total number of atoms is limited to $3n$ per $(D_2O)^+_n$ cluster, the total number N_t of atoms in a cluster–beam with a current of $\sim 1nA$ is large, $N_t = 0.625 \times 10^{10}(3n)/s (\approx 2 \times 10^{12}/s$ for $(D_2O)^+_{100})$, so one has a reasonable statistical system moving into the target.

II.1. Shock–Wave Generation and Energy Loss

Previously, it was thought that macroscopic projectiles were necessary to produce shock waves. However, we have shown[16] that atomic or molecular clusters can produce shock waves due to inertial confinement of the compressed fluid. In

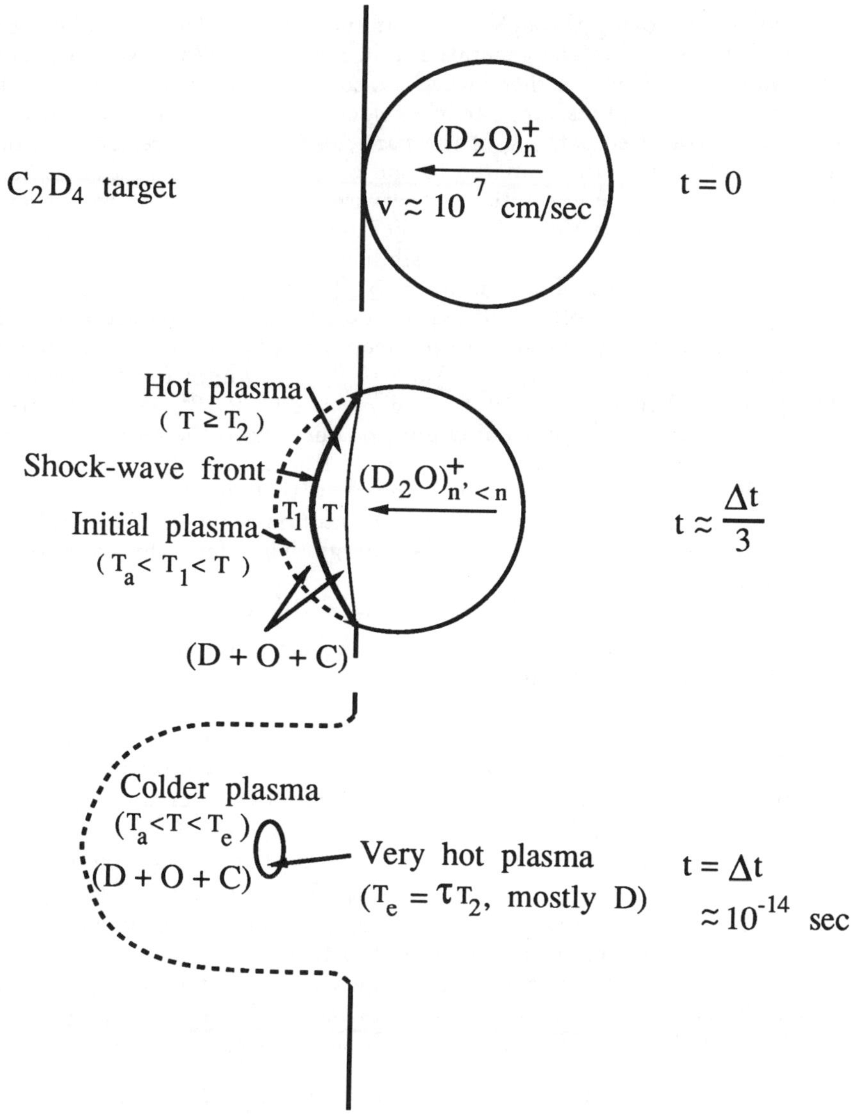

Fig. 1. Schematic illustration of shock wave generated by cluster impact as a function of time.

addition, we have also demonstrated[16] that energy loss mechanisms such as radiation, thermal conductivity, and adiabatic expansion are not significant during confinement.

A projectile moving through a fluid at supersonic velocity produces a shock wave when the speed of the generated fluid motion is greater than the speed of sound (supersonic flow). A shock wave is a compression wave of large amplitude, across which density, pressure, particle velocity, and temperature change drastically. If the mean free path, λ, of the particles in the compressed fluid in front of the projectile is small compared to the projectile's cross–sectional diameter, d, a shock wave can be generated. Otherwise, the compression would normally be unloaded around the sides of the projectile. Let us see if the condition $\lambda < d$ is met, before considering another possibility.

The mean free path of a D is $\lambda_D = \sum_i [(n_x \sigma_{Dx})_i]^{-1}$, where n_x is the compressed number density of constituent x and σ_{Dx} is the cross–section for $D - x$ scattering. Winterberg considered the mean–free path in a fully ionized hydrogen plasma, (eq. (3.1), Ref. 1), $\lambda = 1.2 \times 10^4 T^2/n_x$, where T is the temperature of the plasma. For $T = 3 \times 10^7 K$, and $n_x = 0.92 \times 10^{23}$ cm^{-3}, Winterberg[1] obtained $\lambda = 1.2 \times 10^{-4}$ cm, leading him to conclude to the necessity for micron size particles. Although it is not explicit in Ref. 1, eq. (3.1) appears to come from the Rutherford scattering cross section with long range Debye shielding, $\sigma_R \propto (Z_1 Z_2/E_{12})^2$, where Z_1 and Z_2 are the atomic numbers of the fully ionized colliding species and E_{12} is their relative energy. Therefore, this equation can be generalized to

$$\lambda = 1.2 \times 10^4 T^2/[n_x (Z_1 Z_2)^2]\text{cm}. \tag{1}$$

For comparison, we will also take $T = 3 \times 10^7 K$, as this is close to the temperature we find for cluster–impact fusion.

Equation (1) neglects short–range screening effects on the Coulomb potential, and hence gives an optimistic lower limit for λ_D, assuming full ionization Z. For a $(D_2O)_n^+$ cluster impacting on $ZrD_{1.65}$, with $Z_1 = 1$ and $Z_2 = 40$, we obtain the most optimistic estimate of $\lambda_D \sim 6$ Å. The cluster diameter is $d \sim (4 \text{ Å})n^{1/3}$, so that the shock wave condition requires that $n > (\lambda_D/4 \text{ Å})^3$. In this case, $n > 3$, so $n = 10$ will easily fulfill the condition $\lambda < d$. For $(D_2O)_n^+$ on $TiD(Z_1 = 1$ and $Z_2 = 22)$, $\lambda_D \sim 17$ Å, requiring $n > 81$. For $(D_2O)_n^+$ on $C_2D_4(Z_1 = 1$ and $Z_2 = 6)$, $\lambda_D \sim 94$ Å, requiring $n > 10^4$. Therefore, one has the greatest difficulty in producing a shock wave with the lowest Z target.

The above calculations give lower bounds for λ. Let us estimate an upper bound λ by using the Thomas–Fermi cross section[18] which includes the effects of short–range electron screening on the Coulomb potential. We obtain for λ,

$$\lambda^{TF} = 4.05 \times 10^{10} (Z_1^{1/2} + Z_2^{1/2})^{2/3}\, T/(Z_1 Z_2 n_x)\text{cm} \tag{2}$$

Therefore, using eq. (2) for λ_D, we get $\lambda_D^{TF} \sim 70$ Å for a $(D_2O)_n^+$ cluster impacting on $ZrD_{1.65}$, requiring $n > 6 \times 10^3$. Similarly for $(D_2O)_n^+$ on TiD, $\lambda_D \sim 100$ Å, requiring $n > 10^4$. For $(D_2O)_n^+$ on $C_2D_4, \lambda_D \sim 160$ Å, requiring $n > 6 \times 10^4$.

The mean free path for the higher atomic number atoms in the target is less than λ_D, but that would not greatly change our conclusion **based upon the $\lambda < d$ criterion**: that for $n < 1000$, it is difficult for clusters to generate shock waves. However, there is a second possibility even when $\lambda > d$, that for a very high velocity projectile ($\sim 10^7$ cm/sec) the time scale may be sufficiently short so that the fluid cannot escape laterally around the projectile (**inertial confinement**), and **a shock wave can still be produced**. In the following, we argue that this is the case for cluster impact.

The cluster radius is $r = [3M/(4\pi\rho N_A)]^{1/3} n^{1/3} = r_1 n^{1/3}$, where M is the gram molecular weight of the molecules, n is the number of molecules in the cluster, ρ is the mass density, and N_A is Avagadro's number; $r_1 \approx 2$ Å for a $(D_2O)_n$ cluster. Therefore the cluster diameter, $d = 2r$, increases from ~ 10 Å to 40 Å as n increases from 10 to 1000. The time scale for producing the shock wave is $t_s \sim r/v_o$, where v_o is the initial cluster velocity upon impacting the target. Hence, t_s ranges from 10^{-15} sec to 10^{-14} sec as n ranges from 10 to 1000 for a 300 keV cluster beam.

The general shock region temperature is $\sim 10^7$ K corresponding to a deuteron average velocity of $v_D \sim 3 \times 10^7$ cm/sec. The particle flux in the lateral direction will be less than $n_D v_D/4$. Thus, the lateral displacement of D to the sides of the cluster would be $d_\perp < (v_D/4)t_s \sim 1$ Å to 10 Å$<< d$. The lateral displacement of the heavier atoms such as Zr, Ti, O, and C would even be less due to their smaller velocities, and so most target atoms cannot escape around the projectile. **Therefore a shock wave can be created and sustained for the time scale 10^{-15} to 10^{-14} sec due to inertial confinement.**

We have demonstrated elsewhere[16] that energy loss mechanisms such as radiation, thermal conductivity, and adiabatic expansion during the short impact period do not put a significant limit on the fusion yield.

II.2. Dynamical Shock–Wave Heating

Upon impact, the leading edge of the cluster creates a plasma $\sim$ few Å in thickness consisting of target and cluster ions and electrons which is compressed between the target cavity surface and the trailing cluster atoms. As the trailing cluster atoms begin to plow through the plasma, they produce a shock–wave which initially heats the plasma to temperature T_2 (see Fig. 1).

The plasma can be treated as an ideal gas to good approximation, even at high densities. The sound (or acoustic) velocity v_a in an ideal gas can be written as[19,20]

$$v_a = [\gamma R T_a/m_a]^{1/2} = 1.18 \times 10^4 [T_a(K)/m_a]^{1/2} \text{cm}/s \tag{3}$$

where γ is the ratio of the specific heats ($\gamma = C_p/C_v = 5/3$ for an ideal monoatomic gas), T_a is the initial gas temperature in front of the shock wave discontinuity, m_a is an average atomic weight and R is the universal gas constant ($R = 8.314 \times 10^7 erg/K$). For a cluster with total kinetic energy E_t and rest mass nM_m the shock–wave velocity v (4/3 of the cluster velocity[19]) is

$$v = 4(2E_t/nM_m)^{1/2}/3 = 4\sqrt{2} \times 10^{10}[E_t/(nM_m c^2)]^{1/2} \text{cm}/s. \tag{4}$$

The Mach number M_1 is then

$$M_1 = v/v_a = 4.79 \times 10^6 [E_t(\text{keV}) m_a / n T_a(K) M_m c^2(\text{keV})]^{1/2}. \tag{5}$$

Typically, for a $(D_2O)^+_{100}$ cluster ($M_m c^2 = 1.865 \times 10^7$ keV and $m_a = 20/3$) at $E_t = 300$ keV and $T_a = 300$ K, $M_1 = 2.87 \times 10^2$. Therefore, the Mach number will be very large, thus creating a strong shock wave.

The temperature behind a strong shock–wave front ($M_1 >> 1$) in an ideal gas is given by[19,20]

$$T_2(n) = \frac{2\gamma(\gamma-1)}{(\gamma+1)^2} M_1^2 T_a = 7.17 \times 10^{12} \frac{m_a}{n} \frac{E_t}{M_m c^2} K \tag{6}$$

For example, for $(D_2O)^+_{100}$ with $m_a = 20/3$ (average atomic weight of O and $2D$) and $E_t = 300$ keV, we obtain $kT_2 = 0.67$ and 0.067 keV for $n = 100$ and 1000, respectively (k is the Boltzmann constant).

Although the mean free path of deuterons at T_2 can be comparable to or larger than the cluster diameter of ~ 15 Å– 40 Å for $n = 50 - 1000$, most of them will still be in the region (a few Å thick) between the shock–wave front and trailing part of the cluster during the impact period. A novel ingredient of our model[14–16] is enhanced heating due to dynamic interaction of the back–flow gas with the leading edge of the remaining cluster. As a result of back–scattering, nearly half of the deuterons in an expanding gas at T_2 will flow into the back hemisphere toward the trailing cluster atoms. This back–flow of deuteron gas will be heated even further by the shock wave generated by the trailing cluster atoms since the cluster velocity relative to the back–flow deuteron gas is greater than that relative to the target. Since the acoustic velocity remains the same, the Mach number and T_2 increase (cf., eq. (6)). Such dynamical and continuous shock–wave heating of the plasma will further enhance the temperature during the impact period. We parameterize the resultant enhanced temperature $T_e(n)$ as

$$T_e(n) = \tau(n, E_t) T_2(n) \tag{7}$$

where $\tau(n, E_t)$ is expected to be a smooth function of n and E_t, and is to be determined from the experimental data, as described in sections IV.1 and V.

Dynamical shock–wave heating will be more efficient if magnetically insulated confinement (MIC)[21,22] can occur for the case of cluster–impact fusion. A toroidal impulse magnetic field can be created as a result of the current loop produced by hot ablation electrons that are ejected by the impact of cluster ions on the innermost surface of the crater as shown in Fig. 2. Such a toroidal magnetic field can help to keep the motion of heated deuterons along a one–dimensional direction (parallel or antiparallel to the cluster incident direction), thus enhancing the dynamical shock–wave heating by increasing the back–flow probability for deuterons. The magnetic field will also provide increased thermal insulation by reduction of the electron thermal conductivity. Furthermore, a very strong pinch off of the plasma into a sausage instability is likely to occur, thereby causing further local heating of the deuterons in the plasma. A more complete description of MIC as it relates to cluster–impact fusion will be given elsewhere.[23]

III. SHOCK–WAVE MODEL OF CLUSTER–IMPACT FUSION

When a new unexpected experimental observation (or discovery) occurs in nuclear physics, it is at first usually difficult or even impossible to explain it from

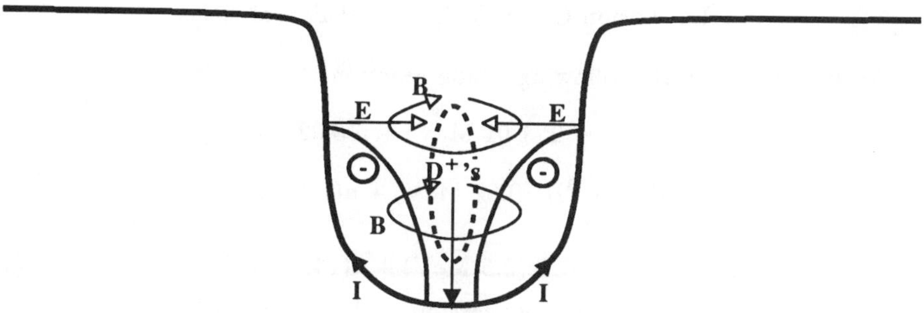

Fig. 2. Process of magnetic field generation in a crater created by cluster impact. Ablation electrons knocked out from the inner surface of the crater by cluster ions tend to drift along the inner surface and produce the current which creates and maintains the magnetic field. Because of the field, the ions in the plasma are squeezed and constrained to move along the original direction of motion of the cluster.

first principles because of the complexity of the new system which has not been investigated before and which is different from the well known conventional cases. Therefore, we are often forced to develop physically reasonable models that have a minimal number of physical parameters. If a physical model is successful in describing the new experimental data, fundamental theoretical justification of such a model and of its parametric values are often sought from realistic (but often not exact) calculations from first principles, which may provide further fundamental understanding of the new phenomenon. Therefore, with this in mind, we have developed a theoretical model[14–16] for cluster–impact fusion based on physically reasonable ideas, predictions of which can be tested experimentally. The shock–wave model[14–16] of cluster–impact fusion is described in the following.

The deuterium gas heated to a high temperature T_e is expected to be confined to a very small volume (a fraction of the cluster volume) with high density up to 4 times ($(\gamma+1)/(\gamma-1) = 4$ for an ideal gas[19]) the initial density during the impact period before it loses its energy by radiation and thermal conduction. The energy losses due to these processes are shown to be insignificant during the short impact period.[16] These deuterons with an average kinetic energy of $3kT_e/2$ can undergo nuclear fusion reactions, $D(D,p)T$ and $D(D,n)^3$He, (or $T(D,n)^4$He) either with the target deuterons (or tritons) as they expand and slow down in the target or to a lesser degree amongst themselves due to the very small volume they occupy.

III.1. Fusion Cross–Sections and Probabilities

We shall focus on the following fusion reactions:

$$D + D \rightarrow T(1.01\ \mathrm{MeV}) + p(3.02\ \mathrm{MeV})\ , \tag{8}$$

$$D + D \rightarrow {}^3\mathrm{He}(0.82\ \mathrm{MeV}) + n(2.45\ \mathrm{MeV})\ , \tag{9}$$

and

$$D + T \rightarrow {}^4\mathrm{He}(3.5\ \mathrm{MeV}) + n(14.1\ \mathrm{MeV})\ . \tag{10}$$

Experimental values of the cross section, $\sigma(E_{cm})$, for reactions (8)–(10) have been conventionally parameterized as (E_{cm} is the energy in the center of mass (CM) frame)[24]

$$\sigma(E_{cm}) = \frac{S(E_{cm})}{E_{cm}}\ e^{-(E_G/E_{cm})^{1/2}}\ , \tag{11}$$

which is based on the assumption that reactions (8)–(10) are non–resonant charged particle reactions; note that this assumption is generally ignored for reaction (10), which is actually dominated by a resonance at $E_{cm} \sim 108$ keV. The parametric values of $S(E_{cm})$ are determined from the experimental data and represent the purely nuclear properties of the fusion reaction. The exponential factor in eq. (11) is proportional to the Coulomb barrier penetration factor ("Gamow factor") which here does not include the screening effect due to the presence of electrons surrounding the target nucleus; this effect should be included for very low energies ($E << 1$ keV))[13]. E_G is the "Gamow energy" given by $E_G = (2\pi\alpha Z_i Z_t)^2\ Mc^2/2$ where Z_i, Z_t, and M are the incident nuclear charge, the target nuclear charge, and the reduced mass, respectively. The numerical values for the Gamow energy are $(E_G)^{1/2} = 31.39$, 31.39, and 34.37 $(\mathrm{keV})^{1/2}$ in the CM frame, (or $(E_G)^{1/2} = 44.40$, 44.40, and 44.40 $(\mathrm{keV})^{1/2}$ in the laboratory (LAB) frame), respectively for reactions (8)–(10).

Since the usual two– or three–parameter fits for $S(E_{cm})$ (cf., Ref. 24) are valid only for low energies ($E << 1$ MeV), we have extracted our own wide–energy fits to $S(E_{cm})$ from the experimental values of $\sigma(E_{cm})$ for $E_{cm} \lesssim 10$ MeV($E \lesssim$ 20 MeV) using the following parameterization where E is in the LAB frame

$$S(E_{cm}) = (S_0 + S_1 E + S_2 E^2)e^{-b\sqrt{E}} \tag{12}$$

for reaction $D(D,p)T$, eq. (8), and

$$S(E_{cm}) = (S_0 + S_1 E + S_2 E^2)e^{-b\sqrt{E}} + \frac{S_3}{(E - E_r)^2 + \Gamma^2} \tag{13}$$

for reaction $T(D,n)^4\mathrm{He}$, eq. (10). The parameters in eqs. (12) and (13) have been obtained[25] by a minimization fitting procedure and are given in Table I. The resultant $S(E_{cm})$ are plotted as a function of E in Figs. 3a and 4a for reactions $D(D,p)T$ and $T(D,n)^4\mathrm{He}$, respectively; the respective cross–sections (eq. 11) are plotted in Figs. 3b and 4b.

Table I. Parameters for S-factor fit for $D(D,p)T$ and $T(D,n)^4$He

Reaction / Parameters	$D(D,p)T$	$T(D,n)^4$He
S_0 (keV–barns)	54.84	1059.
S_1 (barns)	0.212	0.0
S_2 (10^{-4}barns/keV)	0.206	1.706
b (10^{-2}keV$^{-\frac{1}{2}}$)	1.943	3.603
S_3 (barns–keV^3)	—	1.152×10^8
E_r (keV)	—	82.74
Γ (keV)	—	67.12
N_0	4	7
N	160	277
$\chi^2/(N - N_0)$	10.7	4.0

For a given deuteron LAB kinetic energy $E_d(= m_d v_d^2/2)$, the probability $P(E_d)$ for a deuteron to undergo $D - D$ (or $D - T$) fusion while slowing down in the deuterated (or tritated) target is given by[26,27]

$$P(E_d) = n_{D(T)}^t \int_0^{E_d} \frac{\sigma(E_{cm})}{|dE/dx|}\, dE. \tag{14}$$

Here the target deuteron (triton) number density $n_{D(T)}^t$ is 5.68×10^{22} cm^{-3}, 7.05×10^{22}cm^{-3}, 8×10^{22}cm^{-3}, and 5.05×10^{22} cm^{-3} for the targets TiD, $ZrD_{1.65}$, C_2D_4 (C_2T_4) and $D_2(T_2)$, respectively, and dE/dx is the stopping power of the target for a deuterium projectile as given in Refs. 16 and 28.

III.2. Fusion Yield

The total fusion yield per cluster is given by $Y = Y_{(1)} + Y_{(2)}$ where $Y_{(1)}$ is the non–shock wave contribution and $Y_{(2)}$ is the shock wave contribution. Since we expect $Y_{(1)} << Y_{(2)}$ for $n > 10, Y_{(1)}$ will be neglected, i.e. $Y \approx Y_{(2)}$. For a three–dimensional Maxwell–Boltzmann velocity distribution $f(v)$ with $T_e = \tau(n, E_t)T_2$ and $v_e = \int v f(v) d^3v$ (i.e., $mv_e^2/2 = 3kT_e/2$),

$$Y_{(2)}(n, E_t) = 2ng\nu(n, E_t) I(kT_e) \tag{15}$$

where

$$\begin{aligned} I(kT_e) &= \frac{1}{v_e}\int v f(v) P(v) d^3v = (8/3\pi)^{1/2}(kT_e)^{-2}\int_0^\infty dE\; E e^{-E/kT_e} P(E) \\ &= n_D^t \left(\frac{8}{3\pi}\right)^{1/2} \int_0^\infty dE\; \frac{S(E_{cm})(1 + E/kT_e)}{E_{cm}|dE/dx|}\; e^{-(E/kT_e + \sqrt{E_G}/\sqrt{E})} \end{aligned} \tag{16}$$

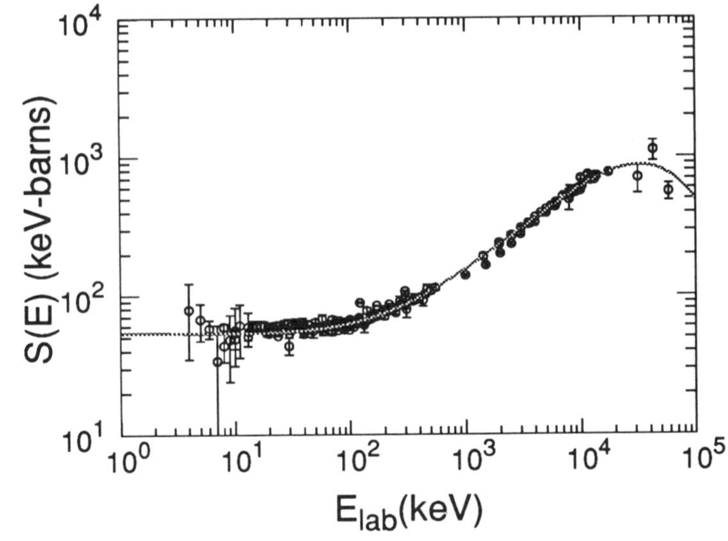

Fig. 3a. The $S(E)$ factor (eq. (12)) extracted from the experimental values of the cross–section for reaction $D(D,p)T$. The extracted values are shown as circles with error bars. The fitted values for $S(E)$ using parameters given in Table I are shown as a shaded line.

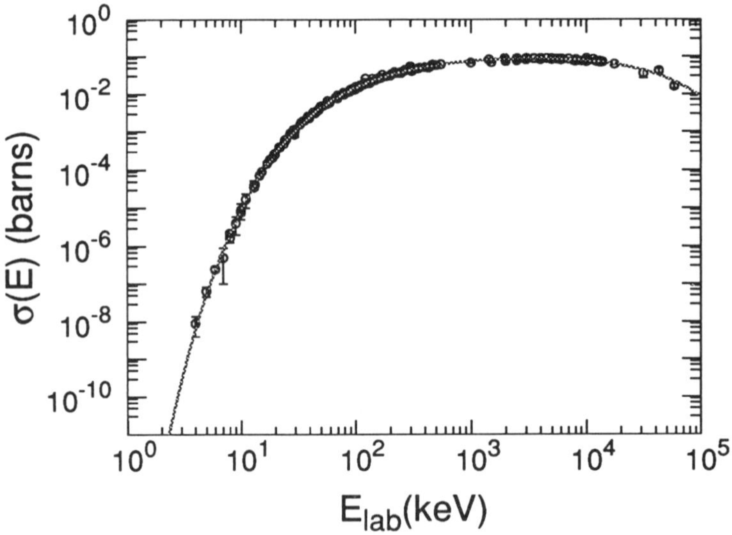

Fig. 3b. The fusion cross–section for the reaction $D(D,p)T$ as a function of deuteron laboratory energy. The experimental data are plotted as circles with error bars, and the gray line is the fit obtained from using eq. (12) for $S(E)$ in eq. (11) for $\sigma(E)$.

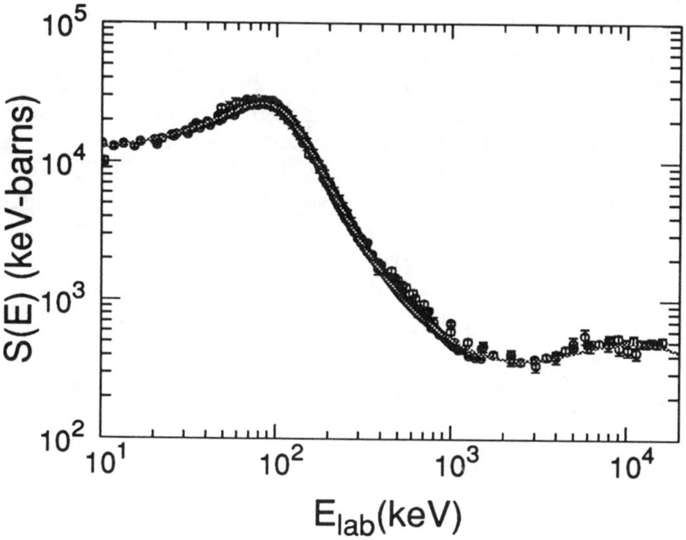

Fig. 4a. The $S(E)$ factor (eq. (13)) for the reaction $T(D,n)^4$He. See Fig. 3a for further explanation.

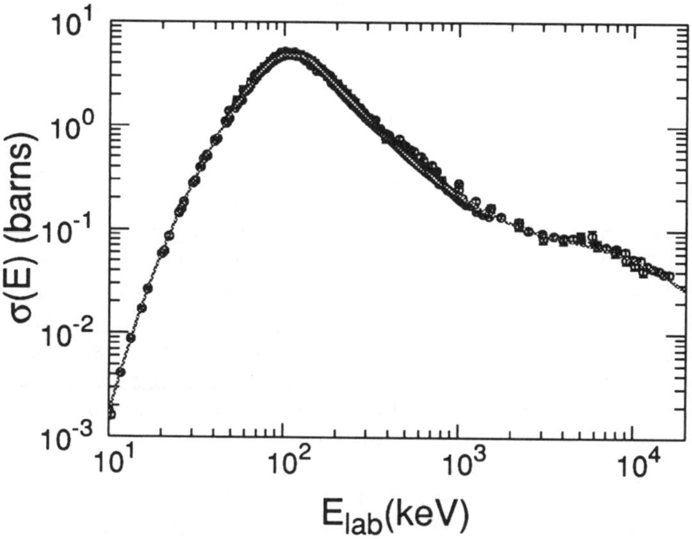

Fig. 4b. The fusion cross–section for the reaction $T(D,n)^4$He as a function of deuteron laboratory energy. Same as for Fig. 3b, but with eq. (13) used for $S(E)$.

with $P(E)$ and dE/dx as described above. Here we use for $S(E_{cm})$ the wide-energy parametric fit[25] to the $D-D$ and $D-T$ fusion cross-sections described by eqs. (12) and (13) in subsection III.1. The parameter $\nu(n, E_t)$ in eq. (15) is defined as $\nu(n, E_t) = \alpha(n, E_t) + \beta(n, E_t)$ where $\alpha(n, E_t)$ and $\beta(n, E_t)$ are defined as $\alpha(n, E_t) = n_s^p(E_t)/2n$ and $\beta(n, E_t) = n_s^t(E_t)/2n$ with n_s^p and n_s^t representing the average numbers of deuterons originating from the projectile cluster and the deuterated target, respectively, that are heated to an average temperature T_e given by eq. (7). Conservation of energy requires $3(n_s^p + n_s^t)kT_e/2 \leq E_t$. Note that $n_s^p = 0$, and therefore $\alpha = 0$, for the case of $(H_2O)_n^+$ cluster projectiles. The factor g (set to 0.5) in eq. (15) is included to account for the fraction of $(n_s^p + n_s^t)$ that is lost in the back hemisphere away from the target.

If we now use a one-dimensional Maxwell-Boltzmann velocity distribution, $f(v_x) = \frac{1}{\sqrt{\pi}}\left(\frac{m}{2kT}\right)^{1/2} e^{-mv_x^2/2kT}$, $I(kT_e)$ in eq. (15) is replaced by

$$\begin{aligned} I^{(1)}(kT_e^{(1)}) &= \frac{1}{v_e^{(1)}}\int_{-\infty}^{\infty} P(|v_x|)|v_x| f(v_x) dv_x = \frac{1}{kT_e^{(1)}}\int_0^{\infty} dE e^{-E/kT_e^{(1)}} P(E) \\ &= n_D^t \int_0^{\infty} dE \, \frac{S(E_{cm})}{E_{cm}|\frac{dE}{dx}|} \, e^{-(E/kT_e^{(1)} + \sqrt{E_G}/\sqrt{E})} \end{aligned} \tag{17}$$

where $v_e^{(1)} = \int_{-\infty}^{\infty} |v_x| f(v_x) dv_x = \sqrt{\frac{2kT}{m\pi}}$ with $\int_{-\infty}^{\infty} f(v_x) dv_x = 1$. Note that $\frac{1}{2}m\overline{v_x^2} = \int_{-\infty}^{\infty} \frac{1}{2}mv_x^2 f(v_x) dv_x = kT/2$.

Our shock-wave model of cluster-impact fusion is expressed in the form of a universal scaling equation (eq. (15) with eqs. (16) or (17)) with parameters $\tau(n, E_t)$ (cf., eq. (7)) and

$$\nu(n, E_t) = \alpha(n, E_t) + \beta(n, E_t) \tag{18}$$

which (cf., eq. (15)) are to be determined from the existing and/or future experimental data, in a manner somewhat analogous to the determination of $S(E_{cm})$ in $\sigma(E_{cm})$ from the experimental values of cross-sections. In fact, the parameterization of $\tau(n, E_t)$, $\alpha(n, E_t)$, and $\beta(n, E_t)$ in similar forms to those given by eqs. (12) and (13) enable us to determine these parameters from a n- and E_t-dependent finite data set of the experimental fusion yields. Once these parameterizations are accomplished, we can extend our calculations to the cases of different targets and a wider range of E_t. If our model is successful in describing new experimental data and functions as a working theoretical model, further calculations based on first principles are to be carried out to justify or understand the working model and its parametric values.

Note that the parameters τ and ν are not independent. By conservation of energy and number of particles, ν is expected to decrease as τ increases. Thus our theory effectively has only one fitting parameter ν (or τ). It is also important to note that, if we treated $T_e(n, E_t)$ (instead of $\tau(n, E_t)$) and $\nu(n, E_t)$ as our parameters, eq. (15) with eq. (16) or eq. (17) can be regarded as a more general working model of cluster-impact fusion than even the specific physical (shock-wave) model described in subsection II.2.

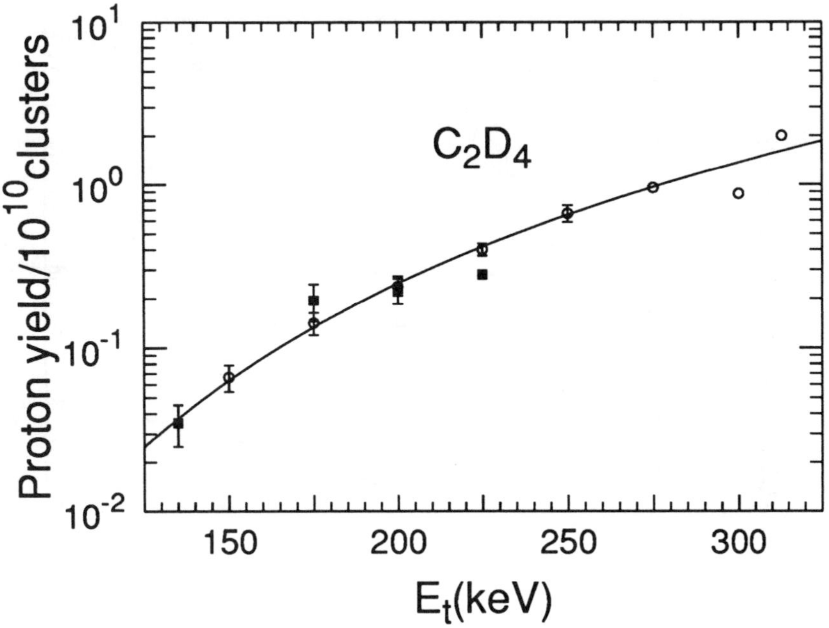

Fig. 5. Proton yield for a $(D_2O)^+_{100}$ cluster beam impacting on a C_2D_4 target as a function of cluster total energy, E_t. The experimental data are from Ref. 4 (open circles) and Ref. 6 (solid squares). The solid line is the fit to the data using eqs. (15) and (16) with $\tau(100, E_t) = 9.7$ and $\nu(100, E_t) = 0.75 \times 10^{-2}$.

IV. RESULTS AND PREDICTIONS

IV.1. $(D_2O)^+_{100}$ Cluster on C_2D_4, TiD, and $ZrD_{1.65}$ Targets

For a given value of n, the two parameters $\nu(n, E_t)$ and $\tau(n, E_t)$ can be determined uniquely by fitting the experimental data for the proton yield $Y_{(2)}(n, E_t)$ as a function of E_t using eqs. (15) and (16), assuming that ν and τ are independent of E_t. Such a fit to the $(D_2O)^+_{100} - C_2D_4$ data of BCFFK,[4] which is shown in Fig. 5, yields $\nu(100) \approx 0.75 \times 10^{-2}$ and $\tau(100) \approx 9.7$(i.e., $kT_e \approx 6.5$ keV at $E_t = 300$ keV). The recent data of BLY[6] for $n = 115$ are also shown in Fig. 5. Using the same values of ν and τ, theoretical proton yields for $(D_2O)^+_{100}$ on TiD and on $ZrD_{1.65}$ are calculated as our predictions and compared with the corresponding experimental data[4] in Fig. 6. Our predicted results are in good agreement with the experimental data for both the TiD target and $ZrD_{1.65}$ target. Note that the predicted slope for $Y_{(2)}$ vs. E_t, being a function only of T_e, is nearly independent of the target composition.

Since the average initial kinetic energy of the hot plasma is the same whether we consider three–dimensional or one–dimensional thermalization, $kT_2^{(1)}/2 = 3kT_2/2$, or $T_2^{(1)} = 3T_2$. Thus $T_e = \tau T_2 = \tau^{(1)} T_2^{(1)}$, so that $\tau^{(1)}(100) = \tau(100)/3 \approx 3.23$. Therefore the temperature enhancement parameter for the one–dimensional model is much smaller than that for the three–dimensional model. Thus, from a

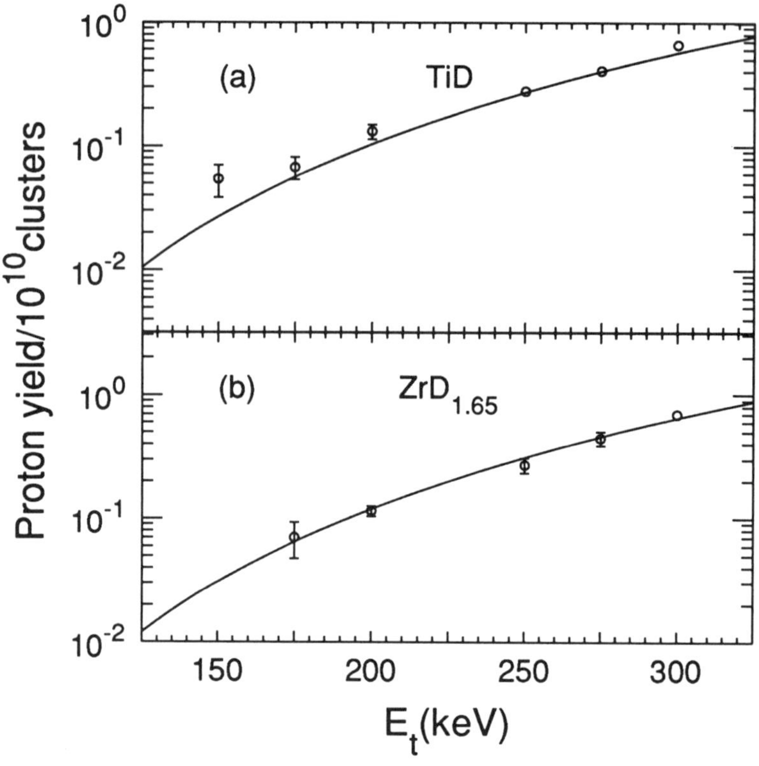

Fig. 6. Proton yields for a $(D_2O)_{100}^+$ cluster beam impacting on a TiD (a) and $ZrD_{1.65}$ (b) target as a function of total cluster energy, E_t. The data (open circles) are from Ref. 4. The solid lines are the calculated yields using eqs. (15) and (16) with $\tau(100, E_t) = 9.7$ and $\nu(100, E_t) = 0.75 \times 10^{-2}$ as for the $(D_2O)_{100}^+ - C_2D_4$ experiment.

fit to the $(D_2O)_{100}^+ - C_2D_4$ data of BCFFK[4] using eqs. (15) and (17) for the one-dimensional distribution, we obtain $\tau^{(1)}(100) = 3.23$ and $\nu^{(1)}(100) \approx 4.5 \times 10^{-2}$, which is about six times larger than $\nu(100)$.

If the MIC mechanism[21–23] presented in subsection II.2 occurs in cluster-impact fusion, the one-dimensional model with eqs. (15) and (17) may be more appropriate than the three-dimensional model with eqs. (15) and (16).

IV.2. $(D_2O)_{100}^+$ Cluster on Other Targets

Using our three-dimensional theoretical model (eqs. (15) and (16)) with the two parameters, $\nu(100) = 0.75 \times 10^{-2}$, and $\tau(100) = 9.7$, we extend our calculations of the proton yields to include several deuterated targets (liquid deuterium (D_2), LiD, $ZrD_{1.65}$ and TiD) over a wider range of cluster beam energies, 100 keV $\lesssim E_t < 2$ MeV. The predicted yields as a function of E_t are shown in Fig.

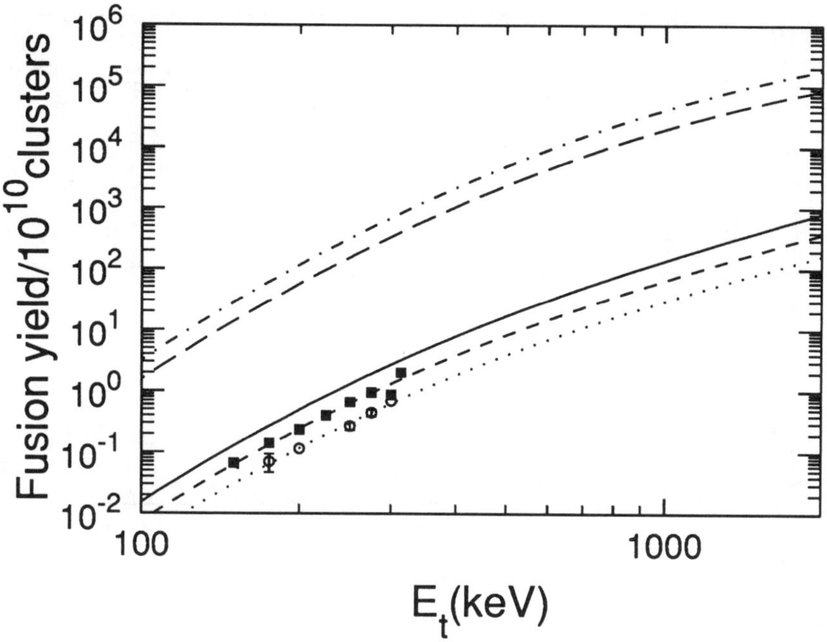

Fig. 7. Predicted fusion yields for the impact of $(D_2O)^+_{100}$ clusters on various deuterated or tritated targets for total cluster energies 100 keV $\lesssim E_t \lesssim 2$ MeV calculated using eqs. (15) and (16) with $\tau(100, E_t) = 9.7$ and $\nu(100, E_t) = 0.75 \times 10^{-2}$. The results are for T_2 (dot–dashed line), C_2T_4 (long–dashed line), D_2 (solid line), C_2D_4 (short–dashed line), and TiD (dotted line) targets. The predicted yields for LiD and $ZrD_{1.65}$ targets (not shown here) are very similar to those for C_2D_4 and TiD, respectively. The $(D_2O)^+_{100} - C_2D_4$ (solid squares) and $(D_2O)^+_{100} - TiD$ data (open circles) of BCFFK[4] are shown for comparison.

7 for the case of $(D_2O)^+_{100}$ clusters impacting on the targets D_2, C_2D_4, and TiD. We find that the predicted yields for LiD and $ZrD_{1.65}$ are almost identical to that for C_2D_4 and TiD, respectively; hence, they are not shown here. In order to test these predictions, it is desirable to measure the proton yields as a function of E_t for $(D_2O)^+_{100}$ cluster beams impacting on various deuterated targets.

The cross–section for the $T(D,n)^4$He reaction is substantially larger than that for $D(D,p)T$ for $E_d \leq 100$ keV. Therefore, we have computed the $T(D,n)^4$He fusion yields for $(D_2O)^+_{100}$ cluster beams impacting on two types of tritated targets (C_2T_4 and T_2), using eqs. (15) and (16) with the same values for the two parameters $\nu = 0.75 \times 10^{-2}$ and $\tau = 9.7$, and the multi–parameter fit[23] (cf., eq. (13)) of $S(E_{cm})$ for $T(D,n)^4$He extracted from most of the available experimental cross–section data for $E_d \lesssim 16$ MeV. The calculated results are plotted in Fig. 7 for comparison with the deuterated target results. As expected from the larger values of the $T(D,n)^4$He cross–section, the calculated $T(D,n)^4$He yields are some two orders of magnitude larger.

There are sufficient data[3,4] only at $n = 100$ for an approximate determination

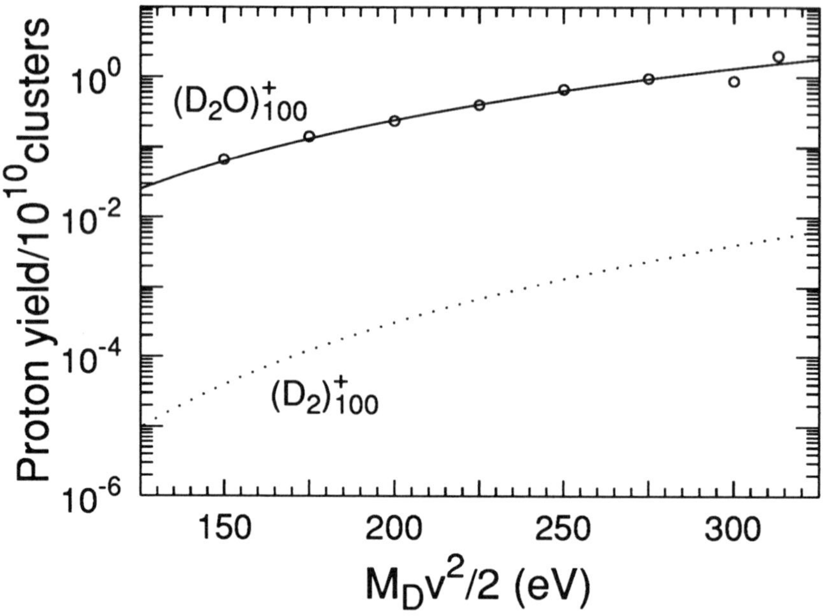

Fig. 8. The calculated proton yields as a function of average deuteron energy for $(D_2O)^+_{100}$ (solid line), and $(D_2)^+_{100}$ (dotted line), projectiles impacting on a C_2D_4 target. The calculations were done using eqs. (15) and (16) with $\tau(100, E_t) = 9.7$, and $\nu(100, E_t) \doteq 0.75 \times 10^{-2}$; here we further use eq. (19) for T_2 to adjust for the mean atomic weight of the plasma, m_a, which we assume here to be the same as that for the projectile—i.e., $m_a = 20/3$ for D_2O and $m_a = 2$ for D_2.

of the parametric values of $\nu(n, E_t)$ and $\tau(n, E_t)$ for 100 keV $\lesssim E_t \lesssim$ 300 keV. More experimental proton yield data $Y_{(2)}(n, E_t)$ are needed as a function of E_t for $n \neq 100$ to determine uniquely the dependence of ν and τ on both n and E_t for $(D_2O)^+_n$ clusters.

IV.3. $(H_2O)^+_{100}$ and $(D)^+_{200}$ Clusters on C_2D_4 Target

The much higher yield observed for both D_2O and H_2O clusters[3,4,6] as compared to D clusters[5] indicates the importance of a heavy atomic partner such as O in the shock–wave mechanism. From eqs. (3)–(6), T_2 can be written as

$$T_2 = \frac{2(\gamma - 1)v^2 m_a}{(\gamma + 1)^2 R}. \tag{19}$$

For the same cluster projectile velocity v, the shock–wave heating mechanism is less efficient for a $(D)^+_n$ cluster as compared to the case of $(D_2O)^+_n$ or $(H_2O)^+_n$ clusters because of lower values of T_2 (eq. (19)) due to a smaller value of $m_a = 2$ for $(D)^+_n$ as compared to $m_a = 20/3$ for $(D_2O)^+_n$ and $m_a = 18/3$ for $(H_2O)^+_n$.

Using eqs. (15), (16), and (19) with the same value of $\tau^{H_2O}(100) = \tau^{D_2O}(100) = 9.7$ and $\nu^{H_2O}(100) = \beta(100) = 0.75 \times 10^{-2}/20 = 3.8 \times 10^{-4}$, the proton yield

$Y_{(2)}^{H_2O}(n, E_t)$ can be calculated as a function of E_t for the case of $(H_2O)_{100}^+$ cluster – C_2D_4 target. From eq. (19) we find that T_2 is about the same for both $(H_20)_{100}^+$ and $(D_2O)_{100}^+$ for the same value of E_t, so that $Y_{(2)}^{D_2O}(100, 300)/Y_{(2)}^{H_2O}(100, 300) =$ 20 which is the experimentally observed value for this ratio for approximately this cluster size (precisely, $n = 115$) at roughly this energy.[4,6] If $\nu^{H_2O}(100) = \beta(100)$ and $\nu^{D_2O}(100) = \alpha(100) + \beta(100)$, then $\nu^{D_2O}(100) = \alpha(100) + \nu^{H_2O}(100)$, so that we find $\alpha(100)/\beta(100) = 19$.

For the case of $(D_2)_{100}^+$ cluster – C_2D_4 target, the proton yield $Y_{(2)}^{D_2}(n, E_t)$ is calculated from eqs. (15), (16), and (19) using the same values of $\tau(100) =$ 9.7 and $\nu(100) = 0.75 \times 10^{-2}$ as for the case of $(D_2O)_{100}^+$cluster – C_2D_4 target. The calculated proton yield for $(D_2)_{100}^+$ is plotted as a function of E_d in Fig. 8 for comparison with the results for $(D_2O)_{100}^+$ with the same C_2D_4 target. Our results for the $(D_2)_{100}^+$ cluster case are orders of magnitude smaller than those for the $(D_2O)_{100}^+$ cluster case and are consistent with the null results of Ref. 5.

IV.4. Contaminant Effect

In order to investigate the contributions from contaminants, we calculate the proton yields Y(contaminant) for $D^+, (D_2O)^+$, and $(D_3O)^+$ using

$$Y(\text{contaminant}) = 2nf\ P(E_d) \tag{20}$$

where $2n$ is the number of deuterium in a contaminant particle, f is the fraction of a given contaminant in the cluster beam, and $P(E_d)$ is the fusion probability given by eq. (14). For the cases of D^+, $(D_2O)^+$, and $(D_3O)^+$, we have $2n = 1, 2,$ and 3, and $E_d = E_t, E_t/10$, and $E_t/11$, respectively. f has been determined by BCFFK[4] to be $f \leq 10^{-5}$ for D^+ and $f \leq 2 \times 10^{-4}$ for $(D_2O)^+$. The values of Y(contaminant) due to $D^+, (D_2O)^+$, and $(D_3O)^+$ impacting on a C_2D_4 target are calculated using eq. (20) and compared in Fig. 9 with the (experimental and calculated) proton yields for $(D_2O)_{100}^+$ clusters impacting on the same target. As can be seen from Fig. 9, the possibility of the contaminants mimicking the $(D_2O)_{100}^+$ proton yields can be ruled out on the following two grounds: (1) the calculated values of Y(contaminant) are substantially smaller than the experimental $(D_2O)_{100}^+$ yields, and/or (2) the calculated slopes for Y(contaminant) as a function of E_t are different from that for the $(D_2O)_{100}^+$ proton yield. Note that heavier contaminants (e.g. $(D_2O)_2^+$), which have smaller average energies per deuteron for a given E_t, will have much smaller calculated yields at contamination levels similar to that for $(D_2O)^+$. Even if the contamination level of $(D_2O)_2^+$ were significant, the calculated slope of the yield as a function of E_t is steeper than that for the experimental data.

IV.5. Efficiency for Energy Production

The efficiency for cluster–impact $D-D$ fusion for the experimentally detected reaction $D(D,p)T$ (eq. (8)) is

$$\epsilon\ (D - D) = Y_{(2)}(n, E_t)(4.03\ \text{MeV})/E_t(\text{MeV}), \tag{21}$$

and for the $D - T$ fusion reaction (10), the efficiency is

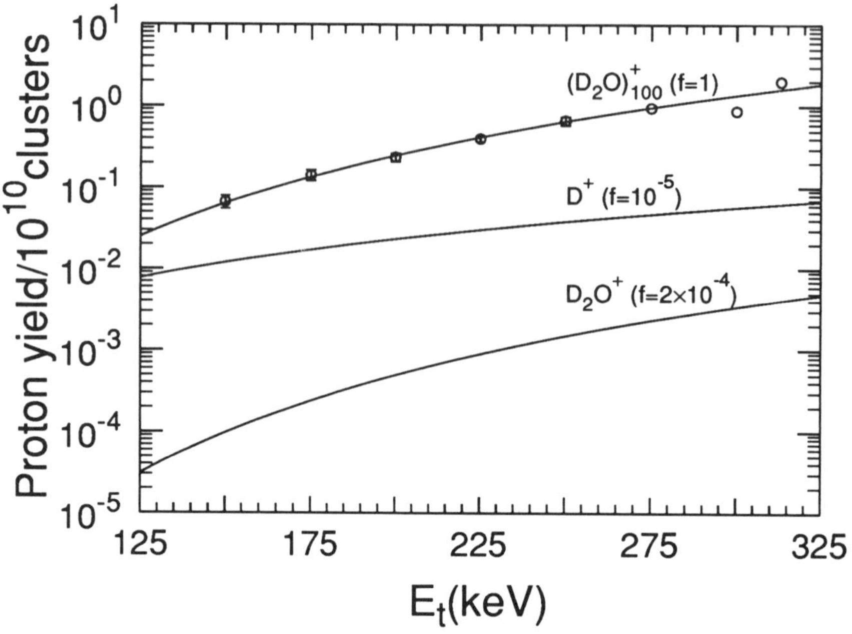

Fig. 9. Calculated contaminant yields, eq. (20), for D^+ and $(D_2O)^+$ impacting on C_2D_4 as a function of energy, E_t. The upper limits for the presence of these contaminants, f, in the $(D_2O)_n^+$ cluster beam as determined in BCFFK[4] are indicated on the figure. These contaminant yields are to be compared with the experimental (open circles) and the calculated (solid line) results for a $(D_2O)^+_{100}$ cluster beam.

$$\epsilon\ (D-T) = Y_{(2)}(n, E_t)(17.6\ \text{MeV})/E_t(\text{MeV}). \tag{22}$$

In eqs. (21) and (22), the accelerator efficiency and the efficiency for generating and delivering the electricity to run the accelerator have not been included.

Using the same energy–independent parameters, $\nu(100) = 0.75 \times 10^{-2}$ and $\tau(100) = 9.7$, which fit the $(D_2O)^+_{100} - C_2D_4$ data[4] for $E_t \lesssim 300$ keV, we have calculated $\epsilon\ (D-D)$ and $\epsilon\ (D-T)$ for $n = 100$ as a function of E_t. The calculated results for $\epsilon\ (D-D)$ and $\epsilon\ (D-T)$ are plotted in Figs. 10 and 11, respectively. For comparison, the efficiencies for a D^+ beam incident on C_2D_4 and C_2T_4 targets are also shown in Figs. 10 and 11, respectively.

As can be seen from Figs. 10 and 11, in the range 10 keV $\leq E_t \leq$ 100 MeV the highest cluster–impact efficiency for $(D_2O)^+_{100}$ is $\sim 10^{-4}$ for the case of $\epsilon\ (D-T)$ at $E_t \approx 3$ MeV in Fig. 11. However, if the energy–dependence of $Y_{(2)}(n, E_t)$ for $n > 100$ turns out to be much steeper for $E_t > 300$ keV than that predicted by the energy independent parameters $\nu(100) = 0.75 \times 10^{-2}$ and $\tau(100) = 9.7$ which were used for the calculation of $\epsilon\ (D-T)$ shown in Fig. 11, then $\epsilon\ (D-T)$ in Fig. 11 is an underestimate. Therefore it is important to measure $Y_{(2)}(n, E_t)$ for $n > 100$ and $E_t > 300$ keV, in order to determine more precisely $\nu(n, E_t)$

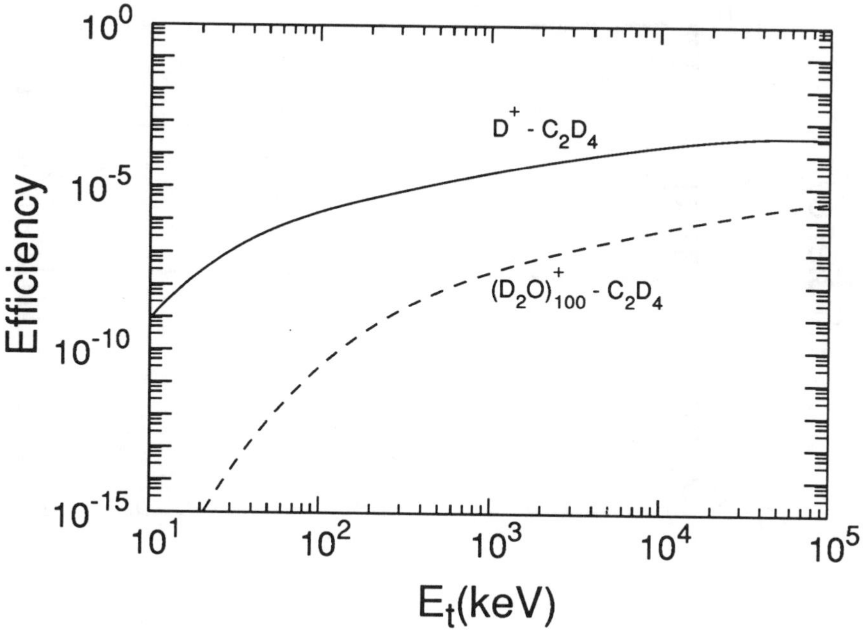

Fig. 10. Energy efficiency as a function of beam energy for $D(D,p)T$ fusion, eq. (21), for the $(D_2O)^+_{100} - C_2D_4$ experiment (dashed line) using eqs. (15) and (16) with $\nu(100, E_t) = 0.75 \times 10^{-2}$ and $\tau(100, E_t) = 9.7$ for the yield. We compare this to the expected energy efficiency for the D beam–C_2D_4 experiment (solid line) where we have used eq. (20) for the yield in eq. (21) with $f = 1$. The total efficiency for $D - D$ fusion for both experiments, which would include the $D(D,n)^3$He reaction channel, is approximately twice that shown in this figure.

and $\tau(n, E_t)$.

The purpose for this exercise in the calculation of energy efficiencies is two-fold. First, it demonstrates the crucial need for many new cluster–impact fusion (CIF) experiments as a function of cluster–size and energy (as well as projectile and target types) for the refinement and verification of our model which predicts that CIF is a very high temperature phenomenon. Second, it emphasizes the critical need to investigate the optimum conditions under which shock–wave cluster–impact fusion rates can be maximized as a function of cluster size (n) and cluster energy (E_t) for many different nuclear fusion reactions ($p + D, D + D, D + T$, etc.).

V. EXPERIMENTAL CONFIRMATION OF TEMPERATURE

It is important to note that the CIF temperature T_e in eqs. (7) and (16) can be extracted independently from the experimental proton energy spectrum as in the case of $D - D$ fusion in a hot plasma for which the neutron energy spectrum has been used as a diagnostic of the temperature.[29,30]

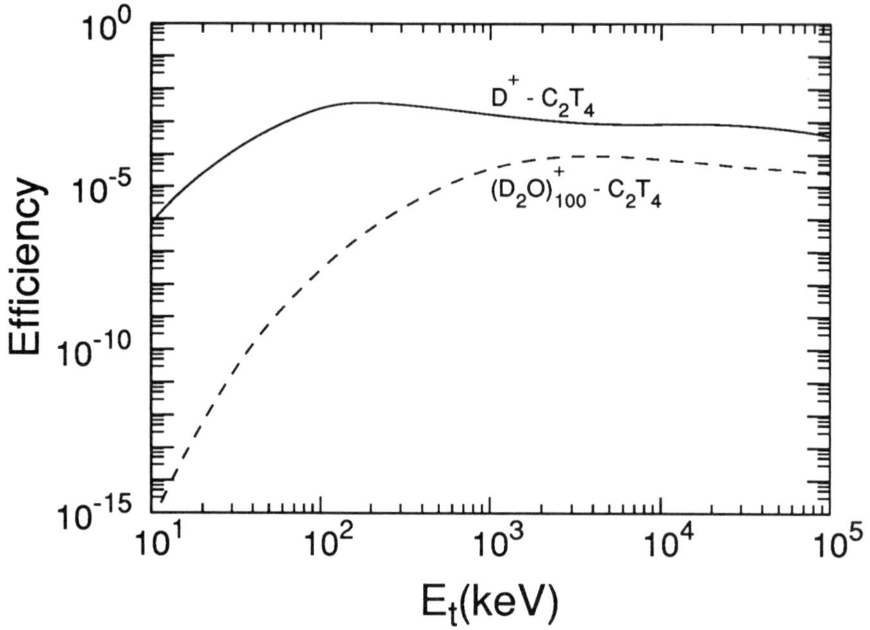

Fig. 11. Energy efficiencies for $T(D,n)^4$He fusion for the $(D_2O)^+_{100} - C_2T_4$ (dashed line) and D beam–C_2T_4 (solid line) experiments as a function of beam energy, E_t. Same as in Fig. 10, but we use eq. (22) here instead of eq. (21).

For a double Maxwell–Boltzmann velocity distribution, the proton energy distribution is given by (see eq. (36) of reference 30)

$$f(E_p)dE_p = dE_p \exp\left[-(E_p - \langle E_p\rangle)^2(m_p + m_t)/4m_p\langle E_p\rangle kT\right] \; . \qquad (23)$$

For a single Maxwell–Boltzmann velocity distribution corresponding to our case of the target deuteron at rest,[13] $kT = \frac{1}{2}kT_e$ in eq. (23) (see eqs. (9) and (10) of reference 30). For our predicted parametric value of $kT_e \approx 6.0$ keV at $E_t =$ 275 keV, we obtain from eq. (23) using $\langle E_p\rangle \approx 3$ MeV, a FWHM of $\Delta E =$ $2(E_p - \langle E_p\rangle) = [8kT_e m_p\langle E_p\rangle \ln 2/(m_p + m_t)]^{\frac{1}{2}} \approx 158$ keV compared to $\Delta E \approx$ 300 keV extracted from Fig. 5 of BCFFK[4] for the case of $E_t = 275$ keV. We will give elsewhere[31] a detailed description of a more careful extraction of kT_e from the proton spectrum which also considers additional line broadening due to instrumental and kinematic effects and consideration of dE/dx for both the projectile deuteron and fusion proton.

VI. SUMMARY AND CONCLUSIONS

We have obtained a universal scaling equation for the fusion yield of cluster impact fusion experiments which depends mainly on two inter–dependent adjustable parameters: $\tau(n, E_t)$, the temperature–enhancement factor, and $\nu(n, E_t)$, the fraction of target and projectile deuterons that reach the enhanced temperature. There is effectively only one fiting parameter, as these are not independent parameters. Most importantly, the theoretical value we obtain for $kT_e \approx 6$ keV is in excellent agreement with the experimental value extracted from the line broadening of the emitted 3 MeV protons at $E_t = 275$ keV. Our theory explains the known data, reconciles apparently contradictory experiments, and makes testable predictions. Although our high temperature model is obtained from a shock–wave analysis, it is sufficiently general to encompass other heating processes such as pinch instability heating. Further refinement of our model is readily achievable through its present form as we improve the definitions of $\tau(n, E_t)$ (or $T_e(n, E_t)$) and $\nu(n, E_t)$ with the acquisition of additional data, and also through the consideration of such processes as magnetically insulated confinement. Higher efficiencies than those presently obtained are shown to be readily achievable. Although break-even may not be achievable for $n = 100$, larger clusters and mechanisms such as multiple simultaneous impacts may allow CIF to exceed break-even. Finally we have shown that contaminants are an unlikely artifact in the experimental data.

ACKNOWLEDGMENT

We wish to thank Dr. S. M. Lee for securing financial support for one (YEK) of us to present this paper as an invited talk at this symposium.

REFERENCES

1. F. Winterberg, Z. Naturf. **19a**, 231 (1964); Case Inst. Tech. Plasma Res. Prog. Tech. Report A-21 (1963).

2. E. R. Harrison, Phys. Rev. Lett. **11**, 535 (1963).

3. R. J. Beuhler, G. Friedlander, and L. Friedman, Phys. Rev. Lett. **63**, 1292 (1989).

4. R. J. Beuhler, Y. Y. Chu, G. Friedlander, L. Friedman, and W. Kunnmann, J. Phys. Chem. **94**, 7665 (1990).

5. J. Fallavier, R. Kirsch, J. C. Poizat, J. Remillieux, and J. P. Thomas, Phys. Rev. Lett. **65**, 621 (1990).

6. Y. K. Bae, D. C. Lorents, and S. E. Young, SRI Report MP91-023 (February 1, 1991) submitted to Phys. Rev. Lett.

7. M. Rabinowitz, Mod. Phys. Lett. B**4**, 665 (1990).

8. Y. E. Kim, Fusion Technology **17**, 507 (1990).

9. C. Carraro, B. Q. Chen, S. Schramm, and S. E. Koonin, Phys. Rev. A**42**, 1379 (1990).

10. P. M. Echenique, J. R. Manson, and R. H. Ritchie, Phys. Rev. Lett. **64**, 1413 (1990).

11. S. N. Yang, Y.-C. Chen, W.-Y. P. Hwang, Proc. of First Ann. Conf. on Cold Fusion, Salt Lake City, Utah, March 28–31, 1990, pp. 335–343 (National Cold Fusion Institute, Salt Lake City, Utah, 1990).

12. M. Rabinowitz, Y. E. Kim, R. A. Rice, and G. S. Chulick, in AIP Proceedings of Int. Work on Anom. Nucl. Eff. in Deuterium/Solid Syst., Oct. 22–24, 1990.

13. Y. E. Kim, M. Rabinowitz, G. S. Chulick, and R. A. Rice, Mod. Phys. Lett. B**5**, 427 (1991).

14. Y. E. Kim, G. S. Chulick, R. A. Rice, M. Rabinowitz, and Y. K. Bae, "Shock–Wave Impact Fusion with Cluster Beams," Purdue Nuclear Theory Group Report PNTG-90-3 (February 1991), to be published in Chemical Physics Letters.

15. Y. E. Kim, G. S. Chulick, R. A. Rice, and M. Rabinowitz, "Shock–Wave Theory and Predictions for Impact Fusion with Cluster Beams," PNTG-91-4 (March 1991).

16. Y. E. Kim, M. Rabinowitz, Y. K. Bae, G. S. Chulick, and R. A. Rice, "Cluster–Impact Nuclear Fusion: Shock–Wave Statistical Analysis," PNTG-91-5 (March, 1991), to be published in Mod. Phys. Lett. B.

17. M. W. Matthew, R. J. Beuhler, M. Ledbetter, and L. Friedman, Nucl. Inst. & Meth. Phys. Res. **B14**, 448 (1986).

18. L. C. Feldman and J. W. Mayer, *Fundamentals of Surface and Thin Film Analysis*, North Holland, N.Y. (1986).

19. Ya. B. Zel-dovich and Yu. P. Raiser, *Physics of Shock Waves and High Temperature Thermodynamic Phenomena*, 2 vols. (Academic Press, New York, 1967).

20. L. D. Landau and E. M. Lifshitz, *Fluid Mechanics*, Pergamon Press, London (1959).

21. A. Hasegawa et al., Phys. Rev. Lett. **56**, 139 (1986).

22. Y. E. Kim in *Laser Interaction and Related Plasma Phenomena* (International Conference Series), edited by H. Hora and G. H. Miley, Plenum, New York (1989).

23. Y. E. Kim et al., to be published.

24. W. A. Fowler, G. R. Caughlan, and B. A. Zimmermann, Ann. Rev. Astr. Astrophys. **5**, 525 (1967); **13**, 69 (1975).

25. G. S. Chulick, et al., to be published.

26. Y. E. Kim, Fusion Technology **19**, 558 (1991).

27. Y. E. Kim, "Surface Reaction Theory of Cold and Warm Fusion," paper presented at Int. Work on Anom. Nucl. Eff. in Deuterium/Solid Syst., Oct. 22–24, 1990 (to be published in AIP proceedings).

28. H. H. Anderson and J. F. Ziegler, *Hydrogen Stopping Powers and Ranges in All Elements*, Pergamon Press, New York (1977).

29. G. Lehner and F. Pohl, Z. Phys. **207**, 83 (1967).

30. H. Brysk, Plasma Physics **15**, 611 (1973).

31. Y. E. Kim et al., to be published.

MAXIMUM EXCITATION ENERGY AND ANGULAR MOMENTUM OF HOT NUCLEI

HOT NUCLEI : HIGH TEMPERATURES, HIGH ANGULAR MOMENTA. TOWARDS NUCLEAR DISASSEMBLY.

D. Guerreau

GANIL, BP 5027, Caen Cedex, France

ABSTRACT

A review is made of the present status concerning the production of hot nuclei above 5 MeV temperature, concentrating mainly on the possible experimental evidences for the attainment of a critical temperature, on the existence of dynamical limitations to the energy deposition and on the experimental signatures for the formation of hot spinning nuclei. The data strongly suggest a nuclear disassembly in collisions involving very heavy ions at moderate incident velocities. Furthermore, hot nuclei seem to be quite stable against rotation on a short time scale.

I - INTRODUCTION

The development of heavy ion facilities has offered to the experimentalists the opportunity to investigate the domain of hot nuclei. Intermediate energy heavy ion beams appeared indeed to be the best tool to produce and study nuclei under extreme conditions of temperature and pressure[1,2)].

What is the highest excitation energy that a nucleus can accomodate ? Is the concept of a critical temperature meaningful ? Is it possible to reach a point where either a liquid gaz phase transition or mechanical instabilities could occur ? Are there some dynamical limitations to excitation energy deposition ? How stable is a hot nucleus against rotation ? What is the influence of the angular momentum on the decay process ?

From a large set of experiments, one gets now partial answers to these exciting and fundamental questions and some of these points will be debated in this paper. The emphasis will be put on the possible evidence for the existence of a critical temperature above which the system becomes unstable as well as on the possibility to form hot spinning nuclei for large and intermediate impact parameters.

How to define a hot nucleus ? the word nucleus means that one has to deal with a self bound system, even for a short while, at high temperatures (T>3 MeV i.e. a region where shell effects can be neglected). Furthermore, as we are going to use the concept of temperature, this implies a relaxation of the intrinsic degrees. These hot nuclei should then be in thermal equilibrium and it raises immediately the problem of using the concept itself of a hot nucleus at high temperature. The difficulty relies on the time needed for the two colliding ions to form a composite object and reach an equilibrium state for which a temperature can be defined[2,3)].

A comparison is made in Fig. 1 between the neutron decay half-life of a Pb compound nucleus as a function of temperature and some characteristic time associated with the relaxation of various degrees of freedom (the intrinsic degrees, the giant dipole resonance and the nuclear fission). It stresses the very sharp decrease of the half life with increasing temperature and it is easily seen that for $T \approx 6$ MeV, the thermalization time becomes similar to the neutron decay time ! At such high temperatures, the characteristic time for energy dissipation and the decay time start to overlap and consequently, one would be unable to decouple the probability of formation of the compound system from its decay properties. This automatically

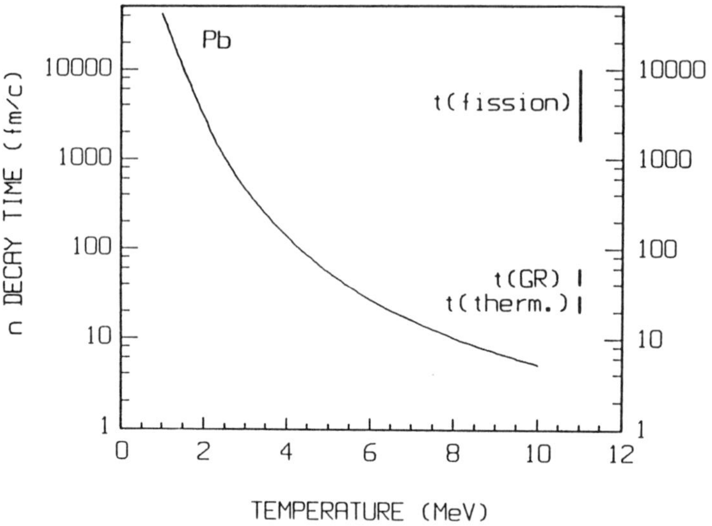

Fig. 1 : Evolution of the neutron decay half-life of a ^{208}Pb nucleus as a function of its temperature. A comparison is made with typical relaxation times (thermalization time, the giant resonance mode and nuclear fission).

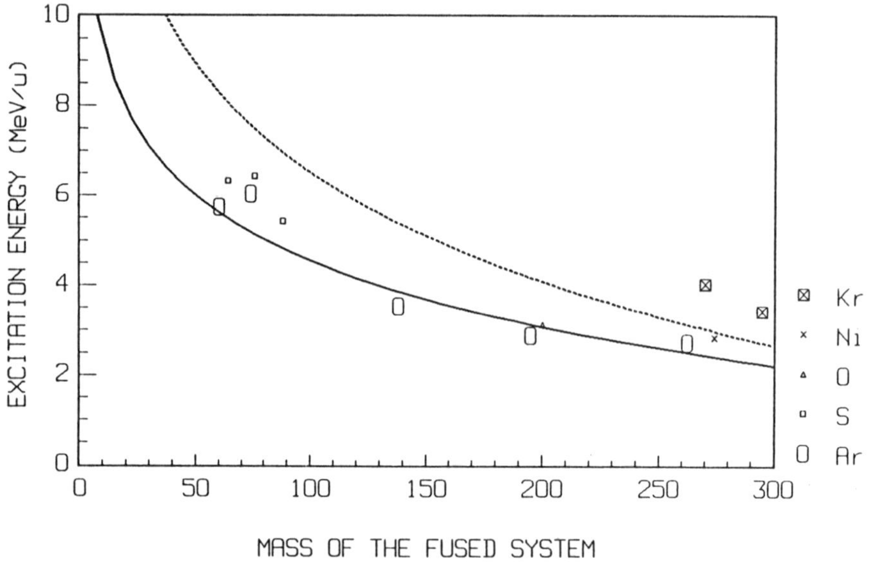

Fig. 2.: Systematics of the highest excitation energy per nucleon reached so far in heavy ion collisions for a given mass of the fused system. The data are compared to static calculations[4)] using the hot liquid drop model and a soft nuclear EOS. The two curves correspond to different surface tension coefficients.

introduces the role of the dynamics and of non equilibrium processes. Moreover, as shown also in Fig. 1, the collective degrees of freedom have quite large relaxation times, indicating that the hypothesis of equal life-times for the different decay channels (light particles, fission) which is contained in the standard statistical model and the Weisskopf formula is no more fulfilled. Large consequences are then expected for the decay of these hot systems.

This paper is intended to focuse on two main aspects of the production and the decay of hot nuclei. The first one has to deal with the attainment of some critical temperature above which the system becomes unstable. This problem has given rise to extensive experimental and theoretical studies[1,2]. The main questions are dealing with the experimental signatures of the existence of such critical temperatures and obviously with the optimum experimental conditions. The influence of the entrance channel will be discussed and it will be shown that there are some dynamical limitations to the energy deposition. Evidences for a complete disassembly of the Pb + Au system will be given when triggering on the most violent collisions. The second aspect which has not been tackled very often is concerning the angular momentum that a nucleus can accomodate. It will be shown that such hot nuclei can also store quite large angular momenta and that consequently, they appear to be quite stable against rotation on a short time scale.

II - EVIDENCE FOR THE EXISTENCE OF A LIMITING TEMPERATURE ?

How hot can be a self bound system in thermal equilibrium ? A partial answer to this question, can be obtained from Fig. 2 which shows the highest excitation energy per nucleon (E*/A) reached so far experimentally for a given mass of the composite system. These values are displayed over the mass range 60-300 and they evolve from 6.5 MeV/u for the light systems down to about 3 MeV/u for the heaviest ones. This would correspond to a temperature decreasing from almost 7 MeV down to 4.5 MeV. This trend seems to be well predicted by static calculations as those of Levit and Bonche[4] which are shown in the figure. In such calculations, the limit of stability is influenced by the balance between Coulomb, Bulk and surface energies, inducing a significant dependence of T_{lim} on the nuclear charge. Both curves in Fig. 2 result from a calculation using a soft nuclear equation of state with an incompressibility modulus K = 220 MeV (different surface tension coefficients have been used).

Does it mean that one has a comprehensive picture of the problem and that static approaches are quite successfull to account for the existing data ? This might not be so sure if one notices the large differences observed for heavy systems when using different projectiles as Ar and Kr[5,6]. (excitation energies are differing by more than 1 MeV/u). Obviously, the entrance channel plays a very important role and it might well be that the values indicated in Fig. 2 result from such limitations and have nothing to do with the intrinsic properties of the nucleus itself.

Furthermore, experimental evidences are needed before to ascertain that we have reached some critical point where the system becomes unstable. In the next section, we shall concentrate on these two points.

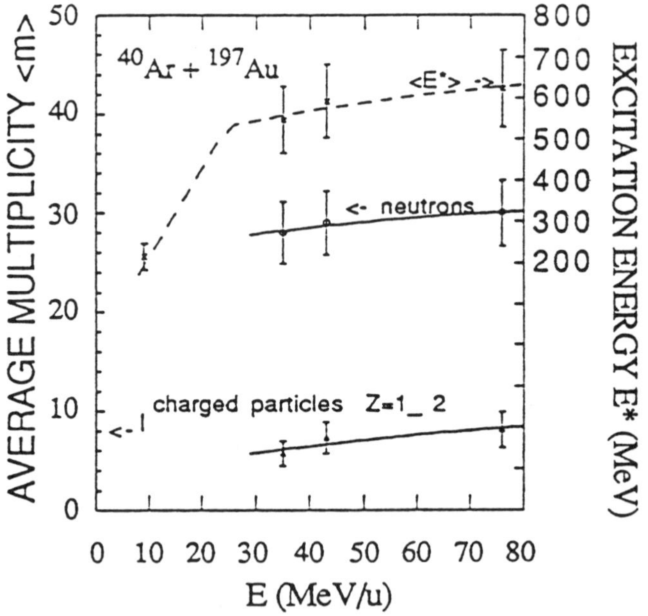

Fig. 3. : Evolution with bombarding energy of the total number of neutrons (circles) corrected for detector efficiency and evaporated charged particles (triangles) summed over Z = 1,2 released from the most dissipative collisions. From these multiplicities, the excitation energies (crosses) have been estimated[5].

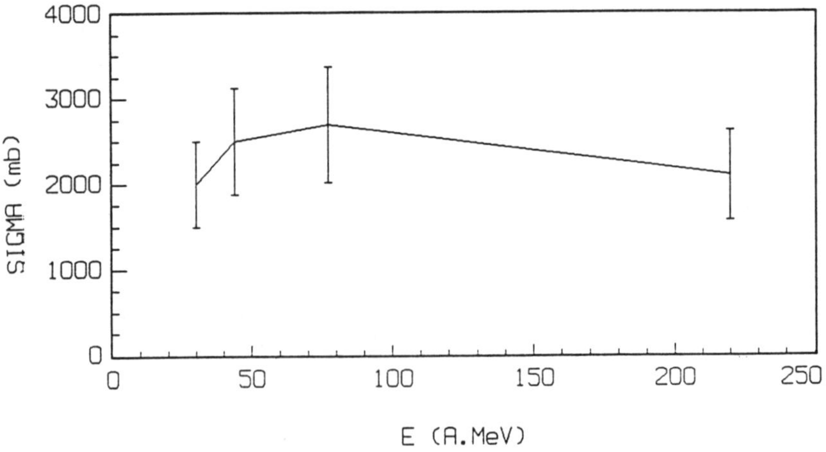

Fig. 4. : Evolution with the bombarding energy of the IMF cross section in the reaction Ar + Au. Data at 30 and 220 A.MeV are extracted from Ref. 8, those at 44 and 77 A.MeV from Ref. 7.

III - THE EFFECT OF THE ENTRANCE CHANNEL ON THE ENERGY DEPOSIT

In order to shed light on the influence of the dynamics of the entrance channel, essentially two parameters can be varied, the bombarding energy and the mass asymmetry of the system.

III.1. The influence of the bombarding energy

The influence of the bombarding energy on the thermal energy deposit has been carefully studied by Jiang et al.[5) for the systems Ar + Au and Ar + Th in the energy range 27 - 77 A.MeV. The results are summarized in Fig. 3 for the system Ar + Au. In such experiments, the most probable multiplicity of neutrons evaporated by the hot composite system has been measured using a 4 π neutron detector. The associated light charged particles have been detected in the backward direction. A soft saturation of the number of evaporated light particles (≈ 35 neutrons and 5 light charged particles) is observed above a very moderate value of the incident energy, i.e. 30 A.MeV.

Assuming a sequential emission of these particles, the primary thermal energy deposited in the system has been estimated. It is shown to increase very weakly from 620 up to 700 MeV in this wide incident energy range (30- 77 A.MeV). When increasing the bombarding energy, preequilibrium processes seem to pump most of the additional available energy. Would it mean that such a soft saturation indicates that the limiting temperature has been reached ? Would it be the case, one should then observe some change in the decay process. The saturation observed in the light particle emission is obviously a clear indication that the hot system has not undergone a complete vaporization ! Looking at the excitation function for the emission of intermediate mass fragments (IMF) may provide the key to our problem. Fig. 4 shows the evolution of the IMF cross section in the energy range 27-220 A.MeV for the same system Ar + Au[7,8)].(It has been checked at 44 and 77 A.MeV that these fragments were associated with the very dissipative collisions). This cross section remains more or less constant around 2 - 2.5 barns, a value corresponding to an emission of one IMF per dissipative collision on the average.The fragment emission is clearly not becoming a dominant decay mode with increasing energy and it may rule out their formation through a dynamical process.

Consequently, from these experiments, there is no sign for having reached a critical temperature for the Ar + Au system. Most dissipative collisions have suffered a dynamical limitation of the thermal energy deposit. It does not exclude the possibility, for the most central collisions, associated with small cross sections, to get a sizeable increase of the maximum energy deposited in the system. This has been predicted by recent calculations and indeed observed in very exclusive experiments[9)]. But again, this, by no mean becomes a dominant process over a significant range of impact parameter.

III.2 - Mass asymmetry effects in the entrance channel

It has been shown in the previous section that the most probable deposited thermal energy remains almost independant of the bombarding energy (≈ 2.6 MeV/u for the Ar + Au system). Another manifestation of entrance channel effects is to look at the dependance of the deposited energy on the projectile-target combination. Such a dependance is displayed in Fig. 5 which shows the amount of excitation energy for reactions involving various projectiles, from nitrogen to krypton, impinging very

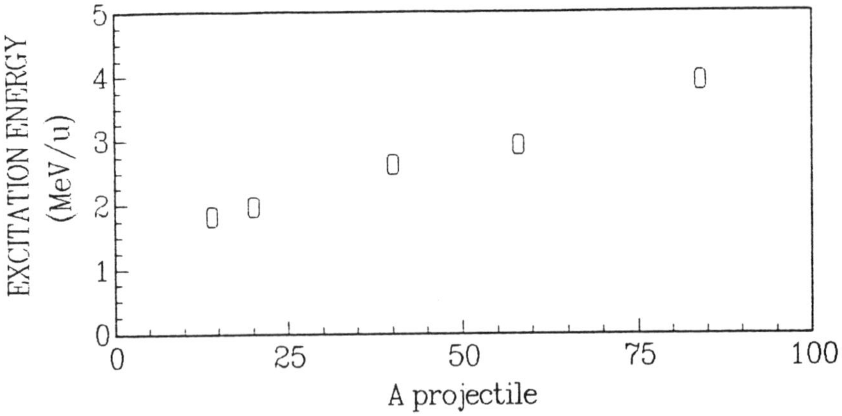

Fig. 5.: Experimental systematics of the excitation energy deposit for heavy ion projectiles impinging a heavy target (Au, Th or U) at 30 A.MeV. Data are extracted from Ref. 5, 6, 10.

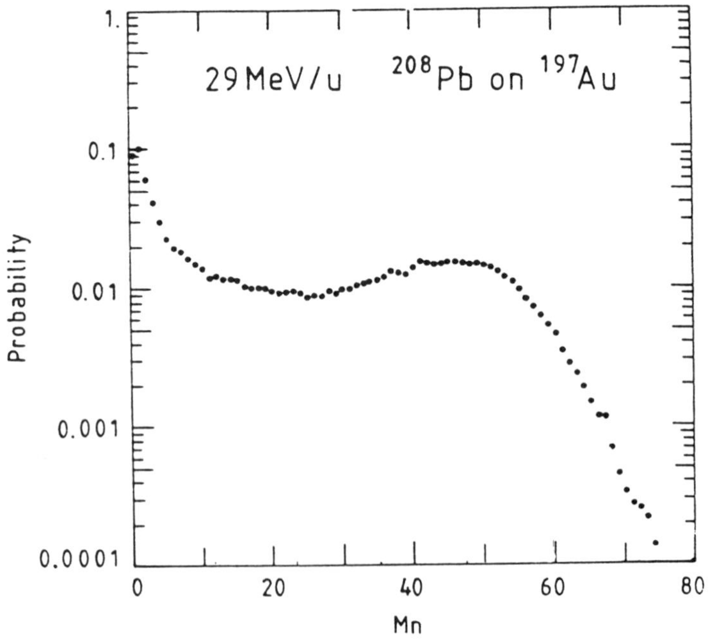

Fig. 6. : Inclusive neutron multiplicity distribution, as measured (no efficiency correction) for the system $^{208}Pb + ^{197}Au$ at 29 A.MeV. (Ref. 12).

heavy targets (Ar, Th, U) at a very moderate incident velocity, ≈ 30 A.MeV. A steep increase of the most probable excitation energy per nucleon is observed with increasing projectile mass, from 1.8 MeV/u for N induced reactions up to 4.5 MeV/u for Kr induced reactions[5,6,10]. It is then quite clear that the limits reached in these experiments are not those characterizing the intrinsic properties of the nuclear systems. They are more likely to reflect the influence of the collision dynamics on the efficiency of heat generation mechanism. Keeping a moderate incident velocity allows to minimize direct processes as preequilibrium emission. It has been shown already that around 30 A.MeV, the most probable linear momentum transfered to the target amounts to about 75 % - 80 % of the initial projectile momentum. Extrapolating our present knowledge on the systematics of linear momentum transfer for much more massive projectiles would indicate that even for such very heavy systems, temperatures close to 7 MeV are within the reach of the experimentalist.

IV - THE Pb + Au SYSTEM TOWARDS NUCLEAR DISASSEMBLY

It was when tempting to extend this systematics to the heaviest projectile-target combination and the almost symmetric system Pb + Au has been recently studied at 29 A.MeV[12]. For such a system, the available excitation energy per nucleon is close to 7.5 MeV. Extrapolating from the previous systematics would indicate that an average energy deposition exceeding 6 MeV/u could reasonably be expected. The aim of such an experiment was to search for changes in the decay process which could be clearly correlated with the attainment of a critical temperature. In the experiment, the charged distribution of the products has been essentially measured with a forward angle hodoscope ($6° \leq \theta \leq 20°$) and the filter on the degree of violence of the collision (strongly correlated with the impact parameter) was provided by the 4 π detector ORION giving an event by event measurement of the neutron multiplicity M_n with a high efficiency[13]. A global information on the energy dissipation is already provided in Fig. 6 by the inclusive neutron multiplicity distribution. The shape of this distribution is quite similar in character to those measured previously for the Ar + Au and Kr + Au systems at comparable beam velocities[5,6]. The very dominant feature is the presence of two components which can roughly be related to peripheral (low $M_n$values) and more central collisions (high M_n values). This broad bump at a high multiplicity is centered at a significantly higher M_n value ($M_n \approx 50$) than with lighter systems. Moreover, this value corresponds to approximately 78 neutrons when corrected for detection efficiency. This represents almost one-third of all neutrons of the system, a much larger fraction than observed earlier in the Ar + Au and Kr + Au experiments. A more quantitative comparison is made in Fig. 7 between the three systems Ar + Au, Kr + Au and Pb + Au, where the fraction of the neutron excess of the system observed as free neutrons (as determined at the centroid of the high multiplicity bump) is plotted versus the neutron excess of the system. The striking feature is the observation of a fraction approaching 100 % for the system Pb + Au. This means that all charged products must have an average N/Z ratio close to unity, which is already a very strong indication for the system to have disassembled into rather light fragments. It is then very likely that higher temperatures are thus attained with increasingly symmetric beam-target combinations.

Exclusive measurements of charged products shed more light on the drastic evolution of the decay processes according to the degree of violence of the collision, the later being selected through the measurement of the associated neutron multiplicity M_n with the 4 π detector ORION. Fig. 8 displays the distribution of these reaction products in the Energy Versus Atomic Number plane gated by 6 different neutron multiplicity

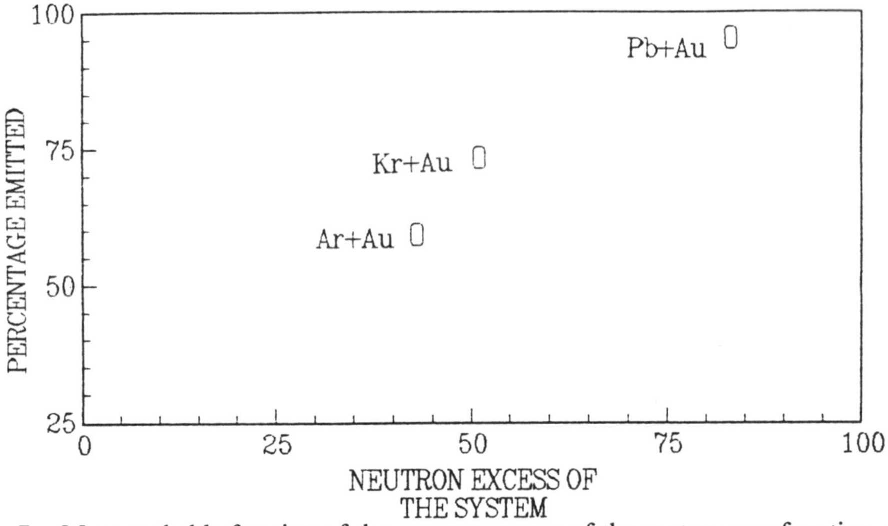

Fig. 7. : Most probable fraction of the neutron excess of the system as a function of the neutron excess contained in the (Ar, Kr, Pb) + Au system, studied at E_{lab} close to 30 A.MeV (Ref. 5, 6, 12).

bins[12,14]. For low M_n values (gate 1), the picture is dominated by elastic and quasi-elastic reactions. For the intermediate impact parameters (M_n = 10,19 and 20,29), two components are now present : the first one consists of heavy projectile-like residues originating from partially damped collisions. The second component corresponds to the sequential binary fission of the excited projectile. The double hump which is observed reflects the two kinematical solutions which are beeing observed when the recoil velocity of the fission fragments.

When selecting larger M_n values, the distribution undergoes important changes. The former components, the projectile-like fragments and the fission fragments progressively vanish and at the same time the light fragment production becomes the dominant process. Fig. 9 demonstrates even more clearly the change in the decay process according to the value of the associated neutron multiplicity (i.e. the impact parameter). It shows the emission probability as a function of M_n for a typical intermediate mass fragment (Z = 42). As IMF emission is growing continuously with decreasing impact parameter, the binary fission is a dominant process only for intermediate impact parameters and it disappears for the most violent collisions. In that case, one has to deal either with successive binary fission decays or a prompt multifragment production. An attempt to distinguish between those two different processes is out of the reach of such an experiment and would require the use of a 4 π charged product detector. For these small impact parameters, the nuclear system appears indeed to disassemble in a large number of neutrons (more than 80 after detector efficiency correction) and a copious number of intermediate mass fragments, their yields decreasing exponentially with increasing atomic number. Massive fragments did not survive for these most dissipative central collisions.

This disassembly is by no means a rare process as it represents about 25 % of the total reaction cross section. If one assumes this process to occur for the most central collisions, this would imply that all impact parameters below b = 6 fm are leading to a disassembly. The available energy (≈ 7.3 MeV/u) is obviously efficiently used in the system Pb + Au, at such a very moderate incident bombarding energy, to

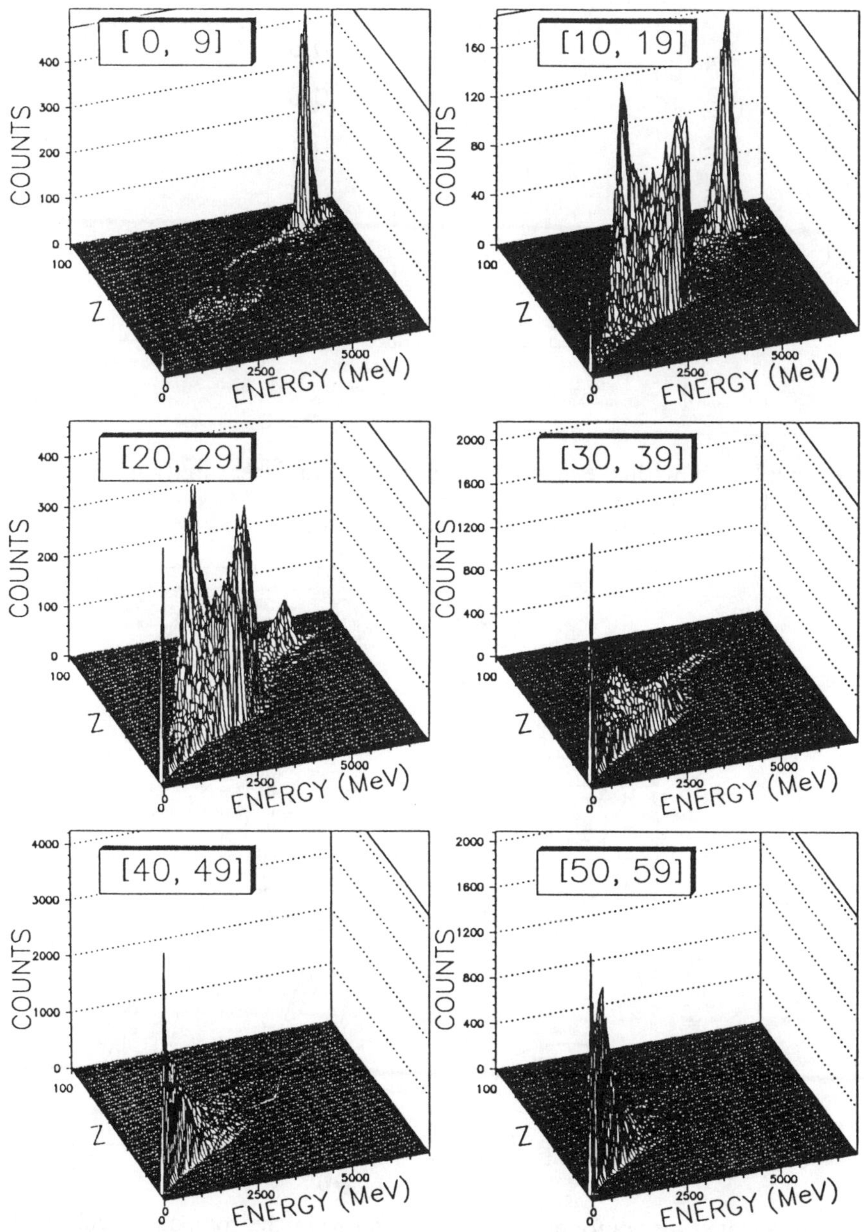

Fig. 8. : Distribution of reaction products detected between 6° and 20° for the Pb + Au system, as a function of their atomic number and kinetic energy for several neutron multiplicity gates (Ref. 12, 14).

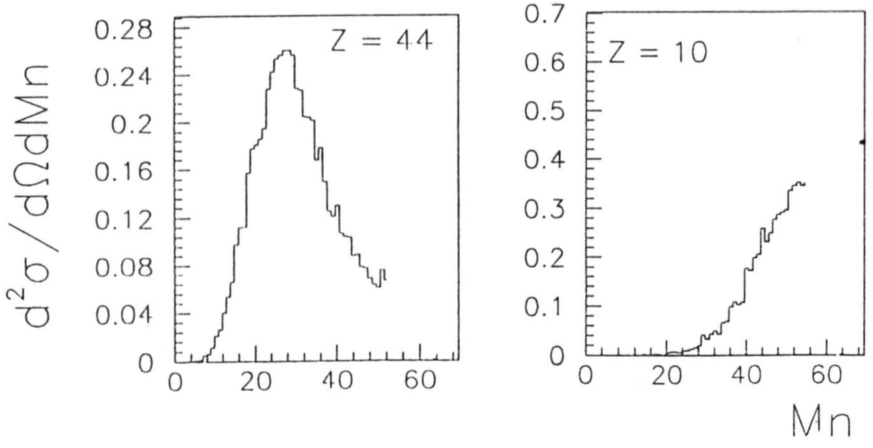

Fig.9.: Evolution with the neutron multiplicity (as measured) of the emission probability for a typical IMF (Z = 10) and a typical fission fragment (Z = 42) in the reaction Pb + Au at 29 A.MeV (Ref. 14).

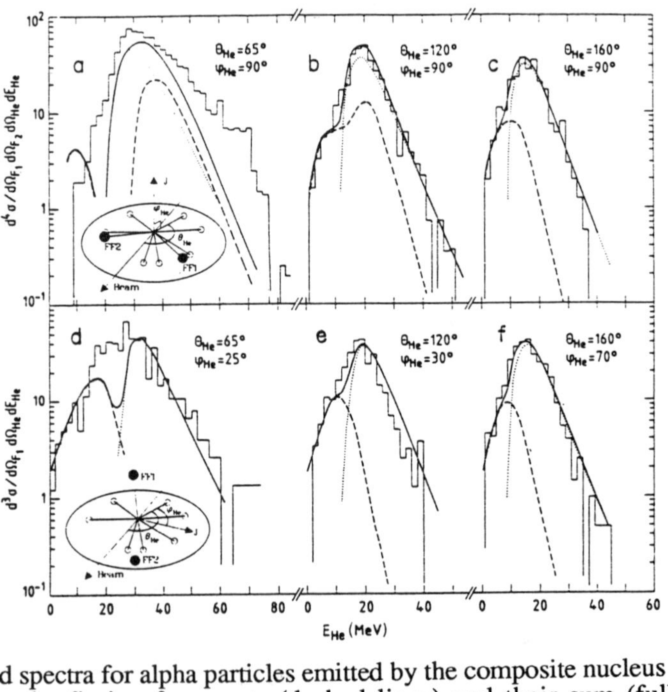

Fig.10.:Calculated spectra for alpha particles emitted by the composite nucleus (dotted lines), by the fission fragments (dashed lines) and their sum (full lines) compared to in-plane (φ = 90°) and out of plane (φ = 25°, 30°, 70°) experimental spectra. The reaction is 27 A.MeV Ar + U (Ref. 17).

heat-up the system (presumably up to the critical temperature) and lead to the observed nuclear disassembly. This would strongly favour the hypothesis of a thermal disassembly. A dynamical disassembly seems less likely to occur as the same available energy of 7.3 MeV/u for the Ar + Au system (implying a bombarding energy of 53 MeV/u) does not lead at all to such a high cross section for the IMF production (see Fig. 4).

For this Pb + Au reaction, the limit of the existence of a real nucleus has probably been reached and it seems quite probable that one has to deal with a hot blob of nuclear matter for these very central collisions.

V - THE PRODUCTION OF HOT SPINNING NUCLEI

At first sight, it seems straightforward that, in any dissipative phenomenon, kinetic energy dissipation has to be more or less strongly connected with angular momentum transfer. It is then worth trying to determine the amount of angular momentum transferred during the formation of hot nuclei and study their effect on the decay properties of such systems. The ultimate question would be obviously to determine how much angular momentum the nucleus can accomodate without undergoing a multifragmentation process.

Unfortunatly, very few experiments have been dealing with such measurements. We shall try to show, through a few examples, that the nucleus is indeed quite stable against rotation.

V.1 - Experimental evidences

An estimate of the angular momentum deposited in these incomplete fusion reactions can be made in an indirect way either by detecting γ-rays[15] or by measuring the azimuthal angular distributions of light charged particles or fission fragments or looking at correlations between emitted particles[16-18]. However a quantitative estimate of the intrinsic spin of the emitter can be made only using the statistical model which implies that one has to deal with a fully thermalized composite nucleus.
The first example is concerning the system Ar + U at 27 A.MeV for which the in plane and the out of plane (OOP) distributions of the α-particles have been measured, the reaction plane being defined by the detection of the fission fragments[17]. Typical experimental energy spectra for He are shown in Fig. 10 together with calculated ones assuming two sources for these particles, the composite nucleus and the fission fragments. The simulations demonstrate clearly that the major contribution arises from the emission by the composite nucleus (> 85 %) before scission, which is obviously a necessary condition to extract an angular momentum information from the OOP distribution. This distribution is essentially sensitive to the ratio J_o^2/IT where J_o is the spin of the parent nucleus, I the moment of inertia and T the temperature[19]. The extraction of J_o from the experiments implies an attempt to single out the effects of these different parameters : temperature, spin and deformation. The method is detailed by Jacquet et al.[17]. A precise analysis indicates that the experimental results (anisotropy, energy spectra) are well reproduced assuming a temperature T close to 4.5 MeV, a quite large deformation (axis ratio 2 : 1) and a r.m.s value of the angular momentum for the hot emitter $J_{r.m.s} \approx 120\ \hbar$.This value is a minimum value as it is an average over the whole deexcitation chain of the composite nucleus from the moment it is formed up to scission. If one takes into account this fractional loss of the angular momentum, the value which has been extracted is then in agreement with an estimate

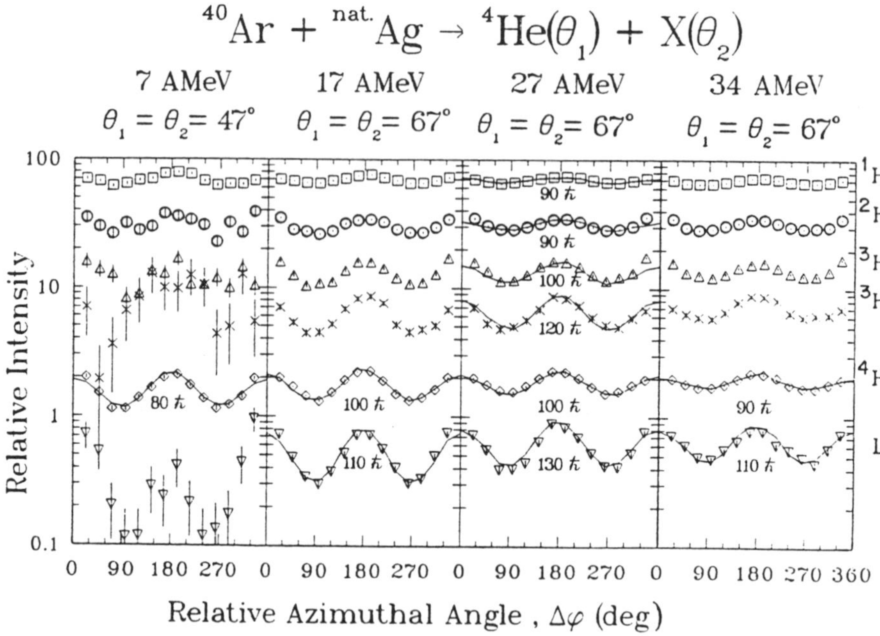

Fig.11.: Azimuthal angular correlations for ^{4}He-X pairs. The curves are statistical model calculations. Values of the angular momentum indicated in the figure are the J_{max} values used in the calculation. (Ref. 18).

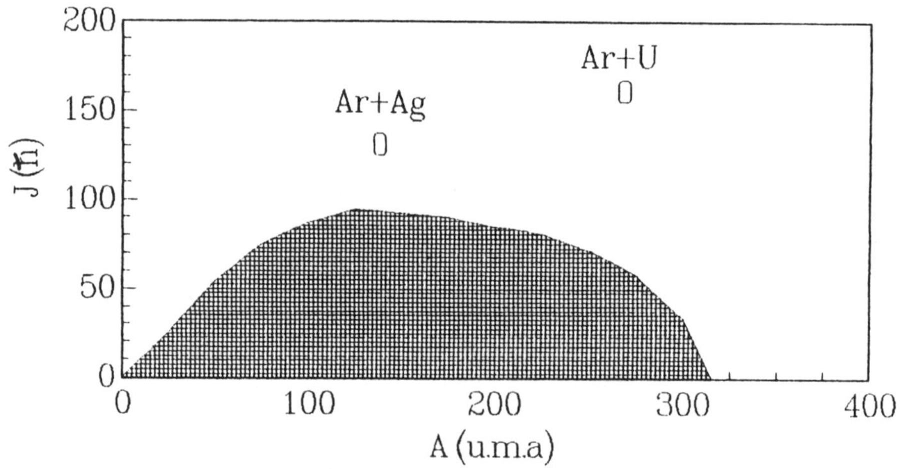

Fig.12.: The values of J_{max} found for the Ar + Ag and Ar + U systems are compared to the limits predicted by the rotating liquid drop model.

of $J_{r.m.s}$ from the measurement of the fusion cross-section. The striking result is that such hot and heavy composite systems seem to accomodate large values of angular momentum. For the Ar + U case, maximum values of 160-200 ħ have been reached.

Another example has been recently discussed by Ethvignot et al.[18]. The role of angular momentum has been studied by measuring correlations between emitted particles. The system Ar + Ag has been investigated in the bombarding energy range 7-34 A.MeV. Examples of these azimuthal distributions are shown in Fig. 11 for various pairs of emitted particles. The largest anisotropies are observed for heavier ejectiles, which have more spin-off energy.The curves in the figure result from simulations within the statistical model framework and are able to provide a very satisfactory description of all experimental data. Again, one has here a clear evidence that very highly excited nuclei ($T \approx 5$ MeV) have been formed with spins as high as 130 ħ. Moreover, due to the averaging over the whole deexcitation chain, the maximum spin values have been obviously underestimated. (Cross sections for central collisions at 27 A.MeV are comparable with incident partial waves extending up to 200).

V.2 - Hot nuclei are quite stable against rotation

It may appear quite amazing that hot nuclei with high thermal excitation ($T \approx 5$ MeV) can be so stable against rotation. The large angular momenta involved in the above reactions are indeed much larger than the ones corresponding to the limit of stability against fission[20]. Such a comparison with the rotating liquid drop model (RDLM) is shown in Fig. 12. While the RDLM limit is 80 ħ for an U nucleus, the data for the Ar + U system are fully compatible with $J_{max} \approx 160$ ħ.

Why such high angular momenta ? The reason lies probably in the fact that we are dealing with different timescales. It is now well known that the fission process is strongly hindered in the early stage of the collision as compared to the predictions of static calculations. This is explained through dynamical models where the collective motion of the nucleons towards the scission point is a very slow process as compared to one evaporation step[21,22].

Coming back to Fig. 1, one sees that the scission time is almost two orders of magnitude shorter than the time for particle evaporation at $T = 5$ MeV. The limit of stability for fission is a limit on a very long timescale ($t > 2000$ fm/c) as in the experiments described previously, the α particle emission allows to get a snapshot of the hot composite system at a very early stage of the deexcitation. Consequently, it tells us that the nucleus is quite stable against rotation over a very short timescale ($T \leq 100$ fm/c). This is confirmed by recent calculations showing clearly that a hot rotating nucleus can be very stable against fission and that particle evaporation at high angular momentum can stabilize the nucleus so strongly that binary fission may even be totally suppressed[24].

The final question would be to know how much angular momentum can the nucleus accomodate. Would high angular momenta help to destabilize the system and induce more easily a multifragmentation process ? De Paula et al.[25] have performed calculations for Au nuclei based on molecular dynamics and restructured aggregation. It indicates that at T = O, the critical angular momentum for multifragmentation is close to 450 ħ ! As mentioned by these authors, it means that a lot of excitation energy would then be stored into this collective mode at the expense of the thermal energy which again might tend to stabilize the system. Anyhow, such high values of the spin should play a significant role in the decay process. This has to be taken into account seriously in any calculation

VI - CONCLUSION

This paper was intended to give a short review of the experimental data concerning both the production of hot nuclei up to some critical temperature and the possibility to induce very large angular momentum transfer.

It has been shown that in most cases the apparent limitations in the temperature of the system do not reflect the intrinsic properties of the nuclei but are essentially related to the dynamics of the entrance channel. The highest thermal energies are reached for symmetric systems at moderate incident bombarding energies ($\approx$ 30 A.MeV). For the Pb + Au system, the central collisions ($b < 6$ fm) lead to a nuclear disassembly ; the system disintegrates into a large number of nucleons and small fragments.

What is the stability of a hot nucleus against rotation ? Recent experiments demonstrate clearly that nuclei can easily accomodate very large angular momenta ($J > 150\ \hbar$), much larger than the limit of stability predicted by the rotating liquid drop model. The nucleus appears then to be remarquably stable against rotation at least on a short time scale ($t \approx 100$-300 fm/c).

In the future, it would be interesting to investigate very properly the effect of both the temperature and the spin on the decay process. Recent measurements indicate that peripheral reactions are most suited to undertake such studies[26)]. Pretty hot nuclei could be obtained in these collisions for which a precise determination of the impact parameter can be achieved experimentally at variance with fusion like reactions which have been discussed in this paper. This would allow to prepare uncompressed hot nuclei with well defined values of the temperature and the spin.

REFERENCES

1. E. Suraud et al., Prog. Part. Nucl. Phys. 23, 357 (1989)
2. D. Guerreau, in : Nuclear Matter and heavy ion collisions, (Plenum, New York, 1989),p. 187.
3. D. Brink, Nucl. Phys. A519, 3c (1990).
4. S. Levit and P. Bonche, Nucl. Phys. A437, 426 (1985).
5. D.X. Jiang et al., Nucl. Phys. A503, 560 (1989).
6. E. Crema et al., Phys. Lett. 258, 266 (1991).
7. A. Sokolov, Ph.D Thesis (Paris, 1990) and to be published.
8. A. Milkau et al., GSI Preprint, GSI 91-08, 1991.
9. M. Louvel et al., Contribution to this Symposium.
10. M. Fatyga et al., Phys. Rev. Lett. 55, 1376 (1985).
11. E.C. Pollaco et al., Nucl. Phys. A488 (1988) 319c.
12. E. Piasecki et al., Phys. Rev. Lett. 66, 1291 (1991).
13. J. Galin, XXIth Summer School on Nucl. Phys., Mikolajki (Poland, 1990), Ganil Preprint P 90-17 ; X.M. Wang et al., to be published.
14. S.Bresson et al., to be published.
15. K. Hagel et al., Nucl. Phys. A486, 429 (1988).
16. R. Lacey et al., Phys. Rev. C37, 2540 (1988).
17. D. Jacquet et al., Nucl. Phys. A511, 195 (1990).
18. T. Ethvignot et al., Phys. Rev. C43, R2035 (1991).
19. N.N. Ajitamand et al., Phys. Rev. C34, 877 (1986).
20. Cohen et al., Ann. Phys. 82, 557 (1974).

21. P. Grangé et al., Phys. Rev. C27, 2063 (1983).
22. C. Grégoire et al. Z. Phys. A329, 497 (1988).
23. D. Hilscher et al. Phys. Rev. Lett. 62, 1099 (1989).
24. F. Garcias et al. Phys. Lett. B255, 311 (1991).
25. L. De Paula et al., Phys. Lett. B258, 251 (1991).
26. J. Galin , First Biennal Workshop on Nucl. Phys., Megève (France, 1991). GANIL Preprint P 91-10 ; S. Bresson et al., to be published.

DECAY OF EQUILIBRATED VERY HOT NUCLEI VIA MULTIFRAGMENT EMISSION

M. Louvel, N. Prot, G. Bizard, R. Bougault, R. Brou, H. Doubre, D. Durand, Y. El-Masri, H. Fugiwara, A. Genoux-Lubain, K. Hagel, A. Hajfani, F. Hanappe, S. Jeong, G.M. Jin, S. Kato, J.L. Laville, C. Le Brun, J.F. Lecolley, S. Lee, T. Matsuse, T. Motobayashi, J.P. Patry, A. Péghaire, J. Péter, R. Regimbart, F. Saint-Laurent, J.C. Steckmeyer, B. Tamain

LPC ISMRA- GANIL Caen FRANCE , Universities of Tsukuba -Rikkyo JAPAN, Bruxelles - Louvain la Neuve BELGIUM

ABSTRACT : Coïncidences between at least 3 fragments emitted at large angle have been used to study central collisions in the Ar+Au reactions at 30 and 60 MeV/u incident energies. For both energies, the formation of equilibrated very hot nuclei is evidenced . Excitation energy as high as 1 GeV are reached in the reaction at 60 MeV/u .

Recently, limitations to the formation of very hot nuclei have been claimed for the Ar+Au (Th) system. For such heavy systems [1,2], at bombarding energies lower than 35 MeV/u , the central collision mechanisms were studied with the help of the standard decay channels leading to the formation of heavy residues and the subsequent emission of fragments [3,4]. At higher incident energies, the detection of evaporation residues or fission fragments selects peripheral and dissipative processes but these experimental methods have not allowed to observe the most violent collisions [5].

We have performed at GANIL the study of the Ar+Au system at 30 and 60 MeV/u bombarding energies with an apparatus which allows to investigate complex phenomena such as sequential decay of highly excited nuclei, multifragmentation or even emission from a non equilibriated subsystem. These energies were chosen to surround the 44 MeV/u energy for which the disappearence of the central collision folding angle bump was observed. The main purpose of this paper is to show evidence of the formation of equilibrated very hot nuclei decaying by multifragment emission.

The experimental set up (Fig. 1) allows us to cover the whole impact parameter range in order to follow the dynamical evolution of the decay processes with excitation energy. As an example, we were able to observe the onset of multifragment emission in the reaction Ar + Au at 60 MeV/u [6].

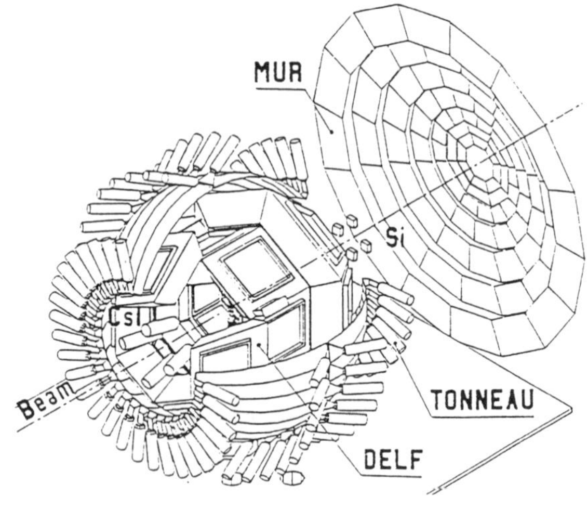

Fig. 1 : *Experimental set up*

The detection system consists in three multidectors, DELF [7] for fragments with Z larger than 8 and a velocity larger than .5cm/ns , emitted in the 30°-150° angular range, the MUR [8] and TONNEAU [9] for light charged particles (LCP) and light fragments in the 3°-150° polar angle range. In addition, 4 large area silicon telescopes[10] were located at forward angles to detect projectile-like fragments. Each detected fragment or particle is identified by its atomic number and its velocity vector is determined with a good accuracy.The total geometrical efficiency for the fragments and LCP emitted at large angle is 2π sr.

The present data were recorded using a trigger requirement of at least 3 fragments in coincidence. These 3 or more body events can be roughly divided into two classes the characteristics of which are presented on figure 2.

Fig. 2 : *Bi-dimensional velocity (cm/ns) - Z spectra for the fragments detected at large angles for the two different triggers and for the reaction Ar+Au at 30 MeV/u and 60 MeV/u*

a) "peripheral " trigger at 30 MeV/u
b) "central" trigger at 30 MeV/u
c) "peripheral" trigger at 60 MeV/u
d) "central" trigger at 60 MeV/u

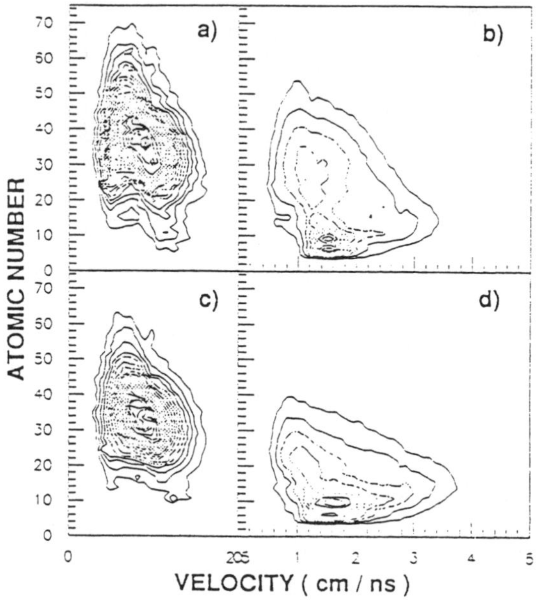

The upper part concerns the 30 MeV/u bombarding energy , the lower part the 60 MeV/u one. In each figure, are presented the two dimensional spectra : velocity - atomic number for the fragments detected at angles larger than 30° .

On the left hand side, for both incident energies we have selected peripheral collisions: one of the 3 detected fragments is a projectile remnant emitted at forward angles. In the analysis we have imposed as conditions for the projectile like fragment a Z larger than 8 and a velocity larger than 64% of the beam velocity. The two fragments detected in DELF exhibit the characteristics of fission fragments, the mean atomic number $<Z>$ is around 35 and the average velocity is ~1 cm/ns .

On the right hand side , for each 3 body recorded event, the 3 fragments plotted on the figures are emitted at angle larger than 30° . We observe at both energies the presence of light or intermediate mass fragment and a clear decrease of the atomic number of the heaviest fragment with bombarding energy.

Let us now focus on events shown in Fig. 3 in which 3 or 4 fragments have been detected at large angle. At 30 MeV/u, the ratio of the numbers of events with three detected fragments to four detected fragments (Nd=3/Nd=4) =137 is very high and in addition, the large value ~73 of the mean total atomic number of the fragments indicate that the bulk of the events are true 3 body events.

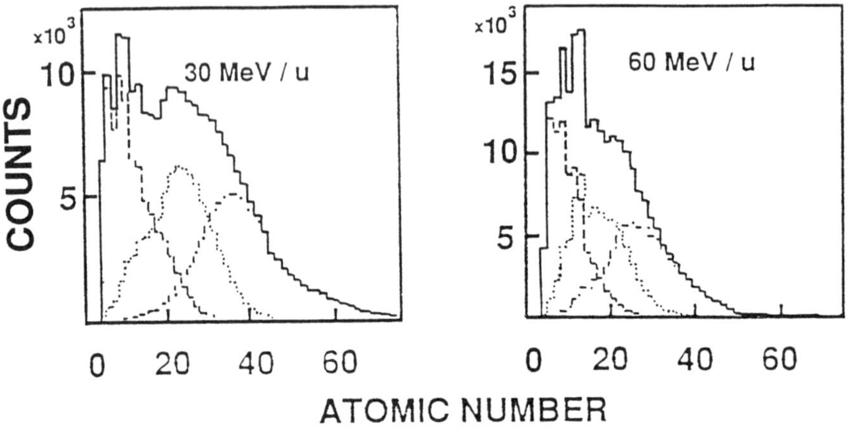

Fig. 3 : *The full curve is the Z fragment distribution which includes the light fragment (dashed curve), the medium one (dotted curve) and the heavy one (dot-dashed) curve.*

At 60 MeV/u , the decrease of the ratio Nd=3 to Nd=4 and of the mean total Z are strong clues that the average fragment multiplicity is higher than in the previous case which can be interpreted as an increase of the violence of the collision. This behaviour can be understood by relating the violence of the collisions to physical quantities such as the linear momentum transfer and excitation energy deposit. These two quantities will be deduced from the recoil velocity measurement.

At 30 MeV/u , binary fission was usually observed in central collisions and the measurement of the recoil velocity is deduced from the most probable value of the folding angle between fission fragments. In the present analysis, we will justify that the 3 or 4 coïncident fragments detected at large angle, are issued from a unique source in thermal equilibrium. Then we identify the recoil velocity of the hot source to the center of mass velocity of the detected fragments. The recoil velocity is calculated event by event with the formula :

$$\vec{V}_R = \sum_{i=1}^{Nd} Z_i \vec{V}_i \,/\, \sum_i Z_i$$

The sum is extended to all the detected fragments.

In Fig. 4 the recoil velocity distributions are shown. The values corresponding to the complete fusion are indicated by arrows.

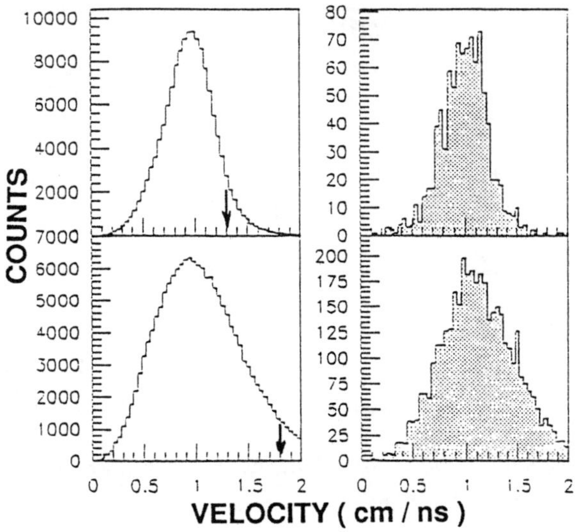

Fig. 4 : *Recoil velocity distributions for three fragments (unfilled histograms) and four fragments (dashed histograms) detected at large angles in the Ar+Au system at 30 and 60 MeV/u. Upper curves : 30 MeV/u, lower curves : 60MeV/u.*

At 30 MeV/u , whatever the multiplicity of the detected fragments, the maximum of the V_R distribution is about 1 cm/ns which represents a linear transferred momentum of 75% of the incident momentum.

At 60 MeV/u , all the emitted fragments are not detected and we need a simulation [11)] to study the influence of the experimental filter. The effect of the apparatus is a widening of the distribution without modifying the mean value as long as the recoil velocity is lower than 1.5 cm/ns .The linear momentum transferred is reduced to 50% of the incident linear momentum. At both energies it corresponds to about 170 MeV/c per incident nucleon .

Proving the equilibrium of the composite system requires a great deal of care. Near the detection threshold, the efficiency of each PPAC is variable so that we cannot draw conclusions from the shape of the angular distribution of the lightest fragments. Several independent features indicating the formation of an equilibrated hot system are presented in figures 5 to 7. On each figure, the data are presented for two ranges of the light fragment (LF) detection angle : crosses for LF emitted in the angle range from 30° to 60° and triangles for the backward 120°-150° angles. Indeed, if the damping of the kinetic energy is not complete [12)], we could expect the collision to be more or less violent whether the light fragment is forward or backward emitted.

First, the velocity distributions of the light fragment, calculated in the center of mass frame of all the detected fragments are presented on Fig. 5. The most probable values are centered at the coulomb barrier whatever the LF emission angle. This is a necessary condition for the composite system to be a nucleus.

Now, let us compare the multiplicities of light charged particles for both incident energies. Whatever the selected particles, fast particles detected in the MUR (Fig. 6) or

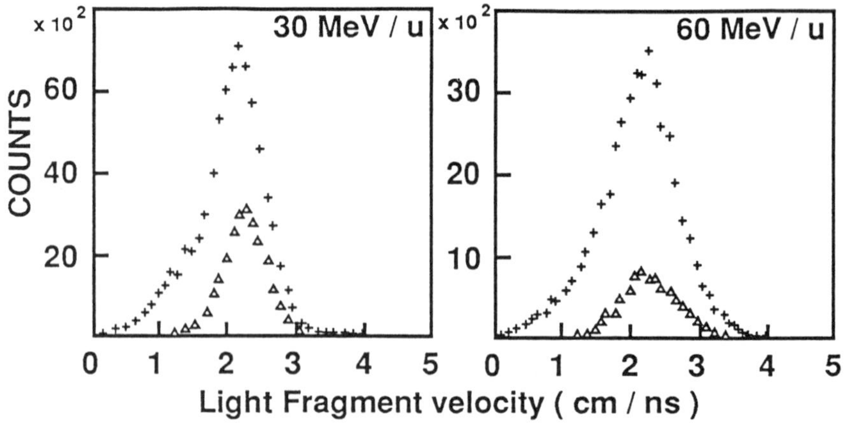

Fig. 5 : *Light fragment velocity distributions in the center of mass frame, gated by two ranges of the LF detection angle.*

evaporated particles detected backward in the TONNEAU (Fig. 7), the largest multiplicities are observed at 60 MeV/u . In each figure, the superposition of the + and Δ symbols indicate that the multiplicities do not depend on the light fragment emission angle.

Fig. 6 : *Fast particle (Z=1,2) multiplicity distributions gated by light fragment emission angle as a function of the recoil velocity. (+) the light fragments are detected between 30° and 60°, (Δ) the LF are detected at backward angles (θ_{LF} > 120°). The largest multiplicities are obtained at E/A= 60 MeV.*

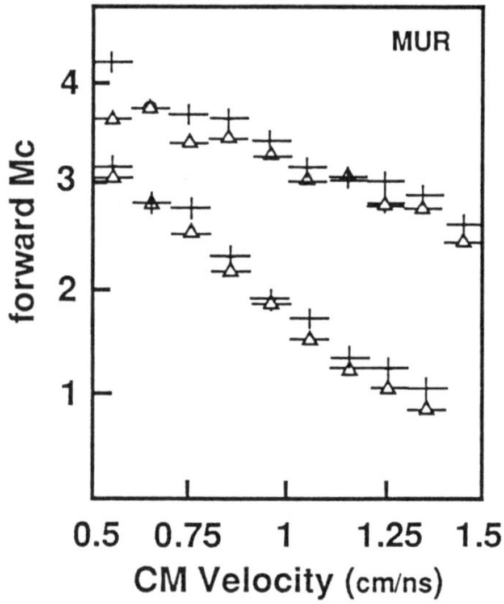

Thus, each multidetector can provide useful information about a specific mechanism which is independent of the fragment angles: DELF for the emission of fragments at the coulomb barrier, the MUR for the nuclear transparency (pre-equilibrium emission) and the TONNEAU for excitation energy deposit. All these observations characterisc the formation and the decay of an equilibrated hot nucleus.

The evolutions of the multiplicities of fast light particles (Fig. 6) as well as evaporated particles (Fig. 7) with increasing recoil velocity are consistent with an incomplte fusion mechanism. For a given E/A , largest recoil velocities are associated to lowest multiplicities in the MUR and largest multiplicities in the backward hemisphere of the TONNEAU. This agreement between these 3 observables establishes that the recoil velocity calculated event by event - and not only the mean value - is a good global variable.

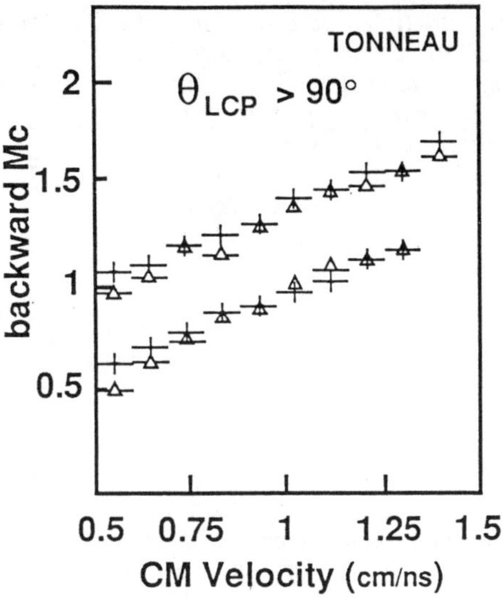

Fig. 7 : *Evaporated particle multiplicity distributions as a function of the recoil velocity. The particles detected at angles larger than 90° are selected to avoid pre-equilibrium contamination. The symbols are the same as in Fig. 6.*

These are clear proofs that an increase of V_R selects more complete fusion events. The effects of an increase of the incident energy are more complex. For a given V_R , the jacobians are the same at 30 and 60 MeV/u so we can compare in the two figures 6 and 7
the multiplicities measured in the Laboratory frame. The highest values obtained at 60 MeV/u confirm that the linear momentum transfer is less effective than at 30 MeV/u as shown in Fig. 4. However it remains sufficient for the excitation energy to increase since the multiplicity of evaporated particles is in average 50% higher at 60 MeV/u than at 30 MeV/u .

It is now possible to compare the corresponding excitation energy with the total multiplicity of evaporated light charged particles.

From linear momentum ccnservation we estimate event by event the number of nucleons of the projectile which fused with the target and deduce within the framework of massive transfer mechanism the excitation energy, which depends on the recoil velocity and on the beam velocity. The average total multiplicity, for evaporated particles Z=1 and Z=2 , is calculated for each recoil velocity bin, assuming an isotropic emission in the frame of the hot nuclei. No correction is applied for the experimental velocity threshold of 2.8 cm/ns .

In figure 8, the multiplicity is plotted at 30 MeV/u (filled symbols) and 60 MeV/u (openned symbols) as a function of the excitation energy.

The data associated with central collisions are presented by triangles and the ones associated with less dissipative phenomena leading to binary fission by circles

The four sets of points lie on the same curve indicating that the multiplicites of evaporated particles are only depending on the excitation energy deposit. This coherence between various data is a strong indication that the method used to extract excitation energies is reliable.Values exceeding 1 GeV are obtained in the 60 MeV/u

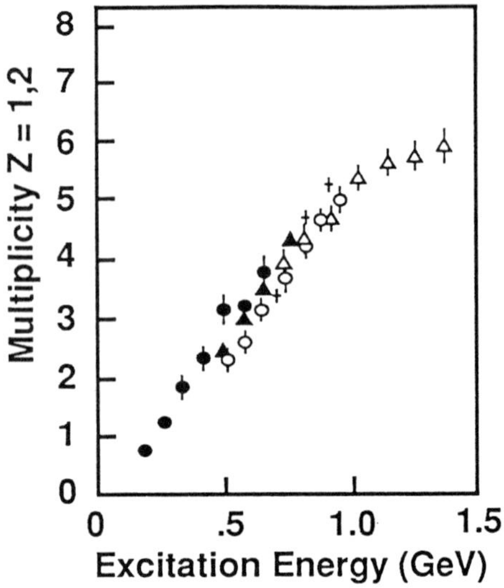

Fig. 8 : *Multiplicity of light charged particles as a function of the excitation energy deposited in the composite system. The symbols represent two ranges of impact parameter at the two energies : filled (30 MeV/u) open (60 MeV/u) peripheral or intermediate (circles) and central collisions (triangles).*

case. The estimated central collision cross sections for these 3 and 4 body events are of the order of 300 mb at 60 MeV/u and 3 times lower at 30 MeV/u.

In summary, the detection of a large fraction of all charged products allowed to select and characterise central collisions in the Ar+Au reactions at 30 and 60 MeV/u . At both incident energies, we have shown the occurence of the formation of an equilibrated system decaying by emission of particles and fragments. At the highest incident energy, in spite of a smaller proportion of linear momentum transfer, the data reveal larger multiplicities both for fragments and evaporated particles. The estimated average excitation energies are respectively 650 MeV and 1 GeV for the 30 and 60 MeV/u cases.

Thus, we have shown that when selecting the most violent collisions with the help of fragment characteristics, we could reach higher excitation energies when increasing the beam energy.

REFERENCES :

1) S. LERAY, Journal de Physique Fr. (1986) C4-275
2) D.X. JIANG et al., Nucl. Phys. A503 (1989) 560
3) D. JACQUET et al., Phys. Rev C32 (1986) 1594
4) G. BIZARD et al., Nucl. Phys. A486 (1986) 178
5) E. POLLACO et al., Phys. Let. 146B (1984) 29
6) A. BADALA et al., XXIII Int. Meeting on Nucl. Phys., Bormio (1990)
7) R. BOUGAULT et al., NIM A259 (1987) 473
8) G. BIZARD et al., NIM A244 (1986) 483
9) A. PEGHAIRE et al., NIM A295 (1990) 365
10) T. MOTOBAYASHI et al., NIM A284 (1989) 526
11) D. DURAND, Preprint LPCC 91-03 submitted to Nuclear Physics.
12) B.BORDERIE et al., phys. lett. B205 (1988) 26

EQUILIBRIUM AND NON-EQUILIBRIUM FEATURES OF NUCLEAR FRAGMENTATION

M.Colonna, M.Di Toro
INFN-LNS Catania, Italy

N.Colonna, L.G.Moretto, P.Roussel-Chomaz, G.J.Wozniak
LBL Berkeley, USA

ABSTRACT

One of the most interesting problems in intermediate energy heavy ion collisions is the complex fragment production. We concentrate on statistical multifragment emission from projectile-like, target-like and quasi-fused sources. This implies the need of a coupling between a dynamical and a statistical picture at some critical freeze-out time, when all the promptly produced primary sources can be considered as equilibrated. Applications are shown to the reactions Kr + Al at 34.4 MeV/A, La + Al and La + Cu at 55 MeV/A beam energies. Several features of the complex fragment production can be understood from this coupling of a dynamical to a subsequent statistical stage. In the energy range discussed here dynamical fluctuations seem not to play a relevant role in fragmentation. Some discrepancies revealed for high multiplicity events, in cases with large excitation energy available, could be a signature of the onset of a direct multifragmentation mechanism.

1. - Introduction.

In this report we will show that a carefull analysis of the entrance channel dynamics coupled to a complete description of decay channels can account for most of the complex fragment production in medium energy heavy ion collisions. We follow a mean trajectory in phase space for the time evolution of the collision: dynamical fluctuations which lead to a direct multifragmentation are not included. The fact that most data are well reproduced is an indication that in such energy range we are only marginally enter the mechanical instability region of the nuclear matter phase diagram.

For the dynamics we solve the Boltzmann-Nordheim-Vlasov (BNV) equation within a test particle approach [1,2,3]. The collision simulation is based on the concept of nucleon mean free path [4]. Fragment production is obtained from a clustering procedure of the mean distribution in phase space, with a suitable definition of a "cluster". At some "freeze-out time", when all the primary prompt produced fragments can be considered equilibrated, we follow a deexcitation stage with the possibility of complex fragment emission through sequential binary decays.

2. - Dynamical stage: the freeze-out problem.

The BNV equation, given below, describes the time evolution of the Wigner function in phase space :

$$\partial f_1/\partial t + \vec{p}_1/m\ \nabla_r f_1 - \nabla_r U \nabla_p f_1 = I_{coll.}$$

$$= \int d\vec{p}_2 d\vec{p}_3 |\vec{p}_1 - \vec{p}_2|/m\ \sigma_{NN} d\Omega [f_3 f_4 (1-f_1)(1-f_2) \tag{2.1}$$

$$- f_1 f_2 (1-f_3)(1-f_4)]\delta(\vec{p}_1 + \vec{p}_2 - \vec{p}_3 - \vec{p}_4)$$

The two fundamental ingredients that enter the BNV equation are: i) the self - consistent nuclear mean field $U_{nucl} = U(\rho)$ and the elementary nucleon-nucleon cross section. In our calculations the Coulomb potential U_{coul} is explicitely taken into account and the neclear part of the mean field is approximated by a density dependent Skyrme - like form:

$$U(\rho) = A(\rho/\rho_0) + B(\rho/\rho_0)^\sigma + C(\rho_n - \rho_p)\tau_z/\rho_0 \tag{2.2}$$

Here ρ, ρ_n and ρ_p are the local nucleon, neutron and proton densities respectively; τ_z is the isospin operator. The parameters A, B, C, σ are chosen such as to reproduce nuclear matter saturation properties and to give a compressibility coefficient K=200 MeV (soft nuclear equation of state). The nucleon - nucleon cross section $\sigma_{NN}(E,\theta)$ is assumed to be the energy and angular dependent free one.

The equation is solved within a test particle approach until a freeze-out time, when some statistical equilibrium has been reached for each primary fragment, defined with a suitable clustering procedure. We stress again that the BNV equation is an equation for the average one body distribution function and therefore it will be able to reproduce only the most probable distributions of primary fragments generated by the dynamics. Moreover, even in a simplyfied situation of a fusion reaction, when the dynamical fluctuations are not expected to play an important role, such kinetic description can be inadeguate on long time scales because in the BNV approach a subsequent evaporation represents the most likely way to account for fragment deexcitation. Other decay channels, with more complex fragment emission, are absent in such a picture. A coupling of the dynamical description to a statistical decay of equilibrated primary sources will overcome such difficulties.

The freeze-out time is defined by the requirement that all primary excited fragments have reached statistical equilibrium without having started the de-excitation stage. A proper determination of the freeze out time is of crucial importance in the evaluation of primary observables: a too late choice would imply the loss of important informations. We consider different criteria:

i) change in the slope of emitted nucleon spectra. In case of incomplete fusion as dominant mechanism this slope permits to distinguish between pre-equilibrium emission and evaporation from an equilibrated source,

ii) saturation of the number of collisions,

iii) vanishing of the quadrupole moment of the momentum distribution in each cluster,

iv) restoration of normal density values in the compressional oscillation.

In fig.1 we show, for the reaction $La + Al$ at beam energy 55 MeV/A and impact paramenter 1 fm, the emitted neutron mean energy vs. time. We deduce a freeze-out time around 100 fm/c, confirmed by the analysis of the other criteria ii) - iv).

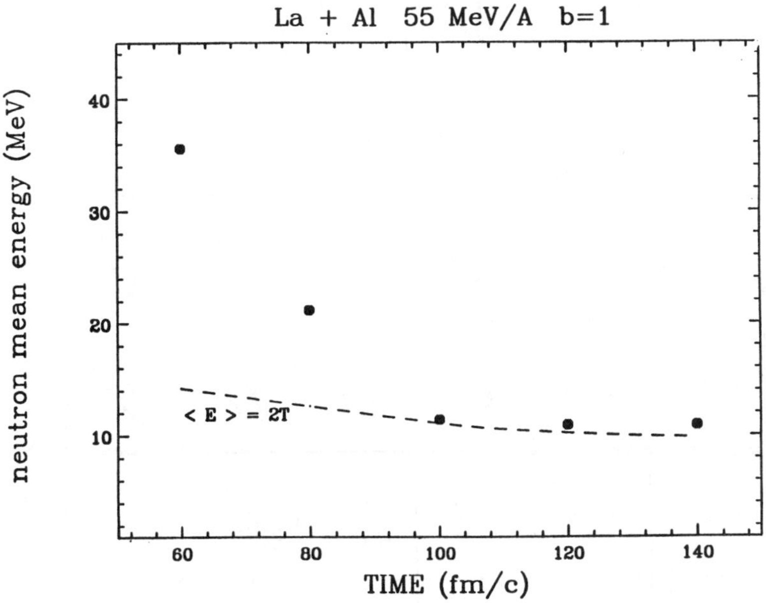

Fig.1

In fig.s 2, 3 we report some density contour plots on the reaction plane at different times for the reactions $La+Al$ and $La+Cu$ at 55 MeV/A beam energy.

Different reaction mechanisms appear quite evident, with a clear dependence on the energy and angular momentum available in the center of mass. We note also the presence of an intermediate source, of "fireball" type, for larger impact parameters in the Cu target case.

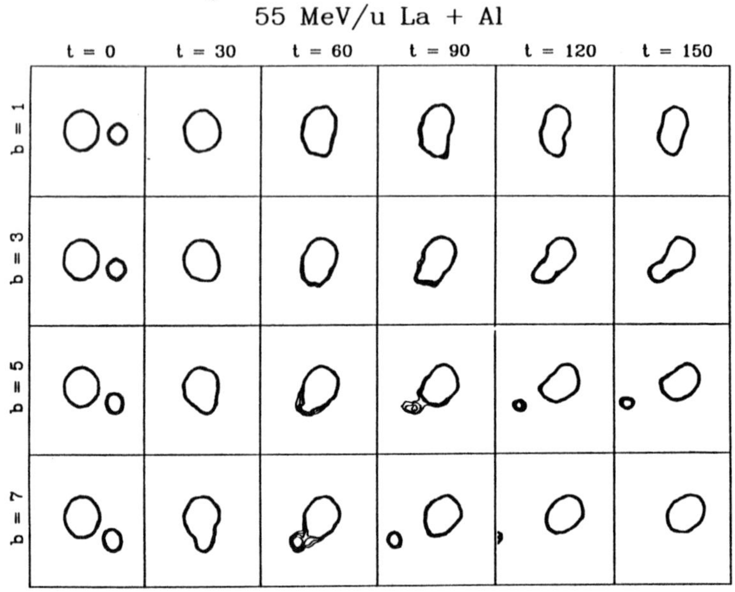

Fig.2

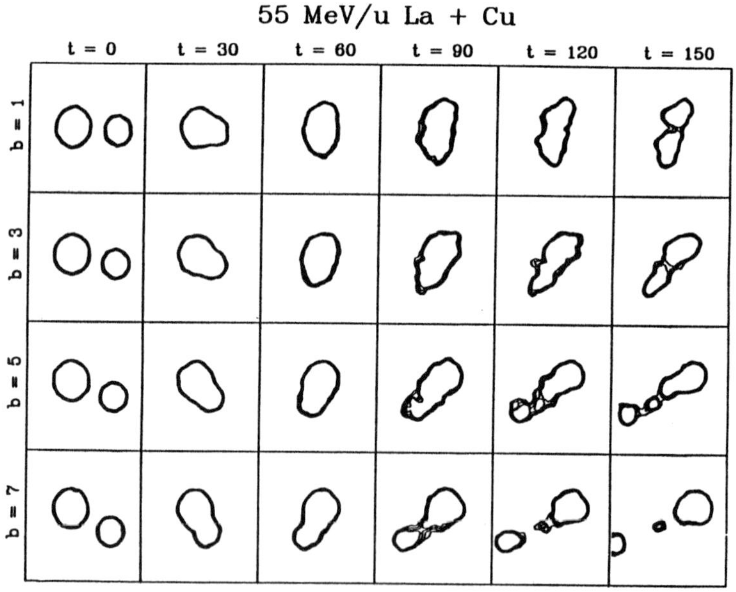

Fig.3

3. - Clustering and de-excitation stage.

Once the freeze-out time is determined the clustering procedure goes as follows[5]. Two test particles belong to the same cluster if they are sufficiently connected in $\vec{r}$ space, that is:

$$|\vec{r_i} - \vec{r_j}| < D \tag{3.1}$$

The key parameter D is strictly linked to the average distance among test particles in a cluster, that it is a functional of the fragment local density. At normal density we can univocally fix $D(\rho = \rho_0) = D_0$ by requiring that the initial condition of the collision is reproduced. In fact the mass and charge distribution in the entrance channel, obtained by the clusterization procedure, must show, up the touching point, a δ-function behaviour peaked on the mass and charge of projectile and target. With this procedure we have found a $D_0 \sim 1.5 fm$. Then D is scaled with density according to the following formula:

$$D/D_0 = (\rho_0/\rho)^{1/3} \tag{3.2}$$

In this way it is possible to calculate mass numbers, atomic numbers , the velocities, deflection angles, excitation energies and angular momenta of the clusters. The intrinsic angular momentum is calculated as :

$$\vec{I} = \sum_i \vec{r_i} \wedge \vec{p_i} \tag{3.3}$$

in the cluster center of mass. The excitation energy is computed as:

$$E^* = E_{kin} + E_{m.f.} + E_{coul} - E_{g.s.} \tag{3.4}$$

where E_{kin} is the total kinetic energy (in the cluster center of mass), $E_{m.f.}, E_{coul}, E_{sym}$ are respectively the mean field energy, the Coulomb energy and the symmetry energy; $E_{g.s.}$ is the ground state total energy calculated from the static solution used as initial condition for the BNV equation.

In practice we cannot always apply the clustering procedure at $t = t_{freeze-out}$ because the fragments are not yet well separated. In the case of $La + Al$ at 55 MeV/A we had to run the BNV calculations up to $t = 140 fm/c$ in order to clearly identify the clusters. Once the characteristics of the fragments are known at $t = 140 fm/c$ and above, the time evolution of these observables is parametrised and the fragment properties at $t = 100 fm/c$ are obtained by extrapolation.

With this procedure we get the most probable distribution of primary fragments for each impact parameter with all their properties at the freeze-out.

4. - The statistical stage.

As already discussed, from the dynamical evolution of the system we get all the informations on primary fragments at some freeze-out time as a function of the impact parameter: mass, charge, intrinsic angular momentum, excitation energy, velocity and deflection angle. As an example we show in fig. 4 the mentioned properties for the largest fragment that we observe in the case of the $La + Al$ reaction at 55 MeV/A. Note the sudden change in intrinsic angular momentum when a transition in the reaction mechanism occurs between $b = 3$ and $b = 5fm$ (see fig. 2), from incomplete fusion to deep inelastic.

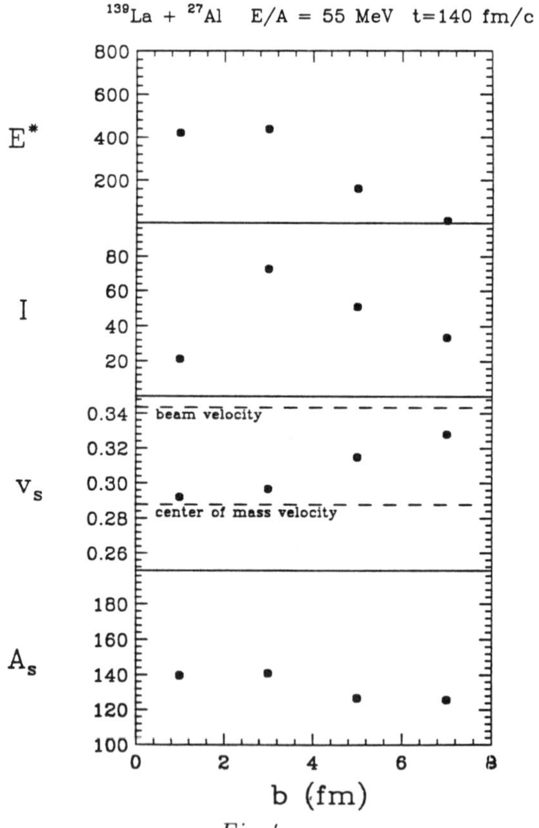

Fig.4

We remark that almost half of the energy available in the center of mass is dissipated in pre-equilibrium processes, but still the primary fragments are at quite high "temperature", between 5 and 7 MeV.

The next step consists into accounting for the de-excitation stage of such equilibrated sources. We use the statistical decay code $GEMINI$ [6]. All decay channels are considered, from light particle emission to symmetric division. A Monte-Carlo technique is used to follow the de-excitation chains of individual nuclei through sequential binary decays until the resulting products are unable to undergo a further decay. The evaporation widths of fragments with $Z < 3$ are calculated using the Hauser-Feshbach formalism, whereas for binary divisions corresponding to heavier fragment emissions the decay widths are calculated using the transition state formalism of L.Moretto [7].

5. - Results.

We have applied the model described before to the analysis of complex fragment production in the reactions $Kr + Al$ at 34.4 MeV/A, $La + Al$ and $La + Cu$ at 55 MeV/A beam energies.

In the Kr case our calculations predict for the most central impact parameters ($b = 1, 2, 3 fm$) the occurrence of incomplete fusion accompanied by pre-equilibrium emission. For more peripheral reactions we have mainly evidence of binary processes of deep-inelastic type. To stress the importance of the coupling between dynamical and statistical stage in fig.5 we show the inclusive production cross section obtained in both cases: BNV alone and $BNV + GEMINI$ coupling. In the first case (dashed line) we see in the calculated yield a Projectile-like a Target-like and a Fusion peak: it turns clearly out that the BNV approach is inadeguate to describe by itself the intermediate fragment production (IMF). The complete lack of cross section in the intermediate mass region has to be ascribed to the fact that the kinetic approach takes into account only nucleon emission to dissipate energy (classical dynamics). Here we have also a lot of angular momentum dissipated into the primary fragments (see fig.4) and so the fission channels are well in competition with nucleon emission [8]. The "fused" systems which occur in the case of small impact parameters are therefore highly unstable with a large probabitily of fissioning with emission of IMF's. The IMF's production can be simply understood by a sequential decay of such excited primary sources.

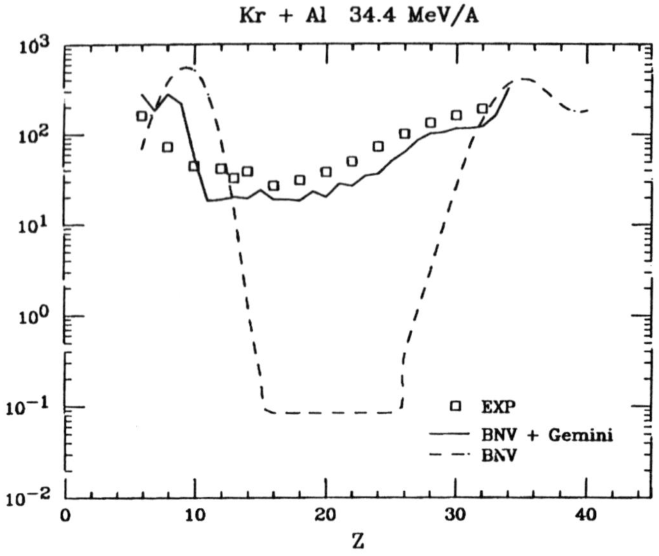
Kr + Al 34.4 MeV/A
10^3
10^2
10^1
10^0
10^-1
10^-2
EXP
BNV + Gemini
BNV
0
10
20
30
40
Z

Fig.5

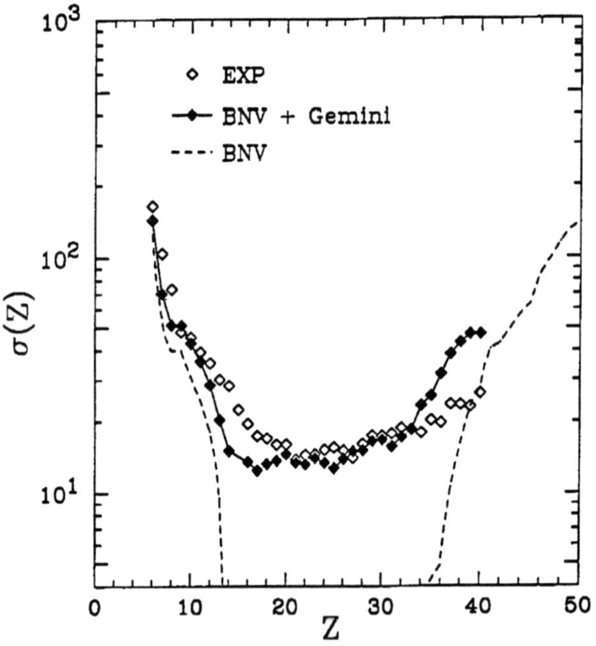
10^3
EXP
BNV + Gemini
BNV
10^2
σ(Z)
10^1
0
10
20
30
40
50
Z

Fig.6

The same calculation has been performed for the system $La + Al$ at 55 MeV/A. In this case we observe the occurence of a fusion (incomplete) mechanism for the most central impact parameters ($b = 1, 2fm$). The observed mechanism at $b = 3fm$ can be characterised as fast fission. The more peripheral reactions lead to deep inelastic like processes. A freeze-out time of 90 fm/c was chosen for this system. As already mentioned, this time corresponds approximately to the "kink" observed in the energy spectra of emitted neutrons, see fig.1. Fig. 6 shows the inclusive fragment production cross section. Diamonds represent the data and the solid line is the result of our calculations. Also in this case a whole BNV calculation severely underestimates the IMF production rate (dashed line).

For this reaction we have also compared the results of our method with more exclusive measurements in order to extract more informations on the origin of the detected fragments.

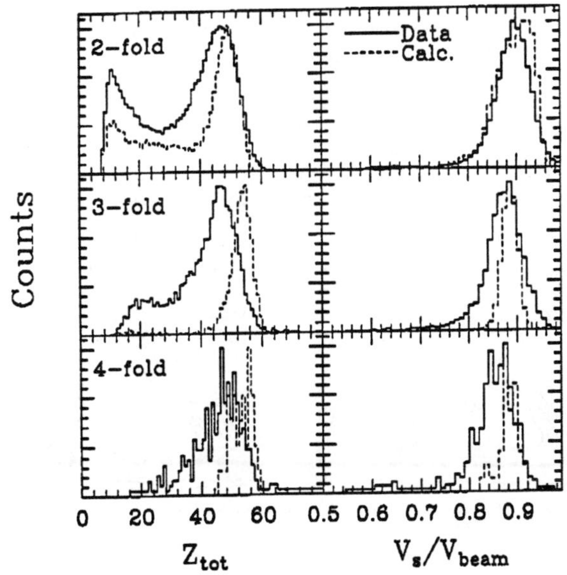

Fig.7

In fig.7 we report total charge and source velocity distributions (for $Z_{tot} > 30$) for different multiplicity events. The calculations are filtered through the experimental detection efficiency. The overall agreement is quite good. The tail at low total Z observed for the 2 fold events clearly indicates the occurence of a predominant 3-body mechanism, easily understood in our approach: for impact paramenters $b > 2fm$, which give a large contribution to the cross section, a light fragment comes from the dynamics of the target residue and the other two from the fission of the largest primary fragment. Three-body decays are reported by means of Dalitz plots: each 3-fold event is represented as a point for which the distances to the sides of the triangle indicate the values $Z_1/Z_{tot}, Z_2/Z_{tot}, Z_3/Z_{tot}$[9]. Fig. 8 compares the data with the results of the $BNV + GEMINI$ calculation before and after filtering (for $Z_{tot} > 30$). We remark that at the end of the dynamical stage the Dalitz plot would be empty, since only either an incomplete fusion residue or a target and projectile remnant are produced. After the de-excitation stage, the simulated Dalitz plot is very similar to data, with a dominance of events with one heavy and two light fragments.

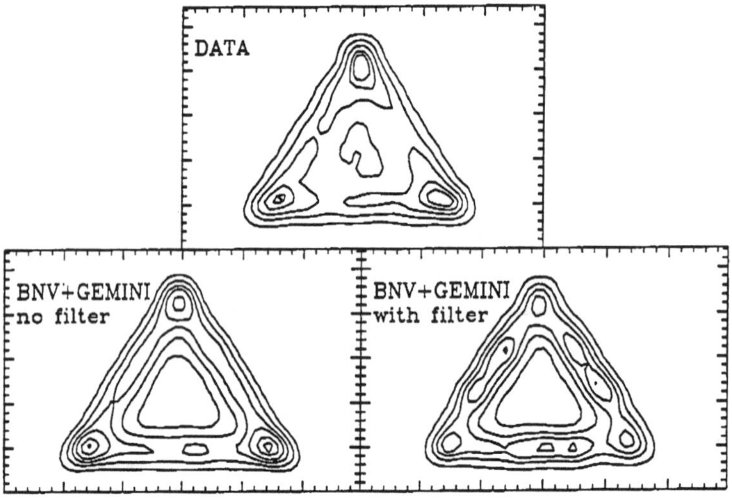

Fig.8

The underestimation of the tail for both 2- and 3- fold events in fig.7 can be ascribed to the fact that the calculations underestimate the number of 4- and higher n-fold events, which actually represent a few percent of the total fragmentation cross section, as clearly indicated in Table 1. This could be an indication of possible contributions from a direct multifragment emission.

Table 1 : Comparison of the experimental and calculated rates of multifragment events. A gate has been set on the total detected charge (Ztot>30).

	2-fold/total	3-fold/total	4-fold/total
Data	0.869	0.122	0.0088
Simulation	0.855	0.144	0.0017

Indeed some hints for an onset of multifragmentation processes related to dynamical instabilities can be further supported from the analysis of the $La+Cu$ results at the same beam energy, where actually more energy is available in the center of mass.

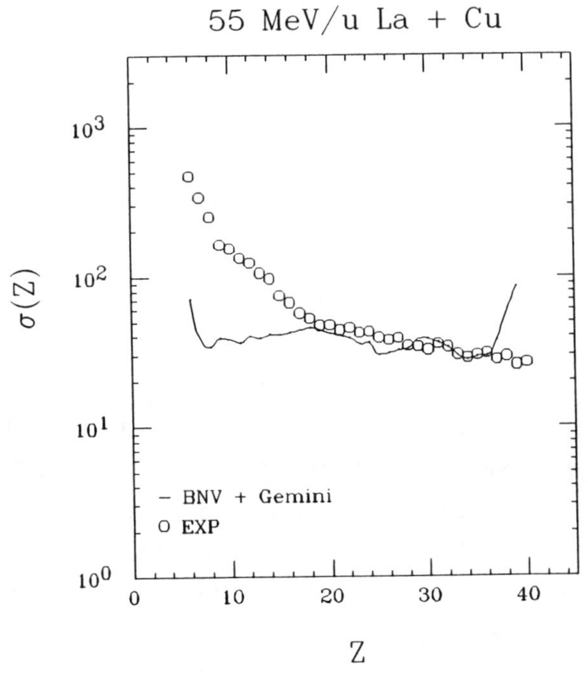

Fig.9

Fig. 9 shows the comparison of calculated (solid) and measured inclusive cross section for fragment production. The discrepancy is clear for light fragments. These are probably coming from more central impact parameters where in any case our clustering procedure of the average space space distribution presents several uncertainties. E.g. for $b = 1fm$ at $t = 150fm/c$ the density distribution on the reaction plane (see fig.3) seems to show two main primary fragments while the same density plot on the plane orthogonal to the beam axis (fig. 10) clearly indicates more complicated structures not easy to disentangle. Moreover in this case the number of 4- and higher n-fold events is much larger [10].

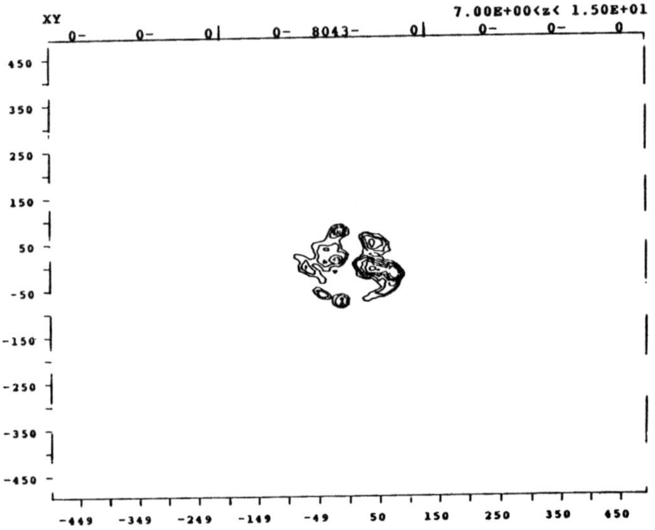

Fig.10

6. - Conclusions.

We have shown that, due to the very different involved time scales, we can understand the complex fragment production from heavy ion reactions in the energy range 30-60 MeV/A just considering non-equilibrium and equilibrium features by coupling a kinetic description of the dynamical stage of the collision with a subsequent statistical decay of primary sources. The dynamical evolution has been simulated by solving the BNV equation for the average phase space distribution up to a freeze-out time where the characteristics of the primary fragments have been extracted and introduced as inputs in the statistical code $GEMINI$. It is very essential to carefully take into account all possible decay channels in the de-excitation stage.

An overall agreement is obtained not only for inclusive production yields , but also for charge distributions, source velocities and Dalitz plots. Some discrepancies are observed for high multiplicity events that could be ascribed to the onset of direct multifragmentation processes, related to dynamical instabilities [3,11]. However the corresponding cross section seems to be only a few percent of the total fragmentation yield. This means that at this relative velocity for the projectile-target systems studied here we are just marginally entering the spinodal region of the nuclear matter phase diagram. The analysis of collisions at the same beam energy but with more energy available in the center of mass could be of great interest.

We warmly thank A.Bonasera (Catania), F.Gulminelli and M.Cavinato (Milano) for many helpful discussions on the clustering procedure.

REFERENCES

1) H.Stöcker and W.Greiner Phys.Rep. 137(1986)277.
2) Ch.Grégoire, B.Rémaud, F. Sebille, L.Vinet and Y.Raffray, Nuclear Physics A465(1987)315.
3) G.F.Bertsch and S.Das Gupta, Phys. Rep. 160(1988)190.
4) A.Bonasera, G.F.Burgio and M.Di Toro, Phys.Lett. B221(1989)233 A.Bonasera, G.Russo and H.H.Wolter, Phys.Lett. B246(1990)337
5) A.Bonasera,M.Colonna,M.Di Toro,F.Gulminelli and H.H.Wolter, Phys. Lett. 244B(1990)169.
6) R.J.Charity et al., Nucl.Phys. A483(1988)371 and Nucl.Phys. A511(1990)59
7) L.G. Moretto, Nucl.Phys. A247(1975),211.
8) A.Adorno, M.Colonna, M.Di Toro and G.Russo in "The Nuclear Equation of State" Ed.s W.Greiner and H.Stöcker. Plenum Press N.Y.1990 p.413-420, and A.Adorno et al., Nucl.Phys. (1991) in press

9) M.Colonna, P.Roussel-Chomaz, N.Colonna, M.DiToro, L.G.Moretto and G.J.Wozniak, "Role of dynamics and statistics in multifragment production", LBL preprint 1991 submitted to Phys.Rev.Lett.

10) P.Roussel-Chomaz, private communication

11) S.Ayik and Ch.Gregoire, Phys.Lett.B212(1988)269 and Nucl.Phys. A513(1990)187,
J.Randrup and B.Remaud, Nucl.Phys. A514(1990)339
E.Suraud, S.Ayik, J.Strjewski and M.Belkacem Nucl.Phys. A519(1990)171c
Ph.Chomaz, G.F Burgio and J.Randrup, Phys.Lett B254(1991)340

A SYMMETRY TERNARY FISSION SHAPE AND ITS POTENTIAL

Guangxi Dai
Institute of Modern Physics, Academia sinica,
P.O.Box 31, Lanzhou, 730000, P. R. China

ABSTRACT

Symmetry Ternary Fission (STF) becomes a important out-going channel in intermediate energy heavy ion reactions. In the paper a shape with spherical residue added 3-rose-leaf curve rotating bodies are assumed as the shape of STF. The shape has one deformation parameter δ and an adjustable parameter for choosing better shape. The general function, such as Ec/Eco and Es/Eso aganist δ, (here Eco and Eso are Coulmb and surface energies for spherical nucleus, respectively) have calculated. In comparing with binary fission, the TKE of STF is $0.2335Eco$, which is large than the TKE of binary fission by 30%, but is slight smaller than that of $0.254Eco$ calculated by cascade binary fission process, that is resonable. However, the more deformation for new born fragment in the STF is present. There are 3 times of the deformation energy in the STF large than that in the binary fission for each fragment. The deformation of new born fragment in the STF corresponds to a axes length ratio by roughly 2:1. The temperature dependence of surface tension energy is introduced in the STF potential calculation and discussion. For hot nuclei the STF barrier decreases and saddle point shrinks with increasing nuclear temperature.

I. Introduction

Many experiments on symmetry ternary fission (STF) have been performed since W. Swiatecki[1] proposed ternary fission is possible if fissionable parameter $\chi > 0.44$, In ^{40}Ar induced fission experiment[2], with ^{232}Th, ^{238}U, ^{208}Pb and ^{197}Au targets, the probability ratio of STF to binary fission, P_{32} can be reach to 3% when the excitation energy goes up to 240 MeV. However For ^{252}Cf spontaneous fission, STF is a rare event, the ratio P_{32} has a up-limit, 8×10^{-8}, that is quite small; still, the mass and kinetic energy distributions for three massive debrises of ^{252}Cf has been obtained [3]. In the work for Kr+Au at 43MeV/u systemthe events with 3- or 4- folds massive debrises have large yeilds; among them the prompt STF events with relative velocity of 2.15cm/ns exist with considerable cross-section[4]. In work[5] the direct ternary fission was proposed in prolate and oblate configurations. They found a preference for prolate over oblate saddle point shapes and a higher barrier in prolate than that of BF by three times.

Indeed, Cassinian ovaloid is a simple shape with good approximation for binary fission description[6,7]. Using a curve described by biquadratic equation to form a rotating body, the variation calculation demonstrates that the fission barrier of the ovaloid is quite close to that of the best shape, for which the fission barrier is minium[8]. Similarly, in present paper, we would like to choose an oval-like curve as a scission profile in STF shape description too.

In the current liquid drop model, the surface tention is temperature-independent. According to nuclear thermodynamics, the tension coefficent decreases against nuclear temperature. From sphere to scission point, the surface area of STF changes larger than that in the binary fission(BF). Therefore, the barrier height of STF is quite large than that of BF. In order to explain the large P_{32} in measured experiments[2], it may be useful to consider the tension dependence on the temperature. A similar consideration of a limit temperature in hot nucleus has been done in a work[9] and was comfirmed by some experiments[10]. Therefore, we would use the tension coefficient with the temperature-dependent factor.

II. The Shape Geometery and Potential Calculation in STF

In similar with the ovaloid, a shape changing from sphere to scission point with one parameter for STF deformation process has been proposed. At scission point of the STF the shape profile looks like a 3-rose-leaf curve, that is a oval-like curve with 3 branches. A so-called spherical residue plus the 3-rose-leaf-like rotating bodies (SRRL) is defined to describe the STF deformation. The spherical residue is defined by a sphere removed 3 capes with hight of 0.5 radius, as shown in Fig.1. The radius of the sphere $r = r_0(1-\delta)^{1/n}$ depended on the deformation parameter

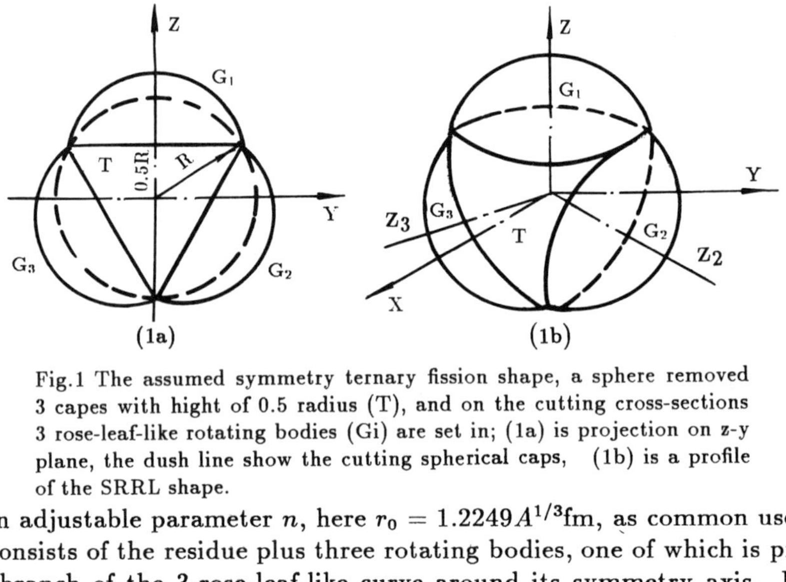

Fig.1 The assumed symmetry ternary fission shape, a sphere removed 3 capes with hight of 0.5 radius (T), and on the cutting cross-sections 3 rose-leaf-like rotating bodies (Gi) are set in; (1a) is projection on z-y plane, the dush line show the cutting spherical caps, (1b) is a profile of the SRRL shape.

δ and an adjustable parameter n, here $r_0 = 1.2249A^{1/3}$fm, as common used. The SRRL consists of the residue plus three rotating bodies, one of which is produced by one branch of the 3-rose-leaf-like curve around its symmetry axis. In polar coordinates, the curve reads,

$$\rho^n(\theta,\delta) = \delta\cos(3\theta) + \sqrt{1-\delta^2\sin^2(3\theta)} \tag{1}$$

where n is a adjustable parameter, we set it by 2 or 3 . When $\theta = 60°$, $\rho(\pi/3,\delta) = (1-\delta)^{1/n} = r/r_0$.

The SRRL body is formed by such a way: on the cutting cross-section of the spherical residue, a rotating body associated with curve (1) can be joined with. A special case of SRRL is for $\delta = 1.0$, namely at scission point, in which the residue is vanish, and the scission profile with 3-rose-leaf-like curve where only $\theta < 30°$, the $\rho(\theta, \delta)$ has non-zero value.

With the shape SRRL, the STF process from sphere, where $\delta = 0$, and pass through the saddle point then to scission point, where $\delta = 1.0$, can be simply and smoothly described.

Considering volume conservation, here we introduces a factor $f_v(\delta, n)$, which is equal to $[Vo/V(\delta, n)]^{1/3}$, here the Vo is sherical volume and the $V(\delta, n)$ reads

$$V(\delta, n) = \frac{17}{24}\pi(1-\delta)^{3/n} + 3\pi\int_0^{\theta_o} \rho^3 \sin^2\theta[\frac{3\delta \sin 3\theta \cos\theta}{n\sqrt{1-\delta^2\sin^2 3\theta}} + \sin\theta]d\theta \qquad (2)$$

and the area of SRRL can be expressed as following

$$S(\delta, n) = f_v^2(\delta, n)\{\pi(1-\delta)^{2/n} + 6\pi\int_0^{\theta_o} \rho^2 \sin\theta\sqrt{\frac{1+\delta^2\sin^2 3\theta(9/n^2-1)}{1-\delta^2\sin^2 3\theta}}d\theta\} \qquad (3)$$

here $\theta_o = \pi/3$, where $\delta < 1$; $= \pi/6$, when $\delta = 1$; As example, the ratios of S(1,2) and S(1,3) at scission point are 1.5575 and 1.5017, respectively. When the $S(\delta, n)$ is calculated, the ratio of surface energy Es over spherical one, Eso, can be obtained from $S(\delta, n)/So$.

The SRRL is 120° rotating symmetry; for each fragment inside the 120°, the position of centre of mass to original point $Zc(\delta, n)$ can be expressed as

$$Zc(\delta, n) = \frac{Zg + Zt}{V_o/3} f_v^4(\delta, n) \qquad (4)$$

where Zg and Zt are associated with 3-rose-curve expressed in (1) and one third of the spherical residue, respectively;

$$Zg = \pi\int_0^{\theta_o} \rho^4 \sin^2\theta\cos\theta(\frac{3\delta\cos\theta\sin 3\theta}{n\sqrt{1-\delta^2\sin^2 3\theta}} + \sin\theta)d\theta \qquad (5)$$

$$Zt = \int_0^{.5\rho_o} [(\pi - 2\psi)(\rho_o^2 - z^2) + 2z\sqrt{3\rho_o^2 - 12z^2}]zdz \qquad (6)$$

where $\rho_o = (1-\delta)^{1/n}$ and

$$\psi = \arcsin\sqrt{(\rho_o^2 - 4z^2)/(\rho_o^2 - z^2)} \qquad (7)$$

At scission point, the $Zc(1,2)$ and $f_v(1,2)$ have values of $1.424r_o$ 1.687 respectively; in comparing with binary fission the corresponding values are $1.2046r_o$ and 1.4816, respectively. In Fig.2, the $Z_c(\delta, n)$ as function of δ for $n = 2$ and 3 are shown. Considering the shift of the centre of mass from one third sphere to fragment at scission point, the shift value is $0.96r_o$, that is large than $0.83r_o$ associatted with binary fission.

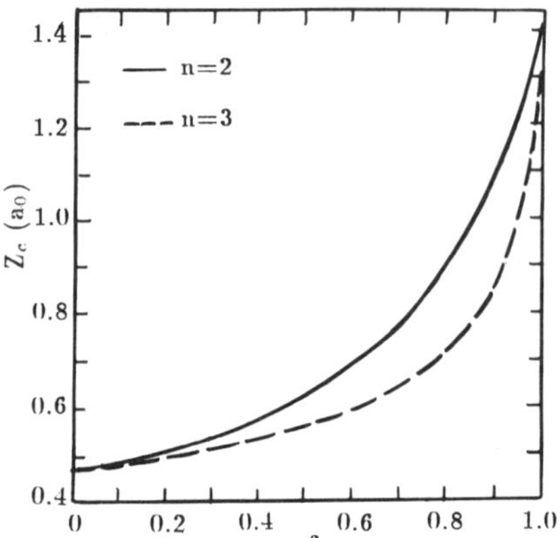

Fig.2 The dependence of center of mass Z_c associated with one fragment at scission point on δ.

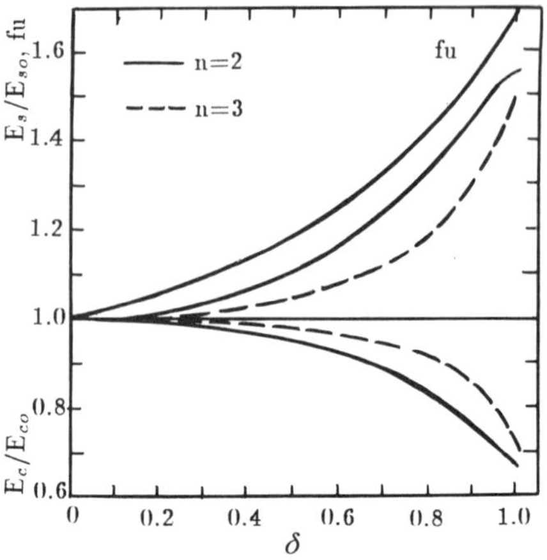

Fig.3 The dependence of E_c/E_{co}, E_s/E_{so} and $f_v(\delta,2)$ on the deformation parameter δ for n=2 and 3 (dash line), in STF with assumed shape, the SRRL; here E_{co} and E_{so} are Coulomb and surface energy associated with spherical shape.

Integrating whole volume of the SRRL, we can get Coulomb energy E_c/E_{co} functions on the parameter δ and n. In Fig.3, these general functions such as $E_c(\delta,n)/E_{co}$ and $E_s(\delta,n)/E_{so}$ for $n = 2$ and 3 cases and $f_v(\delta,2)$ are shown.

At scission point, a test to the SRRL by the mutual Coulomb energy among the 3 fragments associated with total kinetic energy (TKE) for the STF has been performed. As listed in table 1, the TKE's in the SRRL shape for $n = 2$ and 3 have been calculated; in addition the TKE for binary fission and for a STF by double cascade binary fission have been also calculated by Viola systematics[11]. From the Table I, the TKE/Eco for prompt STF is slight smaller than that for cascade binary fission; because the fragment center of mass in STF move more distance than that in BF.

Table I. the comparation of TKE and deformation of new born fragments in STF, binary fission and cascade double BF

	STF(n=2)	STF(n=3)	BF+BF	BF
TKE/E_{co}	0.234	0.245	0.254	0.1780
$\Delta Es/Eso(\%)$	11.6	5.95	2.6	2.6

ΔEs – the difference of the surface energy between new born fragments and its spherical shape, in which the surface energy is Eso.

In comparing with binary fission , the TKE of STF is larger than the TKE of binary fission by 30%, as shown in Table I. The STF considered here is a prompt process by the SRRL, in which the calculated value of TKE is $0.2335E_{co}$ that is slight smaller than that of $0.254E_{co}$ calculated by cascade binary fission process, that is resonable and confirmed by work[4]. However, the more deformation for new born fragment in the STF is present. As shown in table I, there are 4 times of the deformation energy in the STF larger than that in the binary fission. The deformation of new born fragment in the STF corresponds to a ratio of axes length by roughly 2:1. If we want to decrease the deformation, we can set larger power index n in eq.(1) expression by 3. As listed in table I, the deformation (n=3) of new born fragment deviating from spherical becomes less, however there are a great change when δ are near scission point (see Fig. 3). However it seems not resonable. The n value of SRRL could be given only by fitting experimental data; however it seems that n=2 is a initial value as likely the Casinian ovaloid in BF.

III. The Temperature Dependence of Surface Tension Energy

The STF is a rare event in spontaneous fission or in low excitation fission processes. It becomes considerable only in high excitation[2]. From the calculated results on the shape of SRRL,Table II lists the deformation parameter values and fission barrier values calculated with the shape of SRRL at saddle point for various nucleus in different nuclear temperatures. Here we adopt the surface tension temperature factor by following expression,

$$E_s(T) = E_s(1 - T/T_c)^2 \tag{8}$$

where the T_c is critical temperature for infinite nuclear matter, 17.22 MeV[9], where another temperature-dependence expression has been introduced, by which the dependence of barrier and δ at saddle point is weaker than expression (8). However, the prediction of limit nuclear temperature by (8) has been confirmed by some experiments.

From calculated results listed in Table II, for various nuclear system it seems that there is a relationship between deformation δ and barrier value at saddle point in average.The barrier values decrease as decreasing δ. At saddle point, the deformation of STF and barrier value become smaller with nuclear temperature increasing. For easy fissioning nucleus, like ^{238}U, at $T > 3.5$ MeV the fission barrier is vanishes. Therefore in hot nuclei the probabilities of STF increase and reaction time become faster because the barrier is lower and the shape at saddle is closer to sphere.

Table II. The dependences of δ and fission barrier B in MeV at saddle point on the nuclear temperature T for various nuclei

T	^{197}Au		^{209}Bi		^{238}U		^{252}Cf	
(MeV)	δ	B	δ	B	δ	B	δ	B
0	.95	83.21	.95	78.68	.95	57.97	.85	45.01
0.5	.95	66.55	.95	61.59	.85	41.82	.80	31.70
1.0	.95	50.39	.90	45.27	.80	29.26	.70	21.45
1.5	.90	35.94	.85	31.93	.75	19.56	.65	13.78
2.0	.80	24.72	.75	21.57	.65	12.41	.55	8.00
2.5	.70	16.12	.70	13.92	.55	7.08	.50	4.02
3.0	.65	10.00	.60	8.28	.50	3.42	.40	1.41
3.5	.55	5.49	.55	4.28	.40	1.10		
4.0	.45	2.46	.45	1.70				
4.5	.35	0.63	.30	.20				

IV. Discussion

In the paper we proposed a simple deformation shape SRRL with one deformation parameter is used for describing STF process. The SRRL with n=2 or 3 could predict the TKE. However, at saddle point, the shape should be tested by fitting experimental data,such as ratio of yeild binary fision over prompt STF, and the ratio changing with Z^2/A and excitation energy, etc. From the Ec/Eco and Es/Eso against deformation parameter, these shapes for n=3 has a drastic changing near scission point, in comparing with the case for $n = 2$. Therefore the saddle pointsn for various nuclei are crowed near the scission point for $n = 3$ case, that is not reasonable. However, it seems that the shape $n = 2$ is more reasonable than the shape $n = 3$. Therefore, in fitting process n value should be choosed around 2.0, and it is not necessary with integer values. In our calculation to $^{298}114$ nuclei shown in Fig.4, the barrier height and separation distance between two fragments among the three are 22.9 MeV and 11 fm; in comparing with the results in[5], those value are 55 MeV and 10 fm, respectively. It seems that the SRRL shape with lower barrier than the oblate is preferential one. The folding angles between two fragment in STF have 120° in work[2], that supports the oblate shape and does not the prolate. In the prolate case, when the two necks are broken simultaneously, the middle fragment gets only small momentum; then the folding angle is near 180°. Indeed, the prolate has more lower barrier, that is a special case of cascade fission with short interval time.

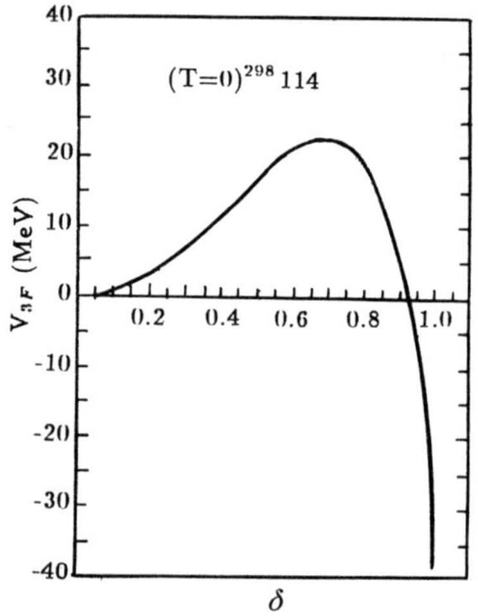

Fig.4 The STF potential $V_{3f}(\delta)$ calculated by current liquid drop model with the SRRL deformation

The prompt STF is associated with a new deformation type other than binary fission. In hot nuclei produced in intermediate heavy ion reactions, the prompt STF may become a considerable process. It is may important to consider the surface tension depending on nuclear temperature. Only preliminary consideration has been made in the paper. Because of the large changing surface area in STF, it could be sensitive for testing the temperature-dependent factor in the tension. The temperature-dependent factor is a important parameter for thermodynamics describing the finite nucleus. There are two different kinds of factor; one is common used as in text, it is stronger dependence on temperature than other[9]. from the testing, one may get some conclusions.

References

[1] W.J.Swiatecki, et.al.,Second UN Intern. Conf. on the Peaceful Uses of Atomic Energy (Geneva,1958), vol.15, p.248

[2] V.P.Perelygin, et.al., Nucl. Phys., A127 (1969) 577

[3] P.Schall, P.Heeg, et.al., Phys. Lett., 191 (1987) 339

[4] R.Bougault, LPCC 90-02, 'Winter Meeting on Nucl. Phys.'(Bormio 22-26 Jan. 1990)

[5] H.Diehl and W.Greiner, Nucl. Phys., A229 (1974) 29

[6] V.S.Stavinsky, et.al., Nucl. Phys.(USSR), 71 (1968) 1051

[7] G.Dai, X.Liu and G.Liu, Chinese Phys., 8 (1988) 427

[8] G.Dai, W.Li, High Ener. Phys. & Nucl. Phys., 12 (1988) 399

[9] S.Levit, P.Bonche, Nucl. Phys. A437 (1985) 426

[10] E.Crema, et al. , GANIL P-90 19b (1991)

[11] V.E.Viola Jr., Nuclear Data Sect., A1 (1966) 391

SHAPE PARAMETRIZATION OF FISSION NUCLEI BY CASSINIAN OVALOIDS

Guangxi Dai[1]
Institute of Modern Physics, Academia Sinica,
P. O. Box 31, 730000 Lanzhou, P. R. China

Hartwig Freiesleben
Institut für Experimentalphysik I, Ruhr-Universität Bochum
D 4630 Bochum-1, F.R.Germany

ABSTRACT

Based on well-defined fission shape, the Cassinian ovaloid, the PES (potential energy surface) for asymmetry fission with shell revision has been calculated. Two halves of Cassinian Ovaloids smoothly joined together are utilized as asymmetry fission shapes. Therefore there are only two collective variables are needed in the approach: an eccentricity, ε_1, and a mass asymmetry parameter, p;necking-in is an inherent property of Cassinian ovaloids and does not require an additional parameter. When neck appears, the shell revision is treated to two parts of nucleus on both sides of the neck, separately. For each part there is a inner lemniscatoid as a core, which developes to whole fragment, when ε_1 increases from 0.7 to 1.0. By semi-emprical folmula due to Swieatecki, the shell revision energy is calculated only on the core instead of whole fragment. Examplary results for ^{208}Pb, ^{233}Th and ^{252}Cf qualitatively feature the expected properties of static fission barrier and saddle points.

I. Introduction

The full quantum mechanics treatment of fission is one of the most challenging many-body problems of nuclear physics. Various attempte were undertaken in the past to describe the dynamics in a proper way [Pauli, Breack, Dietrich, Negele, et.al.]. A less ambitious attempt is to calculate the potential energy surface (PES) of a fissionning nucleus and infer from such a static property the fission path a nucleus might take [Sweatecki, Brosa, et.al. ,]. Howevere, the calculation of a PES itself is not a simple task. In principle, a selfconsistent microscopic calculation is required. The inhernt problems are not overcame yet. An alternative method is a macroscopic-microscopic approach where the PES is derived from a macroscopic liquid drop or droplet model with its various refinement [Swiatecki, Myers, Nix, Moller, Krappe, et.al. ,], on top of which microscopic corrections are added. Here, the shape parametrization of the liquid drop is of fundamental importance: the short-range nuclear interaction gives rise to the surface tension which counteracts the Coulomb repulsion; their interplay determines the stability of a liquid drop against fission. Both potential energies are strongly shape dependent. In turn the

[1] A.v.Humbolt-fellow at Ruhr-Universität Bochum.

the Coulomb repulsion; their interplay determines the stability of a liquid drop against fission. Both potential energies are strongly shape dependent. In turn the nuclear shape determines the mean field in which nucleons move. Hence, shell and pairing corrections become shape dependent, too.

Various shape parametrization were advanced in the past (see e.g. the review article of Nix[1], and of Myers[2]). For small deformation the approach of Nilssion[3] of expanding the surface in terms of spherical harmonics was extremely successful. However, it is not well suited for a scission configuration with large deformations. Therefore other parametrizations were introduced in order to render possible the treat of large deformations such as perturbed spheroids, generalized spheriods, perturbed Cassinian Ovaloids[7], two-center models, three quadratic surfaces of revolution. Common to these approaches is the need of essentially three collective variables, namely a deformation along the fission axis, a mass asymmetry parameter, and a necking degree of freedom.

The macroscopic PES can be calculated in all cases in a straight forward manner once the shape is specified. The microscopic part of the PES need to be calculated for each nuclear shape. first the single particle Schrödinger equation for the potential related to the macroscopic shape under consideration must be solved and second, the microscopic corrections have to be derived by the Strutinsky method[4], which is based on the Nillson diagram with an ellipsoid potential. In X. Scharnweber work[5], the two center potential calculation shows at intermediate values of deformation where the neck develops, the lower levels first become degenerate, and reflecting the fact that a barrier is developeing between the two fragments. It is also that the states near Fermi surface still show an appreciable splitting. From the calculation, a concept about how to establish the shell structure of the two fragments by increacing the deformation parameter, becomes practical. In the III.3 of the paper, we use the concept to perform the shell correction in PES calculation.

In the paper we introduce a shape parametrization which is based on Cassinian ovaloids introduced by Stavinsky et.al.[6]. Our simple but quite effective approach uses only two collective variables, the eccentricity along the fission axis and the mass asymmetry parameter; the neck-in is an inherent propery and needs no further consideration. Two halves of Cassinian ovaloids are smoothly joined together, thus additional parameters are avoided which were needed in the work of Pashkevich[7] who expanded Cassinian ovals by a series of spherical harmonics. We develop the model in some detail in sect. II. The microscopic well defined cores of the fission path and become, eventually, the two fission fragments . In sect. III we present and discuss the results, namely the liquid drop properties of nuclei with our shape parametrization and the full PES for ^{208}Pb, ^{233}Th and ^{252}Cf.

II. Definition of Model

II.1 Shape Parametrization

The shape parametrization of our model starts out with the Cassinian oval. It is the geometrical locus of all points for which the product of the distances R_1, R_2, from two fixed points F_1, F_2, separated by twice the Cassinian distance $2c$, is a constant, a^2:

$$R_1 R_2 = a^2 \tag{1}$$

In the following we call a the Cassinian parameter. In Cartisian coordinates this equation reads

$$(z^2 + y^2)^2 - 2c^2(z^2 - y^2) = a^4 - c^4 \tag{2}$$

It reduces to the equation of a circle with radius a for $c = 0$. With the definition of dimensionless quantities

$$\varepsilon = c/a, \qquad \eta = y/a, \qquad \zeta = z/a \tag{3}$$

The eq. (2) can be recast in to

$$(\zeta^2 + \eta^2)^2 - 2\varepsilon^2(\zeta^2 - \eta^2) = 1 - \varepsilon^4 \tag{4a}$$

in polar coordinate with $\rho^2 = \zeta^2 + \eta^2$.

$$\rho^2 = \varepsilon^2 \cos 2\theta + \sqrt{1 - \varepsilon^4 \sin^2 2\theta} \tag{4b}$$

Cassinian ovals have characteristic shape for certain ranges of the parameter ε :

a) $\varepsilon = 0$ circle,
b) $0 < \varepsilon < 1/\sqrt{2}$ ellipse-like shapes,
c) $1/\sqrt{2} < \varepsilon < 1$ ovals with neck,
d) $\varepsilon = 1$ Bernoulli lemniscate,
e) $\varepsilon > 1$ two separated "drops"

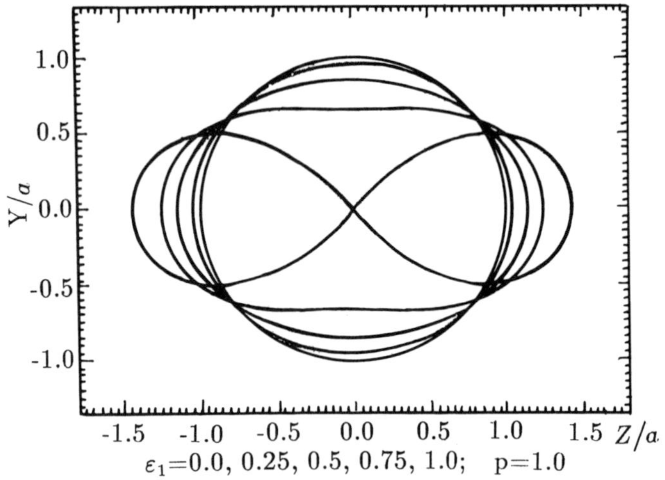

Fig.1 The symmetry Cassinian ovaloid with $\varepsilon = 0.0, 0.25, 0.5, 0.75$ and 1.0.

Some shape for constant Cassinian distance $F_1 F_2 = 2c$ and various values of ϵ are displayed in Fig. 1. If these ovals are rotated about the ζ-axis one obtains

Cassinian ovaloids. They feature for $\varepsilon > 1/\sqrt{2}$ a neck in the symmetry plane $\zeta = 0$ with radius

$$r_n = a\sqrt{1-\varepsilon^2} \tag{5}$$

Apparently, these ovaloids decrease in volume with increasing ε. If volume conservation is properly accounted for, then the Cassinian ovaloids can be visualized to the scission configuration, which is reached for $r_n = 0$, i. e. for $\varepsilon = 1$ (lemniscate).

Volume conservation for a fissioning nucleus of mass number A is incorporated by replacing the Cassinian parameter a in eq.(3) by $a(\varepsilon)$ defined through

$$a(\varepsilon) = a_0 \cdot f(\varepsilon) \tag{6a}$$

where

$$a_0 = r_0 \cdot A^{1/3} \tag{6b}$$

and the volume conservation factor,

$$f(\varepsilon) = [4/\{(1-2\varepsilon^2)\sqrt{1+\varepsilon^2} + 3/\varepsilon \ln(\varepsilon + \sqrt{1+\varepsilon^2})\}]^{1/3} \tag{7}$$

Thus with increasing ε the Cassinian parameter ε changes as well as the Cassinian distance 2c, giving rise to a stretching in the fisssion coordinate (cf. Fig. 1). Hence, ε is an appropriate deformation parameter. For $\varepsilon < 1/\sqrt{2}$ one can compare ε with the eccentricity parameter δ defined via the nuclear quadrupole moment[8] and finds $\delta \approx \varepsilon^2$. The eccentricity parameter δ is related to the eccentricity coordinate en of the non-isotropic harmonic-oscissllator potential of the Nilsson model by $\delta \approx \varepsilon e_N(1+\varepsilon_N/2)$ [9]. We note in passing that the moments of inertia $J_{\|}$ and $J_{\perp}$ can be calculated analytically resulting in[10]

$$J_{\|}/J_0 = \frac{15}{8} f^5(\varepsilon)\{\sqrt{1+\varepsilon^2}[\frac{23}{40} - \frac{\varepsilon^2}{10} + \frac{\varepsilon^4}{5} - \frac{1}{16\varepsilon^2}]$$

$$+ [\frac{1}{16\varepsilon^3} - \varepsilon] \cdot \ln(\varepsilon + \sqrt{1+\varepsilon^2})\} \tag{8a}$$

$$J_{\perp}/J_0 = \frac{15}{8} f^5(\varepsilon)\{\sqrt{1+\varepsilon^2}[\frac{41}{80} - \frac{\varepsilon^2}{60} + \frac{\varepsilon^4}{30} + \frac{1}{32\varepsilon^2}]$$

$$- [\frac{1}{32\varepsilon^3} + \frac{\varepsilon}{2}] \cdot \ln(\varepsilon + \sqrt{1+\varepsilon^2})\} \tag{8b}$$

Using a curve described by biquadratic equation to form a rotating body, the variation calculation demonstrates that the fission barrier of the ovaliod is quite close to that of the best shape, for which the fission barrier is minimum[11]. Asymmetric shapes can be included in a simple manner as indicated in Fig. 2 Halves of two Cassinian ovaloids containing A_1 and A_2 nucleons are joined together in the symmetry plane, $\zeta = 0$. A continuously differentiable connection is guaranteed by equating the neck radii (5)

$$a_1(\varepsilon_1)\sqrt{1-\varepsilon_1^2} = a_2(\varepsilon_2)\sqrt{1-\varepsilon_2^2} \tag{9}$$

while

$$\frac{d\eta}{d\zeta}|_{\zeta=0} = 0$$

is inherently fulfilled for Cassinian ovaloids with $\varepsilon \neq 1$ due to their reflection-symmetry properties. Apparently, these conditions are met by a whole set of Cassinian ovaloids with deferent volume. However, we must choose that pair of mass numbers $A = A_1 + A_2$. We introduce the mass asymmtry parameter p by

$$p = A_1/A_2 \qquad (0 \leq p \leq 1) \tag{10}$$

with $a_i(\varepsilon_i)$ defined in analogy to (6)

$$\begin{aligned} a_i(\varepsilon_i) &= a_{0i} f(\varepsilon_i) \\ &= r_0(2A_i)^{1/3} f(\varepsilon_i) \end{aligned} \tag{11}$$

We obtain from (9) to (11)

$$f(\varepsilon_2) = f(\varepsilon_1)\sqrt{1-\varepsilon_1^2}/\sqrt{1-\varepsilon_2} \cdot p^{1/3} \tag{12}$$

ε_2 is uniquely determined from (9) once $\varepsilon_1$and p are specified and $f(\varepsilon_2)$ can be calculated. The center of mass of each half body is located at a distence z_i from the neck plane at $z = 0$ (cf. Fig. 2) and

$$z_i = \frac{3}{8} a_{0i} f^4(\varepsilon_i)(1 - \frac{1}{3}\varepsilon_i^4) \tag{13}$$

The distance of the total mass center from the neck plane is

$$d = (z_2 - pz_1)/(1+p) \tag{14}$$

It is important to realize that d changes with ε. This means that the neck plane moves during the fission process. This is indicated in Fig. 2 where various

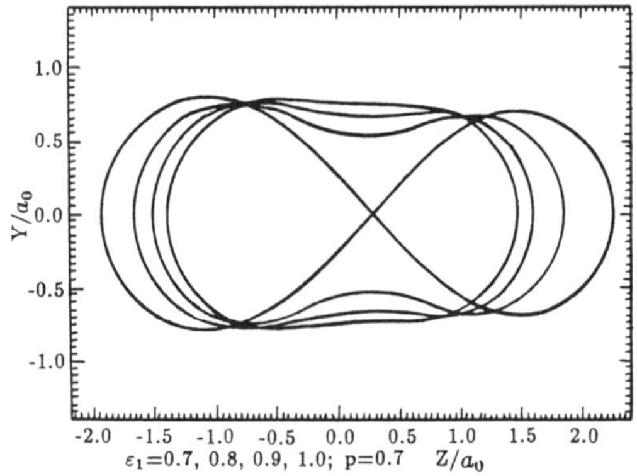

Fig.2 The asymmetry Cassinian ovaloid with p=0.7 and ε=0.7,0.8,0.9 and 1.0

Cassinian ovaloids are depicted for $p = 0.7$ and $\varepsilon_1 = 0.7, 0.8, 0.9$ and 1.0. In order to get a mass independent representation we choose y/a_0 and z/a_0 as coordinates for this plot.

II.2 Shell Correction

Shell corrections can be calculated by the Strutinsky metrod once the single particle energies have been obtained by solving the appropriate Schrödinger equation. It is beyond scope of the present work to solve it for a single particle which moves in a potential determined by mean field of a nucleus displaying a Cassinian ovaloid shape. We rather include shell effects in an average manner. The idea is following. According to the two-center shell model of overlapping spheres performed by Scharnweber[5], in the fission process, the shell property of single nucleus (after $\varepsilon > 0.7$) is breaking by establishing two fragments shell structrue: first in inner nucleons then developing up to outer nucleons, and in final to whole fragment at scission point. Therefore, it seems that when the neck appears, there are two fragment's cores, in which the shell structure has appreciable values. Furthermore we try to revise the shell energies only in the core of fragment; the core is also a ovaloid with $\varepsilon = 1$, namely a lemniscatoid as shown in Fig. 3. However, for ellipse-like shapes ($\varepsilon < 0.7$) we apply the shell correction formula of Myers and

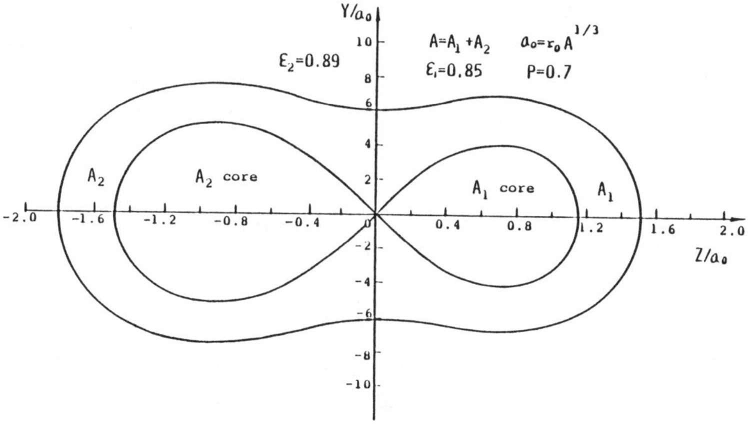

Fig.3 Two halves ovaloid smoothly join together, the ε_2 is function of ε_1 for given p values; here, $p = 0.7, \varepsilon_1 = 0.85, \varepsilon_2 = 0.893$. Inside the halves, two cores are shown too.

Swiatecki[12]. They were derived by bunching of an originally smooth distribution of single-particle levels into one that consists of groups of levels corresponding to the observed magic numbers M_i:

$$S(N,Z) = C\{[F(N) - F(Z)]/(A/2)^{2/3} - cA^{1/3}\} \tag{15}$$

$$F(K) = q_i(K - M_{i-1}) - 3/5(K^{5/3} - M_{i-1}^{5/3}) \tag{16}$$

for $M_{i-1} < K < M_i$, where K assumes either the neutron number N or proton number Z. Both C and c are adjustable parameters with values of 5.8 MeV and

0.26 MeV, respectively[12]. The quantities q_i are defined by

$$q_i = 3/5(M_i^{5/3} - M_{i-1}^{5/3})/(M_i - M_{i-1}) \tag{17}$$

where M_i are the magic number of spherical nuclei and M_0 is defined to be zero. In order to account for the effect of deformation on the shell correction energy Myers and Swiatecki introduce an attenuating function and replaced $S(N, Z)$ by $S(N, Z)\exp(-\theta^2)$. The argument θ is a measure of deformation and will be discussed below.

With the beginning of necking-in ($\varepsilon_1 > 1/\sqrt{2}$) this description is no longer valid because the nucleons do not longer experience a common mean field but begin to feel the two-center potential of the fissioning system. As mentioned above. We assume that, at $\varepsilon_1 = 1/\sqrt{2}$, this two-center potential exists already and can be represented by the cores of the nascent fragments. Only for the cores shell effects need to be considered. The cores are transients; with increasing deformation ε_1 they grow by accretion of nucleons and, eventually, at scission they are identical with the two parts of the lemniscate ovaloid, Lemniscatoid. Then all nucleons within each part contribute to the shell effect of the core changes with deformation ε_1 we observe continuously changing shell correction energies in the ε_1 vs. p plane on top of the liquid PES.

In spirit of our shape parametrization we define the two cores as the two parts of a lemniscatoid which is uniquely obtained for any Cassinian ovaloid with $R_1 R_2 = a^2 = c^2/\varepsilon^2$ by changing a to the value of c, which is equivalent of setting $\varepsilon = 1$ for constant c, thus obtaining

$$R_1' R_2' = c^2 < R_1 R_2$$

(cf. Fig. 1). We show in Fig. 3 the core of a Cassinian ovaloid with $\varepsilon_1 = 0.85$ and $p = 0.7$. The number of nucleons within the core i is

$$A_i^{core} = A_i[\varepsilon_i f(\varepsilon_i)/f(1)]^3 \tag{18}$$

The shape of each core can be approximated quite well by an ellipsoid located at the core's center of mass if the small tipe is neglected. For this ellipsoid an eccentricity parameter (see above) of $\delta = 0.22$ is obtained independent of the volume of the core. It corresponds to an eccentricity coordinate $\varepsilon = 0.47$. Hence, all shell corrections need to be applied for deformed cores, however, thier deformation does not change with increasing ε_1. For these strongly deformed core potentials new magic numbers appear. They are for neutrons: 2, 6, 14, 28[13], and 58, 80, 120,[14]; for larger neutron numbers no significant shell closures are observed at this deformation[14]. Similarly one finds for protons: 2, 6, 14, 28[13] and 58, 88, 120[14].

It is fully in line with the spirit of the derivation of (15)-(17) to use these magic numbers instead of those of spherical nuclei when calculating shell corrections for cores appearing for $\varepsilon > 1/\sqrt{2}$. Since in our approach only the number of nucleons A_i^{core} in core i are determined from (18) we use the hypothesis of an unchanged charge distribution in fission to infer neutronand proton numbers of the cores separately.

The sum of the shell correction energies for the two cores with (N_1, Z_1) and (N_2, Z_2) at the moment of necking-in ($\varepsilon_2 = 1/\sqrt{2}$) must be equal to the value of the attenuated spherical shell correction function $S(N, Z)\exp(-\theta^2)$. From this requirment the damping parameter θ is determined for any mass asymmetry p.

III. Results and Discussion

III.1 Liqid-Drop Moment of Inertia

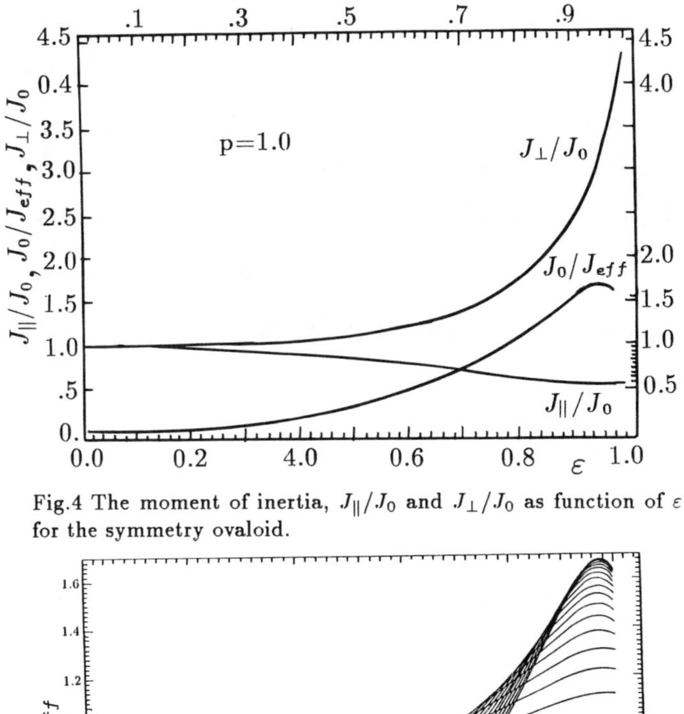

Fig.4 The moment of inertia, $J_\parallel/J_0$ and $J_\perp/J_0$ as function of ε for the symmetry ovaloid.

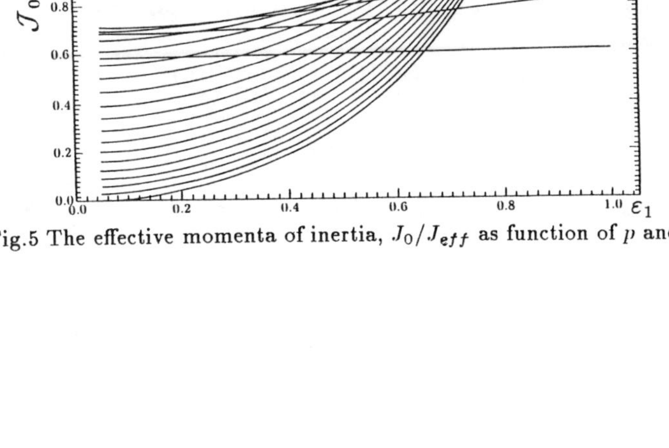

Fig.5 The effective momenta of inertia, J_0/J_{eff} as function of p and ε_1.

The changing distance of the total mass center from the neck plane has interesting effects on the moment of inertia. This is shown in Fig. 4 for a symmetric fission case, i.e. $p = 1$. While $J_{\|}/J_0$ slightly decreases for increasing ε_1 up to $\varepsilon_1 = 0.96$ and then increases again weakly, the $J_{\perp}/J_0$ increases monotonously with ε_1. In Fig. 4 is also plotted the effective moment of inertia.

$$J_0/J_{eff} = J_0/J_{\|} - J_0/J_{\perp} \tag{19}$$

J_0 represents the moment of inertia of a rigid sphere with radius a_o as defined in (5). This figure, hence, is valid for any mass number.

In Fig.5 we depict J_0/J_{eff} as function of ε_1 for various asymmetry parameters p. It is found, that J_0/J_{eff} never exceeds a fission fragment angular distributions[15].

III.2 Liquid-Drop Potential Energy Surface

The liquid drop part of potential energy is derived in the usual way from

$$E_{LD} = [(E_s(\varepsilon_1,p)/E_{s0} - 1) + 2x(E_c(\varepsilon_1,p)/E_{c0} - 1)]E_{s0} \tag{20}$$

where $E_s(\varepsilon_1,p)$ and $E_c(\varepsilon_1,p)$ are surface and Coulomb energies, respectively, of a Cassinian ovaloid specified by ε_1 and p. E_{s0} and E_{c0} are the surface and Coulomb energies of a spherical nucleus and x is the fissility parameter

$$x = 0.5E_{c0}/E_{s0} \tag{21}$$

We follow [12] and use with $I = (N - Z)/A$

$$E_{s0} = 17.944(1 - 1.7826I^2)A^{2/3}MeV \tag{22}$$

$$E_{c0} = 0.7053Z^2/A^{1/3}MeV \tag{23}$$

The calculation of the surface energy is simplified by evaluting that of two symmetric Cassinian ovaloids with mass number $2A_1$ and $2A_2$, their respective deformation ε_1 and ε_2, are connected uniquely by (9) and

$$E_s(\varepsilon_1,p) = 0.5[E_s(\varepsilon_1,2A_1) + E_s(\varepsilon_2,2A_2)] \tag{24}$$

The surface energy E_s is most simply calculated in polar coordinates (4b)

$$E_s = \gamma s \int r^2[1 + \frac{1}{r^2}(\frac{dr}{d\theta})^2]^{1/2} d\Omega \tag{25}$$

which yeilds for each Cassinian ovaloid

$$E_s(\varepsilon)/E_{s0} = f^2(\varepsilon)[1 + \varepsilon^2 \int_0^{\pi/2} d\theta[\cos 2\theta \sin\theta/\sqrt{1 - 4\varepsilon^4 \cos^2\theta \sin^2\theta}] \tag{26}$$

The result of this calculation is shown in Fig. 6 as function of ε_1 for various asymmtry parameters p. For symmetric fission (p=1) the value of E_S/E_{S0} reaches the maximum value of 1.286. This is slightly larger (2.6%) than the surface energy

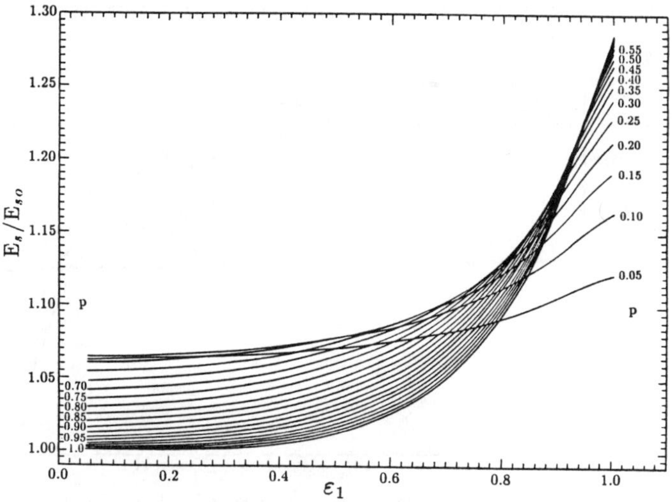

Fig.6 The surface energies, E_s in unit of spherical surface energy E_{s0} as function of ε_1 for various p values.

of two spheres of equal size with mass of $A/2$, since lemniscatoid are less compact. Then in post-fission, the excess of the surface energies become the excitation enegy of fragments. Instead of calculating the Coulomb energy of whole ovaloid

$$E_c(\varepsilon_1, p) = 1/2\rho_e^2 \int\int d\tau_1 a\tau_2/|\vec{r}_1 - \vec{r}_2| \tag{27}$$

where ρ_e is the density of a homogenous charge density distribution, we break up (27) in three terms:

$$E_c(\varepsilon_1, p) = E_c(\varepsilon_1, A_1) + E_c(\varepsilon_2, A_2) + E_{c12} \tag{28}$$

$E_c(\varepsilon_i, A_i)$ are the Coulomb energy of two parts of the ovaloid cut in the neck plane and E_{c12} is the Coulomb repulsion energy of these two parts. this procedure allows to determine, by comparison with experimental values of fission fragment total kinetic energies, the deformation ε_1 at which neck rapture took place. (28) is calculated numerically; the values of $E_c(\varepsilon_1)/E_{c0}$ as function of ε_1 for various asymmetry parameters p are displayed in Fig. 7. For symmtric fission ($p = 1$) the value of $E_c(\varepsilon_1)/E_{c0}$ reaches the minimum value of 0.801. Fig. 8 shows the Coulomb replsion energy E_{c12} of the two parts of the ovaloid as function of ε_1 for various asymmtrry parameters p. For symmtric fission ($p = 1$) this values drops from $=$ 0.4 at $\varepsilon_1 = 0$ (sphere) to 0.178 at $\varepsilon_1 = 1$ (lemniscatoid).

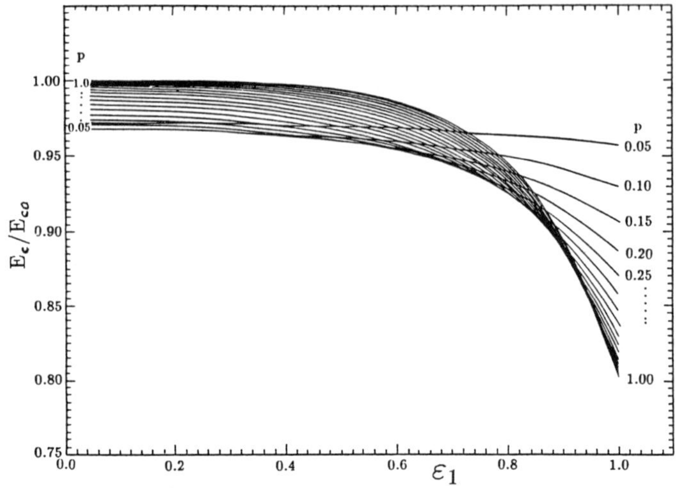

Fig.7 The Coulomb energies, E_c in unit of spherical energy E_{c0}, as function of ε_1 for various p values.

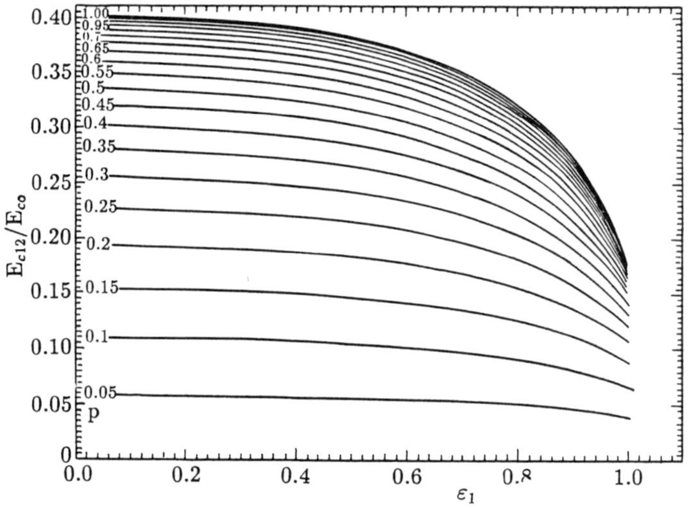

Fig.8 The Coulomb repulsion energy between two parts of the ovaliod cut in the neck plane.

Table 1: Parameters of fission barriers for a Cassinian ovaloid parametrization of the liquid-drop potential energy surface.

		^{208}Pb	^{233}Th	^{252}Cf
Ground	ε_1	0.0	0.0	0.0
State	p	1.0	1.0	1.0
Saddle	ε_1	0.95	0.87	0.75
	p	1.0	1.0	1.0
barrier heigt(MeV)		15.3	7.0	2.3

Finally we show in Fig. 9–11 the liquid drop PES for ^{208}Pb, ^{232}Th and ^{252}Cf, respectively, by contour lines in the ε_1 vs. p plane. Of caurse, the liquid drop part in this or any other parametrization allows only for spherical ground states and for symetric fission, both features are, indeed, observed. The characteristic PES parameters are listed in table 1. The characteristic feature of an increasing compactness of the nuclei at the fission barrier with increasing mass number, which is common to reasonable shape parametrizations, is observed for Cassinian ovaloids, too.

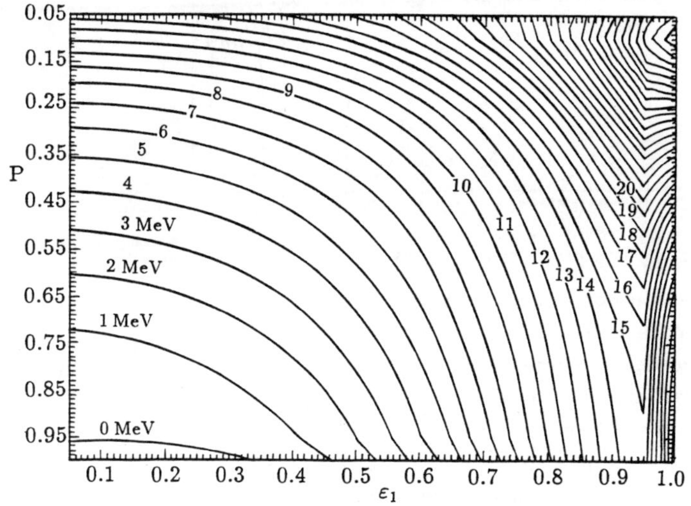

Fig.9 The PES of LDM for ^{208}Pb.

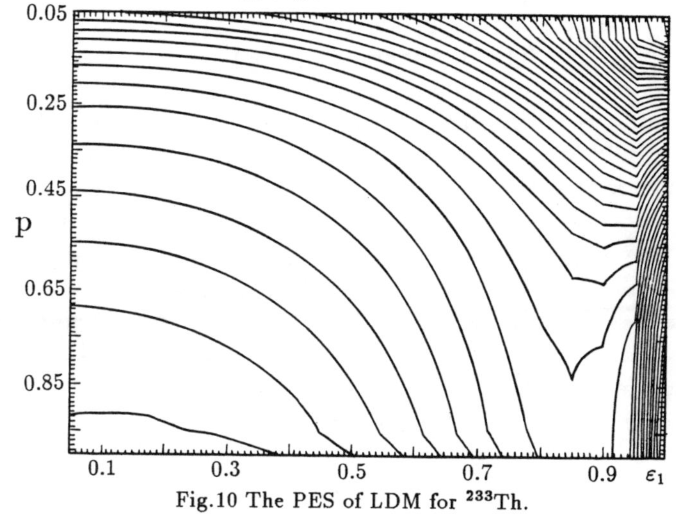

Fig.10 The PES of LDM for ^{233}Th.

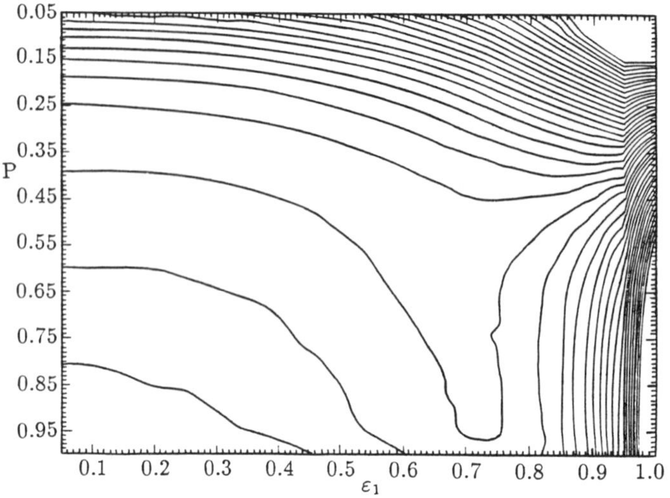

Fig.11 The PES of LDM for ^{252}Cf.

III.3 Potential Energy Surfaces with Shell Corrections

The shell correction energies are known to change the LD potential energy surface quite dramatically when going from lead (single humped barrier), through ^{232}Th (triple humped barrier) to ^{252}Cf (double humped barrier) We der these three nuclei as crucial test cases for our parametrization.

^{208}Pb: Fig. 12 depicts in the ε_1 vs. p-plane the PES for ^{208}Pb. The minimum of -11.5 MeV is located at $\varepsilon = 0$ (sphere) and $p = 1$ (mass symmtry). In is static picture the fission path would proceed on an almost monotonously increasing po-

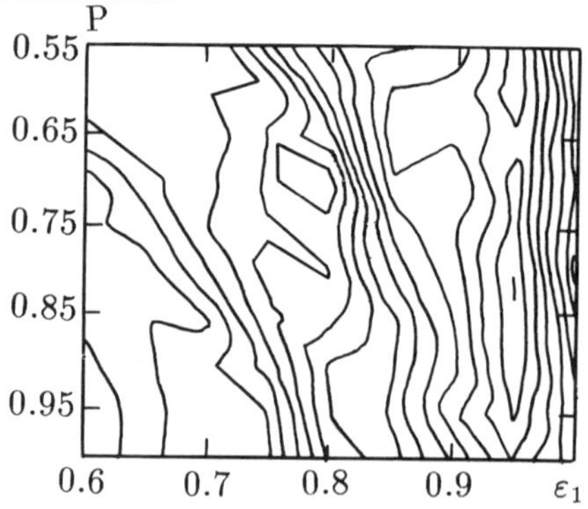

Fig.12 The PES with the shell revision in the core for ^{208}Pb.

tential along $p = 1$ and would pass the barrier of 16.7MeV at $\varepsilon = 1$ (limniscatoid). There is some fine structure to be seen in the shell corrected PES around $\varepsilon_1=$ 0.95–1.0. It pushes the barrier to somewhat larger deformations as in the liquid drop case. The way to asymmtric fission seems to be blocked by local maxima and ridges for that we observe a single humped fission barrier and preferential symmetric fission in case of ^{208}Pb.

^{233}Th: In contrast to the ^{208}Pb-case (Fig. 12) we show in Fig. 13 the shell corrected PES for ^{233}Th for a different range of parameters ε_1 and p in order to display more clearly the details of the potential landscape. There is a well defined local minimum of -3.1 MeV at $\varepsilon_1 = 0.75/p = 1$ which represents the reflection-symmetric ground state of ^{233}Th. A second minimum of -3.9 MeV appears at

$\varepsilon_1 = 0.85/p = 0.95$ and for a larger mass asymmetry a third minimum can of -10.2 MeV be seen at $\varepsilon_1 = 0.8/p = 0.7$. Our simple shape parametrization together with an average shell correction already predicts the existance of a third minimum. The third-minimum hypothesis was advanced by Moller and Nix[16]; further details of the $^{230-233}$Th PES were worked out in refined calculations[17]. In the experimental side there is a quite convincing evidence on the existance of a third minimum in Th-isotopes. A close inspection of the ^{233}Th-PES reveals two furture minima at $\varepsilon_1 = 0.55/p = 0.75$ and $\varepsilon_1 = 0.65/p = 0.6$ of -3.9 MeV and -3.4 MeV, respectively. They might hint at the possibility of excitated states which are due to rotations of such mass-symmetric shapes. Rotational bands of this kind were recently found and extensively inverstigated in many nuclei in this part of the nuclide chart.

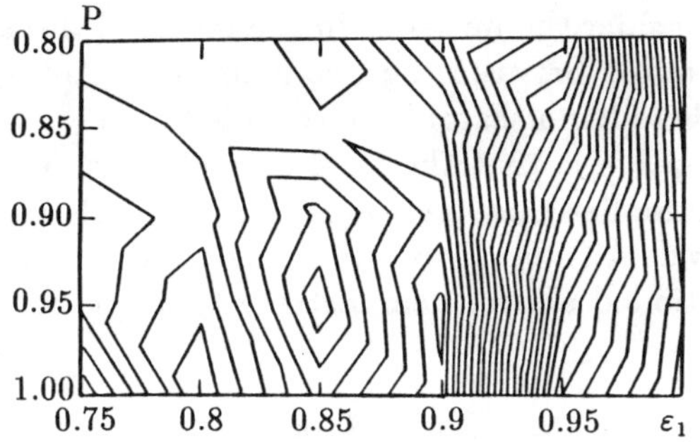

Fig.13 The PES with The shell revision in the core for ^{233}Th.

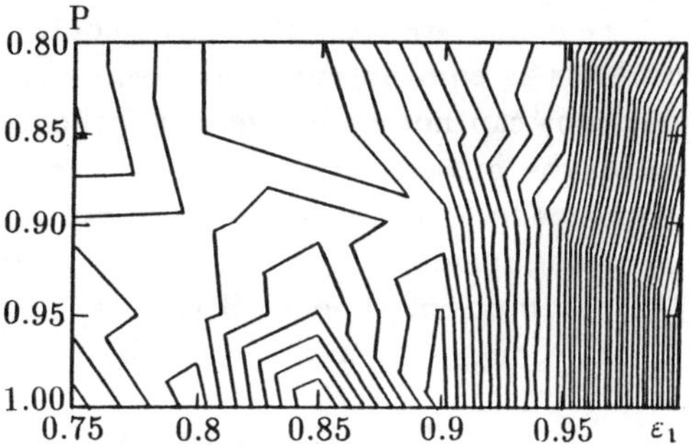

Fig.14 The PES with the shell revision in the core for ^{252}Cf.

^{252}Cf: As in the case of ^{233}Th (Fig. 13) we show in Fig. 14 for ^{252}Cf only that part of the shell corrected PES which is important for our purpose. The ^{252}Cf ground state is observed at $\varepsilon_1 = 0.85\ /p = 1$ with an energy of -6.7 MeV. It is a more compact shape than that of ^{233}Th. The way towards symmetric fission is blocked by a 2 MeV high ridge at $\varepsilon_1 = 0.9$. It can be bypassed by the fissioning system through the saddle point at $\varepsilon_1 = 0.89/p = 0.9$, i.e. in this static picture ^{252}Cf fissions asymmetrically. In our shape parametrization the second minimum found by other calculation is absent. As in the case of ^{233}Th two additional minima are found, namely at $\varepsilon_1 = 0.58/p = 0.78$ and $\varepsilon_1 = 0.75/p = 0.7$ with a depth of -3.4 MeV and -2.3 MeV, respectively. These octupole-deformed nuclear shapes might be bandheads for rotational levels. To our knowlege, they are not known in ^{252}Cf.

IV. Discussions

Comparing with two-center potential results, we introduce the shell correction depended on the deformation by the core. The correction is performed by a well-know formula due to Myers and Swiatecki. Because we face a large asymmetric deformation with neck, we have to get the shell correction for each half (or pre-fragment) instead of a single deformation nucleus. Therefore, anologically introducing the core is based on following reasons: (1) the inner core is a lemniscatoid, which is a scission point shape of fission. (2) from Table 2. the core is accreation of nucleons, vs. ε_1; that means the shell structure first is estabilished in inner nucleons then developed to outer nucleons; that is in agreement with Scharnweber's calculation. (3) There is a correspondence between shape and equi-potential, as shown in Bolsteri's work[18]. If we draw a series ovaloids with $\varepsilon > \varepsilon_0$ inside ovaloid with ε_0, we could consider the nucleons in the layers of these ovaloids as equi-potential nucleons. Therefore between the two lemniscatoids the nucleons can not exchange each other, these lemniscatoid are so called core.

In the present paper, a strong assumption is that the shell correction is only performed in the cores. The assumption implies that the single-particle levels are bunching inside the core and with equivalent interval distributions outside from the assumption. The PES results, mentioned above, demonstrate that the core assumption is effective and available.

Table 2: Percentage of nucleons which belong to core 1 if ε_1 is specified, calculated from [18]

ε_1	0.0	0.1	0.4	0.5	0.7	0.8	0.9	1.0
$A_1(core)/A_1(\varepsilon_1)\%$	0.0	0.03	2.15	4.45	14.95	26.36	48.15	100

when $\varepsilon_1 = 1/\sqrt{2}$ then the ratio of 15.56%

References

[1] J.R. Nix, Ann. Rev. Nucl. Sci. 22(1972)65
[2] P.Müller, J.R.Nix and W.Swiateck, LA-UR-86-2266
[3] S. G. Nilsson, Mat. Fys. Medd. dan. Vid. Selsk 29, No. 16 (1955)
[4] V. M. Strutinsky, Sov. J. Nucl. Phys., 3(1966)449 (Yad. Fiz. 3(1966)614)
[5] D.Scharnweber, W.Greiner and U.Mosel, Nucl. Phys., A169(1971) 257
[6] V. S. Stavinsky, Sov. J. Nucl. Phys., 7(1968)631 (Yad. Fiz. 7(1968)1051)
[7] V. V. Pashkevich, Nucl. Phys., A69(1971)275
[8] A. Bohr and B. R. Mottelson, Nuclear Structure, Vol.II, W. A. Benjamin Inc., New York, 1975
[9] O.Nathan and S. G. Nilsson in α, β, and γ-ray Spectroscopy, North Holland Publishing Company, Amsterdam 1968 Vol. I, P601, K. Siegbahn ed.
[10] G. Dai,X. Liu, and G. Liu, J. High Ener. Nucl. Phys., 11(1987)515 and Chin. Phys., 8(1988)427
[11] G. Dai, and W. Li, High Ener. Nucl. Phys., 12(1988)399
[12] W. D. Myers and W. J. Swiatecki, Nucl. Phys., 81(1966)1
[13] B. R. Mottelson and S. G. Nilsson, Mat. Fys. Ski. Dan. Vid. Selsk 1, No. 8(1959)
[14] C. Gustafson, I. L. Lamm, B. Nilsson, and S. G. Nilsson, Arkiv Fysik, 36(1967)613
[15] R. Vandenbosch, J.R. Huizengen, Nuclear Fission, Academic Press, New York and London,

1973, P312
[16] P. Moller, J. R. Nix, in: Proc. Int. Conf. on Phys. and Chem. of Fission, Rochester 1973 (IAEA Viena 1974) Vol. I, P103
[17] R. Bengtsson, Nucl. Phys., A45(1975)39
[18] M. Bosterli, E. O. Fiset, J. R. Nix, J. L. Norton, Phys. Rev., C5(1972)1050

FROM BINARY DECAY TO MULTIFRAGMENTATION

FRAGMENT DECAY IN ^{12}C + ^{197}AU FROM 30 TO 1100 MEV/NUCLEON

W. Trautmann
GSI Darmstadt, D-6100 Darmstadt, Germany

ABSTRACT

The reaction ^{12}C + ^{197}Au nuclei was studied in several experiments at bombarding energies between E/A = 30 MeV and 1100 MeV. The cross sections for the production of intermediate mass fragments increase by nearly two orders of magnitude within this range. The production mechanism evolves from essentially binary decays at the lower bombarding energies to predominantly multi-fragment processes in central collisions in the relativistic regime of several hundred MeV per nucleon.

INTRODUCTION

The fragmentation of ^{197}Au nuclei in reactions with ^{12}C projectiles was studied in several experiments at bombarding energies ranging from E/A = 30 MeV to 1100 MeV. Earlier experiments[1] performed at the CERN synchrocyclotron at $E/A \leq 84$ MeV were extended to higher energies in a study of inclusive fragment production at SATURNE at bombarding energies E/A = 100, 300, 600, and 1100 MeV.[2] Very recently, in a first experiment with the ALADIN spectrometer, the reaction ^{12}C + ^{197}Au was studied at E/A = 600 MeV with a gold beam from SIS at the GSI.[3,4] There the employed inverse kinematics allowed a nearly complete detection of projectile fragments with atomic number $Z \geq 2$ and thus a reconstruction of the decay patterns of the gold projectiles.

The wide range of bombarding energies gives the opportunity to follow the evolution of the fragmentation process from where it sets in as a major mode of deexcitation in heavy ion reactions up to energies close to where it was observed to saturate.[5] The dramatic increase of the cross section over this energy range will become apparent from the inclusive data. The coincidence studies[3,4,6] provide information on the multiplicities and thus on the strength of the multi-fragment channels which are of particular interest in the present search for the novel mechanism of multifragmentation.[7,8] Furthermore, a variety of correlation techniques has been applied to the coincidence data at the lower energies and has proven rather useful in establishing the sequential binary character of the multi-fragment decays near their threshold.[9,10] It will be important to extend these techniques also to the higher energies.

In this contribution we will first look at the inclusive fragment production as a function of bombarding energy. A brief summary of the results from the earlier coincidence work at $E/A \leq 84$ MeV will follow. Finally, first results from the recent experiment with the ALADIN spectrometer at E/A = 600 MeV will be reported.

INCLUSIVE CROSS SECTIONS

Inclusive cross sections for the production of intermediate mass fragments (IMF's) were measured at the CERN synchrocyclotron (E/A = 30, 48, and 84 MeV) and at the SATURNE facility in Saclay (E/A = 100, 300, 600, and 1100 MeV) with arrays of several detector telescopes. Each telescope consisted of an axial-field ionization chamber, three silicon surface-barrier detectors of 50-μm, 300-μm, and 1000-μm thickness, respectively, and a scintillation detector (BGO at CERN, CsI(Tl) at SATURNE) viewed by a photodiode. The threshold energy for elemental identification of the detected fragments was in the range of E/A = 1 MeV to 2 MeV.

Satisfactory descriptions of the measured cross sections $d^2\sigma/dEd\Omega$ were obtained with fits employing two Maxwellian sources, a slow so-called target source and a faster intermediate-velocity source. The sources move in the beam direction and emit isotropically in their rest frames. The moving-source description was used to generate energy and angle integrated cross sections. The results were found to depend weakly on possible ambiguities in the parameter sets as long as fits of comparable quality were produced by these parameters.

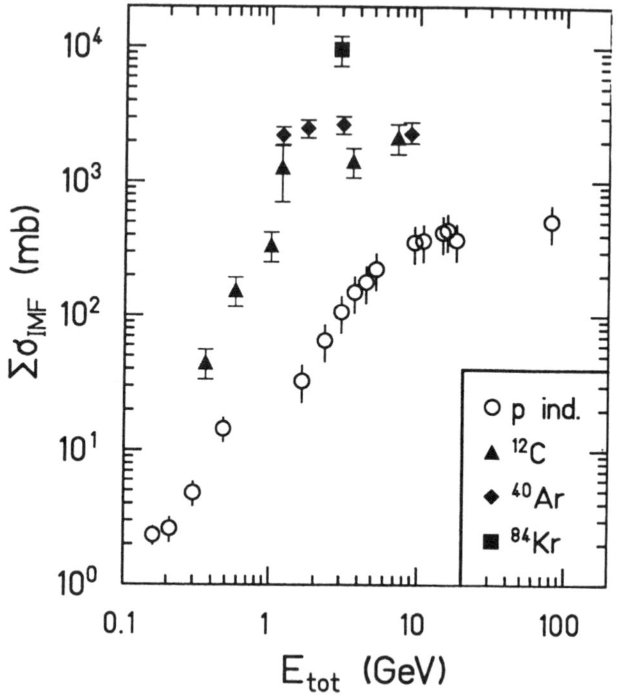

Fig. 1. Cross sections for the production of intermediate mass fragments as a function of the total projectile energy for heavy-ion induced reactions on ^{197}Au and for reactions of protons with silver and xenon targets. The data are from Refs. 2, 11-17.

Total cross sections for the IMF production were obtained by summing the integrated elemental cross sections over the range of atomic numbers $Z = 3$ up to $Z \simeq 15$. The results are shown in Fig. 1. As the beam energy is increased the IMF cross sections rise by nearly two orders of magnitude and reach a value of $\sigma_{IMF} = 2.2$ b at $E/A = 600$ MeV. According to radiochemical data for selected isotopes a saturation of the cross section may be expected at slightly higher energies. In the case of the proton on Ag, Xe reactions the saturation is observed at $E_p \simeq 10$ GeV. For the ^{40}Ar + ^{197}Au reaction, on the other hand, the saturation value is reached already at $E_{proj} = 1.2$ GeV, equivalent to $E/A = 30$ MeV.[16,17] There remains a strong dependence of the IMF cross sections on the mass of the projectile, even if they are viewed as a function of the total projectile energy.

Double-differential cross sections for the production of carbon fragments at $\Theta_{lab} = 90^0$ are shown in Fig. 2. They were generated with the moving source parameters derived from the spectra measured for this element. Besides the dramatic increase of the cross section (note that there are no scaling factors in the figure) we observe that the exponential tails of the spectra fall of much more slowly at the higher energies and that the maxima of the spectra shift to smaller energies. Neither of these two effects is fully understood at present. A shift of the maxima was first observed by Cumming et al.[18] and Poskanzer et al. [19] in proton induced reactions and since then reported for several proton and ion induced reactions at relativistic energies.[17,20–24] It seems to be a general feature of the fragment production at relativistic bombarding energies. It is of particular interest since a multifragment decay of an expanded spectator system, as conjectured by multifragmentation scenarios,[7,8] will result in a reduced Coulomb repulsion of the fragments and thus might cause the observed shift of the maxima.

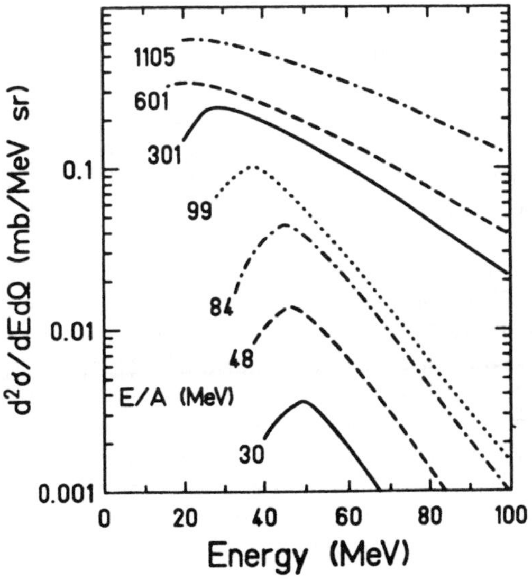

Fig. 2. Double-differential cross sections for the production of carbon fragments at $\Theta = 90^0$ in ^{12}C + ^{197}Au. Note that there are no scaling factors. For $E/A = 1105$ MeV an absolute normalization was not obtained (from Ref. 2).

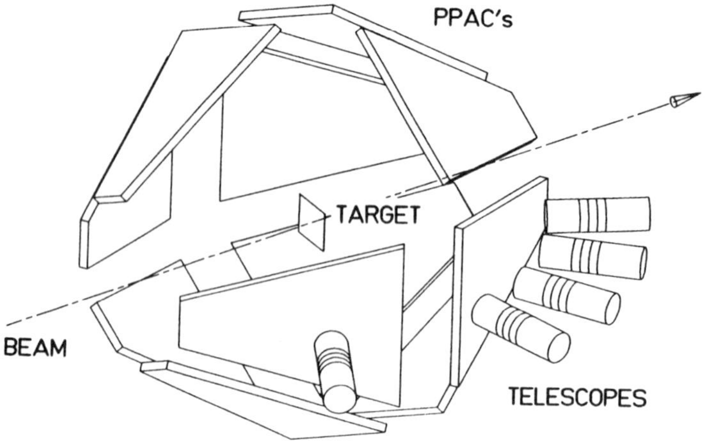

Fig. 3. Schematic view of the experimental setup consisting of five detector telescopes and an array of twelve PPAC's surrounding the target (one element is removed to allow the view inside). The array of plastic detectors backing the PPAC's and designed for light ion detection and the array of phoswich detectors covering the region of forward angles are not shown.

SEQUENTIAL FRAGMENT EMISSION AT E/A ≤ 84 MEV

A schematic view of the setup used for the coincidence measurements at the CERN synchrocyclotron with beams of $E/A \leq 84$ MeV is shown in Fig. 3. The telescopes served as the trigger detectors. Coincident IMF's were detected with an array of twelve parallel plate avalanche counters (PPAC's) which covered $\Delta\Omega = 2.7\pi$ sr between $\Theta_{lab} = 22^{\circ}$ and $\Theta_{lab} = 158^{\circ}$. IMF's were distinguished from the heavier fission fragments and heavy recoil nuclei by their different pulse height ΔE and time of flight. The efficiency of the PPAC's to detect impinging fragments was found to be essentially 100% for carbon and heavier ions. If all four position signals were required, as necessary to determine velocity correlations, the efficiency threshold increased to $Z \simeq 10$. The separation of IMF's from fission fragments via an upper limit in ΔE introduced a smooth cutoff near $Z = 20$.

The PPAC's were backed by an array of ten plastic scintillators of 1-cm thickness and $\Delta\Omega = 1.8\pi$ sr which were used to determine the associated multiplicity of light charged particles. A forward hodoscope[25] of plastic and phoswich detectors extending over the angular range $2.5^{\circ} \leq \Theta_{lab} \leq 20^{\circ}$ was used to detect light particles and projectile fragments in forward direction.

Some of the more detailed analysis was done with the data from the reaction $^{18}O + ^{197}Au$, an asymmetric system similar to $^{12}C + ^{197}Au$.[6,9,10] Some of the basic features of these reactions are illustrated in Fig. 4. Here the distribution of the multiplicities M'_{IMF} of IMF's emitted in coincidence with a $Z \geq 6$ trigger particle is given. Requiring the detection of an IMF with one of the telescopes was found to be an efficient means to select rather violent (central) collisions. In most cases no further IMF was detected with the PPAC array ($M'_{IMF} = 0$). A binary emission process, apart from the light particle emission, is the most likely case. Multifragment decays ($M'_{IMF} \geq 1$) are observed buth with probabilities which rapidly decrease with the IMF multiplicity.

The multifragment decays were studied with the technique of fragment-fragment correlations. The small-angle velocity correlations are strongly influenced by the Coulomb repulsion between the two fragments which depends on their distance and hence on the time interval between their emission. The analysis of the small angle correlation measured for $^{18}O + ^{197}Au$ at $E/A = 84$ MeV has shown that these time intervals of $\tau = 300$ to 500 fm/c are of the same order as the times characterizing the single IMF emission.[9] The latter may be estimated from the facts that the IMF angular distributions are forward peaked and that IMF's are, with some preference, emitted to negative deflection angles.[26] The small-angle correlations thus do not indicate any significant change of time scale for the multifragment decay as compared to the single IMF emission. Similar conclusions were reached by other groups.[27,28]

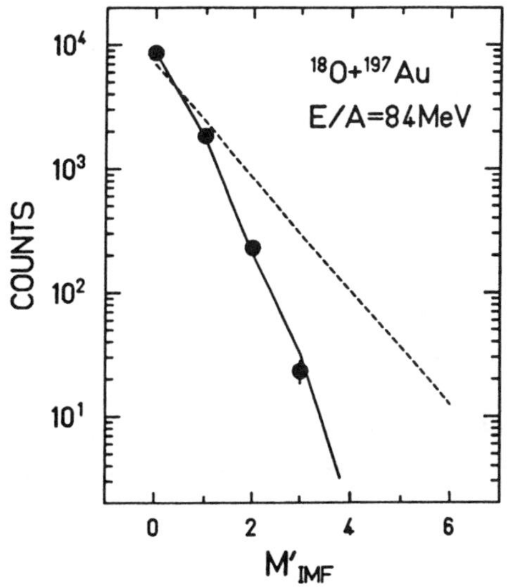

Fig. 4. Relative probability to detect M'_{IMF} fragments with the set of large area parallel plate detectors in coincidence with a fragment of $Z \geq 6$ for the reaction ^{18}O on ^{197}Au at $E/A = 84$ MeV (points). The fit curve (full line) was obtained by assuming an exponential form for the true multiplicity distribution (dashed line) and by applying a Monte Carlo model to account for the efficiency gap in solid angle and in the element range $Z \leq 5$ (from Ref. 6).

The large-angle correlations are governed by the individual Coulomb repulsions of the two fragments from the emitting system and by the recoil effect of the first emission on the second. For a conventional-type sequential emission of two IMF's into approximately opposite directions their relative velocity should be twice the value corresponding to the Coulomb energy governing the individual emissions from the heavy recoil nucleus minus the recoil correction. Smaller values will result if the breakup occurs out of an expanded state, as expected in the case of multifragmentation, where the Coulomb energies are reduced. An increase of the relative velocity is expected if a collective motion of expansion is only partially damped and therefore partially transferred to the fragments.

The measured velocity correlations for the reaction $^{18}O + {}^{197}Au$ are shown in Fig. 5. In the azimuthal coordinate $150^0 < \Delta\Phi < 210^0$ was required. The resulting distributions of relative velocities then still depend on the sum of the polar emission angles $\Theta_{sum} = \Theta_1 + \Theta_2$. For values of Θ_{sum} near 180^0, i.e. for approximately back-to-back emissions the distribution spreads around a value of 4.5 cm/ns. This is close to the expectation for a conventional sequential process if the value 2.3 cm/ns familiar from the Viola systematics[29] is assumed for the individual emissions. The result of a slightly more quantitative estimate[10] for sequential emission is represented by the full line shown in the figure. The data scatter considerably around this line but there is no significant deviation of the mean values. Thus, also the results of the large-angle velocity correlations are fully consistent with a sequential multifragment decay. It turned out, however, that the interpretation of these data is rather complex and that large-angle correlations alone may not be sufficient to distinguish between sequential and other decay modes.[10]

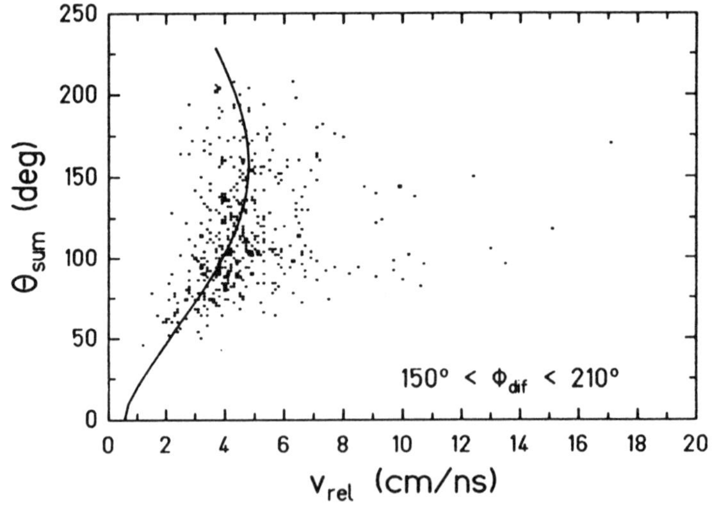

Fig. 5. Scatter plot of two-fragment coincidences from the reaction $^{18}O + {}^{197}Au$ at $E/A = 84$ MeV as a function of $\Theta_{sum} = \Theta_1 + \Theta_2$ (laboratory angles) and the magnitude of their relative velocity, v_{rel}. It was required that the difference of their azimuthal emission angles falls into the range $150^0 < \Phi_{dif} < 210^0$. The full line gives the prediction for a sequential decay (from Ref. 10).

Among the most important results emerging from the coincidence measurements is the governing role of the excitation energy E_x residing in the system at the time of the IMF emission.[6] For this analysis it was assumed that preequilibrium light particles are emitted prior to the IMF's which is supported by the experiment.[30] Consequently, E_x is essentially given by the difference between the total available collision energy E_{cm} and the sum of the separation energies and the kinetic energies carried away by the preequilibrium light particles. The latter quantity was determined with a Monte Carlo model based on the measured light-particle data.

In Fig. 6 the summed cross sections for IMF production ($Z \geq 3$), the power law exponents τ of fits to the inclusive element distributions according to $\sigma(Z) \propto Z^{-\tau}$, and the mean multiplicities $< M_{IMF} >$ of events selected by the observation of at least one IMF are displayed, as a function of the mean $< E_x >$, for several reactions on a gold target. We observe that all three quantities, which may be considered the main observables characterizing IMF emission, are monotonically correlated with $< E_x >$. The bottom panel shows that an excitation energy of about 300 MeV constitutes a threshold for the occurence of multifragment decays. At this energy $< M_{IMF} >$ starts to deviate from one, i.e. from the value characterizing the binary disintegrations. The data suggest, however, that much higher excitation energies will be needed if disintegrations into several fragments are to be observed with large probabilities.

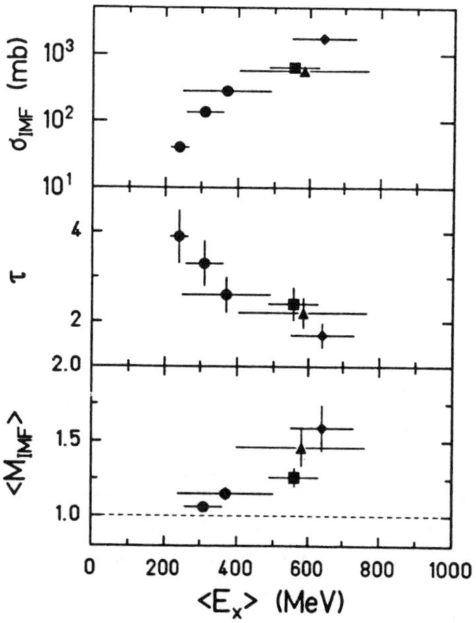

Fig. 6. The production cross section of IMF's (top), the power law exponent τ describing their inclusive element distribution (middle), and the mean multiplicity of IMF's in collisions selected by the observation of one fragment of $Z \geq 6$ (bottom) as a function of E_x for several reactions on ^{197}Au at 30 MeV $\leq E/A \leq$ 84 MeV. The symbols denote the projectiles ^{12}C (points), ^{20}Ne (squares), ^{18}O (triangles), and ^{40}Ar (lozenges). The multiplicity was not determined for ^{12}C of E/A = 30 MeV (from Ref. 6).

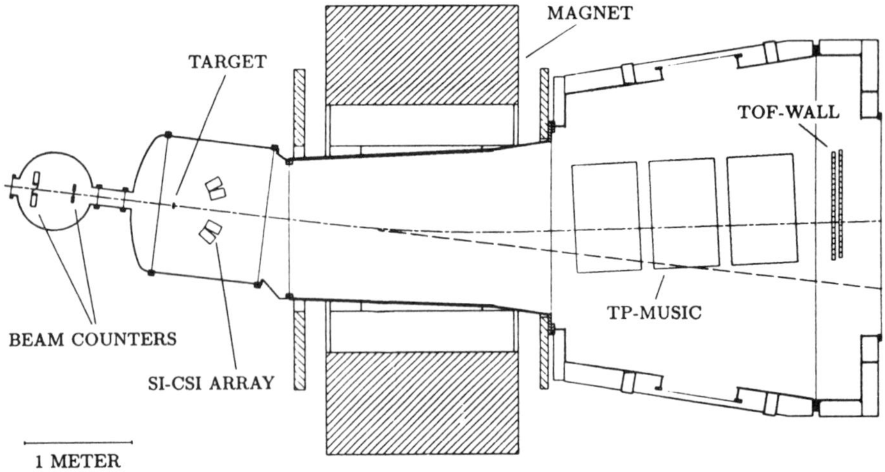

Fig. 7. Schematic view of the ALADIN facility. The dashed line indicates the direction of the incident beam. The dash-dotted line represents the trajectory of beam particles which were deflected by an angle of 8.2°.

MULTIFRAGMENT PRODUCTION AT E/A = 600 MEV

Very recently the reaction $^{12}C + {}^{197}Au$ was studied in inverse kinematics using a ^{197}Au beam with $E/A = 600$ MeV from SIS.[3,4] To allow a systematic study of asymmetric reactions at this bombarding energy also targets of aluminum and copper were used. A schematic layout of the experimental setup is shown in Fig. 7.

The acceptance of the ALADIN spectrometer for beam velocity fragments with $N = Z$ was ± 4.7° in the horizontal and ± 4.5° in the vertical direction. This range proved to be sufficient to detect virtually all of the fragments originating from the decay of the gold projectile (Fig. 8). The atomic numbers Z and the multiplicity of nuclear fragments were determined by means of the time-of-flight (TOF) wall which was positioned at a distance from the target of approximately 6 m. The wall consisted of two layers, with 40 vertical scintillators each, and covered an area of 1 m x 1 m. Individual elements were resolved for $Z \leq 8$, and a charge resolution of ± 2 for heavier fragments was obtained. Fragments with $Z > 10$ were identified and tracked by the multiple-sampling ionization chamber (TP-MUSIC), positioned between the magnet and the TOF-wall. For the data presented here the energy loss measured with the TP-MUSIC was used to calibrate the light-output signals from the TOF-wall. Light particles, predominantly originating from the mid-rapidity source, were detected by the 64 elements of a Si-CsI(Tl) hodoscope, placed near the target at angles between 7° and 40°. The solid-angle coverage of the array was approximately 50% in the range $7° \leq \Theta_{lab} \leq 25°$ and about 15% for $25° \leq \Theta_{lab} \leq 40°$.

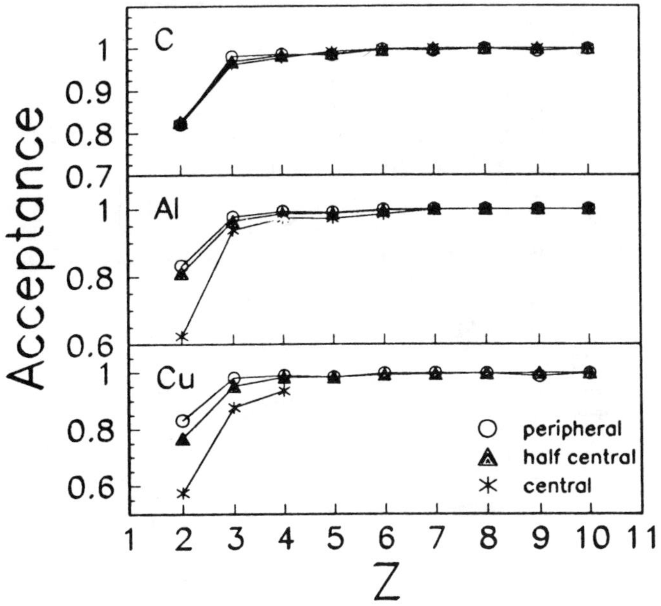

Fig. 8. Geometrical acceptance of the TOF wall as a function of the atomic number Z of the detected fragment and for the three ranges of impact parameter in Au + C (top), Au + Al (center), and Au + Cu (bottom) reactions at E/A = 600 MeV (from Ref. 4).

In the analysis of the fragment decays of the gold projectiles an IMF was defined by $3 \leq Z \leq 30$. Fig. 9 shows contour plots of the distribution of the IMF multiplicity M_{IMF} as a function of the light particle multiplicity M_{lp}. For the C target, M_{IMF} increases with the violence of the collision and reaches a mean multiplicity of three to four for the most central collisions. The reactions on Al behaves similarly in that three to four IMF's are emitted but this occurs over a wider range of M_{lp}. For the Cu target, the region of three to four IMF's is reached at an even lower relative multiplicity of light particles. Where M_{IMF} reaches its maximum, in all three cases, the distribution of M_{IMF} around its mean is nearly symmetric. Here the mean multiplicity is also the most probable multiplicity of intermediate mass fragments. This is a qualitative difference to the multiplicity distribution displayed in Fig. 4 which was found typical for reactions at the threshold to multifragment decay.[6,31] In the reactions at relativistic energies the decay into several nuclear fragments constitutes the dominant decay mode for a whole class of collisions which experimentally may be characterized by a bin in M_{lp}.

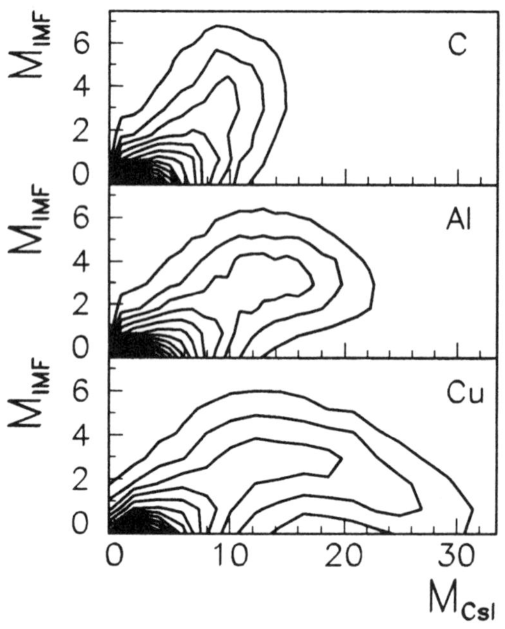

Fig. 9. Measured correlation of the IMF multiplicity with the multiplicity of light charged particles in the CsI array. The results for 600 MeV/nucleon Au on C, Al, and Cu are shown from top to bottom and the contours are drawn using a linear scale (from Ref. 3).

In order to translate the dependence on M_{lp} into an impact parameter dependence we adopted a geometric cross section method first used in Ref. 32. The basic assumption is that our setup and our trigger conditions allowed us to record all reaction events up to a maximum impact parameter given by the total cross section of the event sample. Bins in M_{lp} (starting with the largest values) can then be successively associated with impact parameter bins that represent the same cross section. The same procedure was followed with a second observable Z_{bound} , defined as the sum of the charges of all complex fragments with $Z \geq 2$, which was found equally useful as a measure of the impact parameter.[4]

The results are shown in Fig. 10 for the three target cases. For the carbon target (upper part), the IMF multiplicity increases with decreasing impact parameter and reaches a mean value of three to four for the most central collisions at $b \leq 3$ fm. The reactions on aluminum (center part) show a similar increase of the mean IMF multiplicity up to a maximum at $b \simeq 4$ fm. However, for the most violent collisions $< M_{IMF} >$ begins to decrease. For reactions on Cu (lower part) the maximum value is again about three and is found for even larger impact parameters of 6 to 7 fm.

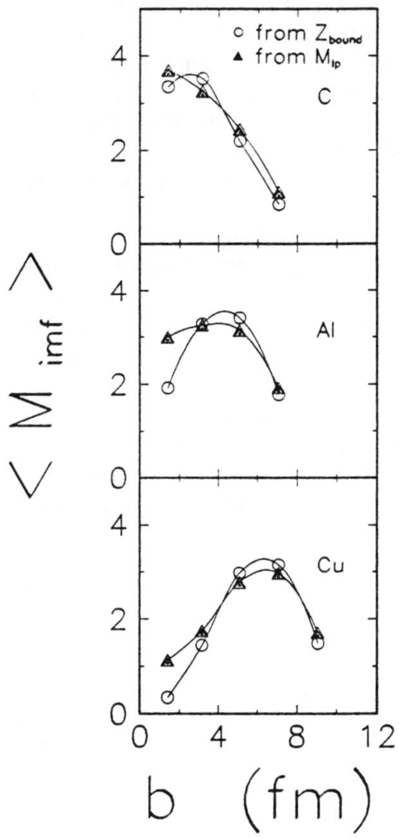

Fig. 10. Mean multiplicity of intermediate mass fragments, $< M_{IMF} >$, as a function of the reconstructed impact parameter (from Ref. 4).

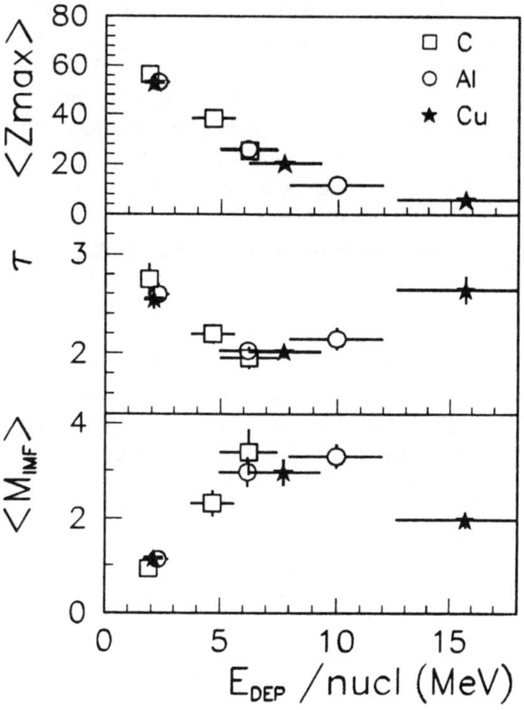

Fig. 11. The average maximum charge within an event, the fitted τ parameter from the charge distribution, and the mean IMF multiplicity as a function of the calculated deposited energy/nucleon. The squares, circles, and stars represent collisions on the C, Al, and Cu targets, respectively. The three points within a target group correspond to peripheral, mid-central, and central collisions with the deposited energy increasing with centrality (from Ref. 3).

The knowledge of the impact parameter allows comparisons with models that include the reaction dynamics. We have used the BUU model[33] in order to calculate the energy deposited into the Au projectile during the initial stage of the reaction.[3] This assumes that the BUU model may give an adequate representation of this first, violent, part of the reaction although it cannot describe the later fragmentation stages. And, furthermore, it is believed that relative dependencies on the deposited energy may be reliably studied, even though its absolute magnitude is clearly model dependent.

The variation of three experimental observables, the mean atomic number $<Z_{max}>$ of the largest fragment within an event (top panel), τ (middle), and $<M_{IMF}>$ (bottom) with the calculated deposited energy $E_{dep}/nucleon$ is shown in Fig. 11. The power law parameter τ has been derived from fits to the elemental distributions of the emitted fragments according to $\sigma(Z) \propto Z^{-\tau}$. Three impact parameter bins were chosen for each reaction. We observe that the data from the three different reactions follow universal curves as a function of E_{dep}. This shows that the deposited energy is a significant parameter also for the multifragment decay processes at relativistic energies. The maximum of $<M_{IMF}>$ is reached near $E_{dep}/nucleon$ = 8 MeV. The τ parameter passes through a minimum at about the same energy, indicating that the maximum multiplicity is associated with the broadest element distributions.

CONCLUSION

We have seen that the rapid increase of the IMF cross section with bombarding energy, the most prominent feature in the inclusive data for $^{12}C + {}^{197}Au$, is accompanied by a transition from binary to predominantly multifragment processes, at least for the more central collisions. It will be interesting to associate the increase of the inverse slope parameters and the decrease of the Coulomb repulsion with this transition in order to understand these inclusive features and to learn about the dynamics of the multifragment decays.

The nature of the multifragment processes at the lower energies, where they occur with small probabilities, were intensively studied with correlation techniques. The emerging conclusion was that they are essentially a sequence of binary-type emissions. The same techniques have not yet been applied to the multifragment decays observed at the relativistic energies. Here it is not clear at present what drives the excited projectiles into the multifragment decay channels and which of the many multifragmentation scenarios may be the most applicable. To find this out will be a challenge for the future.

The more recent experiments at relativistic energies were carried out by a collaboration of J.C. Adloff, M. Begemann-Blaich, P. Bouissou, J. Hubele, G. Imme, I. Iori, P. Kreutz, G.J. Kunde, S. Leray, V. Lindenstruth, Z. Liu, U. Lynen, R.J. Meijer, U. Milkau, A. Moroni, W.F.J. Müller, C. Ngô, C.A. Ogilvie, J. Pochodzalla, G. Raciti, G. Rudolf, H. Sann, A. Schüttauf, W. Seidel, L. Stuttge, W. Trautmann, and A. Tucholski.

REFERENCES

1. R. Trockel et al., Phys. Rev. C38, 576 (1988).
2. U. Milkau, thesis, Universität Frankfurt (1991).
3. C.A. Ogilvie et al., preprint GSI-91-18 (1991).
4. J. Hubele et al., preprint GSI-91-33 (1991).
5. K. Aleklett, L. Sihver, and W. Loveland, Phys. Lett. B197, 34 (1987).
6. R. Trockel et al., Phys. Rev. C39, 729 (1989).
7. D.H.E. Gross, Zhang Xiao-ze, and Xu Shu-yan, Phys. Rev. Lett. 56, 1544 (1986).
8. J. Bondorf, R. Donangelo, I.N. Mishustin, and H. Schulz, Nucl. Phys. A444, 460 (1985).
9. R. Trockel et al., Phys. Rev. Lett. 59, 2844 (1987).
10. J. Pochodzalla, W. Trautmann, and U. Lynen, Phys. Lett. B 232, 41 (1989).
11. S.J. Yennello, K. Kwiatkowski, S. Rose, L.W. Woo, S.H. Zhou, and V.E. Viola, Phys. Rev. C41, 79 (1990).
12. R.E.L. Green and R.G. Korteling, Phys. Rev. C29, 1806 (1984).
13. N.T. Porile et al., Phys. Rev. C39, 1914 (1989).
14. A.S. Hirsch, A. Bujak, J.E. Finn, L.J. Gutay, R.W. Minich, N.T. Porile, R.P. Scharenberg, B.C. Stringfellow, F. Turkot, Phys. Rev. C29, 508 (1984).
15. R. Trockel, PhD thesis, Universität Heidelberg, Report No. GSI-87-17 (1987).
16. A. Sokolov, These de Doctorat de l' Universitè Paris 6 (1990).
17. U. Milkau et al., preprint GSI-91-08 (1991).
18. J.B. Cumming, R.J. Cross, Jr., J. Hudis, and A.M. Poskanzer, Phys. Rev. 134, B167 (1964).
19. A.M. Poskanzer, G.W. Butler, and E.K. Hyde, Phys. Rev. C3, 882 (1971).
20. J.D. Sullivan, P.B. Price, H.J. Crawford, and M. Whitehead, Phys. Rev. Lett. 30, 136 (1973).
21. G.D. Westfall, R.G. Sextro, A.M. Poskanzer, A.M. Zebelmann, G.W. Butler, and E.K. Hyde, Phys. Rev. C17, 1368 (1978).
22. A.I. Warwick et al., Phys. Rev. C27, 1083 (1983).
23. V.V. Avdeichikov et al., Sov. J. Nucl. Phys. 48, 1043 (1988).
24. S.J. Yennello et al., Phys. Lett. B246, 26 (1990).
25. H.J. Rabe et al., Phys. Lett. B196, 439 (1987).
26. M.B. Tsang et al., Phys. Rev. Lett. 60, 1479 (1988).
27. R. Bougault, J. Colin, F. Delaunay, A. Genoux-Lubain, A. Hajfani, C. LeBrun, J.F. Lecolley, M. Louvel, and J.C. Steckmeyer, Phys. Lett. B 232 (1989) 291.
28. Y.D. Kim et al., Phys. Rev. Lett. 67, 14 (1991).
29. V.E. Viola, Nucl. Data Sect. A1, 391 (1966).
30. R. Wada et al., Phys. Rev. Lett. 58, 1829 (1987).
31. Y.D. Kim et al., Phys. Rev. Lett. 63, 494 (1989).
32. W.U. Schröder, J.R. Birkelund, J.R. Huizenga, K.L. Wolf, and V.E. Viola, Jr., Phys. Rep. 45 (1978) 301.
33. G.F. Bertsch and S. das Gupta, Phys. Rep. 160 (1988) 189.

THE NATURE OF MULTIFRAGMENTATION AS EXPLORED WITH A SCHEMATIC MODEL

William A. Friedman

Department of Physics, University of Wisconsin
Madison, Wisconsin 53706. U.S.A.

ABSTRACT

The multifragmentation disassembly mode of highly excited nuclear systems is examined through explorations with a schematic model which involves an expanding and evaporating source. The novel character of the process is investigated. The threshold and demise of the process as a function of excitation energy is studied. A high threshold energy due primarily to dynamics is suggested. Predictions of various features signalling the multifragmentation process are made. The model suggests the sensitivity of this process to the low density nuclear equation-of-state. Observable features associated with the reaction dynamics, i.e., time scales, lowered densities, and the collective velocity of expansion, are discussed. Reference is made to current experiments which are finding features qualitatively similar to the predictions.

INTRODUCTION

In recent years there has been a great deal of interest in a possible new mode of nuclear disassembly associated with highly excited nuclear systems. This mode would differ qualitatively from others, such as simple evaporation or fission, both of which are well known at the lower excitation energies. The new mode has been given the name "multifragmentation", and it is characterized by the copious yields of intermediate mass fragments (IMFs), which we define here to be nuclear fragments having charge greater than or equal to 3, but less than those of typical fission fragments. We suggest that multifragmentation events are those in which, on average, the multiplicity of IMFs is significantly greater than one. This process should be distinguished from events found for lower energies in which there may be 2,3, or more IMFs in a single event as the result of fluctuations about a mean multiplicity which is less than one. To emphasis this point I have shown in Fig.1 a predicted multiplicity distribution for IMFs from a 7 MeV ^{197}Au source for which the mean multiplicity is .61. High multiplicity events of this type should not be designated a true "multifragment" events, even though several IMFs are seen.

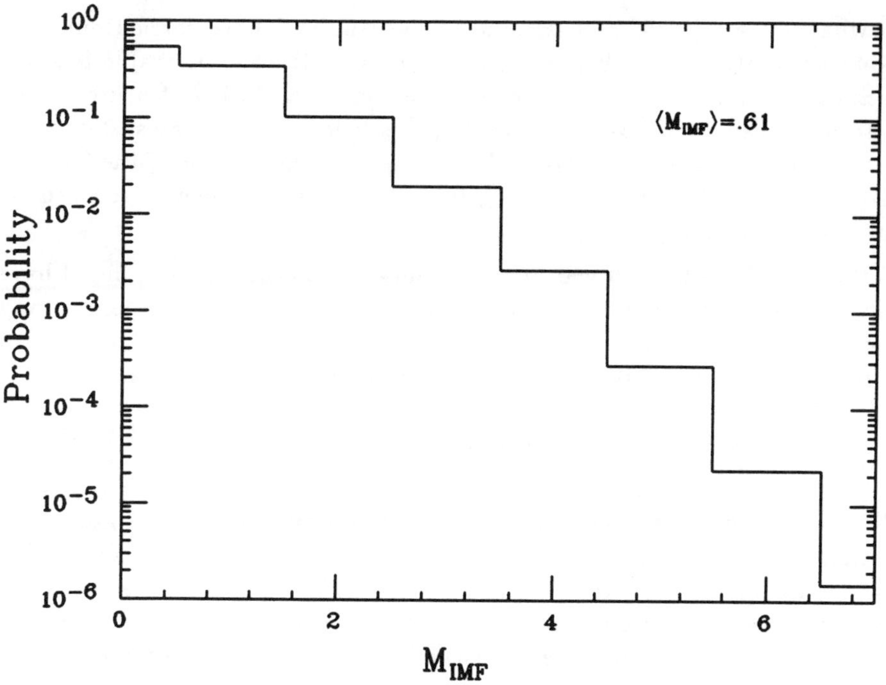

Fig. 1. Calculated probability distribution for the multiplicity of IMFs from ^{197}Au with an initial temperature of 7 MeV.

The nature of the multifragment decay mode, i.e., the underlying physics which governs it, is not well understood. In particular there is little qualitative or quantitative understanding of the conditions for its threshold of onset. Furthermore one expects that in the limit of extremely high excitation energies "multifragmentation" may be superceded by vaporization. This is a process involving the disintegration of the nuclear system into individual nucleons rather than heavier fragments. This transition is also little understood.

In order to facilitate experimental investigation of the multifragment phenomenon it is useful to identify and explore, in a model, the observable features which signal the existence of this mode. To chart both the onset and demise of the phenomenon we will make a systematic study of the evolution of these features with increasing excitation energy. Furthermore, since multifragmentation will be governed by nuclear properties under unusual conditions, namely, high temperatures and low densities, a study of multifragmentation may be useful in exploring nuclear properties in these regimes.

In what follows I will discuss some of the studies I have carried out with a flexible schematic model. These studies suggest both the observable features to be expected and the underlying causes associated with these features. This seems to be an especially exciting period in this field, since the latest data from MSU, GSI, GANIL, and elsewhere, is beginning to uncover some of the features we will discuss, and there is a strong indication that the new mode can be identified and explored.

In the following section I will briefly outline the specific model which I have been using, and then, in subsequent sections, we will discuss both general and specific features of multifragmentation which are suggested by calculations with this model. We will end with a short summary and list of conclusions.

SCHEMATIC MODEL

The model which has been used to explore multifragmentation grows out of a formalism for statistical emission[1] which has been successful in accounting for low energy phenomena. In the spirit of this conference, the model discussed here attempts to unify various aspects of reaction mechanisms. As the excitation energy increases the nuclear system is subjected to increasing thermal pressure which will cause expansion. Our model attempts to include this effect. It couples statistical emission features to dynamical features associated with the density changes of the emitting source. For the lower energies the expansion features have little effect, but with increasing energy they provide dramatic effects. This model has been previously discussed in refs. 2-5. I will, however, very briefly outline its ingredients here.

The model envisions an expanding and evaporating source. This source is characterized by a ground state (T=0) with binding energy which depends on density. The excited states of the source system are assumed to have a level density which is modeled by that of a finite temperature Fermi gas. This can be simply characterized by a Fermi energy that depends on the instantaneous density. The dependence of the ground state energy on density is assumed to rise parabolically for excursions from normal nuclear density, at which the liquid drop binding energy is approximately 8 MeV per particle. The compressibility characterizing the parabolic relation has been taken as a free parameter to explore sensitivity to the equation-of-state. The most important regions for our calculations are for densities less than normal density. With a finite nuclear compressibility (including surface and coulomb effects) of K=144, the binding energy goes to zero at approximately zero density. For K=200 the binding energy vanishes at a density of approximately .15 normal density, and for K=250 it vanishes at .24. For comparison, calculations were performed with no expansion (K=infinity).

The evaporation aspects of the process are governed by rates, dn_b/dt, for each nuclear species, b. These are determined by detailed balance and phase space features provided by level densities which, in turn, depend on the particle density. We include in our calculations all particle-stable species and excited states with charge less than 10. In addition a few unbound resonances for mass less than 6 are included. There is no provision for fission however.

The expansion aspects are very schematic. The density is characterized by a function of time $\rho(t)$ which describes a uniform, sharply cutoff particle density characterized by a radius $R(t)$ and particle number $A(t)$. The density is driven downward by the thermal pressure, and driven toward the normal saturation density by the nuclear attraction characterized by the equation-of-state. It is the influence of the latter on the density dynamics which can be exploited to study the equation-of-state. Apart from the loss of particles from the source the expansion is assumed to be isentropic.

The time evolution of the quantities of the problem is achieved by requiring that total energy be conserved. This energy resides in two places, the residual source and the emitted fragments. The former includes: the collective kinetic energy of the mass of the source, the collective compressional energy, and the thermal excitation energy of the source associated with a temperature. The energy associated with the fragments includes their kinetic energy (assumed to equal the average of their instantaneous spectra), and the separation energy.

The model permits the calculation of the evolution of $\rho(t)$, $Z(t)$, and $A(t)$ of the source, as well as spectra for the emitted particles. The origin of time is taken to be the instant at which the system passes through normal density, presumably following an initial period during which the system may be compressed and during which it receives excitation energy. At the origin of time the source is characterized by a mass A, a charge Z, an excitation energy E^*, and an initial rate of density fall (often taken as zero to reflect the damping of the initial stage). The collective angular momentum has been ignored for simplicity, with the assumption that the collective rotational energy is small compared to the other energies of the problem.

It is found that expansion strongly influences the emission process through: the density dependence of the Fermi energy, the related influence on the instantaneous temperature, and the changes in separation energy associated with a dilute source. Emission, in turn, influences expansion, through the conservation of energy which affects the compressional and the thermal energies.

Calculations track the evolution of the source and accumulate the instantaneous spectra, and particle yields. We have also devised a procedure for statistically calculating fluctuations of the particle yields from the average evolution of the source. This permits the estimation of multiplicity distribution functions, correlations in time between specific emitted fragments, and even the statistical construction of collections of events. All of these features have been explored in examining the multifragmentation process as a function of excitation energy.

EXAMINING AND INTERPRETING RESULTS OF CALCULATIONS

Let us begin by examining the results of a set of 6 calculations performed for a system of mass 190 at initial energies ranging from 10 MeV to 15 MeV, in 1 MeV steps. We have used a compressibility K=144 and have assumed that the density expands, beginning with no initial velocity. In Fig. 2, we show some interesting features in the evolution of each system. We plot the instantaneous mass of the source as a function of the instantaneous density. We have selected the first phase in time for each of the cases. In the cases with T=10 to 14 the density is seen to fall at first and then to return toward normal values; for the case of T=15 no return is found. The lowest 5 cases oscillate in density with decreasing amplitude for times which are longer than those displayed. The feature I wish to emphasize is the sharp drop in the source size when the conditions result in the density falling to the vicinity of .3 normal density. For T=10 MeV no such drop is seen and for T=11 the drop is very

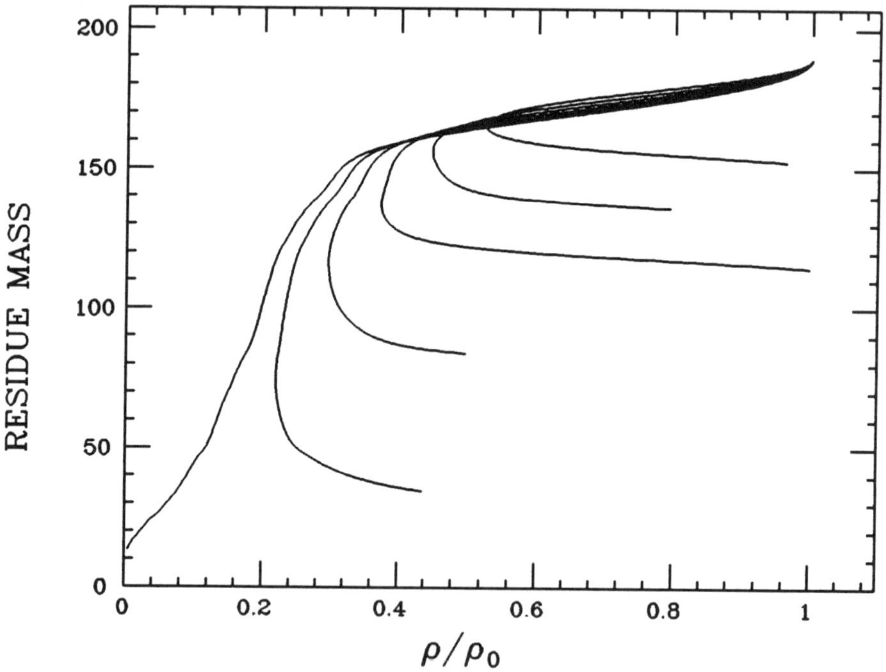

Fig. 2. Instantaneous mass of residual source versus instantaneous density for a source of mass 190 and temperatures ranging in 1 Mev steps from 10 MeV to 15 MeV. Time is the parameter along each curve.

small. For the larger energies it is quite dramatic. The loss in the source mass in these regions occurs over a relatively short period of time (less than 40 fm/c), and it is characterized by an abrupt increase in the mean mass of the emitted fragments. These change to IMFs at lower densities from primarily hydrogen and helium at higher densities. For the cases which return to normal density, the average emitted mass drops again after the excursion into the lower densities. This emission of heavy masses occurs despite the fact that the temperature in the low density region is considerably lower than the initial temperature due to adiabatic cooling.

The model thus suggests that a qualitative change of mechanism attends the low density regions. We wish to associate this with "true" multifragmentation. It is clear that the onset of this behavior is governed by the relative balance of the thermal pressure and the nuclear binding. Thus for a given initial temperature the minimum density achieved is determined by the compressibility, and hence multifragmentation is sensitive to this quantity. The other important point illustrated by Fig. 2, is the fact that substantial initial excitation energy is necessary to overcome the nuclear binding and reach the lower densities. One could estimate that something on the order of 8 MeV/A is required. Even higher values are needed because of the energy carried away by the 20 or 30 mass units (light particles) which leave the source during the expansion phase. These carry away significant energy.

The energy requirements suggest that the threshold for the onset of multifragmentation is quite high, and they explain why there has been such difficulty in observing the phenomenon. The model discussed here was successful[5] in accounting to the the observed mean multiplicities of IMFs observed at GSI with both silver and gold targets. In both cases excitation energies of about 700 MeV were deduced, and even with these energies the observed mean multiplicities were significantly less than 1.

The model also suggests that the onset of multifragmentation should provide sharp features in the excitation function of the mean multiplicity of IMFs, and in the excitation function of the charge distribution dN/dZ. The latter can, for convenience, be characterized by determining the power (negative) of the best power-law fit to the distribution at each energy. The magnitude of this negative exponent is found, with the model, to drop sharply in the same energy region which provides the increase in the IMF multiplicity. This reflects the increased probability for emitting the heavier elements. These two correlated features are shown in Fig 3.

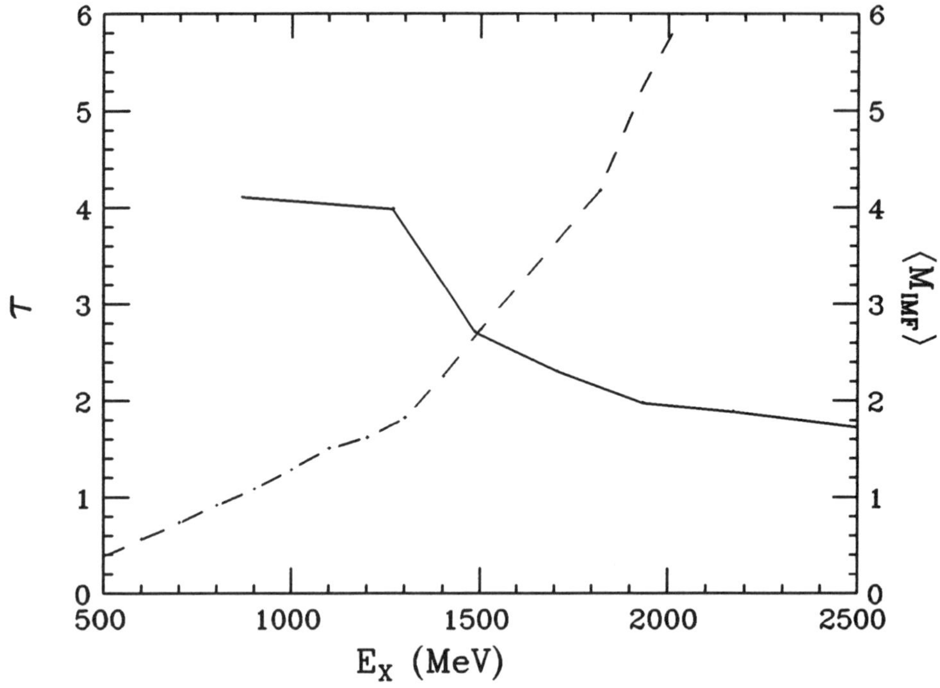

Fig. 3. Excitation function of the mean multiplicity of IMFs (dashed) and the exponent τ which characterizes dN/dZ (solid) for a source of mass 190.

In previous studies[4,5] we explored the physical basis for the rapid emission of massive clusters as predicted by the model. We observed that the conditions under which this occurs correspond to those at which there is a crossover in the values of the free energy for two situations: a homogeneous source, and a source with N of its nucleons formed in a cluster of mass number N. This is shown in Fig.4 which assumes an adiabatic expansion so that a given density is associated with each temperature (abscissa). We associate this transition with a phase transition which leads to cluster formation in the interior of the source. We developed a phase-diagram which plots the locus of the crossover points for different entropies, as a function of temperature and density. Such a diagram shows that the transition, over a large range of initial conditions, occurs for temperatures on the order of 5 MeV, as shown for one case in Fig.4. The corresponding density for Fig.4 is on the order of .25 normal density. Calculations also show that the necessary density for a transition is not reached for lower entropies (initial temperatures). (These would only be reached with equations-of-state which provide substantial binding energies for zero density, a condition clearly unphysical.)

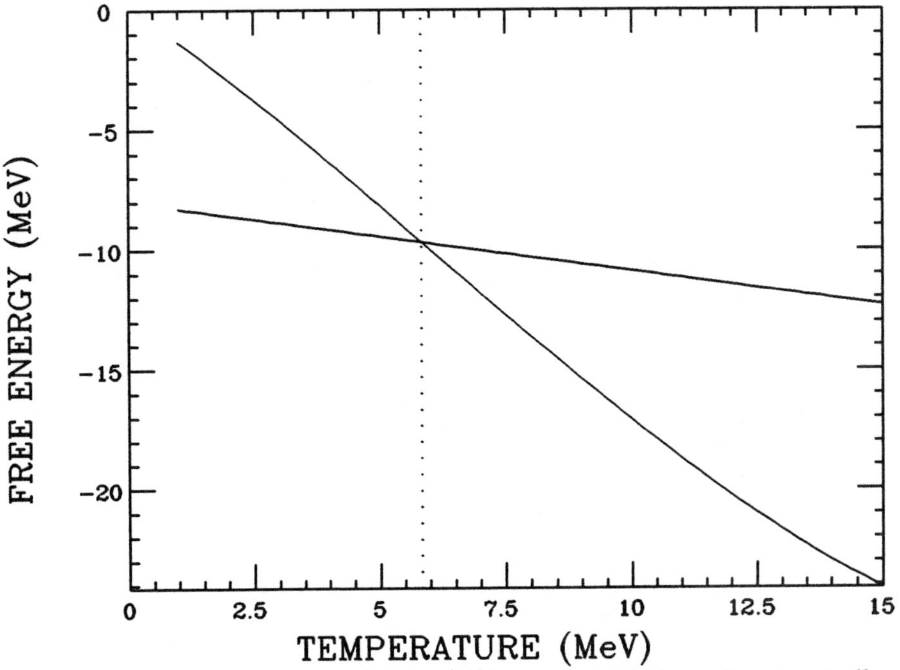

Fig. 4. Free energy per particle versus temperature for an isentropically expanding system: homogeous system (solid), ^{16}O cluster (dashed).

Because the equation-of-state governs the regions of density which are experienced by the system, a plot of the predicted IMF multiplicity versus excitation energy depends on this feature. We find in Fig. 5 four different curves for mean multiplicities of IMFs . These curves result from calculations performed for an initial source which was the size of gold. Fig. 5 shows the predicted threshold behavior for systems with finite compressibility. The curve predicting the lowest multiplicity is a calculation done for comparison, which contains no expansion. The other striking feature of these calculations is the predicted fall of the mean value of multiplicity for very high excitation energy. This fall is associated with systems which expand monotonically. For the K=144 case which reaches zero binding only at zero density, we have assumed that cluster formation ceases for a source more dilute than .1 normal density. Nucleons not already bound in clusters when that condition is reached are freed as light particles. Since this condition is met in decreasing times with increasing energy, the number of IMFs decreases. For the cases of K=200, and K=250, the binding energy of the source vanishes for finite density, and that condition provides a natural cutoff for the fragment formation process.

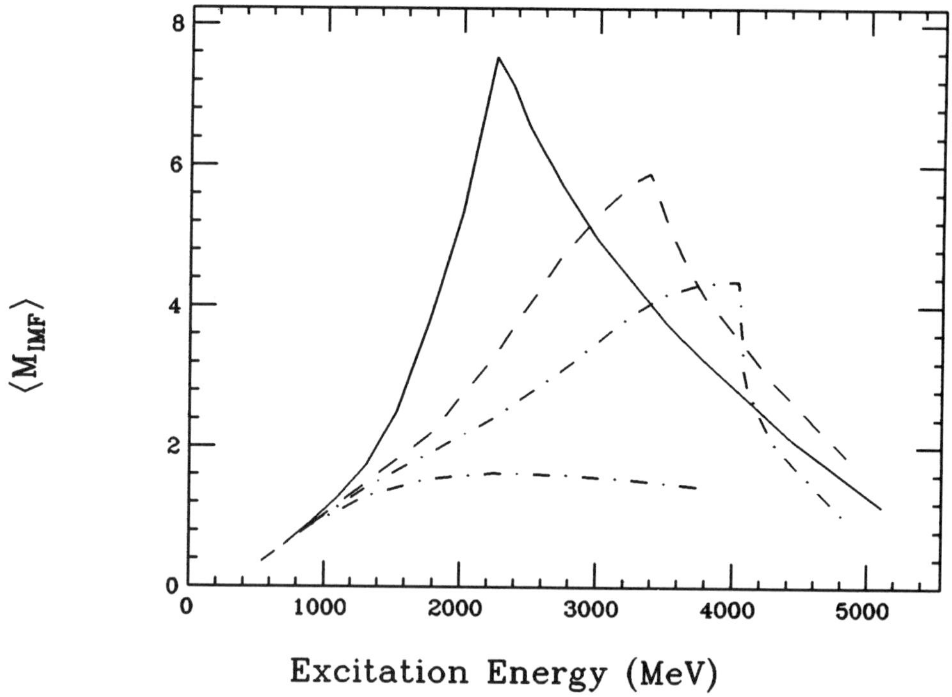

Fig. 5. Mean multiplicity of IMFs versus initial excitation energy for sources of mass 197, with compressibility constants K=144 (solid), K=200 (long dashed), K=250 (dot-dashed), no expansion (short dashed).

IMPORTANCE OF GATES

In the previous section we suggested that the threshold for the multifragmentation process is quite high. It has long been observed the transfer of large amounts of energy from an incoming projectile to systems with a high degree of thermalization is very difficult to achieve with projectiles having energy on the scale of 100 MeV/A. We thus expect that, even when the incoming beam has sufficient energy to achieve the values discussed above (8-10 MeV per thermalized nucleon), only a very small fraction of the events produced in the collisions will involve the deposition of the requisite energy. If one is to examine multifragmentation events it is thus essential to be highly selective in the events which are examined. This can be done by the appropriate gating of the experimental data.

There are currently three types of gates which are being used to examine multifragment reactions: neutron multiplicities, charged particle multiplicities (predominantly light charged particles), and IMF multiplicities. While for the charged particle multiplicities the terminology suggests an association with impact parameter, e.g."peripheral" or "central" trigger, these are also especially sensitive to the total energy deposited. In the context of multifragmentation, it is the link to the excitation energy which may be the most significant. We can assume that most of the reactions involving multifragmentation are "central", but they involve a large range of excitation energies.

We have used the model to estimate the windows in excitation energy associated with the emission of a given number of charged particles. The calculated windows for high numbers of charged particles is shown in Fig. 6 for a very large source. Here, as expected we see a monotonic increase in the energy region selected by increasing multiplicity. To get a qualitative connection one can correlate the value of the peak or centroid of the window with an energy. A more precise treatment would require a weighted integral over the energy region of the window. We have also generated windows associated with a given multiplicity of IMFs. These windows tend to be about twice as wide as the charged particle multiplicity windows. Both of these types of gates can be very effective in screening out the low excitation components of the cross section.

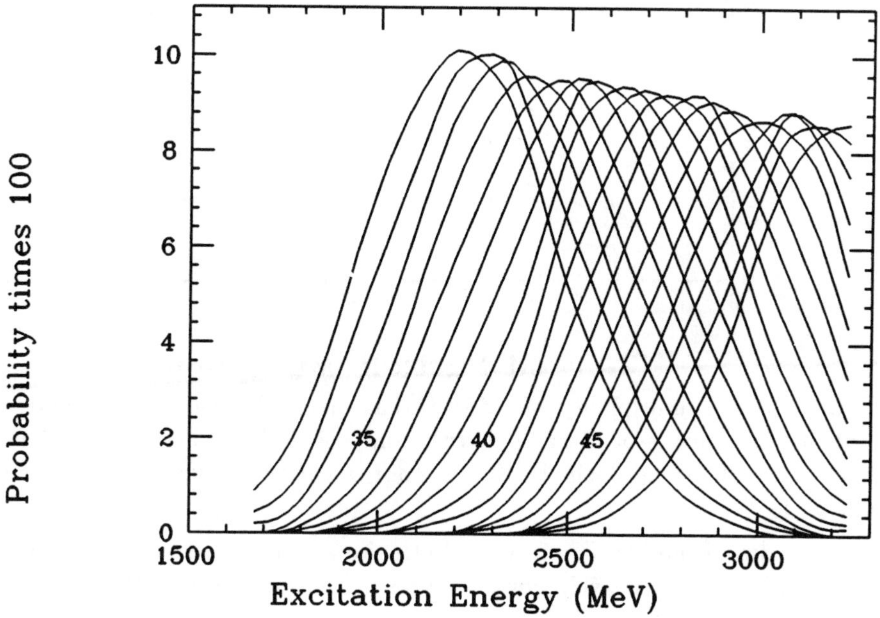

Fig. 6. Calculated excitation energy windows associated with charged particle multiplicities.

In discussing those features which signal multifragmentation, we have dealt with various quantities to be examined as a function of excitation energy. The latter is not an easily observed quantity. Instead, experiments can determine the dependence of these quantities on the values of the gates. To explore the qualitative nature of the correlation between charged particle multiplicity and mean IMF multiplicity we have made calculations for different systems over a range of excitation energies. From each calculation we obtain a mean charged particle multiplicity and a mean IMF multiplicity. We suggest that the correlation between these should be qualitatively similar to what might be seen in a correlation between multiplicity gates and mean IMF. An example of this correlation is shown in Fig.7. This contains the same information as Fig 5. but here we have correlated two experimentally observable quantities.

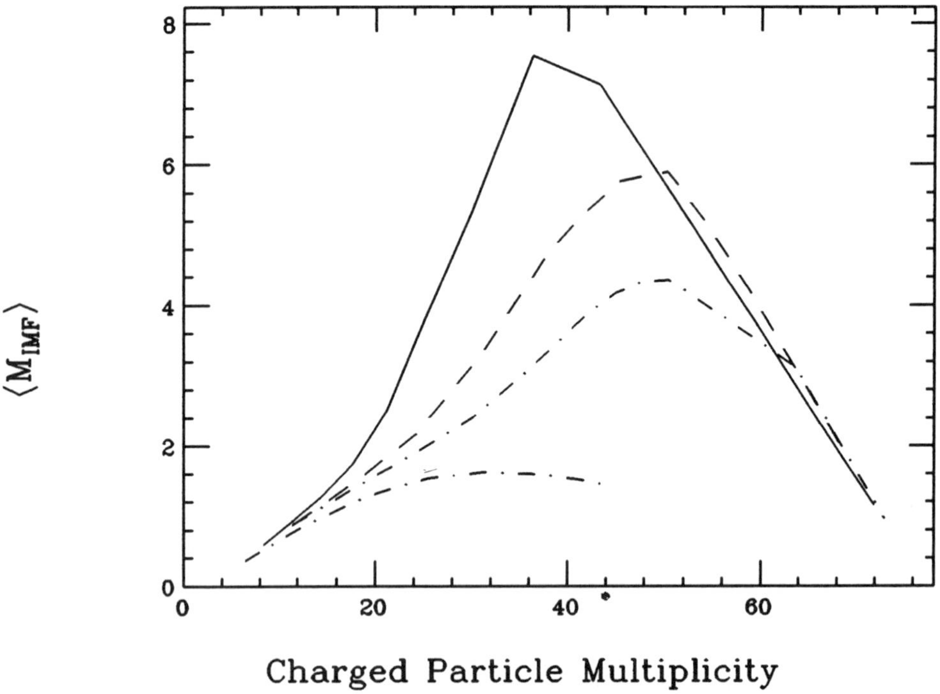

Fig. 7. Mean multiplicity of IMFs versus mean charged particle multiplicity for sources of mass 197, with compressibility constants K=144 (solid), K=200 (long dashed), K=250 (dot-dashed), no expansion (short dashed).

Current data from MSU[6] and GSI[7] both show clear signals of the rise in the mean multiplicity qualitatively consistent with the the curves in Fig.7.

FURTHER SIGNALS

The model has also been used to examine other features which are potential indicators of the development of multifragmentation and of the dynamical aspects of the reactions which lead to multifragmentation. We will discuss a few examples of these in this section.

As mentioned in the discussion of the model, we have developed, through statistical methods, a means for estimating higher order quantities, such as multiplicity distributions and two-particle correlations. Among the features we have examined is the distribution of the total multiplicities and the distribution of IMF multiplicities. These were evaluated for systems with different initial conditions, e.g, different excitation energies. Such calculations were used to establish the windows which are shown in Fig.6 above. We have also studied the variation of the variance (second moment) of the IMF multiplicity distribution with excitation energy . For low energies, where expansion effects and multifragmentation are not important, the distributions are very close to Poisson shapes, and hence the ratio of the variance to the mean value is close to 1.0. In the energy region where multifragmentation becomes important our calculated distributions move away from Poisson forms. This feature is characterized by appearance of variances which are smaller than the mean value. While the variance grows with increasing excitation energy, it does so less fast than the mean value. We have examined this behavior by plotting, as a function of excitation energy, the difference between 1.0 and the ratio of the variance to mean value. In Fig 8. we show the calculations of this quantity for different equations-of-state versus the corresponding total charged particle multiplicity (instead of excitation energy). The calculations cover both the onset and the declining energy regimes. This feature emphasizes the fact that both the mean value and the variance are separate quantities to be examined. Recent data from MSU[8] indeed suggest that the variance may grow at a slightly slower rate than the mean IMF multiplicity in the multifragmentation region. The combined information may provide an increased constraint on the influence of the equation-of-state.

The model has also been used to generate events characterized by individual distributions of charges. These can be used to explore the conditional moment distributions, suggested by Campi[9] as indicators of critical behavior. They also may be used to explore intermittency signals[10]. Both of these studies are currently underway.

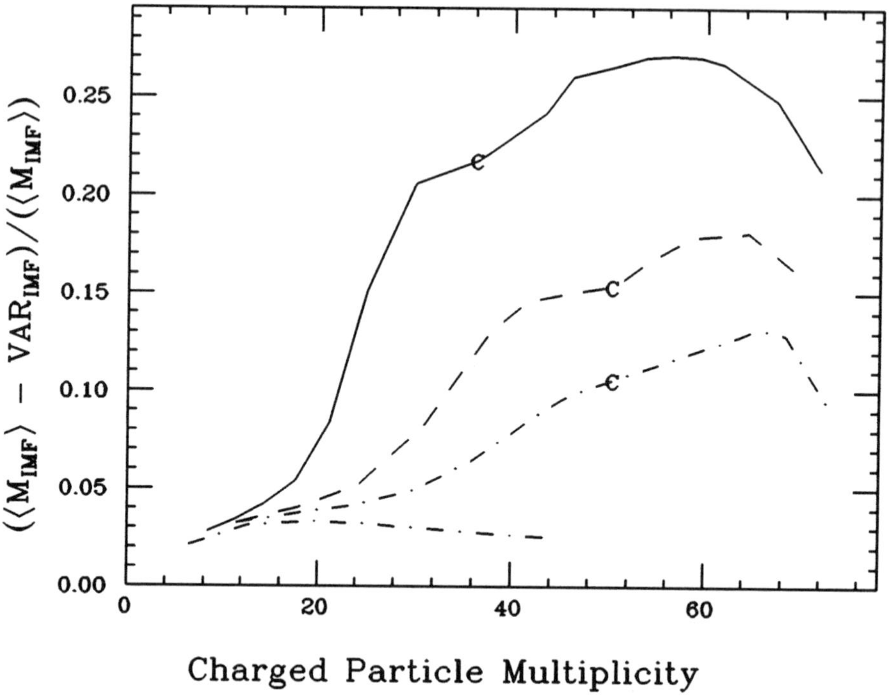

Fig. 8. Influence of equations-of-state on the variance of the IMF multiplicity distributions for associated mean charged particle multiplicity: K=144 (solid), K=200 (long dashed), K=250 (dot-dashed), no expansion (short dashed). The maximum multiplicity is marked by *c*.

Finally I wish to discuss some additional features which may provide specific indications of the dynamics of the reaction. One of these has been discussed at this conference by Y.D. Kim of MSU. This concerns the time correlations between pairs of IMFs emitted at small relative momentum. We have made calculations of the distribution of the time separations between pairs of particles from two sources each with the mass of gold, and with initial temperatures of 8 MeV (sub-multifragmentation) and 13 MeV (multifragmentation). The probability curves for both of these cases are shown in Fig. 9. For the lower energy case, shown in Fig. 9a, the distribution is roughly consistent with an effective mean life time of the order of 300 fm/c. For the case involving multifragmentation, shown in Fig. 9b, the effective mean life time reduces to about 75 fm/c. Hopefully interferometry techniques can be useful in observing this qualitative difference. Care will need to be taken to eliminate the preequilibrium fragments.

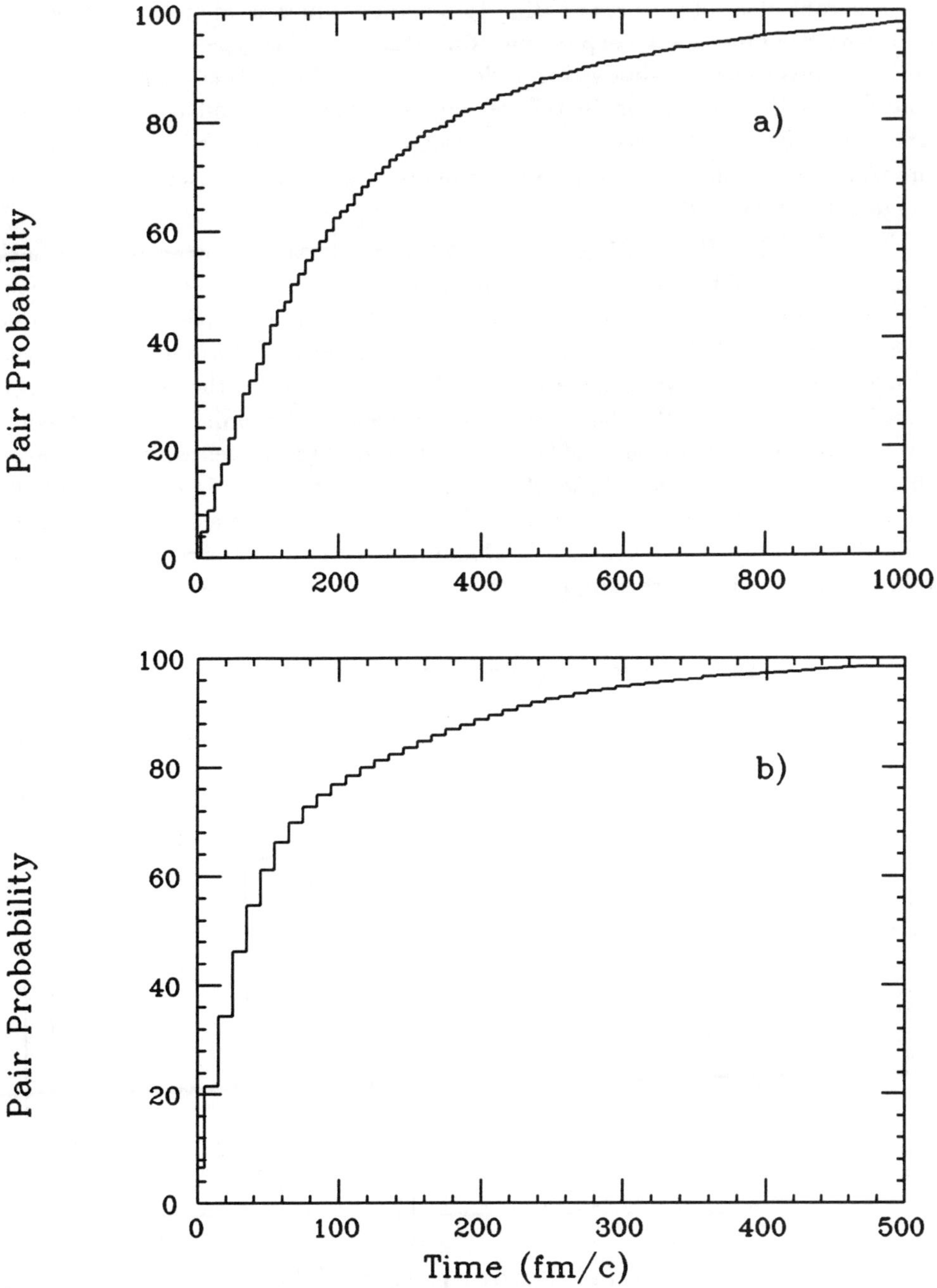

Fig. 9. Time separations of IMF pairs for Au source: a) initial temperature of 8 MeV, b) initial temperature of 13 Mev.

A second class of features relating to reaction dynamics includes those which appear to be influenced by expansion. Two studies of this type have been carried out. One involves a two-body correlation feature and the other a singles feature. The first is the correlation in the velocity of fragments emitted back-to-back from the source. The second is the observed energy spectra of IMFs. Since the treatment of the dynamics is only schematic, the effects discussed here only suggest qualitative features.

In calculating the relative velocities of pairs of IMFs emitted back-to-back from a single event, one takes into consideration the fact that these come from different stages in the evolution of the system. Since the model provides a time characterization of this evolution, we are able to gather information on all sets of pairs. From these we can determine the mean value of the relative energy. This is shown in Fig. 10. There are three features which dominate this behavior. Two are related to the Coulomb field: one of these is associated with the reduced charge of the source due to particle emission; and the other is the effect of the increased radius due to the expansion. The third influence is associated with the collective velocity of expansion. The Coulomb effects cause the relative fragment velocity to fall with increasing excitation energy (temperature).

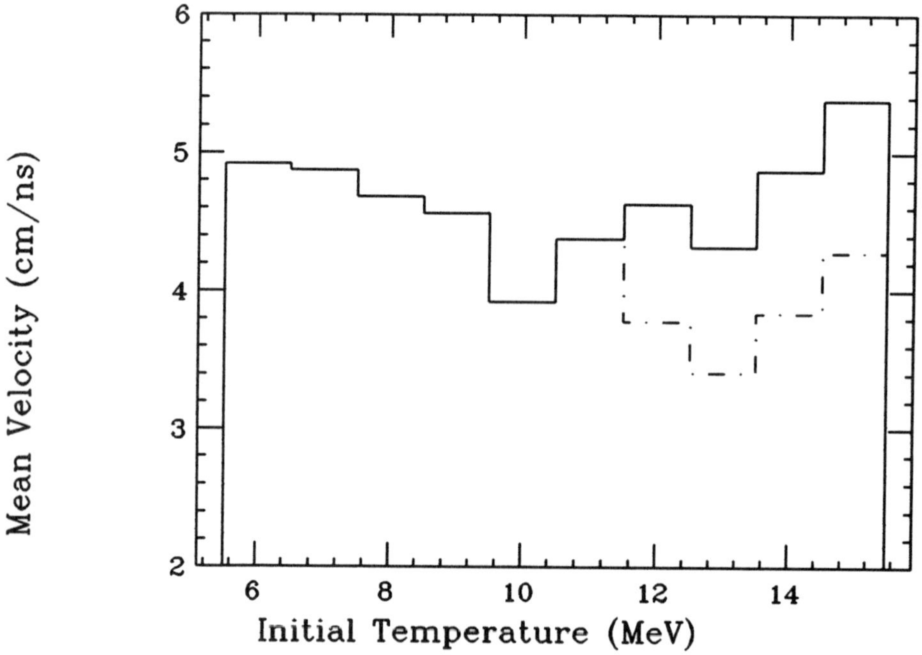

Fig. 10. Mean velocity of back-to-back IMF pairs as a function of initial temperature: surface emission (solid), volume emission for ρ/ρ_0 less than .3 (dot-dashed).

If the emission were to occur only from the surface, then above T=12 Mev the relative particle velocity is raised by the increasing surface velocity. If, on the other hand, there is a transition from surface emission to volume emission when the density falls below approximately $.3\rho_0$, then the effective source velocity and Coulomb energy are both reduced. The relative velocity then follows the reduced curve indicated by the dot-dashed line.

Finally, we suggest that it may be possible to observe the influence of the expansion velocity through measurements of the singles spectra for IMFs. This might be possible because of the boost that the emitted particles receive from the collective expansion velocity of the source. This boost depends on the specific initial conditions of each of the contributing events. In addition to the normal sum over particles emitted at different stages in the evolution of the sources, there is an additional effect which arises from summing over sources with different excitation energies. When expansion is important each contribution provides a different kinematic boost to the particle spectra. In an illustrative calculation we considered the singles spectra of ^{12}C fragments from a mass 100 source, gated on three IMFs. This constraint provides a high energy window for the events.

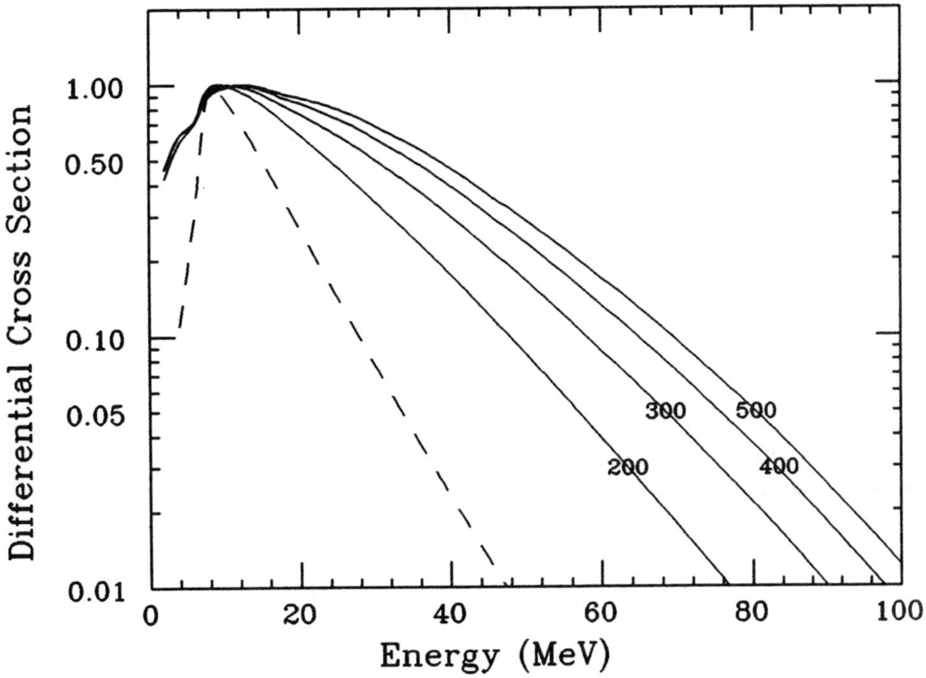

Fig. 11. Calculated spectra for Z=6 fragments, from mass 100 source, with exponentially decreasing distributions of initial excitation energy having mean values indicated (solid curves). Spectrum for a sharp 900 MeV excitation (dashed). Arbitrary normalization.

In addition we assumed that the probability for depositing excitation energy falls exponentially, and we have examined various decay constants for this fall. The results shown in Fig. 11 arise from weighted sums over the calculated spectra for each of the initial excitation energies. The curves show the strong influence of the source expansion. This feature provides spectra which have the appearance of highly enhanced "temperatures". This actually arises from the different distributions of excitation energy and the corresponding velocity boosts.

CONCLUSIONS

By exploring the multifragmentation regime with the aid of the expanding-evaporating source schematic model we have observed several features which may be important in the actual multifragmentation process. First, we suggest that the dynamics, in particular the expansion, plays a very large role in the phenomenon. We find that densities on the order of .3 normal densities are necessary for the process to occur. This feature, in turn, requires initial excitation energies to be quite high, on the order of the total binding energy. Even more energy is required because of the energy carried away by particles emitted during the early expansion stages. We have seen that the rapid and copious emission of massive fragments can be associated with the conditions under which the free energy for the existence of clusters becomes favorable, and that the instantaneous temperature associated with this condition is on the order of 5 MeV. This relatively low temperature is reached from initial temperatures (at normal density) of 12 MeV or more.

Because of the need for high excitation energies the multifragment events often represent a small part of the total cross section. Hence selective gating is required to examine these events. Since the strongest signatures for multifragmentation are to be found in the excitation function of various observables, it is essential to link the excitation energy to some observable quantities. One candidate is the total multiplicity of charged particles, others include the multiplicity of neutrons or IMFs. We have calculated the rise and subsequent decline of the mean multiplicity of IMFs with excitation energy (and charged particle multiplicity), and we suggest this is the clearest sign for multifragmentation. We also found systematic changes in the charge distributions and the variance of the IMF multiplicity distributions. Both of these features vary strongly with excitation energy (charged particle multiplicity) in the region where multifragmentation sets in. In the model we find that the expansion dynamics is sensitive to the finite nuclear equations-of-state, especially for densities lower than normal density. Since multifragment features are very sensitive to this expansion phase of the reaction, there is the possibility of learning about these equations-of-state by studying the decay process.

Finally we discussed the possibility of using experimental signatures to learn about the reaction dynamics. In particular we dealt with the influence of the time scale for the reaction and the influence of the expansion. We have predicted effects to be seen in two-body correlation measurements for time correlations. We have also shown how some of the effects of expansion may appear in the relative velocities between pairs of particles and in the singles spectra of IMFs.

There is new data becoming available at this moment. Some examples of the most interesting of these have been reported at this conference by Guerreau, Kim, Trautmann, Viola, and others. All of this data is related to selected events with expected excitation energy on the order of 8 MeV/A. All show striking features which are qualitatively consistent with the discussions of anticipated signals for multifragmentation that have been discussed above.

This work was supported in part by a grant from the U.S. National Science Foundation.

REFERENCES

1. W.A. Friedman and W.G. Lynch, Phys.Rev.C28, 16 (1983).
2. W.A. Friedman, Proceedings of the Eleventh Oaxtepec Symposium on Nuclear Physics [Notas Fis. 11, 71 (1988).
3. W.A. Friedman, Phys.Rev.Lett.60, 2125 (1988).
4. W.A. Friedman, in *Nuclear Dynamics and Nuclear Disassembly, Proceedings of the Symposium, Dallas, Texas, April 1989*, ed. by J.B. Natowitz (World Scientific, Singapore, 1989), p. 349.
5. W.A. Friedman, Phys.Rev.C42, 667 (1990).
6. D.R. Bowman, *et al.*, MSU preprint (1991).
7. C.A. Ogilvie, *et al.*, GSI preprint 91-18 (1991).
8. R.T. deSouza, *et al.*, MSU *preprint (1991)*.
9. X. Campi, Phys.Letts B208, 351 (1988).
10. M. Ploszajczak and A.Tucholski, Phys.Rev.Letts.65, 1539 (1990).

"MULTIFRAGMENT DISINTEGRATIONS OF THE $^{36}Ar+ ^{197}Au$ SYSTEM AT E/A=35-110MeV"

Y.D. Kim, R.T. de Souza, L. Phair, D.R. Bowman, N. Carlin,†
C.K. Gelbke,W.G. Gong, M.A. Lisa, W.G. Lynch, G.F. Peaslee,
M.B. Tsang, C. Williams,and F. Zhu
National Superconducting Cyclotron Laboratory
and Department of Physics and Astronomy
Michigan State University, East Lansing, MI. 48824, USA

W.A. Friedman and S. Pratt
Department of Physics, University of Wisconsin, Madison, WI 53706

ABSTRACT

Intermediate mass fragment (IMF: 3≤Z≤20) emission for the $^{36}Ar+^{197}Au$ reaction at E/A=35, 50, 80, 110MeV has been studied with a low-threshold 4π charged-particle detector array. Multiplicities of intermediate mass fragments gated on total charged-particle multiplicities have been extracted with an average multiplicity of 4 intermediate mass fragments for the most central collisions at E/A= 110MeV. Two-fragment correlation functions are presented for various gates on the fragment energy, and information about the time-scale of fragment emission is extracted.

INTRODUCTION

Highly excited nuclear systems can be produced in intermediate-energy nucleus-nucleus collisions. As the excitation energy is raised from below 1 MeV per nucleon up to about 10 MeV per nucleon, decays by intermediate mass fragment (IMF: 3≤Z≤20) emission are predicted and observed to become important.[1-18] Model calculations predict that such highly excited nuclear systems can expand either through the action of thermal pressure[1,10] or through a dynamical compression-decompression cycle.[1-3] If nuclear matter expands to sufficiently low densities, it becomes unstable and density fluctuations can grow exponentially as a function of time.[1] Given enough time, the growth of these fluctuations can cause a multifragment breakup.[2,3] Calculations suggest that the probability for multi-fragment breakup may be sensitive to the nuclear equation of state at low density.[4] Therefore, information about the low-density equation of state and the liquid-gas phase transition of nuclear matter may be obtainable from detailed studies of multifragment emission processes.[5,12]

To address this question of multifragment decay of highly excited nuclear systems, and to provide constraints for various theoretical models, we have measured both total charged particle and intermediate mass fragment (IMF: 3≤Z≤20) multiplicities for the $^{36}Ar + ^{197}Au$ reaction at E/A = 35, 50, 80, and 110 MeV. Of particular interest is information about the time scale on which multifragment emissions occur. Since the relative wave-function for two coincident particles is affected by final state interactions and, for identical particles, by quantum statistics, two-particle correlation functions at small relative momenta are sensitive to the space-time evolution of the emitting system.[19-26]

EXPERIMENTAL DETAILS

The experiment was performed in the 92" diameter scattering chamber of the National Superconducting Cyclotron Laboratory at Michigan State University. ^{36}Ar beams of energy E/A=35 MeV from the K500 cyclotron and E/A=50,80, and 110MeV from the K1200 cyclotron bombarded a 1mg/cm^2 ^{197}Au target. A vacuum of better than 10^{-5} torr was maintained in the scattering chamber, and water and hydrocarbon vapor components in the residual gas were strongly reduced by a large cold trap filled with liquid nitrogen. By this means, carbon deposits on the target were reduced to a negligible level.

Light particles and complex fragments were detected using the MSU Miniball phoswich detector array.[27] Fig. 1 shows a half-plane section of the Miniball array in the vertical plane which contains the beam axis. Individual rings are labelled by the ring numbers 1-11 with the number of detectors in parentheses. Since the angular distributions of the emitted particles are strongly forward peaked, the granularity is higher in the forward direction than the backward direction. In this configuration, a total of 188 detectors covered scattering angles of Θ_{lab}=16°-160° and a solid angle corresponding to 89% of 4π. Particles punching through the 4 mg/cm^2 plastic scintillator foils were identified by atomic number up to Z≈20 and by mass number, as well, for H and He isotopes. Representative approximate energy thresholds are E_{th}/A≈2 MeV for Z=3, E_{th}/A≈3 MeV for Z=10, and E_{th}/A≈4 MeV for Z=18 fragments. Particles of lower energy which were stopped in the scintillator foils were recorded but could not be identified by atomic number. All events in which at least two detectors fired were recorded on magnetic tape and analyzed off-line. Random coincidences were negligible due to the low beam intensity.

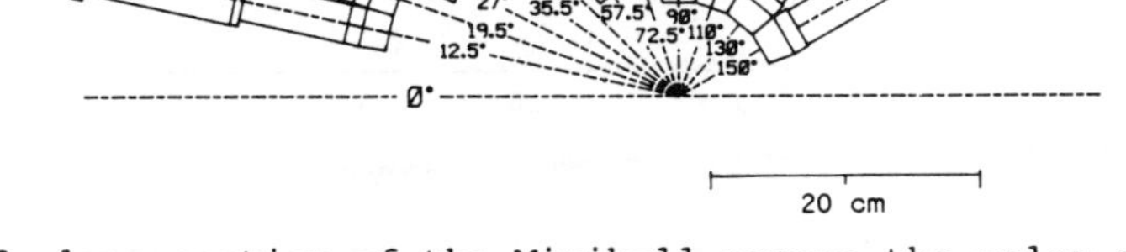

Fig. 1. Half-plane section of the Miniball array; the polar angles for the centers of the rings are indicated.

Energy calibrations, accurate to within 5% were obtained by measuring the elastic scattering of ^{4}He, ^{6}Li, ^{10}B, ^{12}C, ^{16}O and ^{35}Cl beams from a ^{197}Au target at incident energies of E(^{4}He)/A = 4.5, 9.4, 12.9, 16, and 20 MeV; E(^{6}Li)/A = 8.9 MeV; E(^{10}B)/A = 15 MeV; E(^{12}C)/A = 6, 8, 13, and 20 MeV; E(^{16}O)/A = 6, 8, 16, and 20 MeV; and E(^{35}Cl)/A = 8.8, 12.3, and 15 MeV.

MULTIPLICITY DISTRIBUTIONS

To better characterize the various classes of multifragment events, we have extracted the probability distributions $P(N_{IMF})$ of detecting N_{IMF} intermediate mass fragments in a single collision for different gates on the charged particle multiplicity N_C. The four panels in Fig. 2 depict these distributions. At each energy, the IMF multiplicity distributions become wider and shift towards higher average multiplicities as the charged particle multiplicity increases. Peripheral reactions, selected by $N_C \leq 7$, exhibit narrow IMF multiplicity distributions peaked at $N_{IMF}=0$. For these collisions, IMF emission is an unlikely process. On the other hand, for central collisions, selected by large values of N_C, IMF emission is a common process for which the average IMF multiplicity increases from $\langle N_{IMF} \rangle \approx 1$ at E/A=35MeV to $\langle N_{IMF} \rangle \approx 4$ at E/A=110MeV.

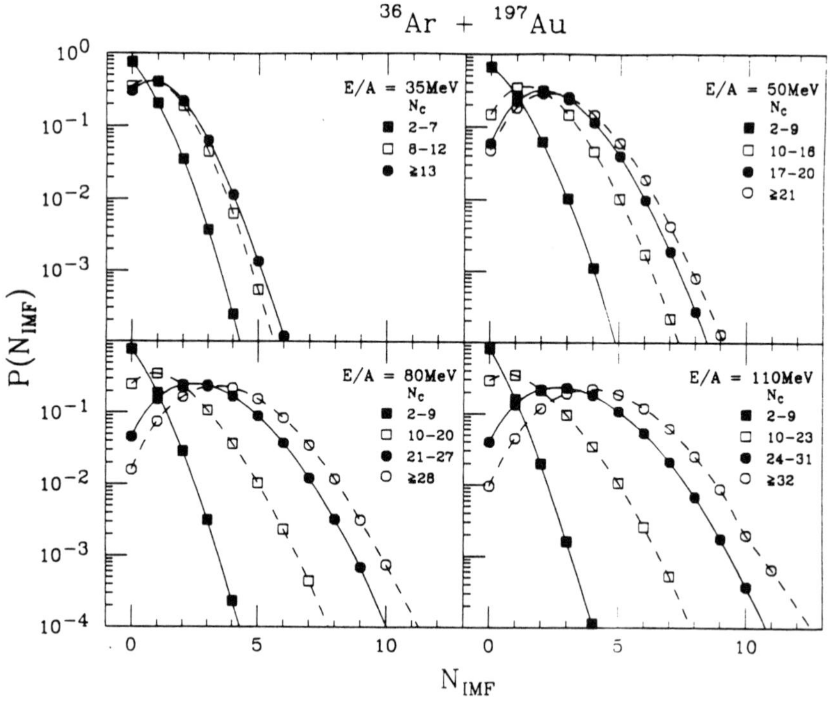

Fig. 2. Measured IMF multiplicity distributions for the indicated gates on charged particle multiplicity N_C. Panels are labeled by incident energy.

At the highest incident energy, events are observed in which more than 10 intermediate mass fragments are detected in the exit channel. The average total detected charge for central collisions increases from 22% of the total charge of the system at E/A=35MeV to 61% at E/A=110MeV, where a number of events were observed for which more than 80% of the total charge of the system was detected. These latter events correspond to a complete disintegration of the system into light particles and intermediate fragments. More detailed comparisons of $<N_{IMF}>$ as a function of N_C with the 'Rapid Massive Cluster Formation' model[10] will be presented in these proceedings by W.A. Friedman.

TWO-FRAGMENT CORRELATION FUNCTION

In this section we explore two-fragment correlation functions to extract time scales for the emission of intermediate mass fragments. These correlation functions will be constructed for data at forward angles, $16° \le \theta_{lab} \le 31°$, and at the incident energy of E/A=35MeV. Over this angular domain, the statistics of the data and the granularity of the array are optimal for the extraction of good-quality correlation functions.

A. Model calculations

A particularly simple expression for two-fragment correlation functions integrated over relative orientations between $\vec{P}$ and $\vec{q}$ can be derived by taking a classical approximation to the original Koonin-Pratt formula[26]:

$$1+R(\vec{P},\vec{q}) = \int d^3r_0 F_{\vec{P}}(\vec{r}_0)\,[1-2\mu\kappa/q^2 r_0]^{1/2} \qquad (1)$$

where $\kappa=Z_1Z_2e^2$ is the product of the charges of the two emitted fragments, $\mu=m/2$ is the reduced mass, and $F_{\vec{P}}(\vec{r}_0)$ is the distribution of relative coordinates between fragments. For fixed $F_{\vec{P}}(\vec{r}_0)$, Eq. 1 depends primarily on the variable

$$v_{red} = q/\sqrt{\mu\kappa} = v_{rel}\sqrt{\mu/\kappa} \propto v_{rel}/\sqrt{Z_1+Z_2}.$$

(In the last step, we have approximated m by 2Z). Thus, according to Eq. 1., one might expect correlations between diferent fragment pairs to display a similar dependence on v_{red}.

For our schematic calculations, we assumed surface emission from a spherical source of radius R_S with a fixed lifetime τ and energy and angular distributions taken to be consistent with the experimental distributions. The emitting source was assumed to move with a velocity of v_S=0.035c consistent with emission from the surface of heavy reaction residues formed in incomplete fusion reactions in which 80% of the projectile momentum was transferred to the fusion-like residue. For emission of the second fragment, we required a minimum initial separation of $r_{min}= 1.2(A_1^{1/3}+A_2^{1/3})$ fm from the first fragment. In all calculations we assumed a fixed source radius of R_S=12 fm. Choices of smaller (larger) source radii would provide slightly larger (smaller) emission time scales. Sensitivities to reasonable parameter variations were discussed in ref. 25; they typically introduce uncertainties of the order of 50% of the extracted emission times. The determination of more accurate emission time scales requires more realistic emission functions and source charge distributions, thus allowing more accurate treatments of the final state Coulomb interactions with the residual system. Here, we will discuss relative variations of fragment emission time scales. Theoretical uncertainties will be illustrated by performing calculations for final-state Coulomb interactions based upon different approximations.

The derivation[26] of Eq. 1 is based upon the assumption that the final-state interaction between the two detected particles dominates, that final-state interactions with all remaining particles can be neglected, that the single particle phase space distribution varies only slowly over particle momenta $p \pm q/2$, and that the correlation functions are determined by the two-body density of states as corrected by the interactions between the two particles. As discussed in ref. 26, these assumptions are only marginally satisfied for realistic IMF emission functions.

In order to evaluate some of the theoretical uncertainties, we have also performed classical trajectory calculations and constructed theoretical correlation functions. Recoil effects were taken into account by assuming a mass of A_S=197u for the emitting system. To provide an estimate of Coulomb distortions due to the interaction of the emitted fragments and a heavy fusion like residue, we have performed three-body Coulomb trajectory calculations for a target-like source of initial charge number $Z_S=79$. In all trajectory calculations, the response of the experimental apparatus was taken into account.

The experimental correlation function, 1+R, is constructed from the measured coincidence and single-particle yields, $Y_{12}(\vec{p}_1,\vec{p}_2)$ and $Y_i(\vec{p}_i)$, respectively

$$\Sigma Y_{12}(\vec{p}_1,\vec{p}_2) = C(1+R(\zeta))\Sigma Y_1(\vec{p}_1)Y_2(\vec{p}_2), \qquad (2)$$

where $\zeta=q$ or $\zeta=v_{red}$. The normalization constant C is determined by the requirement that $\langle R(\zeta)\rangle=0$ at for values of ζ sufficiently large that the final state interaction between the emitted fragments can be neglected. For each gating condition (on P and/or on other quantities), the correlation function is evaluated by summing both sides of Eq. 2 over all momentum combinations corresponding to given values of ζ.

As an example, Fig. 3 shows measured two carbon correlation functions. They exhibit a pronounced minimum as the relative momentum goes to 0 due to the repulsive Coulomb interaction between the two fragments. The curves in the upper and lower panels show correlation functions calculated with the Koonin-Pratt formula and classical 3-body trajectory calculations, respectively, for the indicated emission times. While the agreement between data and calculations is worse for the 3-body trajectory calculations than for the Koonin-Pratt formula, mean emission times of the order of 100-200fm/c can be extracted from both calculations.

B. Combination of different fragment pairs

To study the time scales in more detail by applying some gates, we investigated the possibility of combining different fragment pairs. In Fig. 4, we compared correlation functions for representative pairs of intermediate mass fragments both as a function of the relative momentum q (top panel) and as a function of the the reduced relative velocity v_{red}. Different symbols in the figure indicate correlation functions evaluated for different fragment pairs ($Z_1=6$, and $4 \le Z_2 \le 9$). Due to the increasing strength of the Coulomb repulsion between fragments of increasing charge, the correlation functions 1+R(q) exhibit wider minima at $q \approx 0$ as the charge of the second fragment is increased. On the other hand, the correlation functions $1+R(v_{red})$ are very similar as discussed in the previous section. Hence, little loss in resolution is expected when the correlation functions are summed over different pair combinations and evaluated as a function of the reduced relative velocity. Accordingly, we will construct "mixed-fragment" correlation functions according to Eq. 2, where the sum will also be extended over all charge combinations with $4 \le Z_1, Z_2 \le 9$. By this summation over fragment pairs a sufficient gain in statistics is achieved to allow the exploration of dependences on charged-particle multiplicity and velocity of the emitted fragments.

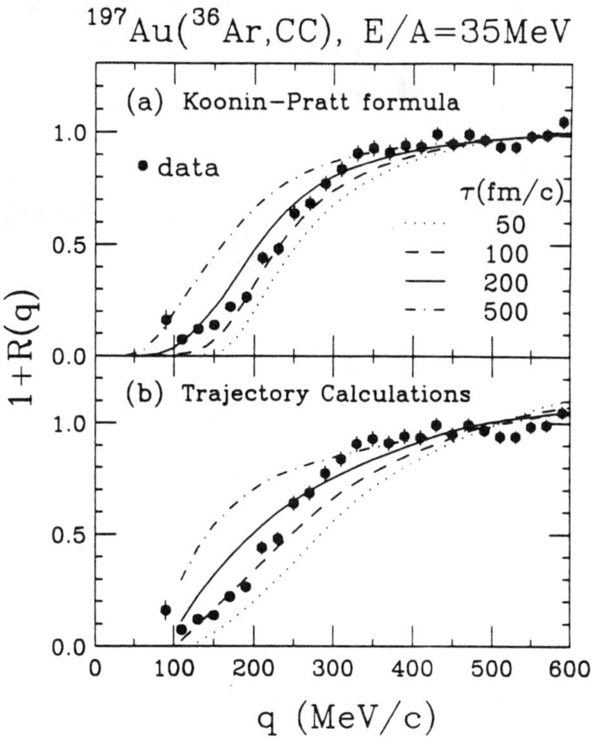

Fig. 3. Experimental C-C correlation function compared to calculations for different emission times

C. Energy spectra

Rather large differences are observed in the shapes of energy spectra gated by different ranges of impact parameter. For illustration, Fig. 5 presents the energy distributions measured for carbon fragments produced in peripheral ($2 \le N_C \le 7$: solid circular points), mid-central ($8 \le N_C \le 11$: open points), and central ($N_C \ge 12$: solid triangular points) collisions. For central collisions, the energy spectra exhibit rather featureless exponential slopes which become steeper at larger emission angles. For peripheral collisions, the energy spectrum at Θ_{lab}=19.5° exhibits a large high-energy shoulder, indicating significant contributions from the decay of projectile-like residues or from strongly damped collisions. Mid-central collisions exhibit an intermediate behavior.

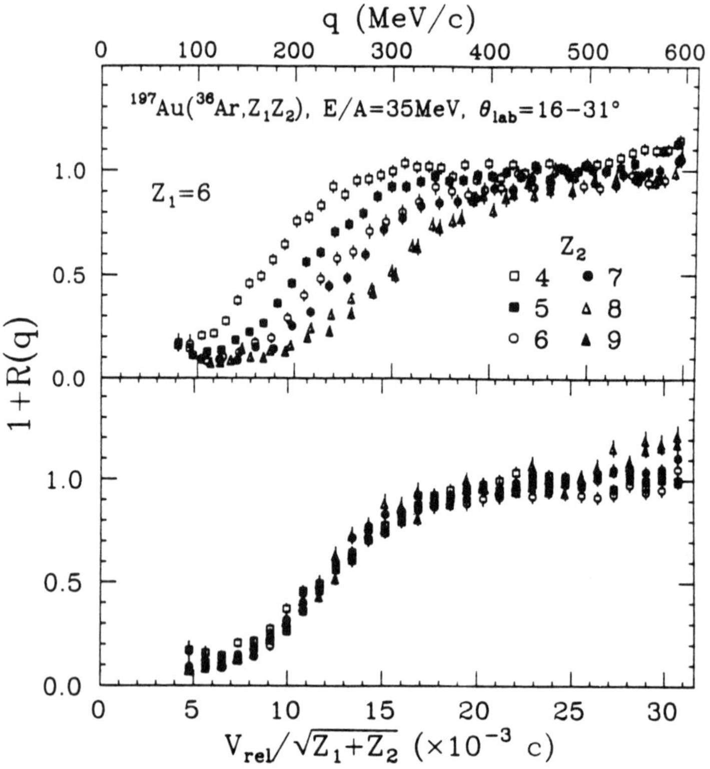

Fig. 4. Dependence of inclusive, energy integrated two-fragment correlation functions on relative momentum q (top panel) and on reduced relative velocity $v_{rel}/\sqrt{Z_1+Z_2}$. One fragment was carbon (Z=6); the atomic number Z_2 of the other fragment is indicated.

The exponential shapes of the energy spectra observed in central collisions suggest emission from a system characterized by a relatively high degree of equilibration. In order to explore whether fragment production in central collisions is consistent with emission from fully equilibrated reaction residues formed in fusion-like processes, we have fitted the differential emission probabilities with a two-source parametrization allowing for contributions from a fusion-like source and an intermediate velocity source. As an example, Fig. 6 shows carbon energy spectra detected at θ_{lab}= 27° gated on central collisions. The solid curve shows the result of a fit with two moving source (which also included data at larger angles, not included in the figures). The dashed and dot-dashed lines represent cross sections of the slow fusion-like and fast non-equilibrium parts respectively. While at higher energy domain the nonequilibrium cross section dominates, we can select different ratios of equilibrium and nonequilibrium cross section by selecting different regions in the energy spectra. While this decomposition into equilibrium and nonequilibrium processes may not be unique, the extracted ratios should, nevertheless, provide

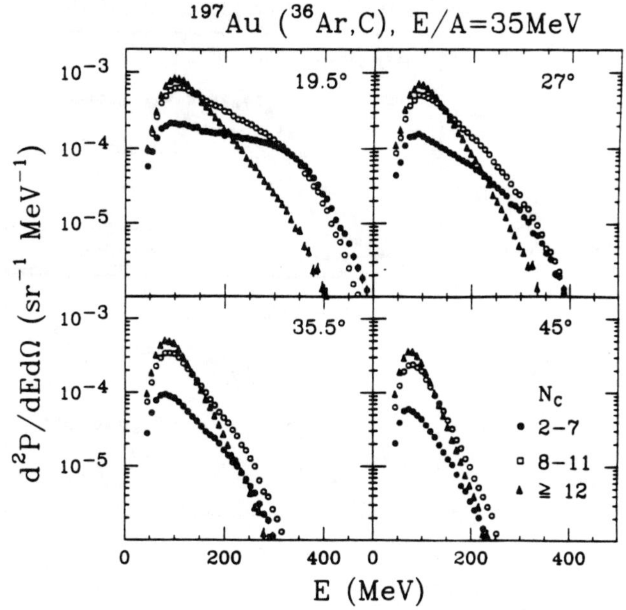

Fig. 5 Energy distributions for carbon fragments produced in peripheral ($2 \leq N_C \leq 7$: solid circular points), mid-central ($8 \leq N_C \leq 11$: open points), and central ($N_C \geq 12$: solid triangular points) collisions. The fragment detection angles are indicated in the individual panels.

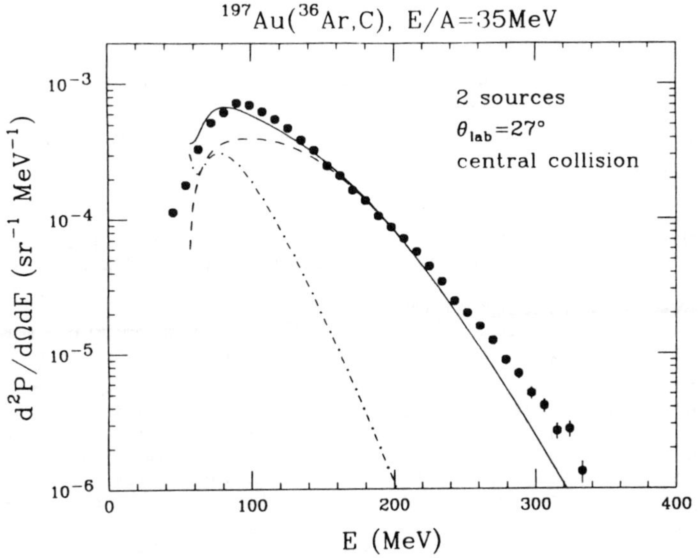

Fig. 6 Carbon energy spectra with 2 moving source fitting, dot-dashed curve is slow fusion-like cross section.

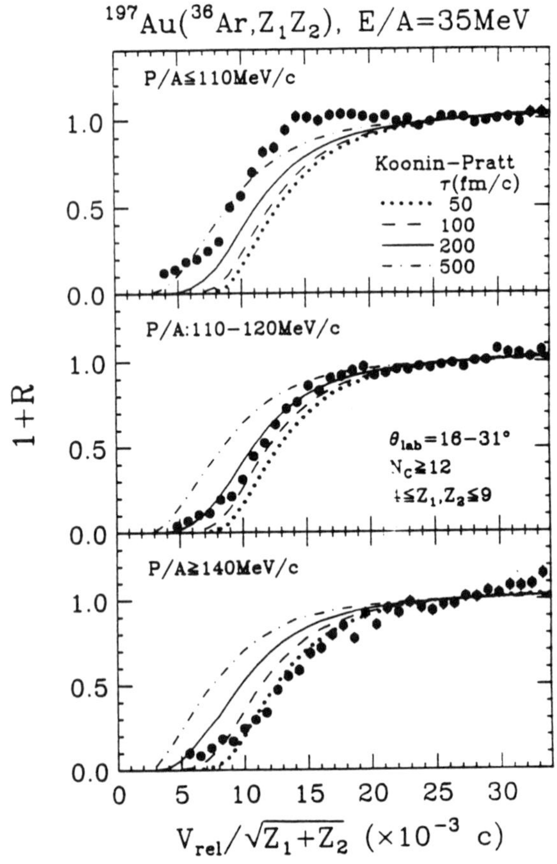

Fig. 7 Two-fragment correlation functions summed over all combinations of Z_1 and Z_2 (with $4 \le Z_1, Z_2 \le 9$) and selected by central collisions ($N_C \ge 12$). The curves represent calculations with the Koonin-Pratt formula for the indicated emission times.

reasonable estimates for the degree of global equilibration characterizing the emission of intermediate mass fragments.

D. Gated correlation function

Figure 7 depicts two-fragment correlation functions gated by central collisions ($N_C \ge 12$) to eliminate contributions from the decay of projectile residues, for three different ranges of P/A, the total momentum per nucleon of the coincident fragment pair (see Fig. 6). The low momentum gate (P/A≤110 MeV/c: top) selects fragment kinetic energies at and below the exit channel Coulomb barrier. Kinetic energies slightly above the Coulomb barrier are selected by the intermediate momentum gate (P/A=110-120 MeV/c: middle). For these two gates, a considerable part of the emission cross section is consistent with equilibrium emission. Kinetic energies significantly above the Coulomb barrier are selected by the high momentum gate (P/A≥140 MeV/c: bottom). In this domain, fragment emission is

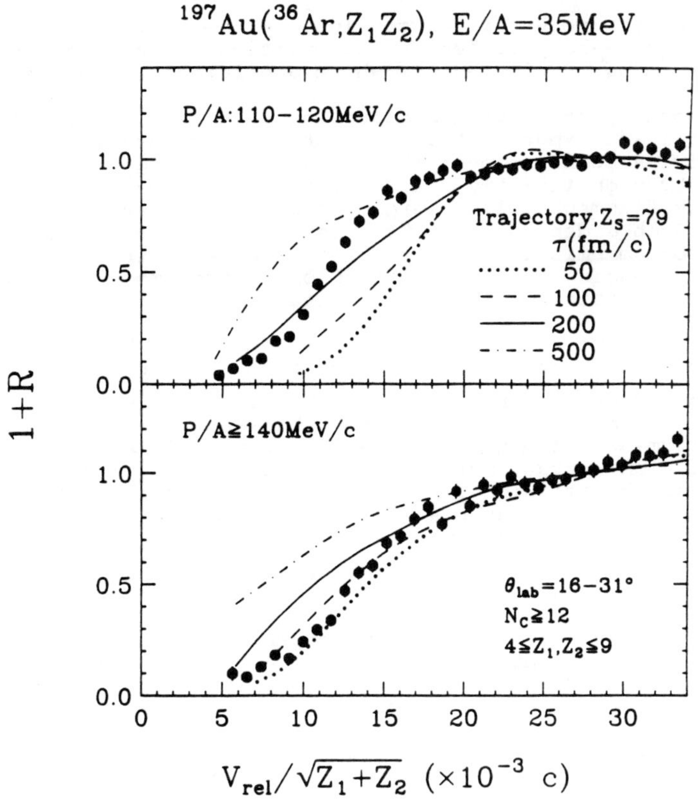

Fig. 8 Two-fragment correlation functions summed over all combinations of Z_1 and Z_2 (with $4 \le Z_1, Z_2 \le 9$) and selected by central collisions ($N_C \ge 12$). The curves represent the results of three-body Coulomb trajectory calculations in which the two fragments are assumed to be emitted from a source of initial charge number Z_S=79.

dominated by nonequilibrium emission processes different from projectile fragmentation. The lines are correlation functions calculated from the Koonin-Pratt formula. From the comparison of measured and calculated correlation function, one can extract emission times of $\tau \lesssim 50$ fm/c for the emission of energetic fragments selected by P/A≥140 MeV/c and considerably larger emission times, $\tau \approx 500$ fm/c, for subbarrier emission selected by P/A≤110 MeV/c. Emission times extracted for P/A=110-120 MeV/c are of the order of $\tau \approx 150$ fm/c.

Fig. 8 shows the correlation functions compared with 3-body trajectory calculations for Z_S=79 which could only be performed for the two higher momentum cuts, P/A=110-120 MeV/c and P/A≥140 MeV/c, as the cut P/A≤110 MeV/c selects mostly energies below the Coulomb barrier.

While qualitatively similar conclusions are drawn from a comparison with this three-body trajectory calculations, they are more consistent with emission time scales which are larger by 25-50%.

For both of the approximations investigated, comparisons between theoretical and experimental correlation functions lead to qualitatively similar conclusions: The emission of energetic fragments is governed by significantly smaller time scales than the emission of low-energy fragments with energies close to the exit channel Coulomb barrier. Such a dependence is consistent with the predominance of nonequilibrium emission processes for energetic fragment emissions and the increasing importance of emission from more equilibrated systems for particles emitted with energies close to the Coulomb barrier.

Emission time scales of the order of several hundred fm/c are consistent with time scales expected from statistical models of compound nuclear decays. For example, the model of ref.[10] predicts average time intervals of $\tau \approx 300$ fm/c between the emission of two carbon fragments from equilibrated heavy nuclei (A=226, Z=93) of 700 MeV excitation energy. Much shorter time scales, $\tau \approx 50-100$ fm/c, extracted for the emission of energetic fragments in central collisions, are incompatible with statistical emission from fully equilibrated heavy reaction residues. These emission times are of comparable magnitude as those predicted[2,3] by dynamical models of fragment production.

This work is based upon work supported by the National Science Foundation under Grant number PHY-89-13815. W.G. Lynch acknowledges the receipt of a U.S. Presidential Young Investigators Award and N. Carlin acknowledges partial support by the FAPESP, Brazil.

† Present address: Instituto de Fisica, Universidade de Sao Paulo, C. Postal 20516, CEP 01498, Sao Paulo, Brazil

REFERENCES

1. G. Bertsch and P.J. Siemens, Phys. Lett. 126B, 9 (1983).
2. G. Peilert et al., Phys. Rev. C39, 1402 (1989).
3. D.H. Boal and J.N. Glosli, Phys. Rev. C37, 91 (1988).
4. E. Suraud, Nucl. Phys. A495, 73 (1989).
5. W. Bauer, Phys. Rev. C38, 1297 (1988).
6. D.H.E. Gross et al., Phys. Rev. Lett. 56, 1544 (1986).
7. J.P. Bondorf et al., Phys. Lett. 162B, 30 (1985).
8. L.G. Moretto, Nucl. Phys. A247, 211 (1975).
9. W.A. Friedman and W.G. Lynch, Phys. Rev. C28, 950 (1983).
10. W.A. Friedman, Phys. Rev. C42, 667 (1990).
11. W. Lynch, Ann. Rev. Nucl. Part. Sci., 37, 493 (1987).
12. A.D. Panagiotou et al., Phys. Rev. Lett. 52, 496 (1984).
13. D.J. Fields et al., Phys. Rev. C34, 536 (1986).
14. D.E. Fields et al., Phys. Lett. B220, 356 (1989).
15. K.G.R. Doss et al., Phys. Rev. Lett. 59, 2720 (1987).
16. D.R. Bowman et al., Phys. Lett. B189, 282 (1987).
17. R. Trockel et al., Phys. Rev. Lett. 59, 2844 (1987).
18. Y.D. Kim et al., Phys. Rev. Lett. 63, 494 (1989).
19. For a recent review, see e.g. D.H. Boal, C.K. Gelbke, and B.K. Jennings, Rev. Mod. Phys. 62, 553 (1990).
20. S.E. Koonin, Phys. Lett. 70B, 43 (1977).
21. W.G. Gong et al., Phys. Rev. C43, 781 (1991).
22. P.A. DeYoung et al., Phys. Rev. C39, 128 (1989).
23. W.G. Gong et al., Phys. Rev. C43, 1804 (1991).
24. W.G. Gong et al., Phys. Rev. Lett. 65, 2114 (1990).
25. Y.D. Kim et al., Phys. Rev. Lett. 67, 14 (1991).

26. Y.D. Kim et al., to be published
27. R.T. de Souza et al., Nucl. Instr. and Meth. A295, 109 (1990).

DYNAMICAL ORIGINS OF INTERMEDIATE VELOCITY LIGHT PARTICLES IN ^{64}Zn + ^{58}Ni at 35 - 79 AMeV

R. Wada, M. Gonin, K. Hagel, J. B. Natowitz
M. Gui, Y. Lou, D. Utley, B. Xiao
Cyclotron Institute, Texas A&M University, College Station Texas 77843

J. C. Angélique, G. Bizard, R. Brou, D. Cussol, A. Kerambrun
J. P. Patry, J. Péter, R. Régimbart, J. C. Steckmeyer, B. Tamain, E. Vient
LPC Caen, ISMRA, IN2P3-CRNS, 14050 Caen, France

G. Auger C. Cabot, E. Crema†, A. Péghaire, F. Saint-Laurent
GANIL, BP5027, 14021 Caen, France

P. Eudes, C. Lebrun
LPN, IN2P3-CRNS et Université de Nantes, 44072 Nantes, France

E. Rosato
Dipartimento di Sience Fisiche, Universita di Napoli, 80125 Napoli, Italy

Y. El Masri
Institute de Physique Nucléaire, Université Catholique de Louvain
1348 Louvain la Neuve, Belgium

†Permanent address: Inst. de Fisica, Univ. de Sao Paulo, CP 20516, 01498 Sao Paulo, Brazil

Abstract

The averaged projected momentum per nucleon and the velocity distribution of light particles are examined in coincidence with projectile-like fragments. A dependence of the left-right asymmetry on the incident energy is observed for the intermediate velocity proton emission. On the contrary the asymmetry of the emission of intermediate velocity α particles shows a similar trend for all incident energies. Possible dynamical models are discussed for the emission of intermediate velocity light particles.

1. Introduction

Intermediate velocity light particles are commonly observed in heavy ion reactions at intermediate incident energies (10 AMeV $\leq E_{beam} \leq$ 100 AMeV). They are characterized by a hard component in the energy spectra at angles around $\sim 20^o \leq \theta_{lab} \leq 90^o$. A detailed study for such particles in coincidence with fragments emitted at small angles in the ^{32}S+Ag reaction at 30 AMeV[1)] has suggested that the dominant emission mechanisms of the intermediate velocity protons and composite particles are different. The emission mechanism of the protons may be explained by a model based on nucleon-nucleon collisions, whereas the composite particles may originate from a classical dynamical mechanism, such as an incomplete deeply inelastic type process. Recently a possible collective flow of the light particles at mid-rapidity has been reported in the reactions of $^{40}Ar +^{51} V$ at 35 AMeV and $^{12}C +^{12} C$ at 50 AMeV [2] and $^{40}Ar +^{27} Al$ at 25 AMeV to 85 AMeV[3]. It has been shown in these experiments that the flow increases when the mass of the emitted particles becomes larger and the impact parameter larger. In this talk we present the results for the peripheral collisions of $^{64}Zn +^{58} Ni$ at 35-79 AMeV in coincidence with projectile-like fragments, where one can define the reaction plane by the fragment direction and the beam axis.

2. Experiment

The experiment has been performed at the GANIL facility, using projectiles of ^{64}Zn incident on a ^{58}Ni target at 35 AMeV, 49 AMeV, 57 AMeV, 69 AMeV and 79 AMeV. The 4π plastic scintillator array MUR +TONNEAU was used for the detection of light particles[4]. MUR is a wall detector array situated at a distance of 2.1m from the target, consisting of 7 rings, having a total of 96 plastic scintillators with thicknesses of 2mm. It covers polar angles from θ=3.2^o to 31.5^o around the beam axis. TONNEAU is a barrel shaped array at a distance of 80 cm from the target, separated into two halves, forward ($30^o \leq \theta \leq 90^o$) and backward ($90^o \leq \theta \leq 150^o$). Each part consists of a total 36 plastic scintillators (2mm thick and 80cm long), each covering an azimuthal angle of $\Delta\phi = 10^o$ and read by photomultiplier tubes at both ends. Particle identification is done by measuring both the energy loss in the 2mm scintillator and the time of flight.

In addition to the light particle detector array, six Si-telescopes are used for measurements of heavier fragments. Each telescope consists of ΔE (100 to 300 μm) and E (3 to 6mm). The telescopes are set at 4.3^o, 7.3^o, 10.9^o 16.0^o, 22.0^o, and two at 28.5^o. A clear Z separation is obtained up to Z=30 for all telescopes. Energy calibration of the telescopes has been performed, using degraded secondary beams from ^{16}O at 95 AMeV.

3. Results

a. Averaged Projected Momentum of Light Particles; $< P_x >/A$

The averaged momenta of protons and α particles projected onto the axis perpendicular to the beam axis in the reaction plane, $< P_x > /A$, were investigated. In Fig.1 the $< P_x > /A$ distributions for protons in coincidence with the fragments of the highest atomic number observed at three forward angles are plotted for the different incident energies. In order to minimize the contribution from the target-like source, the low energy particles are excluded from the $< P_x > /A$ calculation. For all incident energies the $< P_x > /A$

distributions show negative values between the projectile rapidity (y_p) and the center of mass rapidity (y_{cm}) indicating that the emission of protons is enhanced on the opposite side of the beam from the coincident fragments in this rapidity region. At the incident energies below 57 AMeV the distribution was observed to move back to positive values at the rapidity slightly below y_{cm}, indicating enhancement on the same side as the fragments between y_{cm} and the target rapidity (y_t). At 69 and 79 AMeV the distribution stays at negative values.

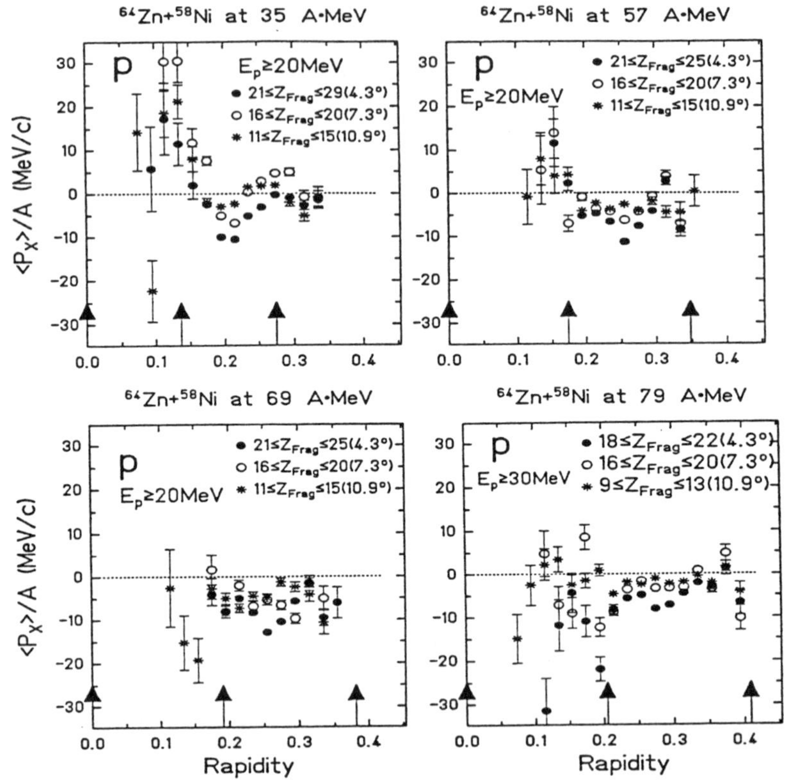

Figure 1. $< P_x > /A$ distributions for protons in coincidence with the fragments of the largest Z group observed at $4.3^o, 7.3^o$ and 10.9^o for the reaction at different incident energies.

Similar plots for α particles are plotted in Fig.2. For all the incident energies, the $< P_x > /A$ distributions show negative values at all rapidity regions. The minimum at the rapidity between y_p and y_{cm} is observed to be smaller when the angle becomes larger, but the basic trend remains the same for the different fragment angles and the different incident energies.

The characteristic trend of the $< P_x > /A$ distributions for protons and α particles show a distinct difference at the lower incident energies and become similar at the higher incident energies. The incident energy dependence of the trend for protons observed above is very interesting and may be a characteristic signature of the emission mechanism.

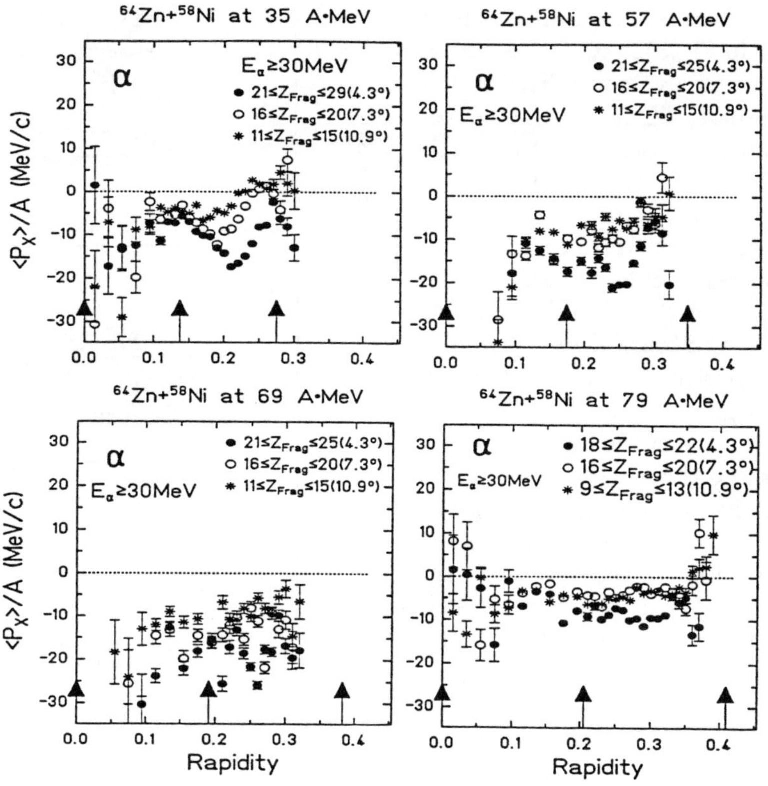

Figure 2. Similar plots to Fig.1 for α particles.

b. Galilean Invariant Velocity Distribution

In order to get a clearer view of the results obtained above, Galilean invariant velocity distributions are studied, in which one may see the contributions from the different sources in detail. In order to make Galilean invariant velocity distributions, particles emitted near the reaction plane are used, namely those with $-30^o \leq \phi \leq 30^o$ and $150^o \leq \phi \leq 210^o$. In Fig.3 velocity distributions for protons (upper) and α particles (lower) in coincidence with the fragments of the largest Z group at 4.3^o are shown. The angles and velocities of coincident fragments are shown by thick arrows. The contribution from the target-like source forms a ridge around the target.

To emphasize the asymmetry, which we observed in the $< P_x > /A$ distributions, the velocity distribution on the same side of the beam from the coincident fragments (the upper half of each figure in Fig.3) is subtracted from the opposite side velocity distribution (the lower half). The subtracted velocity plots for protons and α particles in coincidence with the fragments of the largest Z group, corrected for instrumental offset, are shown in Fig.4 for the different incident energies. The vertical scale of the plots is same for all figures. Contours shown by solid lines have positive values, indicating the enhancement of the emission on the opposite side of the beam from the fragments and those with dashed lines have negative values, indicating the same side enhancement of the emission. For protons at the lower incident energies the subtracted velocity distributions show a maximum at velocities slightly smaller than that of the projectile around $\theta = 15^o$ and a minimum at the intermediate velocities at angles between 40° to 90°. This asymmetry is what we observed in the $< P_x > /A$ distribution in the previous section. When the incident energy increases, the maximum stays more or less the same but the minimum gradually disappears and at 79 AMeV only the enhancement on the opposite side is observed. On the contrary for α particles only the opposite side enhancement is observed and no significant change is seen for the different incident energies.

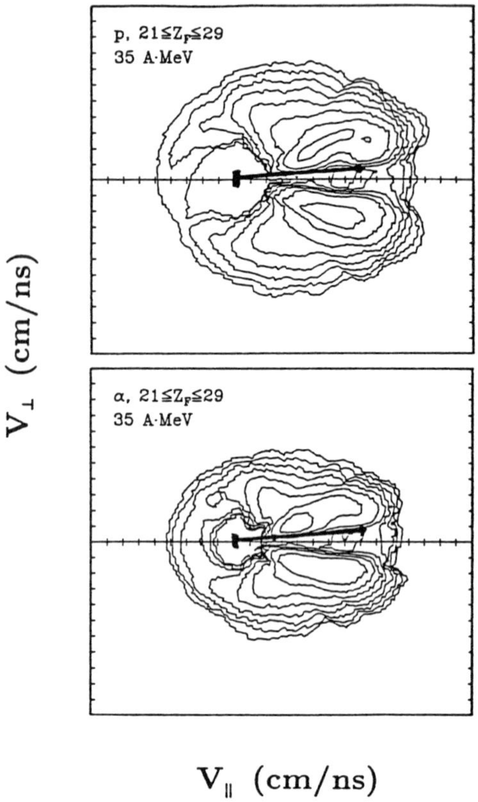

Figure 3. Galilean invariant velocity distributions for protons (upper) and α particles (lower) in coincidence with fragments of Z=21-29 observed at 4.3^o for the reaction at 35 AMeV. The target position is indicated by thick tics on the beam line and units between tics are 1 cm/ns. Vertical scale is logarithmic (factor 2 between contours).

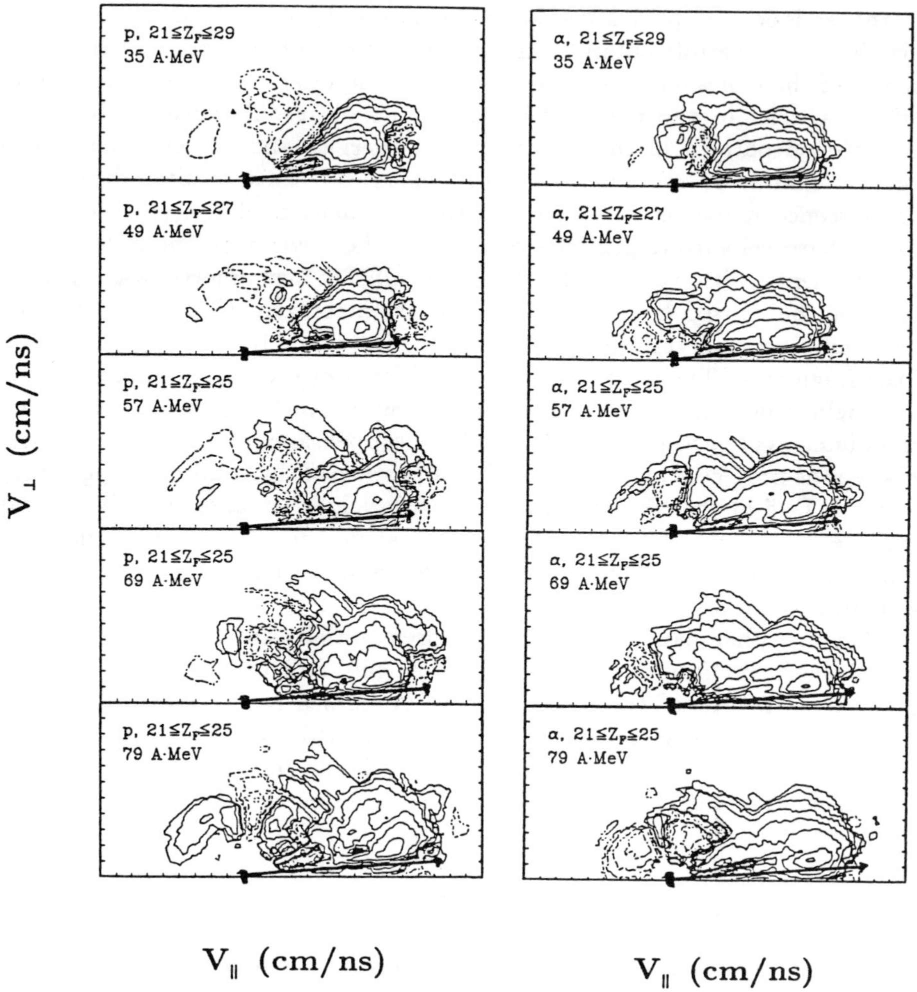

Figure 4. Subtracted velocity distribution for protons (left) and α (right) in coincidence with the fragments of the largest Z group observed at 4.3^o for different incident energies. Solid contures indicate that the emission is enhanced on the opposite side of the beam from the fragments and dashed contours indicate the same side enhancement of the emission.

4. Discussions and Model Calculations

It has already been shown that dynamical mechanisms may be important for the emission of projectile-like fragments and intermediate velocity light particles[1]. The results obtained in the previous section may reveal the dynamic aspects of the emission of light particles. For the intermediate velocity protons and α particles in the peripheral collisions, a clear difference is observed in the emission pattern in the lower incident energies and the emission pattern becomes similar at the higher incident energies. This result indicates that a coalescence model is ruled out for the explanation of the emission of the intermediate velocity α particles at least at the lower incident energies. For the fragments with the smallest Z no significant asymmetry was observed. This may be caused by either the small asymmetry in the central collisions or the ambiguous determination of the reaction plane by the sequential decay of the fragments. The asymmetry observed for protons may originate from the mean field deflection in the compound nucleus[5]. Nucleons in the projectile or target nucleus are accelerated by a sudden disappearance of the potential barrier between the projectile and target at the early stage of the collision going through the mean field of the compound nucleus with velocities large enough to escape from the compound system and be emitted at negative deflection angles. This kind of problem is well treated in the Boltzmann-Uehling-Uhlenbeck (BUU) calculation[6].

On the other hand α particles are preferentially emitted on the opposite side of the beam from the coincident fragments and this emission pattern is almost independent of the incident energies studied here. This opposite side enhancement of the α particles may be explained by a classical dynamical model.[1] Here we first discuss the proton emission in the frame work of the BUU calculation and secondly the α particle emission by the classical dynamical model.

a. BUU Calculation

To investigate the asymmetry of the intermediate velocity proton emission, BUU calculations are performed for peripheral collisions. The intermediate velocity particles are expected to be emitted in the early stage of the collision. In order to minimize the contribution from sequential emissions, the calculation was stopped at an early stage of the collision.

In Fig.5 the $< P_x > /A$ distributions of the emitted test particles are plotted for different incident energies at the impact parameter b=7fm. In the upper part of the figure results for different equations of state are compared at 35 AMeV. The "soft" equation of state (EOS) has the compression parameter of k=200 MeV and the "stiff" EOS has k=375MeV. The $< P_X > /A$ distribution does not depend on the equation of state, as seen in the figure (solid and open symbols). This reflects the fact that the asymmetry of the emission of the prompt emitted particles is mainly determined by the mean field at the early stage of the collision and the effect of the compression becomes important at later stages. In the lower part of Fig.5 results for the reaction at 79 AMeV are shown. One can see a clear evolution of the trend between two incident energies, which is similar to those observed for the proton emission in Fig.1.

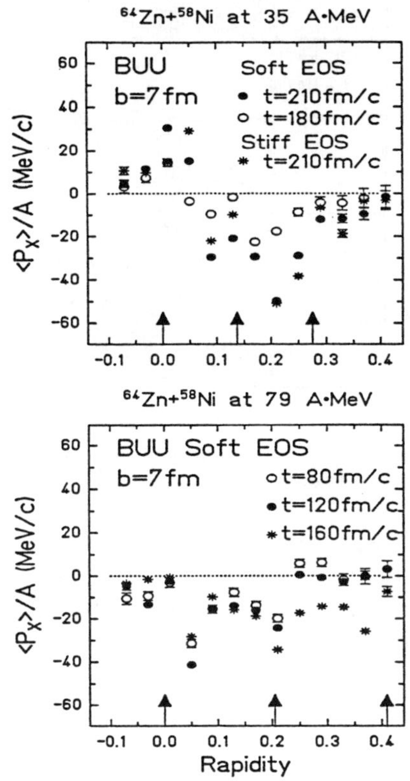

Figure 5. Calculated $< P_x > /A$ distributions for the impact parameter of 7 fm with different conditions indicated in the figure for the reaction at 35 AMeV (upper) and 79 AMeV (lower).

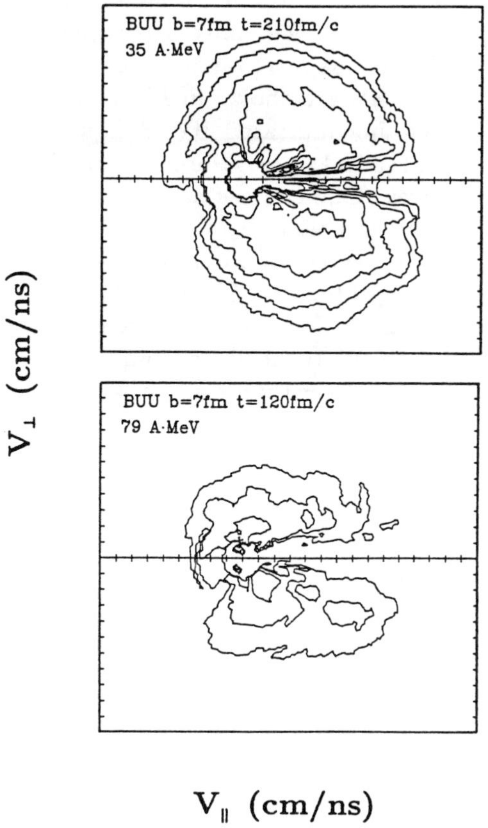

Figure 6. Similar plots to Fig.4 for protons in coincidence with the fragments with Z=6-10 observed at 4.3° for the reaction at 35 AMeV.

In Fig.6 invariant velocity distributions of the emitted test particles are plotted at a given time for b=7 fm. At 35 AMeV one can see a clear contrast in the emission pattern at forward and backward angles in the center of mass system. Most of the test particles observed in the lower half of the figure originate from the projectile and are deflected at negative angles by the mean field. Those in the upper half originate from the target. At 79 AMeV one can still see the asymmetry at the forward direction but not at the backward angles. This is more clearly shown by a subtracted velocity distribution made in the same way as in Fig.4. The subtracted velocity distributions of the emitted test particles for the reaction at 35, 57 and 79 AMeV are shown in Fig.7 at a given time indicated in each figure. The vertical scale is the same for all figures. One can see a clear trend in the emission pattern similar to that observed for intermediate velocity protons in Fig.4, although the angles for the same side enhancement (dashed contours) are slightly larger than those observed in the experiment.

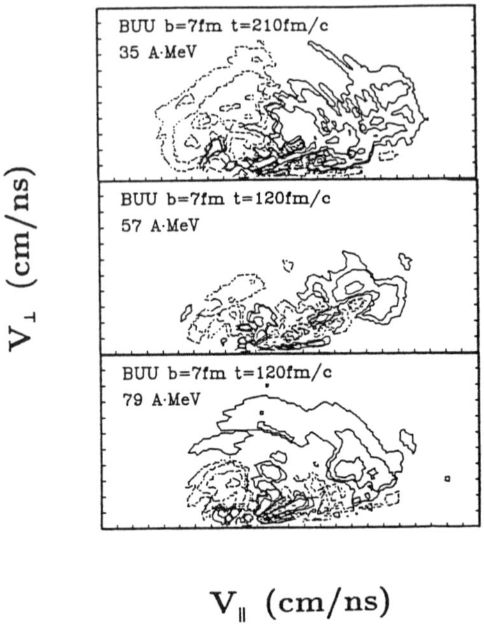

Figure 7. Subtracted velocity distributions for the emitted test particles at different incident energies for b=7fm and a given time.

B. Classical Dynamical model

Classical dynamical models have been applied to the study of preequilibrium α particle emission and fragment energy spectra at forward angles[7-9]. This model has been successfully applied to an intermediate velocity composite particle emission in coincidence with projectile-like fragments for the reaction $^{32}S+Ag$ at 30 AMeV[1]. In this model a projectile breaks into two fragments, 1 and 2, near the target. Then one outer piece from the target is scattered at a positive angle and the other is scattered at a negative deflection angle because of a friction force. The Monte Calro version of the model used here is described in detail in Ref.1. In the calculation three parameters are generated by random numbers, which are the impact parameter b, the initial orientation angle, τ, of fragments 1 and 2 to the beam axis and the angular momentum loss factor F. The angular momentum loss factor is changed from the sticking limit to the rolling limit (F=5/7). The calculations have been done from 35 AMeV to 79 AMeV for different projectile decay channels, such as $\alpha +^{60} NI$, $^3Li +^{58} Co$ and $^{12}C +^{52} Cr$. For all cases the three body decay channel is observed.

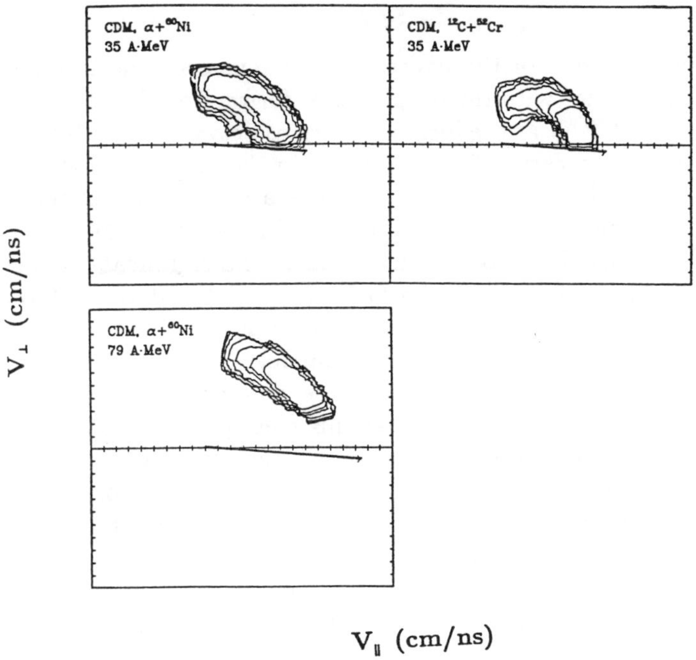

Figure 8. Velocity distributions of the negatively deflected fragments for the classical dynamical calculations for the breakup of projectile into α + ^{60}Ni and ^{12}C + ^{52}Cr at 35 AMeV (upper) and α + ^{60}Ni at 79 AMeV (lower). The angle and velocity of the positively deflected fragments are indicated by arrows.

In Fig.8 calculated velocity distributions of the light fragment are shown for the breakup of the projectile into $\alpha +^{60} Ni$ and $^{12}C +^{52} Cr$ at 35 AMeV and $\alpha+^{60}Ni$ at 79 AMeV. Events which have the larger fragment scattered positively at angle between 3° and 6° are selected. The angular momentum loss factor F is taken as that of the rolling limit. Similar distributions are obtained for the three cases. The calculation does not take into account the sequential decay of the primary deflected fragments, which are normally excited. Since the essential velocity distribution, however, does not depend so much on different breakup channels and the sequential decays are expected to have the same average velocity distribution as the residual nucleus, meaningful comparisons between the calculated results and the experimental ones obtained in coincidence with the fragments of the largest Z group may be made for the intermediate velocity α particles. At 35 AMeV the calculated results seem reasonable but at 79 AMeV the calculated distribution spreads to much larger velocities and larger angles. This may suggest that the angular momentum loss factor is smaller (angular momentum loss increases) for the higher incident energies.

5. Summary

Dynamical aspects of the emission mechanism for the intermediate light particles in coincidence with the fragments at forward angles are investigated for the reaction $^{64}Zn+^{58}Ni$ at the incident energies between 35 AMeV to 79 AMeV. A clear difference is observed for asymmetry in the emission of the intermediate velocity protons and α particles. For protons a dependence of the asymmetric emission on the incident energy is observed in the $< P_x > /A$ distributions and the velocity distributions. On the contrary no significant difference in the asymmetry of the emission is observed for α particles at the different incident energies. A clear contrast observed in the asymmetry of the emissions between the intermediate velocity protons and α particles at the incident energy of 35 AMeV rules out a coalescence model as the dominant emission mechanism of the intermediate velocity α particles at this energy region. The characteristics of the asymmetry in the proton emission for the different incident energies are well explained by BUU calculations, indicating that the observed asymmetry originates from the mean field deflection of the nucleons at the early stage of the collision. A classical dynamical model is applied to describe the observed asymmetric emission for the intermediate velocity α particles. Calculated results obtained are reasonable, although the model may be too simple for detailed comparisons.

Acknowledgments

We thank the staff of the GANIL Accelerator facility for their excellent support during our run. This work was supported by the U. S. Department of Energy under Grant DE-FG05-86ER40256, the National Science Foundation, the Robert A. Welch Foundation.

REFERENCES

1. R. Wada et al. Proc. of XXV Zakopane School on Physics, Zakopane, 1990, p196, World Scientific, Singapore
2. C. A. Ogilvie et al., Phys. Rev. **C40** (1989) 2592.
3. J. P. Sullivan et al., Phys. Lett. **B249** (1990) 8.
4. G. Bizard et al.,Nucl. Inst. & Meth. **A244** (1986) 483
5. K. Niita et al.,to be published.
6. J. Aichelin et al., Phys. Rev. **C31** (1985) 1730, J. Aichelin, Phys. Rev. **C33** (1986) 537
7. K. Moeling et al, Phys. Lett **B203** (1988) 210
8. A. Budzanowski et al, in proceedings of the Symposium on Nuclear Dynamics and Nuclear Disassembly, Dallas , Texas, 1989, p58 editted by J. B. Natowitz (World Scientific, Singapore, 1989)
9. C. Gregoire et al., Phys. Lett **B127** (1983) 308

SUPER HEAVY ELEMENTS

REACTIONS LEADING TO THE SYNTHESIS OF HEAVY ELEMENTS

Yuri Ts. Oganessian

Laboratory of Nuclear Reactions,
Joint Institute for Nuclear Research,
Head Post Office, P.O.Box 79, 101000 Moscow, USSR

1. INTRODUCTION

I have used as the basis of the present report a part of a more general review presented by me at "The Robert A.Welch Foundation Conference on Chemical Research XXXIV Fifty Years with Transuranium Elements", held October 22-23, 1990 in Houston (USA). This part deals with the mechanism of different type reactions bringing to the formation of heavy elements of the end of the Periodic Chart of elements. The problem connected with limitations on the fusion of complex nuclei has been presented in greater detail though still briefly.

The review includes the results of recent experiments carried out at the Laboratory of Nuclear Reactions, JINR and discusses some possibilities of carrying out experiments investigating the dynamics of complex nuclei interaction processes.

The issues connected with studying radioactive and atomic (chemical) properties of heavy elements, present status of these research and prospects have been excluded from the review intentially.

In case of need they can be found in the abovementioned review as well as in some other reviews of the Proceedings of the Houston conference which are to be out of print soon.

2. REACTIONS

Eight transuranium elements were produced by neutron and light charged-particle induced reactions [1]. Moving further along the line of long-lived nuclides one succeeded in synthesizing nuclei with Z through 100 in high-flux reactors. The isotope ^{257}Fm ($T_{1/2}$ = 100 days) is the last one produced by that technique since the ^{258}Fm nuclei undergo spontaneous fission with $T_{1/2}$ = 3×10^{-4} s (ref. [2]). This limit could not be overcome in a high-density neutron flash in nuclear explosions [3].

2.1. Hot fusion with heavy ions

A move into the region of transfermium elements became possible in complete-fusion reactions induced by heavy ions although it was associated with considerable difficulties.

Significant limitations are imposed by the high excitation energy of the compound system ($E_x \sim$ 40-50 MeV), to be removed by the emission of 4 or 5 neutrons. In the process of neutron emission, as a result of the strong competition between fission and neutron evaporation, only a small fraction (10^{-8} - 10^{-10}) of the nascent nuclei de-excite.

At the same time, five transmendelevium elements with Z through 106 were successfully produced by bombarding targets made of the isotopes of actinides ranging from Pu to Cf by ions from carbon to neon.

According to A.Ghiorso et al. [4] the production cross section of the heaviest nuclide $^{263}106$ is equal to about 2×10^{-34} cm^2 in the reaction ^{249}Cf (^{18}O, 4n)$^{263}106$. Attempts to synthesize an element 108 isotope in the reaction ^{249}Cf(^{22}Ne, 4n)$^{266,267}108$ have yielded only the upper limit of its production cross section at a level of $5 \cdot 10^{-35}$ cm^2.

In this situation everything depends on the excitation energy of the composite system.

In fact, within the framework of statistical models it is possible to calculate the probability of the emission of a preset number of neutrons as a function of the compound nucleus excitation energy. These calculations can be carried out taking into account other decay channels including fission which is the dominant mode of decay for heavy nuclei. Following the simple expressions used as early as in the 1960's [5,6], the cross section of the (HI, xn) reaction can be presented in the following form

$$\sigma_{xn}(E^*) = \sigma_{CN}(E^*) \cdot P_x(E^*) \cdot \sum_{i=1}^{x} \frac{\Gamma n_i}{\Gamma n_i + \Gamma f_i} ,$$

where σ_{CN} is the compound nucleus production cross section, $P_x(E^*)$ is the probability of the emission of x neutrons from a nucleus with the excitation energy E* , and Γ_n and Γ_f are the partial widths (probabilities) for neutron emission and fission, respectively.

As (Γ_n/ Γ_f) 0.01 for the transfermium elements, the cross section for producing evaporation residues strongly depends on the excitation energy E* that determines the number of stages in the neutron evaporation cascade process.

Finally, in the reactions of this kind the cross section of evaporation products is defined by the competition of neutron evaporation channels and fission channels at different values of excitation energy and of compound nucleus angular momentum. These processes in essence differ strongly from each other. The probability of neutron evaporation and that of light charged particles according to statistical calculations should grow with growing temperature. Fission probability grows also with the growth of nucleus excitation energy. Nevertheless, in contrast to neutron emission, fission is a strictly collective process during which a nucleus undergoes significant shape changes on all its way from initial configuration which is almost spherical to a strongly deformed figure in the discontinuity point. These substantial changes of the shape require a certain time and like it is demonstrated further below this time period is relatively big (about 10^{-20} sec). Consequently, the value Γ_n increases substantially with growing temperature and at the energies exceeding 50–60 MeV, as follows from recent experiments, the initial stage of nucleus deexcitation is defined to large extent by cascade neutron evaporation. This issue will be considered in greater detail at this symposium in the report of Dr. G.M.Ter-Akopian. At the same time in the case of heavy nuclei the increase in the compound nucleus temperature brings about the increase in the number of cascades and, consequently, the decrease in the yield of the reaction's final product (HI,xn).

On the contrary, the maximum yield can be expected at the minimum temperature of the compound nucleus. But the excitation energy of the compound nucleus in turn cannot be smaller than the value of $E^* = B_c - Q$ determined by the Coulomb barrier and the Q-value of the reaction.

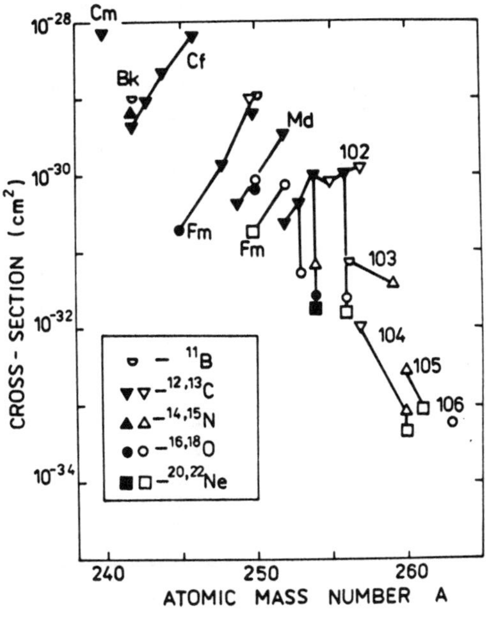

Fig. 1. Cross sections of producing transcurium elements in hot-fusion reactions. Experimental points correspond to the evaporation residues formed as a result of the emission of 4 neutrons and γ-rays. Various symbols stand for various target-projectile combinations (indicated in the figure).

It is clear that a considerable gain in production cross section for evaporation residues can be achieved by lowering the Coulomb barrier of the reaction. This will inevitably lead to strongly asymmetric reactions and to the use of the heaviest possible element as target material and of the lightest possible projectile (Fig. 1). Therefore the accumulation in high-flux reactors of heavy elements now used as target material has long played a significant role for this method of synthesis.

2.2. Incomplete fusion

In principle, the incomplete fusion of the interacting nuclei offers the possibility of reducing the nuclear energy. In this type of reaction the energy is distributed between two (or more) reaction products having proton and neutron numbers somewhat different from those of the initial target and projectile nuclei.

It has been shown experimentally that the interacting nuclei can transfer many nucleons even near the Coulomb barrier [7]. The probability of nucleon transfer is of a stochastic nature and depends on the energy available for the inelastic channel of the reaction. Therefore, as the number of the nucleons transferred from the projectile to the target nucleus increases, the excitation energy of the heavy product also grows.

As a result, the survival probability for heavy nuclei turns out to be extremely low. This seems to account for the fact that, irrespective of the projectile mass, the final heavy nuclei have atomic numbers not far from the target one [8–10] (Fig. 2).

In this situation it is necessary to use the heaviest available nuclei as target material. In this direction experiments were carried out by K.Hulet et al. [11] who succeeded in synthesizing new heavy isotopes with Z = 101, 102 and 103 by bombarding ^{254}Es nuclei by ^{18}O and ^{22}Ne ions. The properties of these nuclides turned out to be rather interesting (bimodal spontaneous fission, the high stability of doubly-odd nuclei with Z = 159, and so forth).

However, the difficulties in accumulating the target material ^{254}Es ($T_{1/2}$ = 276 days) and the drastic cross section decrease in the Z > 103 region are the main obstacles to the synthesis of new elements.

2.3. Cold fusion

Reverting to the complete-fusion reaction between nuclei again we note that the excitation energy of the compound nucleus can be decreased substantially if the Q-value of the reactions is increased. This, in principle, can be achieved only for more symmetric reaction partners. The most advantageous case is when the interacting nuclei lie near the closed nucleon shells. Therefore, instead of actinide isotopes, one can use the magic nuclei ^{208}Pb or ^{209}Bi as target material and ions heavier than argon as projectiles [12].

After it has been demonstrated that reactions of Pb and Bi nuclei with Ar, Ti and Cr ions can result in the production of isotopes with Z = 100–106 [13], this kind of reactions found a wide application in synthesizing heavier elements up to Z = 109 [14,15]. In "cold" fusion reactions the composite system has a low excitation energy so that its de-excitation may be accompanied by the emission of as few as one or two neutrons. However, the advantage indicated above exists until the limitations to the process of fusion of heavy nuclei begin to manifest themselves.

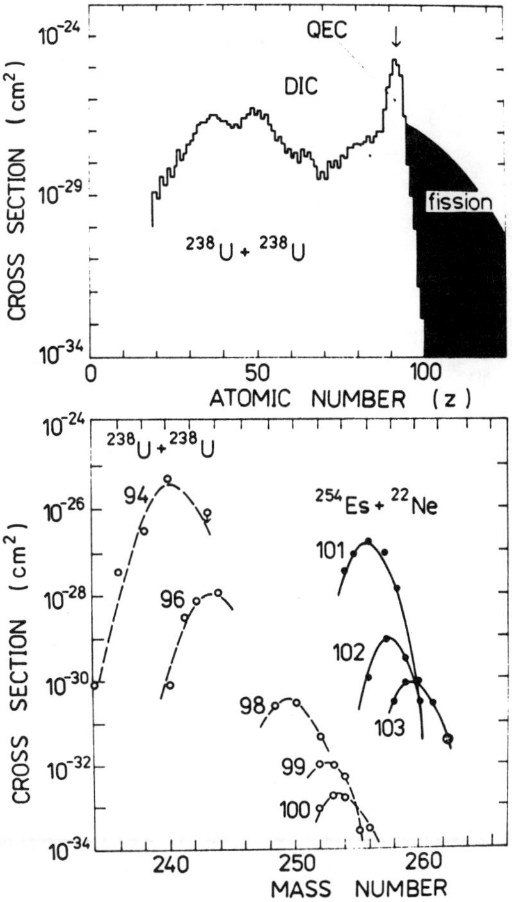

Fig. 2. Cross sections for producing various elements (upper part) and isotopes (lower part) in the reaction $^{238}U + ^{238}U$ near the Coulomb barrier. The closed circles are experimental cross sections of transfermium elements production in the reaction $^{254}Es + ^{22}Ne$. The line is drawn through experimental points to guide the eye.

Unfortunately these limitations appear already in systems with Z = 104 (Fig. 3).

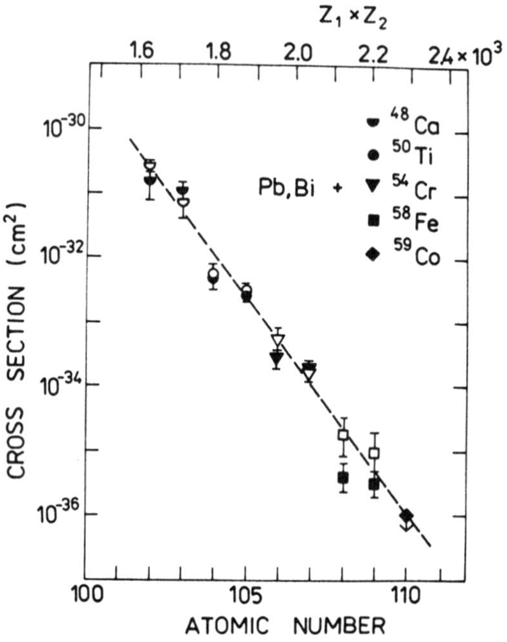

Fig. 3. Cross sections of producing transfermium elements in cold-fusion reactions. Experimental points correspond to the evaporation residues formed as a result of the emission of one neutron and γ-rays. Various symbols stand for various ions (indicated in the figure).

The experimental results obtained at Darmstadt and at Dubna suggest that with the increasing atomic number of the projectile (and consequently, of the composite system) the cross section for forming "cold" fusion products decreases substantially. The heaviest nuclide was produced in the ^{209}Bi(^{58}Fe,n)^{266}109 reaction. The attempts to synthesize element 110 in ^{59}Co and ^{64}Ni bombardments have set only the upper limits of cross sections at a level of 1–4 pb [16,17].

In Table 1 to follow the conclusions drawn from the above analysis of the possibilities of synthesizing elements by different types of nuclear reactions are summarized.

Table I

Reaction	Projectile	Limitation	Heaviest nuclide Z	Heaviest nuclide N
Sequential neutron capture	n_o	break of chain	100	157
High-flux neutron pulse	n_f	spontaneous fission of Fm isotopes $N \geq 158$	100	157
Light-ion induced fusion $Z_1Z_2 \leq 800$	22	low yield of evaporation res. $E_x \sim 40$ MeV	106	157
Incomplete fusion	22	lack of target heavier than ^{254}Es	103	159
Cold fusion $1500 \leq Z_1Z_2 \leq 2200$	50	dynamical hindrance of fusion $E_x \leq 20$ MeV	109	157

As is seen from Table 1, a move from the neutron capture to heavy ion-induced reactions has made it possible to synthesize 9 transfermium elements by increasing Z and almost without changes in the neutron number (N = 157).

According to the calculations of Sobiczewski et al. [18] the transactinide isotopes including even-even ones are expected to be unstable against α-decay ($T_{s.f.} > T_{\alpha}$) over a wide and accessible range of the Z and N values. At the same time the spontaneous fission probability may undergo strong variations with varying Z and N.

One can believe that many experimentally observed "uncommon" properties of heavy nuclei, such as a strong change in the systematics of spontaneous fission half-lives of the element 104 isotopes compared with the actinide elements, the unexpectedly high stability of the even-even nuclides $^{262}102$, $^{260}106$ and $^{264}108$, and others result from the transition from the N = 152 subshell to that with N = 162.

According to calculations, the maximum effect of the N = 162 neutron subshell should manifest itself in the region of fairly heavy nuclei with Z > 108. This prediction needs to be verified experimentally, perhaps, in the first place by using the two methods of elements synthesis in the hot-fusion and cold-fusion reactions each with its own advantages, as our systematics presented in Table I suggests.

However, the intermediate variant associated with the use of actinide targets and of relatively heavy projectiles ($A \sim 40$–50) may prove most suitable in terms of producing isotopes with a maximum neutron excess.

It is unlikely that such combinations can give a substantial increase in the cross section for forming evaporation residues from the compound nucleus. This, however, required more thorough investigation since, as we can see below, the synthesis of superheavy elements can be implemented only in the fusion reactions between relatively heavy nuclei with a maximum neutron excess.

3. LIMITATIONS AND POSSIBILITIES OF FUSION REACTIONS

3.1. Theory

In recent years several theoretical approaches of macroscopic type have been elaborated to investigate the dynamics of the fusion process.

The calculations carried out by W.Swiatecki [19] and subsequently by other authors [20–22] show that as the charges (and, to a lesser extent, masses) of the partner nuclei reach certain values, their further increase leads to a considerable enhancement of the fusion barrier and, consequently, the excitation energy of the composite system. This can serve as a qualitative explanation of the strong decrease in cross sections for forming "cold" fusion reaction products.

In this situation it is natural to try more asymmetric target-projectile combinations in which the Coulomb interaction energy decreases noticeably.

In terms of the synthesis of, say, element 110 this would necessitate the use, along with the reaction $^{208}Pb + {}^{64}Ni \rightarrow {}^{272}110$ ($Z_1Z_2 = 2300$), of more asymmetric systems such as $^{235}U + {}^{40}Ar \rightarrow {}^{275}110$ ($Z_1Z_2 = 1700$) of $^{249}Cf + {}^{26}Mg \rightarrow {}^{275}110$ ($Z_1Z_2 = 1100$).

It is, however, difficult to foretell how effective the proposed systems will prove to be.

A slight variation in the model parameters changes the situation to a great extent. For the U + Ar → 110 reactions, the calculations of W.Swiatecki et al. [19], and of R.Davies et al. [21] suggest that the fusion reaction threshold should not increase whereas the calculated results of J.P.Blocki et al. [22] completely eliminate the formation of evaporation residues with $Z = 110$ ($E_x > 100$ MeV). At the same time, the reaction threshold decreases significantly in the approach used by P.Fröbrich [23]. Clearly, neither of the calculations is qualitatively perfect so far.

On the other hand, the use of targets of uranium isotopes instead of ^{208}Pb leads to a decrease in the Q value and hence to an about 20 MeV increase in the excitation energy of the composite system. In this case fission begins to play a more essential role in the process of composite system de-excitation. It should also be noted that the shell structure effects diminish with growing excitation energy and, as a result, heavy composite systems become "barrierless". The mechanism of the formation and decay of such systems has not been clarified as yet. It can be assumed that in this case the probability of the formation of final products should depend on dynamical factors to a considerable extent.

From the moment of touching up to the formation of a compact mononucleus similar to a compound nucleus the composite system undergoes considerable changes of single particle and collective nature. Moreover, the process as a whole proceeds on the potential energy surface the configuration of which depends on the excitation energy. In the absence of the fission barrier in a heated nucleus the probability for its survival is practically equal to zero. However, this probability may exist if the composite system is cooled in the process of collective motion by emitting neutrons and gamma-rays. Then there will appear structural features due to the nuclear shell effects in a deformed nucleus and the system will cease to be "barrierless" (Fig. 4). The probability for this process to occur depends on the time of neutron evaporation (which can be calculated in the framework of a statistical model) and on the time of collective motion (which is practically impossible to calculate). In other words one needs experimental data on the development of the fusion process from the moment of contact of interacting nuclei to the formation of compact mononucleus.

These data are practically absent since in the experiment one defines, essentially, the end of the fusion process that is the probability of production of separate products, and of its decay.

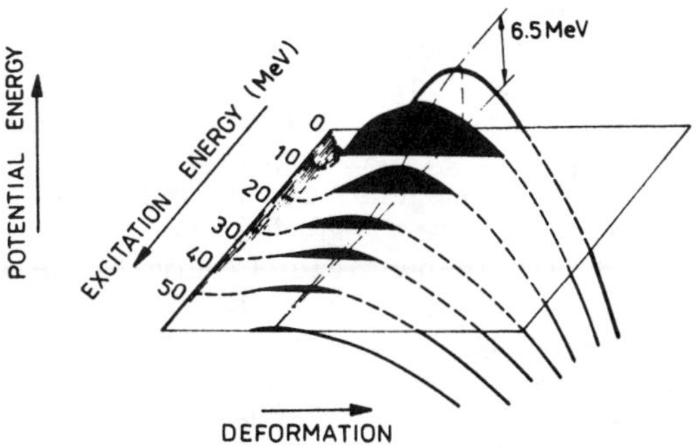

Fig. 4. Calculated fission barriers of the nucleus $^{260}104$ at excitation energies from 0 to 50 MeV (at 10 MeV intervals).

At the same time it is known that the insignificant variation in the nucleon number of interacting nuclei sometimes change significantly the yield of evaporation products (experiments of cold fusion with isotopes of Ca, Ti, Cr). One cannot exclude the possibility that the nucleon content of the partners can produce significant effect at the initial stage of the mononucleus formation process, when strong nucleon exchange takes place in the contacting nuclei. Probably, like it has been mentioned in literature neutron exchange is proceeding in the most intense way (due to the absence of the charge) and this defines the initial conditions of further movement into the region of strong Coulomb interaction (to more compact configurations).

In this connection it would be interesting to define to what extent the probability of neutron exchange correlates with the probability of evaporation products formation. We could use as examples the reactions of ^{209}Bi + $^{44,48}Ca$ in which at ion energies near the Coulomb barrier of the reaction (it's nearly the same for ions of ^{44}Ca and ^{48}Ca) there have been defined cross sections of Bi isotopes formation with masses from 210 to 217 (transfer of 1,2,3,...,8 neutrons from the ion to the target nucleus) and cross sections of evaporation products (in the reaction (HI, 1,2 n), in the given case - of 103 element's isotopes with masses from 251 to 256.

Of course, neutron motion is only the initial stage of the consequent more complicated and longer collective motion. Its development in time will define the end of the fusion story.

In this connection it is of interest to study the nuclear fission process which is estimated according to the emission of pre-fission neutrons [24]. Fig. 5 presents the experimentally obtained average numbers of neutrons emitted in each fission event until the moment of nuclear separation to two fragments[25,26].

From the data presented it follows that pre-fission neutrons are emitted with a high probability even at low excitation energies of highly fissile nuclei. This, in turn, indicates the relatively slow development of deformation. A similar picture can occur in the fusion of complex nuclei then in the process of fusion the system can manage to perform de-excitation via neutron emission even before reaching its final configuration. Really, single-particle motion in the nucleus is a relatively fast process. The produced nucleus may be regarded as an equilibrium system with respect to the single-particle (Thermal) degrees of freedom long time before the completion of a slower collective process of nuclear shape changing. Emission of a neutron decreases the excitation energy by ~ 10 MeV (of 2 neutrons - by 20 MeV) which changes immediately the relief of the potential surface because of the nuclear shell effect. This circumstance should effect substantially the probability of further movement towards the formation of a compact mononucleus. Naturally, the probability of single-neutron emission can have a small value, of 2 neutrons - a square of times smaller value, though we are generally speaking about very small cross sections - of several picobarns. Thus, in constrast to the classical concept of the compound nucleus formation and decay developing in a certain time sequence, here the two stages can proceed simultaneously. Then the probability for evaporation residue production will depend on the initial conditions and it can be determined only experimentally.

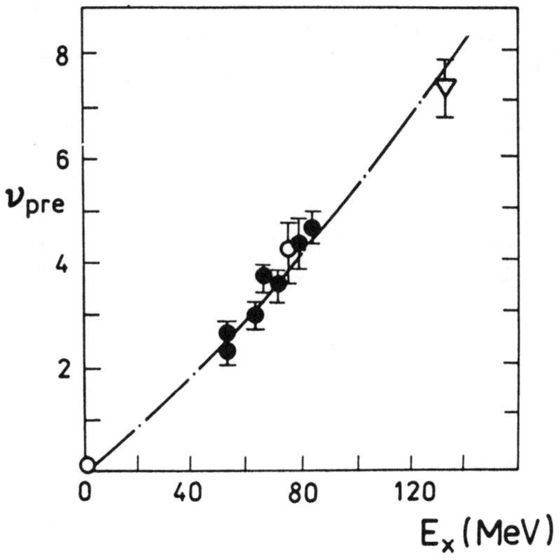

Fig. 5. The number of neutrons emitted prior to nuclear fission into two fragments at different excitation energies. Experimental data correspond to nuclei with Z from 98 to 100. Closed circles, open circles and triangles are data taken from refs. 25,26.

It is very difficult to define experimentally at which exactly moment a neutron is emitted from the composite system since the energy spectrum of neutrons and their angular distribution are practically independent of the nucleus shape.

The emittance probability as well as gamma-irradiations and charged particles (i.e. alpha-particles) characteristics may appear to be more sensitive to the nucleus shape. Since an alpha-particle immediately carriers away out of the nucleus a big portion of the excitation energy, the investigation of this process is of great interest.

Unfortunately, in cold fusion experiments at the irradiation of Pb and Bi nuclei with ions heavier than Ar there have been observed no reactions emitting charged particles. Their boundary cross sections were by two orders of magnitude smaller than the main reaction channel corresponding to the emission of one or two neutrons.

At the same time in the recent experiments of Nomura et al. [27] there have been observed emissions of ^{244}Es nuclei in the reactions of ^{209}Bi + ^{40}Ar with a relatively big cross section (several dozens of nanobarns). According to the authors the ^{244}Es nuclei are produced in the reaction of ^{209}Bi (^{40}Ar; α , n) ^{244}Es since the experiment has been carried out at the conditions corresponding to the maximum cross section of formation and registration efficiency of reaction products with emission of alpha-particles.

Basing on these data and on the earlier data obtained for lighter targets and with lighter bombarding particles the authors have come to the conclusion that the reactions of the type (HI; α, xn) are more promicing than (HI; xn) at going deeper into the area of superheavy elements. There has been undertaken an attempt to test this statement via an independent experiment by Yu.A. Lazarev et al., where the yield of ^{244}Es nuclei has been defined by the radiochemical method over the yield of ^{240}Cm daughter product.

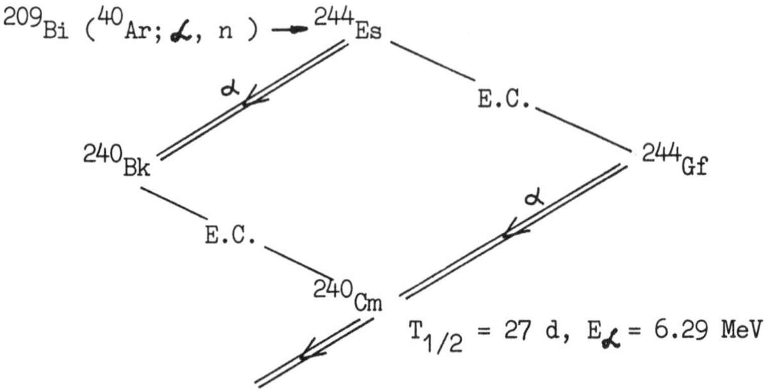

The energy of ^{40}Ar ranged from B_c to 5.75 MeV/nucl.; the beam doze was $\sim 10^{18}$ ions.

In the alpha-spectrum of the Cm fraction there were detected several dozens of alpha-particles from the decay of ^{240}Cm, which, according to preliminary estimations, corresponded to the cross section of $0.4^{+0.3}_{-0.2}$ nb.

By this experimental procedure all reaction products bringing to ^{240}Cm (^{40}Ar; n), (^{40}Ar; α, n), (^{40}Ar; 2α, n) are registered with maximum efficiency, that is why the evaporation value corresponds to the summary cross section of all reaction channels. Basing on the sum of the known data it is most natural to refer it to the reaction ^{209}Bi(^{40}Ar, n) ^{248}Md, that is why the obtained value is the upper cross section limit for the channel (α , n).

Thus it follows that if such a process really exists, it must be detected in experiments with higher sensitivity. For the purposes it is probably more convenient to investigate other reactions, for example $^{206-208}Pb(^{40-44}Ca;\alpha,xn)^{244,246}Fm$ at x = 0,1, 2,..., though they no doubt bring to the formation of highly neutron deficient nuclei.

Another possibility - carrying out direct experiments on synthesis of transuranium elements in reactions with massive ions in the "intermediate region" between the hot and cold fusions. We refer to them the combinations with light actinides and ions with Z $<$ 20.

3.2. Experiment

The experiments carried out in 1988 were aimed at the synthesis of element 110 by the reaction $^{236}U(^{40}Ar,xn)^{276-x}110$. Calculations suggest that the even-even isotope $^{272}110$ (N=162) is expected to have a very high stability against spontaneous fission ($T_{SF} > T_{\alpha}$) [18]. However, in the sequential α-decay of this nucleus we move farther from the N=162 subshell and the situation changes to the opposite: in the daughter nuclei the spontaneous fission probability increases and the α-decay probability decreases ($T_{\alpha} > T_{SF}$). We note that this chain leads to the final product, the spontaneously fissioning nuclide $^{260}104$.

By using a facility capable of selecting evaporation residues according to their characteristics angular and energy distribution during a long bombardment of a ^{236}U target with ^{40}Ar ions rare spontaneous fission events with half-lives of several milliseconds have been detected [28]. Despite a very small production cross section - about 5-10 picobarns - and poor statistics (as few as 16 events have been observed), it was possible to determine the kinematical conditions required for the maximum yield of a given activity to be obtained.These results, as well as the data of control experiments, were close to those expected for the evaporation residues of a composite system with Z=110.

One year later the experiments were reproduced on a more sophisticated facility, a gas-filled kinematic separator for recoil nuclei[29]. In these experiments two spontaneous fission events due to the decay of evaporation residues from the fusion reaction $^{236}U+^{40}Ar$ (206 MeV) were also detected. The future experiments will be aimed at determining the radioactive properties of the nuclei of the whole decay chain, at searching the α-particles emitted prior to spontaneous fission and, possibly, at synthesizing other isotopes with N $>$162. However, these experiments are very complicated because of small cross sections and we are still at the stage of preparing them.

At the same time it is evident that any investigations of reaction mechanism in the field of heavy and superheavy elements demand for a substantial increase in the sensitivity of experiments. This makes it necessary to look for new methods of increasing the beam intensities, ameliorating the separator parameters, developing complicated detecting devices.

4. ADVANCES IN THE DEVELOPMENT OF EXPERIMENTAL TECHNIQUES

4.1. Production of intense ion beams

At Dubna the principle experiments aimed at synthesizing heavy elements and at studying their properties are presently carried out at beams available from the isochronous heavy ion cyclotron with a pole diameter of 400 cm (the U-400). A distinguishing feature of this accelerator is a high level of magnetic induction B = 2.1 T that determines the maximum beam energy

$$E_{max} = 625\ q^2/A\ \text{(MeV/nucleon)}.$$

These parameters allow one to obtain an ion energy exceeding the Coulomb barrier of the reaction even with very heavy targets and at the ion mass-to-charge ratio q/A $\gtrsim$ 0.1.

The ions with relatively low charges, $^{20}Ne^{2+}$, $^{40}Ar^{4+}$, $^{58}Fe^{6+}$, etc., can be effectively produced using the PIG type of plasma ion sources (Fig. 6). We have developed a number of modifications of this ion source which are capable of producing multiply charged ions of rare isotopes of various elements.

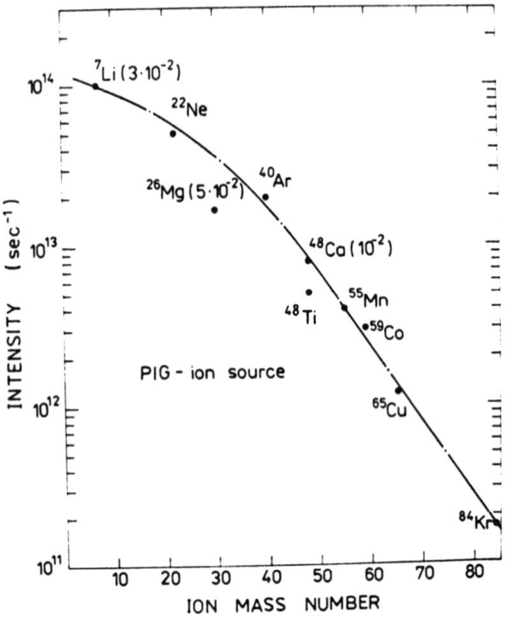

Fig. 6. Average intensities of different ion beams from the U-400 cyclotron, on the target. The figures by the points indicate the ratio of the number of accelerated ions to that of neutral atoms introduced into the ion source plasma.

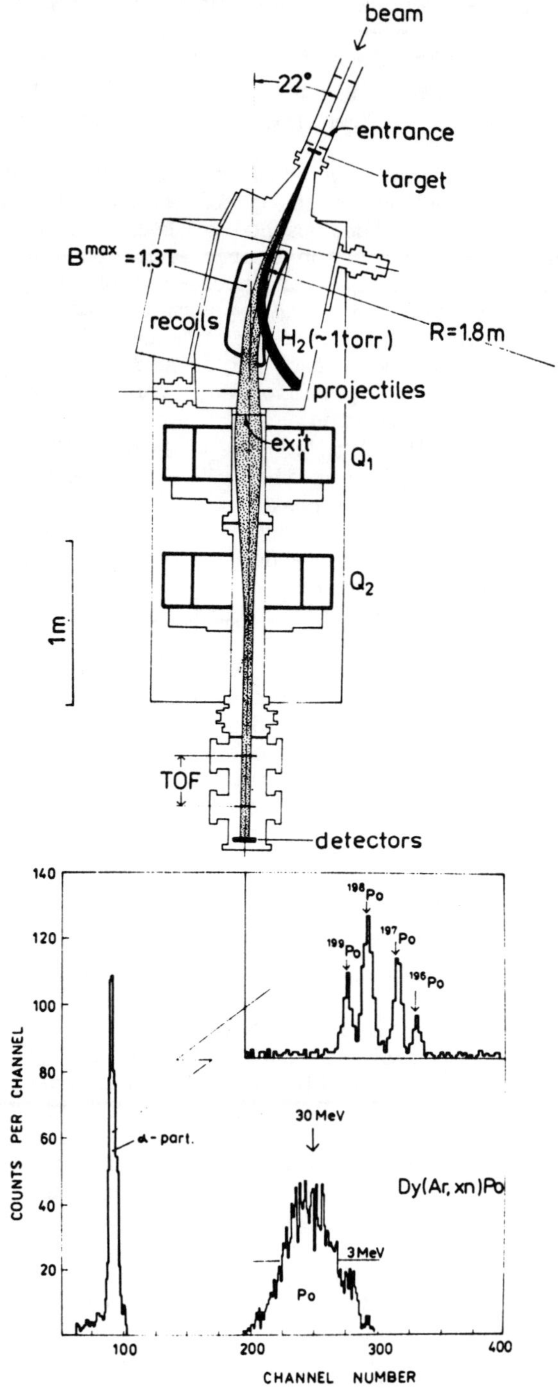

Fig.7. Schematic of the gas-filled separator of recoil nuclei. The lower part shows the energy spectrum of particles recorded by a surface-barrier detector in the spectrometer focal plane. The scale of α-particle energies is given in the upper right-hand corner.

The U-400 accelerator is generally designed for producing intense beams of ions of all the elements located in the initial third part of the Periodic Table of Elements. The average intensity of the ^{22}Ne and ^{40}Ar internal beams is equal to about $5x10^{14}$ pps and $5x10^{13}$ pps, respectively, corresponding to an average beam power of 5 to 19 kW.

The extraction of such an intense ion beam from the accelerator chamber using an electrostatic deflector is an extremely difficult task. Therefore we have developed another method of beam extraction which is based on a sharp change of the ion charge and, consequently, of the particle trajectory radius of acceleration (preset energy). As in this case the radiation and thermal stabilities of a charge exchange target are the factors that determine the limiting beam intensity, theoretical and experimental studies were performed to investigate the operation of various carbon foils prepared using different technologies.

The results obtained by now, indicate the possibility of carrying out prolonged experiments using the ^{18}O, ^{22}Ne and ^{26}Mg ion beam with intensities reaching $5x10^{14}$ pps.

The emittance of an external beam is equal to 20 mm·mrad that corresponds to a cross section of about 3 to 5 mm^2 under real conditions. The maximum beam intensity reaches 0.2 MW/cm^2 taking into account the time structure of the beam (pulse duration 1 ms, repetition rate $\sim$150 Hz).

This enormous thermal load is reduced by a factor of 50 by using forced vertical oscillation (within 8 mm) of the beam with a frequency of 5 kHz and by employing all the facility units (i.e. targets, degraders, the entrance windows separating the gaseous medium and the cyclotron vacuum) having the form of thin foils rotating with a velocity of $3x10^3$ cm/s in the hydrogen atmosphere at a pressure of $\sim$ 1 torr.

4.2. Kinematic separators of recoil nuclei

Earlier it was shown that the yield of the transfermium elements from different types of nuclear reactions is rather small.

The problem, as a whole. consists in the isolation of the very rare events of the formation of new nuclei which, in number, may make up less than the 10^{-10}th fraction of all other reaction products, in their identification and determination of the main properties of their radioactive decay.

Unfortunately, the short life-times of heavy elements block the solution of this problem by using the well-known methods (the chemical separation of elements, mass-separation, and so on).

In should, however, be noted that the nuclei formed in the reaction possess a relatively high velocity, their time of flight being equal to about 10^{-6} s even if the detector is placed rather far from the target. This rapidity of operation can be retained in the case of "in-flight" separation. The principle of separation of nuclei should be based on a difference between their kinematical characteristics determined by the way the nuclei are formed.

In a complete-fusion reaction neutron evaporation slightly changes the compound nucleus momentum. As a result, the evaporation residues retain their strict orientation along the direction of the ion beam. And vice versa, in incomplete fusion reactions the initial momentum is distributed in the exit channel between two fragments that scatter at large angles.

Hence it follows that recoil separation according to the angle and momentum can result in a substantial suppression the background from the nuclei formed in the incomplete-fusion channels. This principle underlies all kinematic separators of recoil nuclei. As the separator should be placed at the angle $O_L = 0°$ it is necessary to carry out the separation of recoil nuclei from the primary beam. The recoil nuclei can be separated only in an electric field by using their difference in energy (velocity) because the magnetic rigidities of evaporation residues and the ion beam are pratically equal. As is known, a similar method is employed in velocity and energy selectors [30,31].

At the same time, as has been shown in works [32-34], it is possible to revert to nuclear separation in a magnetic field if one creates a low-pressure gaseous medium in which the ion and recoil charges depend on velocity and, therefore, greatly differ from each other. Gas-filled recoil separators are based on this principle.

We use both types of kinematic separators of beams available from the U-400 accelerator. As the characteristics of the recoil energy selector (the facility "VASSILISSA") were published earlier[31], we shall give here the parameters of the gas-filled magnetic separator (Fig. 7).

In order to determine the magnetic rigidity of recoil nuclei we varied the equilibrium ion charges of all elements from Ta to Fm in a wide energy range. On the basis of these data we calculated the magnetic rigidities of the complete-fusion and incomplete-fusion reaction products entering the magnetic separator, and the magnetic rigidity of the projectile beam.

As is seen in Fig. 8, the magnetic rigidity of the evaporation residues from different types of reactions differs from that of target-like nuclei and the ion beam. The recoil nuclei that have reached the focal plane are implanted in a mosaic surface-barrier detector about 30 cm^2 in area. The sensitive layer of the detector issues a signal proportional to the total energy of a recoil nucleus. After that the detector is capable of detecting decay products (α -particles or fission fragments) in the 2π-geometry.

At the present time Dr. K.Hulet and his colleagues prepare for operation a more sophisticated detector with high spatial and energy resolutions with respect to the position of the recoil nucleus incidence.

In front of the detector there is a device measuring the velocity and ionization of recoil nuclei. Thus at each act the following measurements are performed:

for recoil nuclei: dE, E, TOF, z, y, t_o

for fission products: E_i, x_i, y_i, t_i.

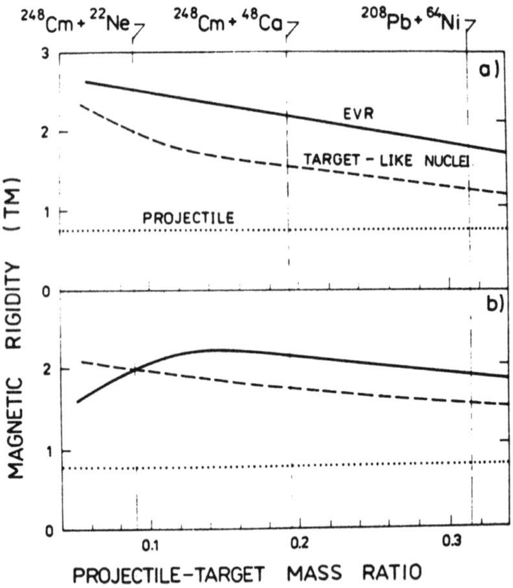

Fig. 8. Ion magnetic rigidity (in Tesla x meter) as a function of the mass ratio of the interacting nuclei:
a) in H_2,
b) in He.

The separating efficiency of the facility was studied in the reactions (^{40}Ar,xn) on ^{164}Dy, ^{207}Pb, and ^{236}U targets.

The results obtained are presented in Table 2, which includes also the data from the reactions induced by the ^{18}O and ^{22}Ne ions. From the table one can see that this design of a gas-filled separator permits a substantial suppression of the background from the incomplete-fusion reaction products and the elimination of the primary beam ions. At the chosen acceptance of the facility the detection efficiency for evaporation residues depends on their angular variance and energy.

In the reactions induced by light ions ($A_I < 20$) the facility efficiency varies in the range from 3 to 20%, strongly depending on the projectile-to-target mass ratio. However, the efficiency rapidly grows to 40-60% as the mass of the bombarding ion increases.

From the above-said it follows that in the experiments aimed at synthesizing new elements and isotopes by the reaction (HI,xn) the modern types of kinematic separators can provide the maximum experimental sensitivity and rapid operation at a level of about 10^{-6}s.

Table 2

Target	^{nat}W	^{238}U	^{164}Dy	^{207}Pb	^{236}U
Projectile	^{22}Ne	^{18}O	^{40}Ar	^{40}Ar	^{40}Ar
Projectile-target mass ratio	0.12	0.08	0.24	0.19	0.17
Evaporation residues	$^{199,200}Po$	$^{250,252}Fm$	$^{199,200}Po$	^{244}Fm	$^{272}110$
Energy of recoil atoms (MeV)	12	6	40	30	28
Collecting efficiency of evaporation residues (%)	27 ± 4	5 ± 2	75 ± 5	44 ± 6	
Collecting and recording efficiency	19 ± 3	3 ± 1	60 ± 4	35 ± 5	30
Suppression of incomp. fusion reaction products	$5\ 10^3$	$5\ 10^3$	10^5	10^5	10^5
Suppression of full energy beam particles	10^{17}	10^{17}	10^{15}	10^{16}	10^{17}

4.3. Future perspectives

It has been noted earlier that the radioactive properties of the isotopes of transactinide elements produced mainly in the cold-fusion reactions indicate the high stability of heavy nuclei against spontaneous fission. As it follows from [18], the shell model predicts a considerable increase in spontaneous-fission half-lives for strongly deformed nuclei in the vicinity of the N = 162 subshell.

At the same time, as it follows from Table 1, the evaporation residues formed in the reactions involving compound nucleus formation followed by neutron evaporation are very neutron-deficient. This deficit can partly be compensated for by an increasing neutron number in the interacting nuclei, and this occurs in the "hot fusion" experiments with strongly asymmetric target-projectile systems.

However, there still exist some possibilities here. In fact, the heaviest isotopes of transuranium elements possess a neutron excess (N-Z) equal to 56 atomic units.

The number of the isotopes used as target material can be increased by decreasing the target area to 0.05–0.07 mm^2, which will result in a drastic decrease in the total amount of target material to 100–200 nanograms (~ 5 x 10^{14} atoms). Although the preparation of such a target requires certain experimental skill.

The effective decrease in the beam size can be achieved by using coolers in which the circulating ion beam is cooled by electrons [35]. As a result of low emittance and high monochromaticity, the "cooled" beam can be focussed onto a small area of about 10^{-2} mm^2. Since in the case of fast beam extraction the beam pulse duration is equal to about 1 μs the separation of recoil nuclei can be done using the time-of-flight technique, too.

Fig. 9 shows a schematic view of the proposed LNR, JINR heavy ion storage ring facility designed for the production of

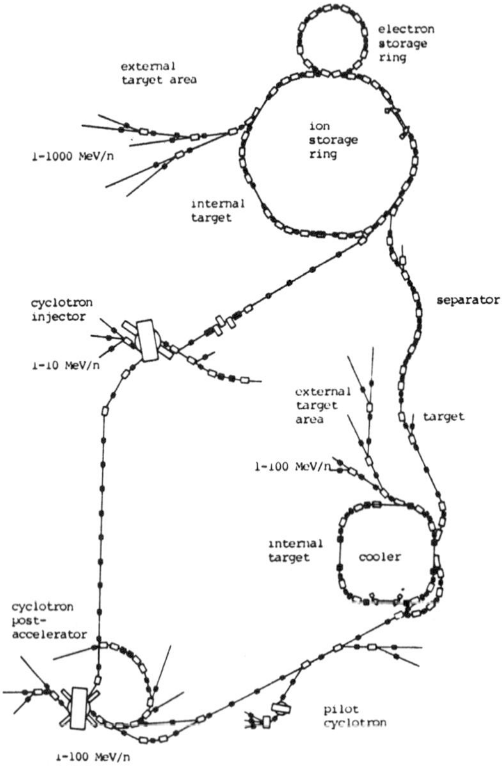

Fig. 9. The heavy ion accelerator facility project of the LNR, JINR.

beams of stable and radioactive isotopes in wide mass and energy ranges. Some data which may form the basis for the future experiments are presented in Table 3. In the Table there are given the target-projectile combinations that lead, as a result of (HI,xn) reactions, to isotopes of heavy elements with Z = 105–110 and with the maximum possible number of neutrons, N = 160–165, respectively. These nuclei are expected to be stable against spontaneous fission ($T_{SF} \sim 1$ d) and to have half-lives of around several tens of second (at least for the $Z \leq 107$ isotopes). In this case a possibility appears for using the chemical separation of elements and, in principle, for investigating heavy transactinides by the methods of classical radiochemistry and nuclear spectroscopy.

Naturally our estimates for properties of new nuclides were based on theoretical hypotheses which themselves need to be verified experimentally. However, the fact that the reactions of fusion between the ^{18}O, ^{22}Ne, and ^{26}Mg projectiles and heavy target nuclei have been studied rather thoroughly [36] permits fairly accurate predictions of the cross sections of forming evaporation residues in these reactions. On the other hand, if the isotopic lifetime turns out to be shorter than expected, the isotopes can be separated by means of kinematic separators which may possess a large angular acceptance as a result of the target having a small size.

Considering that the experiments on ion storage rings will be possible in a rather distant future we note that the given approach can be used already nowadays in connection with a considerable increase in beam intensity and brightness at heavy ion accelerators.

As has been mentioned above, an ion beam density of $10^{13} mm^{-2} s^{-1}$ has been reached at the U-400 accelerator, thus making it possible to carry out, already in the nearest future, very efficient experiments with microamounts of target material.

Table 3

Target material (half-life)	Amount (number of atoms)	Ions	Number of ions on orbit	Max. luminosity ($cm^{-2}s^{-1}$)
^{250}Cm (120 y)	$5 \cdot 10^{14}$	^{18}O	$5 \cdot 10^{10}$	10^{30}
^{252}Cf (2.6 y)	$5 \cdot 10^{14}$	^{22}Ne	$5 \cdot 10^{10}$	10^{30}
^{254}Es (276 d)	$5 \cdot 10^{14}$	^{26}Mg	10^{10}	10^{29}
^{257}Fm (100 d)	10^{13}	^{14}C	$5 \cdot 10^{10}$	

REFERENCES

1. G.T.Seaborg and W.D.Loveland, in: Treatise on Heavy Ion Science, vol. 4, ed. D.A.Bromley (Plenum Press, New York, 1985) p. 255.
2. E.K.Hulet et al. Rev. Phys.Lett. 26, 523 (1971).
3. R.W.Hoff, in "Weak and Electromagnetic Interaction" Proc. Int. Symp. Heidelberg 1986, ed. V.H.Klapdor (Springer-Verlag, 1986) p. 207.
4. A.Ghiorso et al. Phys. Rev. Lett. 33, 1490 (1974).
5. R.Vandenbosch et al. Phys. Rev. 111, 1358 (1958).
6. J.D.Jackson. Can. J. Phys. 34, 767 (1965).
   T.Sikkeland, Arkiv for Fisik 36, 539 (1966).
7. A.G.Artukh et al. Yad. Fis. 28, 1154 (1978).
8. D.Lee et al. Phys. Rev. C25, 286 (1982).
9. W.Schroder and J.R.Huizenga, in: Treatise on Heavy Ion Science, vol. 2, ed. D.A.Bromley (Plenum Press, New Your, 1985).
10. D.Hoffman et al. Phys. Rev. C31, 1763 (1985).
    R.B.Welch et al. Phys. Rev. C35, 204 (1987).
11. R.W.Lougheed et al. LLNL Nuclear Chemistry Division FY87. Annual Report UCAR 10062/87 Livermore (1987).
12. Yu.Ts.Oganessian et al. JINR Preprint D7-8194, Dubna (1974); H.Gaeggeler et al. Z. Phys. A289, 415 (1979).
13. Yu.Ts.Oganessian, in: Classical and Quantum Mechanical Aspects is Heavy Ion Collision, Lecture Notes in Physics, vol. 33 (Springer-Verlag, Heidelberg, 1975) p. 221.
14. Yu.Ts.Oganessian et al. Radiochimica Acta 37, 113 (1984).
15. P.Armbruster. Ann. Rev. Nucl. Part. Sci. 35, 135 (1985).
16. Yu.Ts.Oganessian, in: Proc. Int. School-Seminar on Heavy Ion Physics, Dubna, 23-30 September 1986, JINR (Dubna) 1987., p. 103.
17. G.Münzenberg, in Proc. Int. School-Seminar on Heavy Ion Physics, Dubna, 23-30 September 1986 JINR (Dubna) 1987, p. 127.
18. A.Sobiczewski et al. Phys.Lett. B224, 1 (1989).
    Z.Patyk et al. Nucl. Phys. A502, 591c (1989).
19. W.J.Swiatecki. Phys. Sci. 24, 113 (1981).
20. S.Bjornholm and W.J.Swiatecki. Nucl. Phys. A391, 471 (1982).
21. K.T.R.Davies et al. Phys. Rev. C28, 629 (1983).
22. J.P.Blocki et al. Nucl.Phys. A459, 145 (1986).
23. P.Fröbrich. C. Phys. Rep. 116, 337 (1984).
24. J.O.Newton. Sov. J. Particles and Nuclei 21, 821 (1990).
25. D.J.Hinde et al. Nucl. Phys. A452, 550 (1986);
    Nucl. Phys. A472, 318 (1987); Phys. Rev. C39, 2268 (1989).
26. E.M.Kozulin et al. Yadernaya Fisika 55 (1990).
27. T.Nomura. Int. Symp. on Heavy Ion Physics and its Applications, Lanzhou, 8-12 October 1990.
28. Yu.Ts.Oganessian et al. in: Proc. of the Int. School on Physics "Enrico Fermi" Varenna, Italy 23 June - 3 July 1987.
    Trends in Nucl. Physics (1989) CIII Corso p. 258.
29. Yu.Ts.Oganessian et al. in: Abstr. of the Int. School-Seminar on Heavy Ion Physics, Dubna, October 1989, Dubna, JINR, D7-89-531, p. 145.
30. G.Münzenberg et al. Nucl. Instr.and Meth. 161, 65 (1979);
    B26, 294 (1987).

31. A.V.Yeremin et al. Nucl. Instr. and Meth. A274, 528 (1989).
32. B.L.Cohen, C.B.Fulmer. Nucl.Phys. 6, 547 (1958).
33. V.A.Karnaukhov et al. JINR Comm. P13-4454, Dubna, JINR (1969).
34. A. Ghiorso et al. Nucl. Instr. and Meth., A269, 192 (1988).
35. D.Krämer et al. Nucl. Instr. and Meth. A287, 268 (1990).
36. A.N.Andreyev et al. Preprint JINR E7-89-809, Dubna (1989); Yadernaya Fizika (in press); Yadernaya Fizika 50, 619 (1989); Commun. of the JINR E15-89-683, Dubna (1989).

Heavy Element Synthesis by (HI,αxn) Reactions

T. Shinozuka, T. Nomura[1], K. Morita[2], A. Yoshida[2], T. Inamura[2], M. Koizumi[2], Y. Nagai[3], T. Toriyama[3], H. Miyatake[4], Y. Fujita[4], T. Shimoda[4], M. Fujioka, M. Wada[1], S. Kubono[1], N. Ikeda[1], H. Kudo[5], K.Sueki[6], H. Nakahara[6], Y. Hatsukawa[7], R. Saintola[8], T. Motobayashi[9] and A. Kurokawa[9]

Cyclotron and Radioisotope Center, Tohoku University, Sendai , Japan
Institute for Nuclear Study, University of Tokyo, Tokyo, Japan [1]
Cyclotron Laboratory, RIKEN, Saitama, Japan [2]
Dept. of Applied Physics, Tokyo Inst. of Technology, Tokyo, Japan [3]
College of General Education, Osaka University, Osaka, Japan [4]
Dept. of Chemistry, Niigata University, Niigata, Japan [5]
Dept. of Chemistry, Tokyo Metropolitan University, Tokyo, Japan [6]
Dept. of Radioisotopes, Tokai research Establishment, JAERI, Tokai, Japan [7]
Dept. of Physics, University of Jyväskylä, Jyväskylä, Finland [8]
Dept. of Physics, Rikkyo University, Tokyo, Japan [9]

Abstract

We briefly report the development of the gas-filled recoil-type isotope separator (GARIS), which has recently been constructed at the RIKEN ring cyclotron facility. We then discuss the preliminary results of the (HI, αxn) reactions as a possible new approach to synthesizing superheavy elements with GARIS.

1. Introduction

The (HI,xn) cold fusion reactions near the Coulomb barrier, where the doubly closed nuclei ^{208}Pb or ^{209}Bi are used as target and ions heavier than argon are used as projectile, have been successfully used for synthesizing heavy elements up to Z=109. An advantage of the cold fusion is that it has a large survival probability against fission, because the excitation energy of the compound nucleus is small due to the large negative Q-value of the reaction.

However, if the product of the atomic number of the target and the projectile ($Z_1 \times Z_2$) exceeds 1800, this advantage is offset by the entrance channel limitation, the so called "extra-push energy". As shown in Fig. 1 the fission barrier appears inside of the fusion barrier for heavy system. Due to a strong energy dissipation between the two barrier positions, an extra energy is required for the colliding system to be trapped in the pocket inside of the fission barrier. The experiments at Darmstadt and Dubna suggest that the cross section for making heavy elements rapidly decreases with increasing mass number of the composite system. Hence, the attempt to synthesize the element 110 faces an experimental limitation that the cross section is of the order of pico-barn.

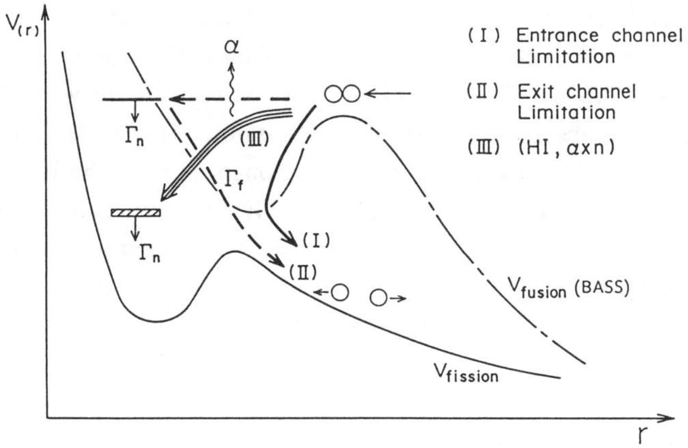

Fig.1 : The cold fusion(1), the hot fusion(2) and the (HI, αxn) reaction (3).

More asymmetric combinations, such as the U + Ar $\rightarrow$ 110 reaction, appear to reduce this dynamic hindrance of the entrance channel. In this case, however, the excitation energy of the compound nucleus becomes higher, to an about 40 MeV. This leads to a reduction of the cross section due to a survival factor $(\frac{\Gamma_n}{\Gamma_n+\Gamma_f})^k$, k being typically 4. Furthermore, the shell stabilized fission barrier certainly lowers or it could even disappear with increasing temperature of the composite system. Several methods have already been proposed to avoid these difficulties in synthesizing the element 110 and/or heavier ones. The use of a neutron rich radioactive beam is one of them [1]. Another possibility is to use (HI,αxn) reactions.

It is well known that the (HI, αxn) reactions have often larger cross sections than those for (HI, xn) reactions when one uses them to produce fissile nuclei like actinide elements [2, 3, 4]. Since the α particle removes a large portion of the excitation energy and angular momenta from a highly excited compound or pre-compound nucleus, the (HI, αxn) reaction will produce the intermediate nucleus at lower excitation energies and with smaller spin than (HI, xn) reactions. This will increase the survival probability against fission. A recent experiment by Morawek

[5] has also shown that the (HI, αxn) reaction channel is the major fraction in the fusion-evaporation process for symmetrical system ^{110}Pd + ^{110}Pd, where a large dynamical hindrance of the fusion cross section should exist at the entrance channel because of the the large $Z_1 \times Z_2 \sim 2100$.

The experimental cross section for the (HI,αxn) reaction cannot be reproduced with statistical model calculations based on a compound nucleus formation. This suggests that the α-particles are emitted at an early stage of the fusion process before a compound nucleus is formed as shown in fig. 1 [2, 5].

So far there are few experimental searches for heavy elements such as Z=110 by (HI, αxn) reactions. No reactions involving charged particles have been observed in the cold fusion experiments, where the Pb or Bi target was irradiated by projectiles heavier than Ar. There could be several reasons for why the previous experiments failed to observe the α-particle emission. One of the problems will be the incident energy. To overcome the extra-push energy, the incident energy should be higher than those expected in usual (HI, xn) reactions. More critical problem is the angular distribution of the residual nuclei produced in this reaction. It is not forward-peaked, but is side-peaked because of the large recoil effect induced by the α emission. This requires a recoil-type separator which has a large angular acceptance.

Here we first introduce the gas-filled recoil separator recently developed at RIKEN, which has a large acceptance angle for recoil products. We then report the results of the first experiment of the (HI,αxn) reactions detected using GARIS. A brief description of the present subject has been given in ref. [6].

2. Gas-filled recoil separator

Since half-lives and production cross sections of transfermium elements than those of transuranium elements are generally short and small, the rapid and high efficient separation technique is indispensable to identify and determine their decay properties of in very rare events of the formation of new nuclei which may make up less than 10^{-10} fraction of all other reaction products. The powerful method which meets the above conditions is the "in-flight" kinematic separation. It should work especially for the case of the recoil nuclei having various velocities, which are relatively high. The gas-filled recoil separator is one of the superior candidates to satisfy the demands mentioned above. The principle of operation is that the recoil products follow an average trajectory in the magnetic field filled with low pressure gas, which make up the equilibrium charge state of recoil ions during many charge exchange processes between recoil ions and atomic gas. Because the average charge is proportional to the velocity of ions in the first approximation, the average trajectory is nearly independent of the initial charge states and momenta of the ions. Therefore, it has a great advantage that the reaction products can be collected at its focal plane with an extremely high efficiency.

The RIKEN gas-filled recoil separator has twofold purposes. One is for the spectroscopic study with high collection efficiency for the recoil nuclei, which are produced with very small cross sections like superheavy elements(SHE). Another is for an injector to an ion-guide isotope separator on-line(IGISOL) developed at Jyväskylä [7]. The IGISOL system coupled with GARIS has an advantage that it does not suffer from plasma effect in heavy-ion reactions. [8].

Figure 2 shows a layout of RIKEN GARIS consisting of a dipole magnet and a quadrupole doublet. The dipole magnet has a bending angle of 45 deg. with ρ=1.2 m; the pole gap is 15 cm and the maximum field is 1.5 tesla. The quadrupole magnet has a bore radius of 15 cm, and a maximum field gradient is 5.2 T/m. The other detailed specifications are given in referece [9]. The GARIS has two different target positions as shown in fig.2 and a large angular acceptance, which amounts to nearly 5 degrees both in horizontal and vertical directions, is realized with relatively large pole gap and bore radius . It should be noted that this characteristic has made it possible to apply the RIKEN-GARIS to search for SHE by using the (HI,αxn) reaction.

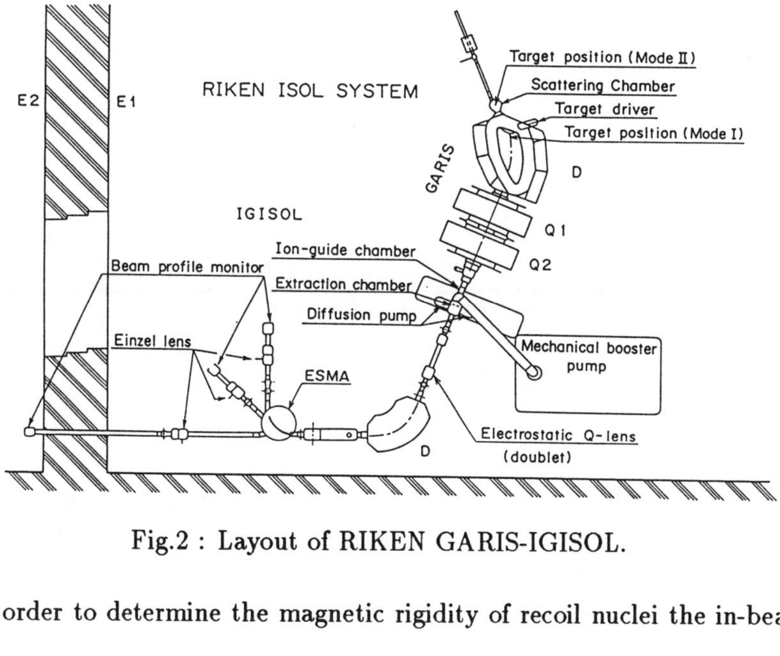

Fig.2 : Layout of RIKEN GARIS-IGISOL.

In order to determine the magnetic rigidity of recoil nuclei the in-beam tunings have been performed by using the energy-degraded primary Kr beams, the ^{169}Tm(^{40}Ar,xn)$^{209-x}$Fr and the ^{181}Ta(^{40}Ar,αxn)$^{217-x}$Ac reactions.

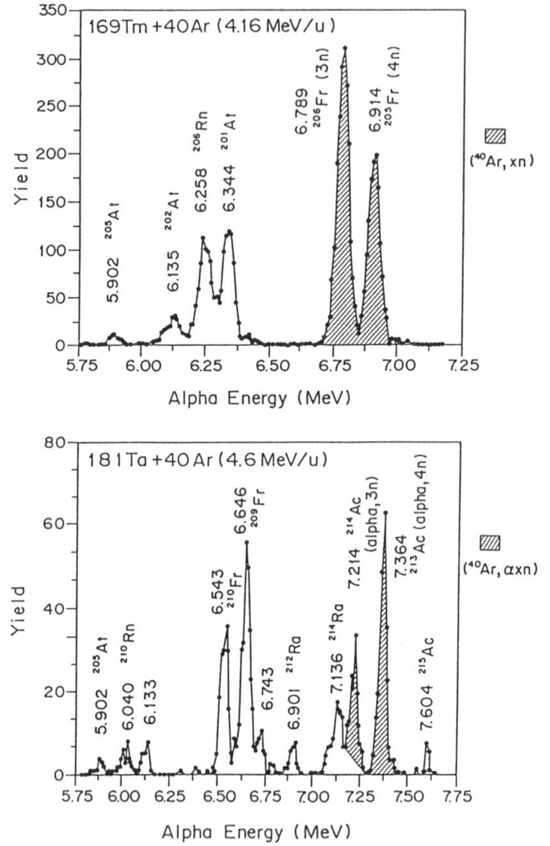

Fig.3 : Typical α-particle spectra from the residual nuclei in (^{40}Ar, xn) and (^{40}Ar, αxn) reaction.

The reaction products were implanted into a large surface-barrier Si detector of about 2000 mm^2 in area and the characteristic α-particles from their decays were detected with almost 2π-geometry during a period of off-beam in a pulsed beam operation. The typical α-decay spectrum of the evaporation residue from the ^{181}Ta(^{40}Ar,αxn)$^{217-x}$Ac reaction is shown in fig. 3. The efficiency of GARIS was measured by comparing the activity on a detector at the focal plane with those on a cather foil located at about 5 cm behind the target. The results show about 30 % collection efficiency for the (HI,xn) channel and 12 % for (HI,αxn), respectively.

The resolution $\Delta B\rho/B\rho$, which is defined in terms of the magnetic rigidity of recoiled particles, was found to be 1/14 in FWHM.

3. (HI, αxn) reaction

Estimation of the cross section

On the basis of the data reported in refs. [2, 3, 4] for lighter targets and with lighter projectiles, and assuming α-emission in the pre-compound stage, we can roughly estimate the production cross sections for heavier elements up to element 110 by the (HI, αxn) reactions. In the zeroth-order approximation, the reaction cross section of the (HI, αxn) can be written as a product of the emission probability of α-particle (σ_α) in an early reaction stage with the survival probability of the intermediate nucleus.

$$\sigma_{\alpha xn} \simeq P_n \sigma_\alpha (\bar{\Gamma}_n/(\bar{\Gamma}_n + \bar{\Gamma}_f))^X \quad (1)$$

$$\sim P_n \sigma_\alpha (\bar{\Gamma}_n/\bar{\Gamma}_f)^X, \quad (2)$$

in which $\bar{\Gamma}_n$ and $\bar{\Gamma}_f$ are the average neutron emission width and fission width, respectively, and the $\bar{\Gamma}_n \ll \bar{\Gamma}_f$ is assumed in eq. (2). P_n is a term expressing an excitation function of present interest, and usually lies between 0.5 and 1 when the incident energy is properly choosen. We therefore omit this term for a further order-of-magnitude estimate. Although there have been no direct experimental measurement on σ_α at the bombarding energies near the Coulomb barrier, it is possible to roughly single out the σ_α values from a few cross section data of $\sigma_{\alpha xn}$ reported in refs. [2, 3, 4] and from the values of $\bar{\Gamma}_n/\bar{\Gamma}_f$ ratio corresponding to the nearly same excitation energies. This $\bar{\Gamma}_n/\bar{\Gamma}_f$ values can be obtained by the experimental data of (HI,xn) reactions and by the following formula

$$\sigma_{xn} \simeq \sigma_{fus.}(\bar{\Gamma}_n/\bar{\Gamma}_f)^X \quad (3)$$

where $\sigma_{fus.}$ is the fusion cross section around the Coulomb barrier. Figure 4 shows the σ_α values versus the bombarding energy per nucleon at the instant of the collision to make a simple comparison with the data in different combinations of projectile and target. The datum for ^{209}Bi+^{40}Ar is the result from our preliminary experiment which we will mention below. From fig. 4, we expect that the value of σ_α should be in a range of 0.1 mb to 1 mb even at the incident energy near the Coulomb barrier.

Here, we can estimate the cross section of (HI,αxn) reaction in comparison with the actinide based (HI,xn) reaction, especially with the (HI,4n), which has a similar excitation energy to the (HI,α2n) having the $U_{CN} \sim 45$ MeV at the bombarding energy near the Coulomb barrier. Since the average kinetic energy of α-particle is about 28 MeV in the region of Z > 100, while Q_α=8-10 MeV, only two neutrons are emitted to reach low-lying levels below the neutron binding energy. However, we should notice that the atomic number of the resultant residual nucleus is by two units smaller than that produced in the usual (HI,xn) reaction when the same projectile- and-target combination is considered.

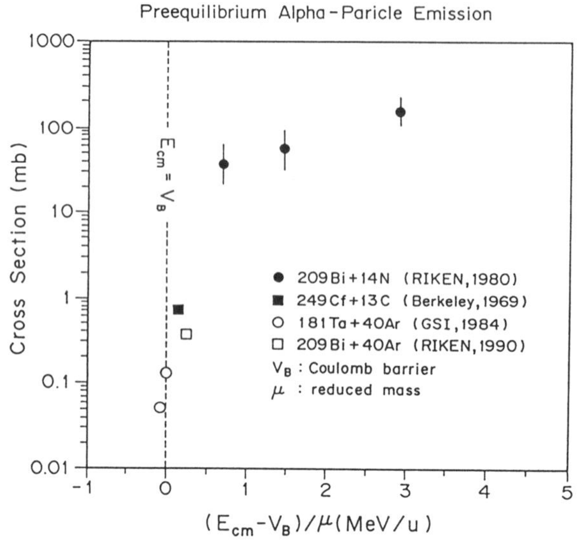

Fig.4 : The estimated cross sections for preequilibrium α-particle emission. see text.

Hereafter, let us consider the two combinations of projectile and target which are slightly different and yield the same residual nucleus in the above two reaction channels. One example is from reference [10],

$$^{246}Cm(^{13}C, 4n)^{255}No; \sigma_{4n}(exp.) = 0.47\mu b \tag{4}$$

$$^{249}Cf(^{12}C, \alpha 2n)^{255}No; \sigma_{\alpha 2n}(exp.) = 1.3\mu b. \tag{5}$$

From eqs. 2 and 3), we obtain

$$\sigma_{\alpha 2n} = \sigma_\alpha(\sigma_{4n}/\sigma_{fus.})^{1/2}. \tag{6}$$

And, if we use the typical values of σ_α and σ_{fusion} as 0.5 mb and 50 mb, respectively, then we obtain,

$$\frac{\sigma_{\alpha 2n}}{\sigma_{4n}} = 0.01(\bar{\Gamma}_n/\bar{\Gamma}_f)^{-2}. \tag{7}$$

Figure 5 shows the estimated cross sections of (HI, α2n) reaction using the known cross sections of (HI,4n), in which the results are normalized by the cross sections at Z=102 using eqs. 4) and 5). Although the estimation was limited to Z = 106, a simple extrapolation of a smooth curve given in fig.5 would yield $\sigma_{\alpha 2n} \simeq 1$ nb, while $\sigma_{4n} < 1$ pb.

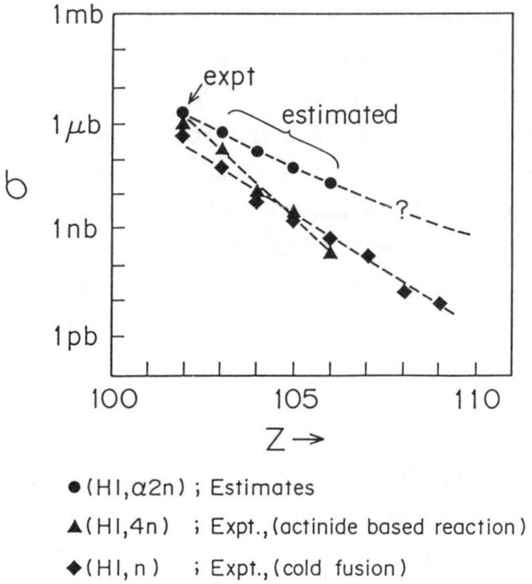

Fig.5 : The estimated cross sections of producing heavy elements in (HI, α2n) reaction from eqs. 6) as well as experimental cross sections in the (HI,4n) and (HI, n) reactions. The cross section data for the latters are taken from ref. [11].

Preliminary experiment for (HI,αxn)

Since in the cold fusion experiment using the symmetric partners with targets and projectiles the (HI,αxn) channel is strongly hindered , it could not easily be compared. It is because the cold fusion reaction at nearly the Coulomb barrier, which is adopted in most experiments, yields the 20 - 25 MeV excitation energy of compound nucleus and thus the α-particle emission is energetically unfavoured. One of the interests is what will happen if one increases the incident energy in the cold-fusion type.

Our first trial experiment by GARIS has been to measure the cross sections of (HI, αxn) channels using the ^{40}Ar + ^{209}Bi reaction at 5.21 MeV/u incident energy of Ar projectile. The experiment has been performed in using almost the same way as in the case of the in-beam tuning of the GARIS described above. That is, the measurement of α decays was carried out during the beam-off period using pulsed beams of 40 sec beam-on and 40 sec beam-off as well as 10 sec beam-on and 10 sec beam-off time structures. Long-lived α decays were also measured after long-period (longer than 1 day) bombardment. Almost all α-peaks could be assigned on the basis of their measured energies, half-lives and α-decay chain relations.

Figure 6 shows delayed α-particle spectra taken 0.8 sec after the termination of the beam-on period. This is to eliminate 0.12 sec $^{212}At^m$ peaks (E_α=7.837 and 7.897 MeV), whose energies are very close to the 7.89 MeV ^{243}Es peak of present interest. $^{212}At^m$ is produced by a transfer reaction on ^{209}Bi, and emits mainly two prominent peaks given above. The possible component of the 7.897 MeV $^{212}At^m$ decay in the region of the shaded area on the right-hand side of fig.6 is estimated to be less than 0.3 events. The half-life of the 7.893 MeV peak shown here could not be measured due to poor statistics, but it turned out that three events fell in the region of t > 5 sec after the termination of beam-on period in the pulsed beam of the 10 sec on and 10 sec off structures. This is consistent with the present assignment of ^{243}Es. Since there is no other known α-decay satisfying the measured energy and time information, we believe that the present assignment is very probable. On the other hand, the assignment of the 7.163 and 7.213 MeV peaks shown on the left-hand side of fig.6 is only tentative, since the assignment is based on their measured energies only.

From our assignment, we obtain 44 $\pm$ 15 nb for (^{40}Ar, αn) and 36 $\pm$ 15 nb for (^{40}Ar, α2n). These values corresponds to $\sigma_\alpha \sim$ 0.36 mb for α-particle emission probability, if we use $\Gamma_n/\Gamma_f \sim$ 0.01 for (^{40}Ar, α2n).

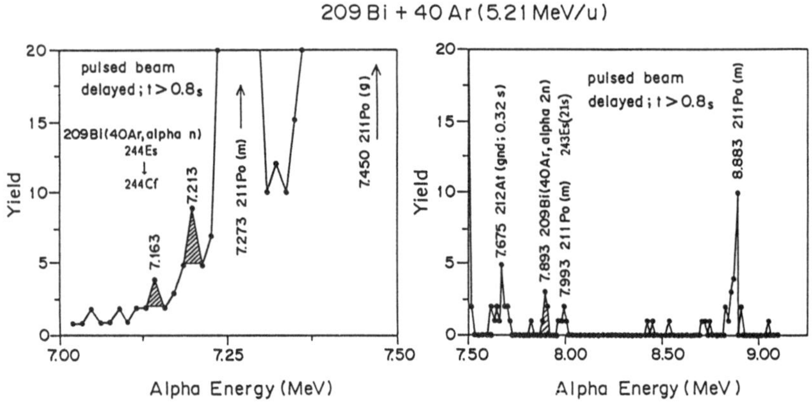

Fig.6 : The measured delayed α -decay spectra in the ^{209}Bi + ^{40}Ar reaction at 5.2 MeV/u. They were taken 0.8 sec after the beam bombardment . See text for more details.

4. Summary and future prospects

Our first experiment has demonstrated the usefulnes of GARIS which has a large acceptance angle of about 5 degrees to collect the side-peaked recoil products after α-particle emission. Furthermore, the detection of the evaporation residues from the ^{209}Bi(^{40}Ar,αxn) reactions has shown that the (HI, αxn) reactions are hopeful in synthesizing heavy elements, 110 etc..

In the recent experiment of ^{209}Bi + ^{40}Ar with the incident energy 5.75 MeV/u of Ar at Dubna [12], they have measured the α-particles from the decay of the ^{240}Cm ($T_{1/2}$=27d), which were sorted with the radiochemical method, in order to obtain the information for the cross section of the (HI,αxn) reaction . Despite the (^{40}Ar,n), (^{40}Ar, αn) and (^{40}Ar, 2αn) reactions are involved, the sum of the cross section for each channel was only $0.4\ ^{+0.3}_{-0.2}$ nb. This is much smaller than the cross section of 44 nb in the (^{40}Ar, αn) reaction in our experiment. It is highly desirable to repeat the same kind of experiments with other combinations in order to resolve these contradictory results.

As the next step at RIKEN-GARIS, we are preparing a detector system, which consists of a 2-dimensional position sensitve silicon detector(60$\times$60) and a parallel plate avalanche counter(PPAC). The aim is to obtain information on the positions, the energies and the time-of -flight of the recoil nuclei. We plan to apply this equipment to study (HI,αxn) reactions in a symmetric ^{65}Cu + ^{208}Pb, and in an asymmetric ^{40}Ar + ^{238}U systems.

We are grateful to the crew of the RIKEN ring cycloron. It is pleasure to thank Prof. N. Takigawa for his stimulate discussions.

References

[1] N. Takigawa, H. Sagawa and T. Shinozuka, Proc. on Nucleus-Nucleus Collision, Kanazawa, 1991. appear in Nucl. Phys.

[2] H. Utsunomiya et al., Nucl. Phys. A334 (1980) 127.

[3] A. Ghiorso et al., Phys. Rev. Lett. 22 (1969) 1317.

[4] D. Vermeulen et al., Z. Phys. A318 (1984) 157.

[5] W. Morawek et al., GSI Scientific report 91-1 (1991) 31.

[6] T. Nomura, Proc. Int. Symp. on Heavy Ion Physics and Its Application, Lanzhou, 1990, ed. W. O. Shen, Y. X. Luo and J. Y. Liu, (World Scientific, 1991) p. 349.

[7] J. Ärje, Nucl. Instr. and Meth. A247 (1986) 431.

[8] T. Nomura et al., Nucl. Instr. and Meth. A269 (1988) 23.

[9] K. Morita et al. Proc. Int. Symp. on Heavy Ion Physics and Nuclear Astrophysical Problems, Tokyo, 1988, ed. S. Kubono, M. Ishihara and T. Nomura, (World Scientific, 1989) p. 271. T. Nomura et al., Nucl. Phys. A217 (1973) 253.

[10] P. Eskola et al., Phys. Rev. C2 (1970) 1058;
A. Ghiorso et al., Phys. Rev. Lett. 18 (1967) 401.

[11] P. Armbruster, Ann. Rev. Nucl. Part. Sci. 35, 135 (1985).

[12] Yu. Ts. Oganessian, private communication.

STUDIES OF FORMATION OF FUSION EVAPORATION RESIDUES WITH Z≥83 IN HEAVY ION REACTIONS.

A. N. Andreyev, D. D. Bogdanov, V. I. Chepigin, Ye. A. Cherepanov, A. P. Kabachenko, O. N. Malyshev, K. V. Mikchajlov, Yu. A. Muzychka, G. S. Popeko, B. I. Pustylnik, R. N. Sagaidak, G. M. Ter-Akopian and A. V. Yeremin.

Laboratory of Nuclear Reactions, Joint Institute for Nuclear Research
141980 Dubna, USSR

ABSTRACT

By employing the kinematic separator VASSILISSA and the beams of ^{20,22}Ne, ^{24,26}Mg, ^{40}Ar and ^{40}Ca we studied 18 projectile-target combinations leading to formation of compound nuclei with atomic numbers 83-102 and obtained xn, pxn and αxn evaporation residues in the range of the projectile energy extending from the Coulomb barrier to the values corresponding to the compound nucleus excitation of about 100 MeV. The systematics is presented of the maximum xn reaction cross-sections which are governed by the fissility of the reaction products. As a result, the values of macroscopic fission barriers were extracted for the neutron-deficient nuclides in the range of Z=83-92 and N=104-134. These values are discussed in terms of contrasting with predictions of the rotating droplet model.

Three isotopes, ^{252,253,254}102, were obtained as deexcitation products of the ^{258}102 compound nucleus formed by ^{22}Ne and ^{26}Mg projectiles. Some conclusions are drawn from these results about the possible use of ^{22}Ne and ^{26}Mg beams for synthesis of Z=106-110 nuclides having neutron numbers close to the predicted island of stability around N=162.

1. INTRODUCTION

During the last decade the considerable progress has been achieved in studies of entrance channel limitations on fusion which arise for heavy colliding nuclei due to predominance of Coulomb repulsion over nuclear forces[1-3]. Conclusions of these theoretical studies have been strongly supported by experiments [4-10]. These findings as well as the results of the successful synthesis of heavy element nuclei[11-15] led us to rather pessimistic views about the prospects of producing even heavier nuclei including superheavy elements. First of all, it is widely admitted that formation of heavy (Z≥100) compound nuclei is strongly prohibited for systems being symmetric in the entrance channel. This conclusion is based on the results of many experiments (see, for example, ref.[4-7]). We mention however that recently the perspectives of such a system as ^{154}Sm+^{154}Sm have been studied[16] in the light of considerable reduction of the nominal Coulomb barrier due to the large static deformation of colliding nuclei. The authors of the work[16] supposed that such reactions which occur as a result of the collision of two deformed nuclei appear to be attractive from the point of view of producing superheavy elements.

The asymmetric cold fusion reaction have been exploited successfully in the work on synthesis of heaviest transfermium nuclei[11-15]. The ability of these reactions to give only slightly excited compound nuclei and significance of this feature for survival of fragile heavy nuclides has been explained due to magic structure of lead and bismuth target nuclei[17-20]. The further advance on this way to heavier (Z≥110) nuclei appears to be, however, very difficult and problematical. Extrapolation of the measured production cross-sections to the element 110 nuclei gives for one of the most favourable reactions $^{208}Pb(^{62}Ni,1n)$ the value close to the subpicobarn region[13-15,19]. Considerations[19,20] based on the conception of the extra-extra push barrier[21] also leaved almost no room for optimistic evaluation of prospects of the new element synthesis by the cold fusion reactions.

The chances of the so-called hot fusion reactions are estimated to be even more pessimistic[19]. Unfortunately, studies of such reactions utilizing the target nuclei of actinide elements and bombarding ions of carbon, oxygen, neon etc. have been ceased since seventies (see ref.[22]) The lack of the experimental results about such asymmetric reactions is an obstacle to the more or less realistic evaluations of the cross sections for reactions which could be used for synthesizing the new heavy nuclides including the nuclei of new elements. Both, fusion cross-sections for asymmetric reactions and the survival probabilities of excited fissile compound nuclei are of interest. Bearing this in mind we initiated systematic studies of evaporation residues formation in the region of compound nuclei with atomic numbers Z≥83. We give in this paper some results of this work.

2. EXPERIMENTAL SET-UP

To carry out the experiments the U-400 cyclotron of the Laboratory of Nuclear Reactions, JINR, Dubna was used. The projectile beams were as follows: ^{20}Ne (120, 140 and 190 MeV), ^{22}Ne (110 and 130 MeV), ^{24}Mg (141 and 172 MeV), ^{26}Mg(136 and 164 MeV), ^{40}Ar (217, 250 and 293 MeV), ^{40}Ca (215, 228 and 270 MeV). The beam intensities passing through targets (12mm in diameter) were $(0.3\text{-}3.0)*10^{12}s^{-1}$ at an energy spread of (1-1.5)% . The beam energy was changed in 3-6 MeV steps using Al and Ti degraders. The energy of the beam was controlled by measuring the energy of ions scattered in a thin (200 $\mu g/cm^2$) gold foil at 30 .

To separate the evaporation residues of heavy-ion fusion reaction from the projectile beams and background products the kinematic separator VASSILISSA[23,24] was used. A schematic view of the separator is shown in fig.1. The evaporation residues knocked out from the target were separated by an achromatic system composed of three electric dipoles. Two triplets of electromagnetic quadrupole lenses provided the focusing on a detector system of recoil nuclei emerging from the target at zero angle within a solid angle of 10 msr. The distance from the target to the focal plane was about 12 m. The detector system consisted of two time-of-flight detectors and silicon detectors. Thin plastic foils emitting secondary electrons and microchannel plates for detecting these electrons were exploited in start and stop time-of-flight detectors. The typical time resolution about 2 ns was obtained for slow (full energy 10-20 MeV) recoil nuclei having mass numbers of about 200. The value 99.95% was achieved for the probability of detection of such recoil nuclei by a single timing detector. After passing the time-of-flight system, recoil nuclei were implanted in silicon detectors which we took in the form of either an array of seven separate detectors or a single crystal divided into eight independent strips. The measurement of the time-of-flight and the energy of recoil nuclei provided their mass determination with an accuracy of about 5%. The whole system provided the uniform detection probability of recoil nuclei within an circular area of 70 mm in diameter.

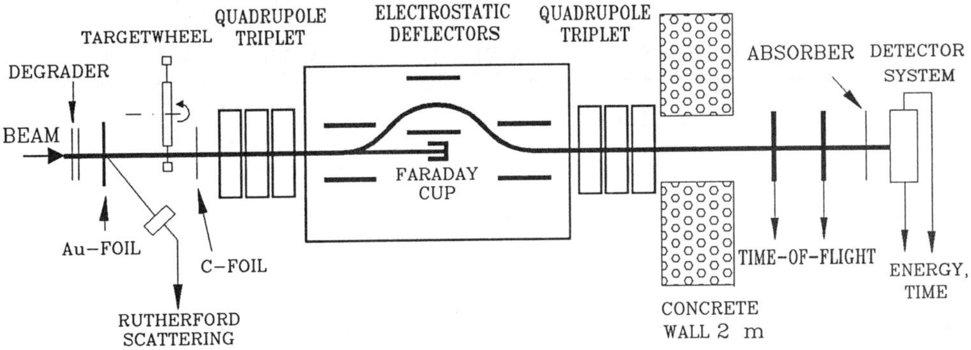

Fig. 1 Schematic view of the experimental set-up.

The heavy-ion fusion reaction products were unambiguously identified by the measured values of their α-decay energy and the life-times. In many cases the identification was performed due to observation of the time correlated α-decays of nuclides belonging to the known α-decay chains. This method was applied in each case when the identification of a previously unknown nuclide or an α-decay line was considered. In fig. 2 we show an example of two - dimensional spectrum of α-α correlation. Such correlations as well as the recoil-alpha time correlations were used for life-time measurements of investigated nuclides. Table I gives the values of measured transportation efficiencies of different evaporation residues from the target to

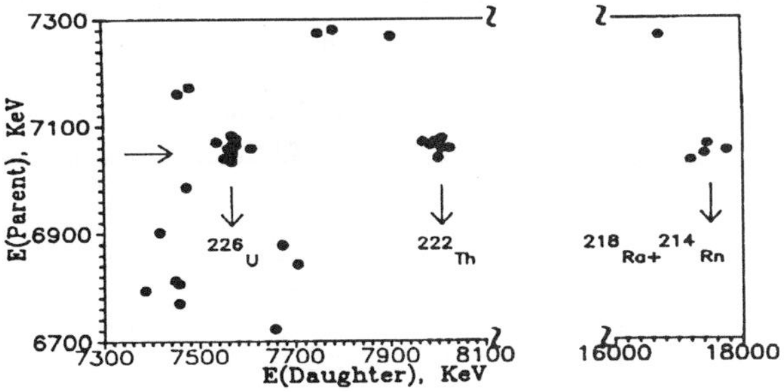

Fig. 2. An example of a two-dimensional α-α correlation plot obtained for reaction $^{26}Mg+^{208}Pb$ at a beam energy $E(^{26}Mg)$=135 MeV for the time window of 100-400 ms[27].

the detector system. We performed the measurements of such values taking targets of different thickness for a set of projectile-target combinations and for evaporation residues produced in xn or pxn as well as in αxn evaporation reactions. The transportation efficiencies of the evaporation residues for properly choosen control reactions were measured in all experiments. Therefore, the relative accuracy of cross section values deduced for a given target-projectile combination was about ±25%. For whole set of the data presented in this work the relative errors in the cross-sections are evaluated to be about ±50%. The errors due to uncertainty and inhomogeneity of the targets used in experiments constituted the negligible part of these errors. Only in few cases the errors were larger due to the poor statistic. The absolute values of the cross sections are probably accurate within factor of two due to errors in measurements of the beam current.

Table I. Experimental results for the separation efficiency

Reaction	Separation efficiency (%) target thickness	
	$0.22\ mg/cm^2$	$0.5\ mg/cm^2$
$^{197}Au(^{16}O,4\text{-}5n)^{208,209}Fr$	3±1	2±1
$^{182}W(^{22}Ne,4\text{-}5n)^{199m,199g,200}Po$	5±1	3±1
$^{166}Er(^{31}P,4n)^{193}Bi$	14±2	
$^{164}Dy(^{40}Ar,4\text{-}5n)^{199m,199g,200}Po$	25±3	19±3

We supposed that the detection efficiency of the α-decay of implanted nuclei was 50% for all reactions studied in this work.

One can evaluate the background conditions of the experiments taking the values of probabilities of passage through the separator to the detector system measured for projectile-like products and products of transfer reactions. These values are given in Table II.

Table II. The suppression factors for multi-nucleon transfer reaction products and for the scattered ions.

Reaction	Target thickness, mg/cm^2	Suppression factors: Scattered ions	Suppression factors: Transfer reaction products
$^{238}U+^{40}Ar$	0.5	$2*10^{10}$	$2*10^4$ (for ^{242}Cm) $8*10^4$ (for ^{227}Th)
$^{208}Pb+^{40}Ar$	0.6	$2*10^{10}$	$7*10^3$ (for ^{211}Bi)
$^{248}Cm+^{22}Ne$	0.33	$3*10^{12}$	$> 4*10^3$ (for ^{254}Fm)

3. SOME FEATURES OF ASYMMETRIC FUSION REACTIONS LEADING TO THE FORMATION OF FISSILE COMPOUND NUCLEI WITH Z≥83

The list of heavy-ion fusion reactions studied in this work is given in Table III. For all of these reactions the xn evaporation channel was investigated in the range of bombarding energies extending from subbarrier values to the region corresponding to excitation energy of compound nuclei close to 100 MeV for the lighter (Z≥91) compound nuclei and about 50-60 MeV for heavier compound nuclei. Several new nuclides (or new α-lines) were obtained in the course of these experiments[25-29] (see Table IV). In addition to the xn evaporation channel, some data were obtained also for cross-sections of pxn and αxn reaction channels (see Table III). The row data were obtained in the form of excitation curves of fusion-evaporation reaction channels. Same examples of excitation curves are given in figures 3 and 4.

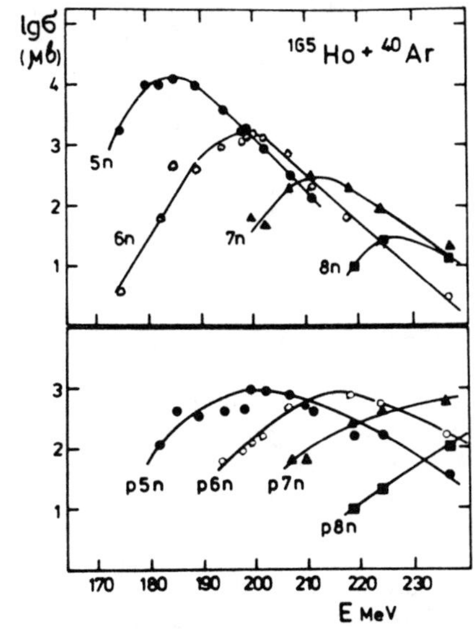

Fig. 3. Excitation functions of the xn and pxn reaction channels in reaction ^{40}Ar+^{165}Ho.

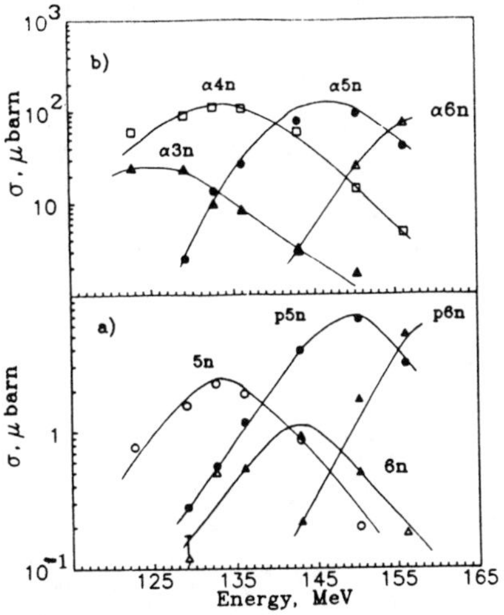

Fig. 4. Excitation functions of the xn, pxn and αxn reaction channels in reaction ^{26}Mg+^{197}Au.

In fig. 5 the systematics is given for the maximum values of xn-reaction cross sections obtained in this work. Similar results for compound nuclei of Ac and Th published by other authors[5,30] obtained with ^{48}Ca and ^{40}Ar bombarding ions are added to the data given in this figure. For comparison, we present in this figure the data for compound nuclei in the range of atomic numbers 98-104 measured by different authors who worked with bombarding ions of carbon, nitrogen, oxygen and neon (see references in the review paper[22]. The points in fig. 5 represent the row experimental results for maximum values of measured cross-sections. We only excluded the data for those reactions for which the maximum values of the cross-sections are strongly affected by the

Table III. The list of heavy-ions fusion reactions studied in this work

Reaction	Z_{CN}	A_{CN}	xn-channel		pxn-channel		αxn-channel	
			x_{min}	x_{max}	x_{min}	x_{max}	x_{min}	x_{max}
$^{40}Ar+^{159}Tb$	83	199	4	10				
$^{40}Ca+^{153}Eu$	83	193	3	6	3	5		
$^{40}Ca+^{151}Eu$	83	191	3	4	2	4		
$^{26}Mg+^{181}Ta$	85	207	4	7				
$^{40}Ar+^{165}Ho$	85	205	4	9	4	9		
$^{24}Mg+^{181}Ta$	85	205	4	9	4	9		
$^{40}Ca+^{159}Tb$	85	199	3		2	5	2	4
$^{22}Ne\ ^{197}Au$	89	219	3	7	3	7		
$^{20}Ne+^{197}Au$	89	217	4	8	4	9	4	9
$^{22}Ne+^{205}Tl$	91	227	3	6	3	5	2	4
$^{26}Mg+^{197}Au$	91	223	5	6	5	6	3	6
$^{24}Mg+^{197}Au$	91	221	3	6	3	7	4	5
$^{22}Ne+^{208}Pb$	92	230	4	5			2	4
$^{20}Ne+^{208}Pb$	92	228	4	5	4	5	2	4
$^{22}Ne+^{209}Bi$	93	231	4	5			2	4
$^{26}Mg+^{208}Pb$	94	234	4				2	3
$^{22}Ne+^{236}U$	102	258	4	6				
$^{26}Mg+^{232}Th$	102	258	4	6				

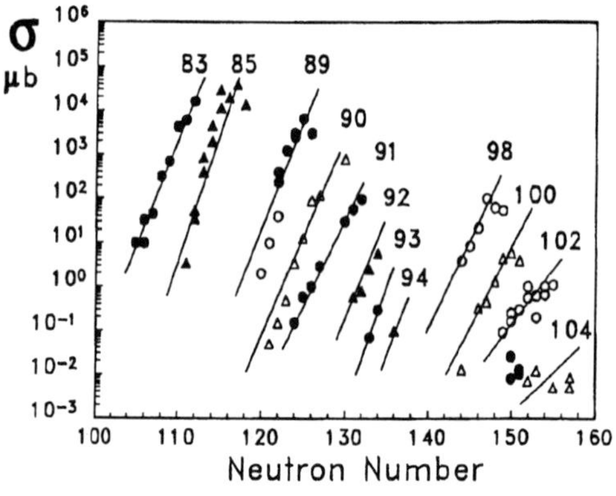

Fig. 5 Systematics of the maximum values of the xn-reaction cross-sections. The circles and triangles alternate in order to facilitate the recognition of data obtained for evaporation residues having different atomic numbers. Our results are presented by closed symbols.

Coulomb barrier. We note that for Z≤ 91 the points are shown in this figure which represent, for given evaporation residues, the maximum cross-sections obtained with projectiles having rather different mass numbers ($^{20,22}Ne$, $^{24,26}Mg$ on the one side and ^{40}Ar, $^{40,48}Ca$ on the other). These point are drawn in figure without correction taking into account the $\pi\lambda^2$ factor. Some of evaporation residues were produced by different bombarding ions (i.e. ^{20}Ne and ^{22}Ne or ^{24}Mg and ^{26}Mg) as a result of evaporation of different numbers of neutrons. The points for all such reactions are presented in fig. 5 without any corrections. Irrespective of these reservation, the points in fig. 5 are grouped around the straight lines corresponding to atomic numbers of evaporation residues. The slopes of these lines show the exponential decrease of xn reaction cross-sections with increasing deficit of neutrons in evaporation residues. This deficit aggravates the competition from the side of fission in the course of deexcitation of compound nuclei. Against the background of this steep exponential decrease the above-mentioned differences between some experimental points look to be of minor importance.

The analysis of experimental data on xn, pxn and αxn evaporation channels was based on the statistical model of the deexcitation process of compound nuclei. Calculations for compound nuclei from Bi to U were carried out using a modified version of the code ALICE. To describe the nuclear level density we used the relations of the Fermi gas model with the phenomenological consideration of shell effects in the level density parameter[31]. We assumed for the fission barrier the form

$$B_f(l)=C*B_f^{cps}(l)+\Delta B_f(Z,A), \qquad (1)$$

where $B_f^{cps}(l)$ is the fission barrier in the model of the rotating charged liquid droplet[32], C is the free parameter, $\Delta B_f(Z,A)$ is

Table IV. The alpha-decay characteristics of some nuclei, investigated in our experiments

	VASSILISSA			other works			
	E_α(keV)	I_α	$T_{1/2}$(ms)	E_α(keV)	I_α	$T_{1/2}$(ms)	
^{223}U	8780±40	100	$0.018^{+0.01}_{-0.005}$				
^{224}U	8470±15	100	$0.7^{+0.9}_{-0.2}$				
^{225}U	7870±20	100	30^{+20}_{-10}	7880±20 7830±20	90 10	80^{+40}_{-20}	[41]
^{226}U	7570±20 7420±20	85±5 15±5	200±50	7430±20	100	500±200	[43]
^{225}Np	8630±20	100	***				
^{226}Np	8000±20 8060±20	50±15 50±15		8044±20	100	31±8	[42]
^{227}Np	7680±20	100	(EC<25%)	7650±20 7677±20		510±60	[42]
^{230}Pu	7050±20	100					

***preliminary data

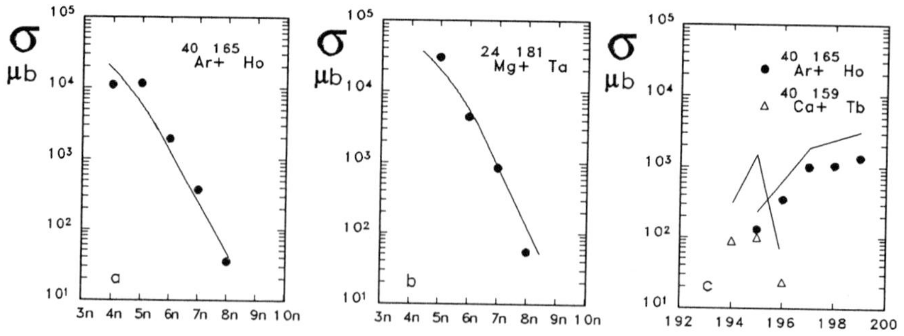

Fig.6. Comparison of the experimental maximum values of the evaporation reaction cross-sections with calculations (solid line) for Z=85 compound nuclei[29].
a),b) - xn-channels, c) - pxn-channels.

the shell correction for the fission barrier of the compound nucleus. The nuclear potential and the choice of the critical value of the angular momentum (l_{cr}) for the compound nucleus formation were considered earlier[33].

The aim of our calculations was the optimum description of the maximum values of cross-sections well above the Coulomb barrier. Therefore, there was no need for additional variation of previously fitted parameters of nuclear potential[33]. More than 90% of cross-section values in their maxima are achieved at $l \ll l_{cr}$. For this reason, the results of calculations were not much sensitive to the value of l_{cr}. We found that calculations could not reproduce experimental results adequately only by variations of shell corrections to nuclear masses and fission barriers, as well as the excitation energy dependence of these values. At this stage we neglected the shell effects in our calculations. We note that there was no need in using the value of level density parameter (a_f) for fission channel different to that for particle evaporation channels (a_ν). Thus, we came to the conclusion that the maximum values of evaporation reaction cross-sections could be fitted rather well by varying only one parameter of the model, i.e. the factor C in

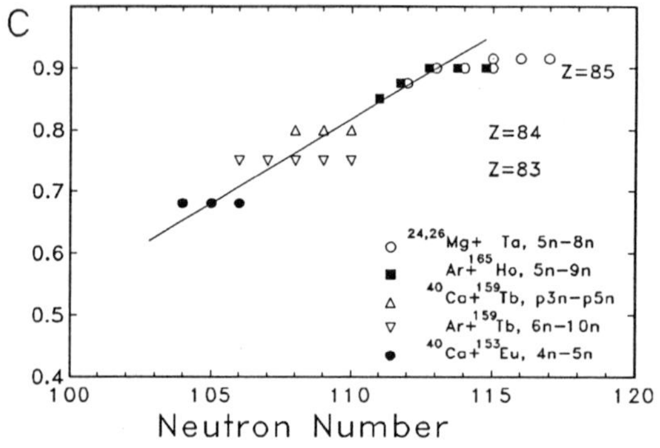

Fig.7 Systematics of the values of parameter C (see, equation (1)) deduced for the neutron deficient isotopes of Bi,Po and At.

equation (1). In other words, these values are rather sensitive to the heights of the fission barriers. By fitting the calculated maximum cross-sections to the experimental results one could eventually determine the fission barriers of the neutron deficient nuclides which participated in the evaporation chains of the compound nuclei studied in this work. Some examples of such a fit are given in fig.6. In fig.7 we show the systematics of the values of the factor C deduced for Z=83 - 85 nuclei. It follows from this figure that the theory[32] overpredicts the fission barrier heights for the neutron deficient isotopes of bismuth, polonium and astatin with neutron numbers N≤112.

Inspection of the data given in fig.5 reveals relatively weak dependence of the maximum cross-section values for compound nuclei lying in two intervals, i.e. Z=83 - 90 and Z=92 - 102 whereas the steep decrease is observed in transition from thorium to uranium. This feature of the cross-sections is seen clearly in fig.8 in which the data are shown for the reactions giving evaporation residues equally displaced from the β-stability line. Similar behaviour was obtained also for pxn and αxn reactions. However, the steep decrease obtained for αxn reactions is shifted in fig.8 to the transition from uranium to plutonium. This leads, in particular, to the fact that xn and αxn reaction channels are similar in the values of their cross-sections for compound nuclei of Ac and Th on one hand and for Pu on the other. In contrast, in the case of the reactions $^{22}Ne+^{208}Pb$ and $^{20}Ne+^{208}Pb$ leading to the compound nuclei of uranium the cross-sections were detected in the microbarn region for 4n and 5n reaction channels whereas for α2n, α3n and α4n reactions channels the obtained cross-section values were close to millibarns[25,28]

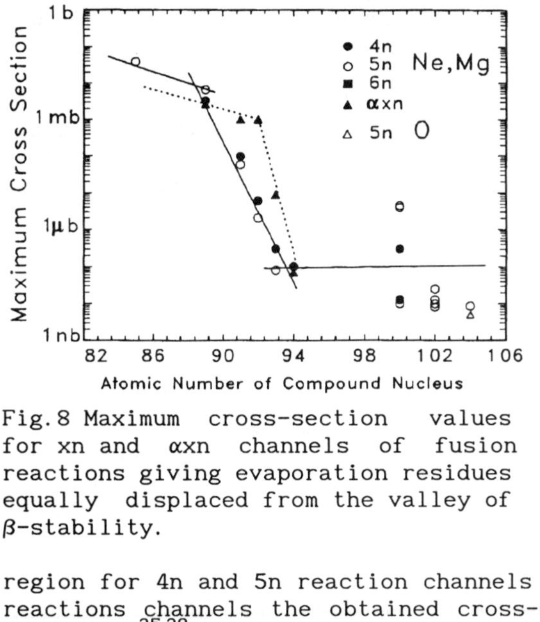

Fig.8 Maximum cross-section values for xn and αxn channels of fusion reactions giving evaporation residues equally displaced from the valley of β-stability.

It is difficult for us to outline any idea which could explain such cross-section behaviour as due to any limitation on the formation of uranium or plutonium compound nuclei. At the same time, the explanation of the obtained steep decrease of the evaporation reaction cross-sections on account of the high fissility of Z>90 nuclides appears to be quite natural. In fact, we could fit in our calculations all set of the results for Ac, Th, Pa and U nuclei presented in fig.5 as well as the data for pxn and αxn reactions by taking the following values for factor C in equation (1) : C=1 for Ac and Th; C=0.7 for Pa and C=0.65 for U nuclei.

4. HOT FUSION REACTIONS $^{22}Ne+^{236}U$ AND $^{26}Mg+^{232}Th$ LEADING TO THE COMPOUND NUCLEUS $^{258}102$.

The aim of the experiments was to obtain the new data about the formation of evaporation residues of the compound nucleus $^{258}102$. Earlier, three reactions, i.e. $^{12}C+^{246}Cm$[34], $^{15}N+^{243}Am$[35] and $^{16}O+^{242}Pu$ [36] have been studied. Addition of two other reactions would complete this sequence of asymmetric hot fusion reactions leading to the same compound nucleus $^{258}102$. The variations of the mean arithmetical fissility parameter[21] along this sequence (see table 5) originates entirely from the difference in entrance channel configurations. Therefore, studies of these two reactions could shed some light on the problem of extra-extra push barrier for hot fusion reactions, which are of interest from point of view of possible synthesis of nuclides in the region of the neutron number 162. In fact, it is not excluded that bombarding ions of ^{22}Ne and ^{26}Mg could be used successfully in combinations with target nuclei of ^{248}Cm, ^{249}Bk and ^{249}Cf for synthesizing nuclides which belong to the predicted[37] stability island centered around this neutron number.

The experimental conditions were essentially the same as was described above. Three evaporation residues, i.e. isotopes of element 102 having the mass numbers 254, 253 and 252 were registered in focal plane of the separator. These isotopes emerging from 4n, 5n and 6n evaporation channels were identified through their characteristic α-decay lines. The 25% spontaneous fission decay branch of $^{252}102$ as well as α-decays of daughter nuclei of $^{252}102$ and $^{253}102$, i.e. ^{248}Fm and ^{249}Fm were detected.

The measured excitation curves of the studied reactions are shown in fig.9. We calculated the xn reaction cross-sections using the method described in Ref.[34,38,39]. Due to the fact that fission and neutron evaporation competition is treated in this approach by exploiting the empirical Γ_n/Γ_f systematics[40], one could simplify the analysis as Γ_n/Γ_f values were essentially the same for all reactions. Therefore, by comparing the experimental data with calculated cross-sections one could count on extracting some information about the fusion probability for different entrance channel configurations. Such a comparison is given in Table 5. We deduced the following conclusions from this table.

Table V. Maximum cross section values for neutron evaporation channels of the compound nucleus $^{258}102$

Reaction	x^{arith}_{mean}	Ref.	Cross section (nanobarn) Experiment			Calculation		
			4n	5n	6n	4n	5n	6n
$^{12}C+^{246}Cm$	0.680	[34]	1000	300		227	107	22
$^{15}N+^{243}Am$	0.694	[35]	80			116	85	20
$^{16}O+^{242}Pu$	0.698	[36]	34	55		111	78	15
$^{22}Ne+^{236}U$	0.721	This work	7	25	15	74	64	9.2
$^{26}Mg+^{232}Th$	0.735	This work	1.5	9	8	42	46	8.2

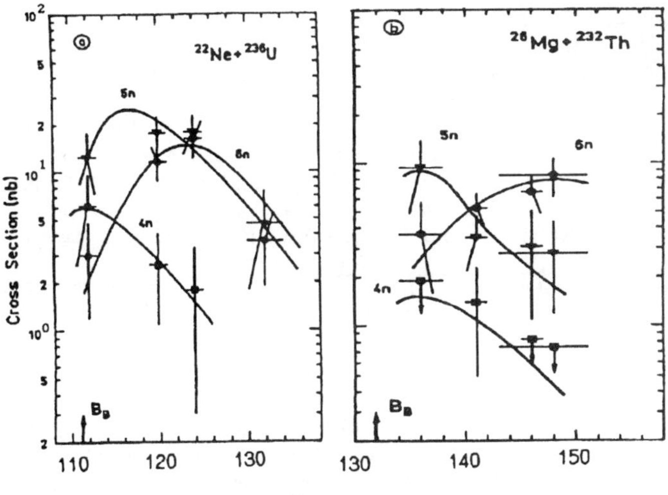

Fig. 9. Excitation curves of neutron evaporation channels for the compound nucleus $^{258}102$.

The experimental maximum values of the cross-sections for the 4n and 5n evaporation channels are reproduced by calculations quite reasonably for the case of the three most asymmetric reactions. Some differences between experimental and calculated values not exceeding the factor of 4 are within the typical limits of the model accuracy[38,39]. The agreement is much worse in the case of the 4n channel obtained for heavier projectiles (^{22}Ne and, especially, ^{26}Mg). One could explain this as an indication of the onset of the extra-extra-push barrier at the transition from $^{16}O+^{242}Pu$ reaction to more symmetric ones, i.e. to $^{22}Ne+^{236}U$ and $^{26}Mg+^{232}Th$. Indeed, the agreement could be improved considerably by adding 5 MeV to the height of the Coulomb barrier for ^{26}Mg. However, we refrain from inferring any conclusion about the extra-extra-push barrier as 4n and 5n channels are strongly subbarrier for $^{22}Ne+^{236}U$ and $^{26}Mg+^{232}Th$ reactions, and we hardly can insist that our accuracy in the calculation of the barrier penetration is high enough.

At any rate, we regard the observation of an anomalously small value of 4n reaction cross-section in the case of ^{26}Mg (^{22}Ne) as an indication that this reaction channel has a little chance to be helpful in the synthesis of new heavy nuclides. Contrary to this, the 6n reaction cross-section does not tend to the drastic decrease for the case of heavy projectiles. Though with some reservations, this is also true for 5n reaction. Therefore, we suppose that these two reactions will be practical in the work on the synthesis of new heavy isotopes of elements 106-110 with the beams of ^{22}Ne and ^{26}Mg.

References

1. J.R.Nix and A.J.Sierk, Phys. Rev. C15, 2072 (1977).
2. W.J.Swiatecki, Phys. Scr. 24, 113 (1981).
3. S.Bjornholm and W.J.Swiatecki, Nucl. Phys. A391, 471 (1987).

4. H.Gaeggeler et al., Z.Phys. A316, 291 (1984).
5. C.-C.Sahm, H.-G.Clerc, K.-H.Schmidt, W.Reisdorf, P.Armbruster, F.P.Hessberger, J.G.Keller, G.Muenzenberg and D.Vermeulen, Nucl. Phys. A441, 316 (1985).
6. J.G.Keller, K.-H.Schmidt, F.P.Hessberger, G.Muenzenberg, W.Reisdorf, H.-G.Clerc and C.-C.Sahm, Nucl. Phys. A452, 173 (1988).
7. W.Reisdorf, F.P.Hessberger, K.D.Hildenbrand, S.Hoffman, G.Muenzenberg, K.-H.Schmidt, W.F.W.Schneider, K.Suemerer, G.Wirth, J.V.Kratz, K.Schlitt and C.-C.Sahm, Nucl.Phys.,A444,154,(1985)
8. R.Bock et al., Nucl.Phys.,A388,334(1982).
9. W.Q.Shen, J.Albinski, A.Gobbi, S.Gralla, K.D.Hildenbrand, N.Herrmann, J.Kuzminski, W.F.J.Mueller, H.Stelzer, J.Toke, B.B.Back, S.Bjornholm and S.P.Sorensen, Phys.Rev.C36,115,(1987)
10. B.B.Back, In : Int. School-Seminar on Heavy Ion Physics, Dubna, 1986; JINR D7-87-68 (Dubna, 1987), p. 465.
11. Yu.Ts.Oganessian, In : Int.School-Seminar on Heavy Ion Physics, Alushta, 1983; JINR D7-83-644 (Dubna, 1983), p.55.
12. Yu.Ts.Oganessian, M.Hussonnois, A.G.Demin, Yu.P.Kharitonov, H.Bruchertzeifer, O.Constantinescu, Yu.S.Korotkin, S.P.Tretyakova, V.K.Utyonkov, I.V.Shirokovsky and J.Estevez, Radiochimica Acta 37, 113 (1984).
13. P.Armbruster, Ann. Rev. Nucl. Part. Sci 35, 135 (1985).
14. G.Muenzenberg, Rep. Prog. Phys. 51, 57 (1988).
15. S.Hoffman, In : Int. School-Seminar on Heavy Ion Physics, Dubna, 1990, JINR, D7-90-142 (Dubna, 1990), p.338.
16. C.H.Dasso, J.Fernandez-Niello and S.Landowne, Phys. Rev. C41, 1014 (1990).
17. P.Moller, J.R.Nix and W.Swiatecki, Nucl. Phys. A469, 1 (1987).
18. P.Moller, J.R.Nix and W.Swiatecki, Nucl. Phys. A492, 349 (1989).
19. P.Armbruster, Varenna Course in Physics, 1987, Preprint GSI-87-9.
20. P.Armbruster, J. Phys. Soc. Jpn. 58, 232 (1989).
21. J.Blocki, H.Feldmeier and W.J.Swiatecki, Nucl. Phys. A459, 145 (1986).
22. G.N.Flerov and G.M.Ter-Akopian, Prog. Part. Nucl. Sci. 19, 197 (1987).
23. A.V.Yeremin et al., Nucl. Instr. & Meth. A274, 528 (1989).
24. A.N.Andreyev, D.D.Bogdanov, V.A.Gorshkov, A.V.Yeremin, A.P.Kabachenko, O.A.Orlova, G.M.Ter-Akopian, V.I.Chepigin, JINR, Rapid Communications, N3[29]-88 (Dubna, 1988), p.33.
25. A.N.Andreyev, D.D.Bogdanov,V.I.Chepigin,A.P.Kabachenko, O.A.Orlova, G.M.Ter-Akopian and A.V.Yeremin, Yad. Fiz. 50, n.9, 619,(1989) (in Russian).
26. A.N.Andreyev, D.D.Bogdanov, V.I.Chepigin, A.P.Kabachenko, S.Sharo, G.M.Ter-Akopian and A.V.Yeremin, Z. Phys. A337, 229 (1990).
27. A.N.Andreyev, D.D.Bogdanov, V.I.Chepigin, O.N.Malyshev, A.P.Kabachenko, S.Sharo, G.M.Ter-Akopian, A.V.Yeremin, Z. Phys. A337,231,(1990).
28. A.N.Andreyev, D.D.Bogdanov, V.I.Chepigin, O.N.Malyshev, A.P.Kabachenko, G.M.Ter-Akopian and A.V.Yeremin, Z. Phys., A338,363,(1991).
29. A.N.Andreyev, D.D.Bogdanov, V.I.Chepigin, A.P.Kabachenko,

Yu.A.Muzychka, O.A.Orlova, B.I.Pustylnik, G.M.Ter-Akopian, A.V.Yeremin, Yad. Fiz. 52, n.3(9),640,(1990) (in Russian).
30. D.Vermeulen, H.-G.Clerc, C.-C.Sahm, K.-H.Schmidt, J.G.Keller, G.Muenzenberg and W.Reisdorf, Z.Phys. A318,157,(1984).
31. A.V.Ignatjuk et al., Yad. Fiz. 21,485,(1975) (in Russian).
32. S.Cohen, P.Plasil and W.Swiatecki, Ann. Phys. 82, 557 (1974).
33. Yu.A.Muzychka and B.I.Pystylnik, In : Int. School-Seminar on Heavy Ion Physics, Alushta,1983. Preprint JINR, D3-83-644, (Dubna, 1983), p.420.
34. T. Sikkeland, A. Ghiorso and M. J. Nurmia, Phys.Rev.172,1232 (1968).
35. E.D.Donets, V. A. Shchegolev and V. A. Yermakov, Atomnaya Energiya 20, 223 (1966) (in Russian).
B. A. Zager, M. B. Miller, V. L. Mikheev, S. M. Polikanov, A. M. Sukhov, G. N. Flerov and L. P. Chelnokov Atomnaya Energiya 20, 230 (1966) (in Russian).
36. V. L. Mikheev, V. I. Ilyushchenko, M. B. Miller, S.M. Polikanov, G. N. Flerov and Yu. P. Kharitonov, Atomnaya Energiya 22, 90 (1967) (in Russian).
37. K. Böning, Z. Patyk, A. Sobiczewski and S. Cwiok, Z.Phys. A325, 479 (1986).
Z.Patyk and A.Sobiczewski, Preprint GSI-90-53, (Darmstadt, 1990).
38. T.Sikkeland et al., Phys. Rev. 169, 1000 (1968).
39. A.S.Iljinov, Yu.Ts.Oganessian and Ye.A.Cherepanov, Yad. Fiz. 36, n.1, 118 (1982) (in Russian).
40. Ye.A.Cherepanov, A.S.Iljinov and M.V.Mebel, J.Phys.G : Nuclear Phys. 9, 931 (1983).
41. F.P.Hessberger et al., Z. Phys. A333, 111 (1989).
42. V.Ninov et al., Z. Phys. A336, 473 (1990).
43. V.E.Viola et al., Nucl. Phys. A217, 372 (1973).

TOWARDS A UNIFIED PICTURE

Distributions of Reaction Strengths and the Quest for a Unified Picture

W. Henning
Gesellschaft für Schwerionenforschung

INTRODUCTION

Heavy-ion reaction dynamics has been studied intensively over the past decades from low to intermediate energies. The theme of this conference reflects the general sentiment that a unified picture for the various phenomena observed will hopefully emerge. Whether this will also lead to a quantitative unified model may be more doubtful. In any case, the overall distribution of reactions strengths into the various reaction channels has been often considered as a zero-order guide for such an attempt. In the following some recent experimental results are summarized which might have a bearing on this discussion. I will not attempt to summerize all information that is known about the distribution of reaction strengths; rather, I will discuss some recent results and possible implications.

The discussion will follow along the most simple and direct variables which span a three-dimensional parameter space: i) excitation energy E_x : near the barrier involving few-nucleon degrees of freedom and subject to the detailed structure of the collision partners; then the transition to the statistical regime, the disappearence of, for example, collective giant modes, and finally the disintegration of the nuclear system via multifragmentation and evaporation; ii) the impact parameter and angular momentum dependence: from peripheral collisions with approximate preservation of the symmetry of the collision system to central highly dissipative rearrangement collisions; and iii) the system-size and Z/A dependence: Coulomb vs. nuclear dominance, surface vs. volume effects, and isospin-dependence, in particular for extreme collision systems involving nuclei far off stability. The real challenge of course comes from the interdependence of these variables which makes the reaction dynamics so complex. Keeping this in mind, we still want to discuss some experimental results along this simplyfied picture.

QUASI-ELASTIC COLLISIONS NEAR THE BARRIER

These have been extensively studied now for three decades, yet not in a comprehensive manner: elastic scattering data for heavy systems, for example, is essentially non-existent. Some results for heavy systems have been recently obtained at GSI from which the elastic scattering potential can be extracted. In the first case, the systems $^{208}Pb + ^{206,208}Pb$ were investigated with the particle-particle gamma-ray coincidence technique[1]. This allowed to extract the angular distributions for inelastic scattering to the low-lying collective states, as illustrated with an example in Figure 1. A coupled-channel model fit yielded parameters with unusual values. In particular, a diffuseness parameter of 0.95 fm for both the real and imaginary part of the potential is quite unexpected. It was not possible, despite the often experienced ambiguities in potential parameters for lighter systems, to significantly deviate from this value while maintaining an acceptable fit to the data.

A very different experiment, designed to study for the system $^{110}Pd + ^{238}U$ the time-delay in deep-inelastic collisions by atomic delta-electron spectroscopy, yielded a similar result[2]. At an observation angle for the elastically scattered nuclei where the trajectory is affected by the nuclear potential, delta-electron yields at high momentum are observed that are one order of magnitude larger than predicted for the pure Coulomb trajectories generally assumed in calculations (Figure 2). A reasonable description of the electron spectrum is only achieved after a nuclear potentail is added, but again only if a diffuseness of about 1.0 fm is used.

There exist model calculations in the literature[3] which may provide an answer to these seemingly unusual scattering potentials. These are calculations for $^{208}Pb + ^{208}Pb$ with a complex energy density functional derived from a Reid soft-core potential. Figure 4 shows an example of the results indicating the increase in diffuseness over a frozen-density folding model. The essential degrees of freedom determining these results are illustrated in Figure 4: the quadrupole degree of freedom leads to oblate stages at large distance when the Coulomb repulsion dominates and to prolate elongation under the attractive nuclear force at short range. The diffuseness increases at short distances due to the energy gain from increased nuclear overlap. Microscopically, the quadrupole deformation reflects the influence of low-lying collectivity modes of the system whereas the increased diffuseness and enhanced neck reflect increased nucleon exchange. Is this behaviour restricted to the heaviest systems? Of course, due to the increased charge the Coulomb excitation of collective modes is more probable in heavy systems. The nucleon transfer, however, should play a similar role in pheripheral collisions whether in light or heavy systems. In fact, recent data[5] for $^{16}O + ^{16}O$ scattering in the nuclear rainbow region indicate the need for an equally large diffuseness (1 fm) for the real potential. So why do the many systematic studies in light and medium systems yield diffusenesses more in accord with results from frozen-density folding models (typically diffusenesses between 0.5 and 0.7 fm)? One possible reason might be that due to the quantum fluctuations in collision distance for lighter systems, the results are only carrying information about a mean depth of the surface potential and not its detailed radial dependence. This also leads to the often observed ambiguities in potential parameter sets; unless (as in the $^{16}O + ^{16}O$ case) measurements are carried to the nuclear rainbow region where interference between short and large collision distances establishes the potential over a wider radial region, or in the heavy systems where the quantum fluctuations are small and the semiclassical trajectories are highly localized in r-space.

As mentioned above, the macroscopic neck has its microscopic counterpart in the nucleon transfer channels. Figure 6 shows the distribution of the quasi-elastic neutron-transfer strengths measured near the barrier ($E_{cm}/V_B \approx 1.2$) for a wide range of systems and as a fraction of the total reaction cross section. For the heavier systems this fraction approaches 100% while for the light systems it is only a few percent (for light systems also charged-particle transfer contributes of about the same magnitude). Qualitatively this behavior is readily understood by considering simple trajectory calculations as illustrated in Figure 7 for a light system ($^{16}O + ^{208}Pb$, $Z_1Z_2 < 1000$) and a heavy system ($^{208}Pb + ^{208}Pb$, $Z_1Z_2 \gg 1000$). For the light system the close impact parameters lead to maximum overlap and absorption, i.e. fusion. Only peripheral trajectories maintain the symmetry of a bi-nuclear system. Quite differently for the heavier system: the strong Coulomb repulsion never allows substantial overlap at these incident energies. Therefore the total reaction cross section (of the order 1-2 barns and thus at this energy already an appreciable fraction of the geometrical cross section) is essentially made up of only peripheral processes, and as results for the $^{208}Pb + ^{208}Pb$ system show[5] dominantly quasielastic ($|Q| < 30$ MeV) neutron transfer processes. Another interesting feature from Figure 7 is the high localization in r-space for the transfer process in $^{208}Pb + ^{208}Pb$; the implications of a highly-localized nucleon-exchange potential contribution to the total potential (pockets ?), as discussed for the Ni + Sn system earlier[6], may become even more important for the heaviest systems.

DISSIPATIVE HEAVY-ION COLLISIONS

The dominance of the quasielastic nature of the transfer processes for the $^{208}Pb + ^{208}Pb$ just discussed, arises partly from the closed-shell nature of the colliding Pb nuclei. But already also for this system, appreciable yield of more inelastic processes (yet still binary) is observed near the barrier. This has been observed even more clearly in collision systems at the barrier involving less tightly bound nuclei. An example is the $^{110}Pd + ^{238}U$ reaction studied near 6 MeV/u incident energy[7]. Processes with very negative Q-values, well below the barrier for spherical shapes of the nuclei in the exit channel are observed. The interesting (and still unanswered) question in these dissipative reaction concerns the distribution of internal excitation energy between the collision partners. Results from the $^{110}Pd + ^{238}U$ collision at the barrier suggested[8] a splitting of excitation energy heavily weighted towards the light reaction partner, and outside the range generally considered acceptable within the transport models: namely equal sharing during the initial phase of the reaction, and sharing according to mass (and thus favouring the heavy nucleus) in the equilibrated phase. The observed behavior at the barrier led us to conclude[8] that single-particle properties (mean level spacing) are the dominant factor. This seemed supported by data in the $^{50}Cr + ^{208}Pb$ system at energies well below barrier[9] (Figure 8), although an alternative explanation was proposed by the authors. At higher collision energies, one expects that the energy dissipation in nucleon transfer becomes dominated by the momentum transfer from the relative motion of the colliding nuclei (window-wall picture).

The complete distribution of reaction strengths into the various channels, including deep-inelastic processes, has been studied for the relatively heavy system $^{58}Ni + ^{124}Sn$[10] from below to well above the barrier. The results are shown in Figure 9. The deep-inelastic and fusion processes follow closely the same behavior. The decrease below the barrier is easily understood in the effective closing of channels, while the quasielastic channels dominate similarly as for the $^{208}Pb + ^{208}Pb$ system discussed above. This behavior suggests a parallel to the relation between imaginary and real optical potential depths deduced from elastic scattering in the $^{16}O + ^{208}$ system[11] and found connected through the dispersion relation[12].

HIGH EXCITATION, FRAGMENTATION AND FLOW IN HEAVY ION COLLISIONS

At the energies discussed in the previous chapters the mean field effects and built-in correlations dominate which probably requires a corresponding description based on a potential approach. When dissipation becomes very large and the excitation energy approaches the limit of stability of the nuclei involved transport models based on the statistical individual nucleon motions, like quantum-molecular dynamics (QDM) codes, should be more appropriate provided long-range correlations, in particular Coulomb effects, are adequately treated.

With increasing bombarding energy a collision system is created with higher and higher excitation energy. Several talks here have addressed the decay of such hot nucleon systems and there is not much that I can add. Clearly a lot of new information has been obtained on intermediate-mass fragment production, which appears to become an important reaction channel at collision energies near the Fermi momentum in nuclei (see various contributions to these proceedings). From the point of view of a statistical model, any corner in phase space (small as it may be) can be populated and the occurence of such events does not necessarily say much about the average reaction behavior. The distribution of strengths needs to be established for such systems on an absolute scale. I would like to point to a few specific aspects. The first concerns the recent data obtained by the ALADIN group at the new SIS accelerator facility at GSI and presented at this meeting[13]. The group succeeded in establishing the intermediate-fragment multiplicity for nuclei highly excited in peripheral collisions at 600 MeV/nucleon with 4π efficiency and individual nuclear-charge resolution. When correlating these multiplicities with light-particle multiplicities and a deduced nuclear excitation energy, the results of Figure 10 are obtained. The rise of intermediate-fragment multiplicity at about average nuclear binding in nuclei is intriguing. The detailed analysis of these data will be a severe constraint on fragmentation models as, for example, QMD.

A second aspect concerns recent results[14] obtained with bremstrahlung photons emitted in energetic nucleus-nucleus collisions. At GANIL the photon-photon correlations for bremsstrahlungs quanta above 25 MeV were measured with the TAPS BaF_2 photon detector array and the quantum intensity interference (Hanbury-Brown and Twiss effect) was observed (Figure 11). Whether this effect can indeed be used to deduce new information on the reaction behavior (space-time extent of the collision system) remains to be seen; yet the observation of the effect itself at intermediate energies is interesting.

Finally, a short comment to the nucleon motion at even higher energy and more central collisions. At several hundred MeV/nucleon, where the collision process is considered to arise from individual nucleon-nucleon interactions, rescattering and multiple interactions lead to a correlated motion, sideward flow[15]. Recently flow has been observed[16] for nucleons from a collision system at much lower bombarding energy, typically around the Fermi energy in nuclei. The fact that the magnitude of flow goes (first) towards zero when increasing the energy and then increases again has led to the interpretation that at higher energies the short-range nuclear repulsion leads to 'positive' flow while at lower energies the mean field attraction and dissipative coupling leads to 'negative' flow. In the same spirit the long-range Coulomb repulsion leads to the well known 'positive' flow of the scattering system at even lower energies, between the barrier and energies a few times the barrier. Here the short-range attraction, of course, binds the nucleons within the mean field and the flow observed is that of the two nuclei. Nevertheless, the gradual transitions from short-range correlated, Coulomb - repelled flow of the mean fields to that of the correlated flow due to statistical rescattering and short-range repulsion of individual nucleon pairs clearly must be an inherent part of any unified picture of heavy-ion reaction dynamics.

SUMMARY

The present discussion focussed on a few examples to illustrate the distributions of reaction strengths under these selected conditions. Various other results were presented at this meeting and are given in the literature. As a zero-order guide from all this information the distribution of reaction strength as a function of incident energy, angular momentum (impact parameter) and for a heavy and light system, respectively, is sketched in Figure 12. Whether these pictures can even be treated in a unified approach is questionable or, if one is more optimistic, represent an enormous challenge.

REFERENCES

1. H.J. Wollersheim et al., submitted to Z. Phys. A.
2. J. Stroth et al., contribution to the Kanazawa meeting on nucleus-nucleus collisions, Kanazawa (1991); J. Stroth, Thesis, University of Mainz (1990).
3. M. Rashdan et al., Z. Phys. A330 (1988) 417.
4. E. Stiliaris et al., Phys. Lett. B223 (1989) 291.
5. R. Holzmann et al., unpublished.
6. A.M. van den Berg, Phys. Rev. Lett. 56 (1986) 572.
7. W. Mayer et al., Phys. Lett. 152B (1985) 162.
8. G. Beier et al., Z. Phys. A336 (1990) 217.
9. H. Keller et al., Z. Phys. A328 (1987) 255.
10. F.L.H. Wolfs, Phys. Lett. 196B (1987) 113.
11. J.S. Lilley et al., Phys. Lett. 151B (1985) 181.
12. M.A. Nagarajan et al., Phys. Rev. Lett. 54 (1985) 1136.
13. C.A. Ogilvie et al., GSI-Preprint GSI-91-18, Darmstadt (1991).
14. R. Ostendorf et al., Nouvelles du GANIL, no. 38 (1991) 16.
15. H.H. Gutbrod et al., Rep. on Progr. Phys. (1989); and references therein.
16. C.A. Ogilvie et al., MSU Preprint; J. Peter, Int. Symp. on Heavy Ion Phys. and its Applications, Lanzhou (1990).

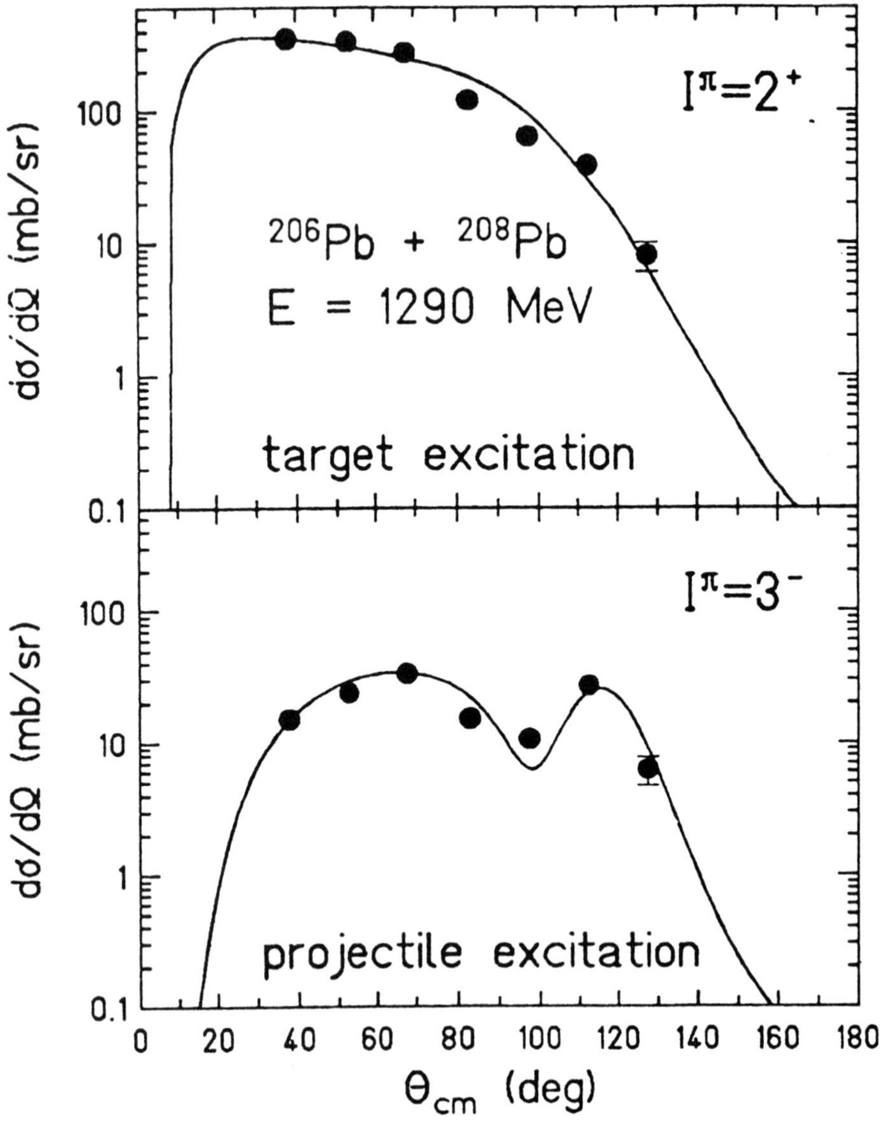

Fig. 1: Inelastic scattering for the system ^{208}Pb + ^{206}Pb [1].

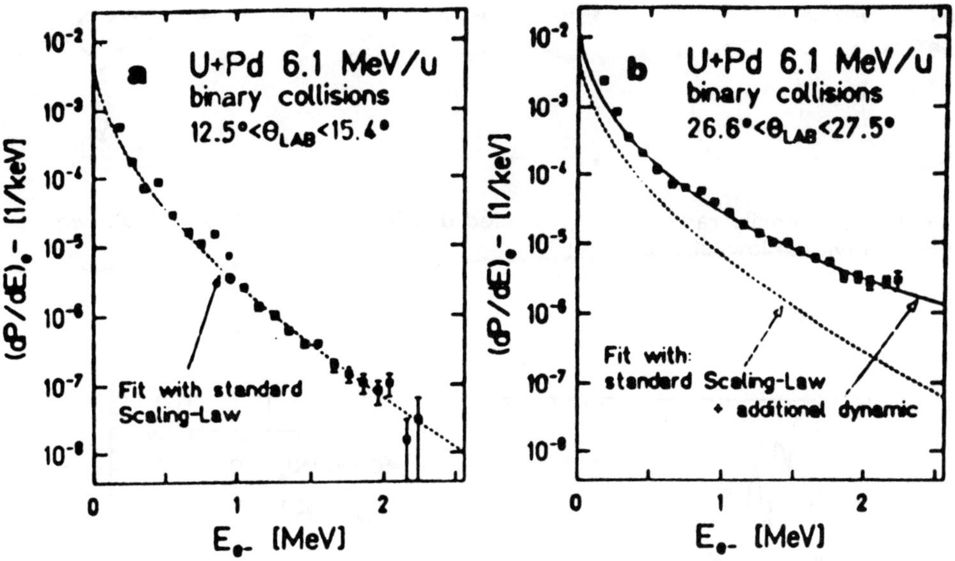

Fig. 2: Delta electron spectra for the collision system Pd + U [2].

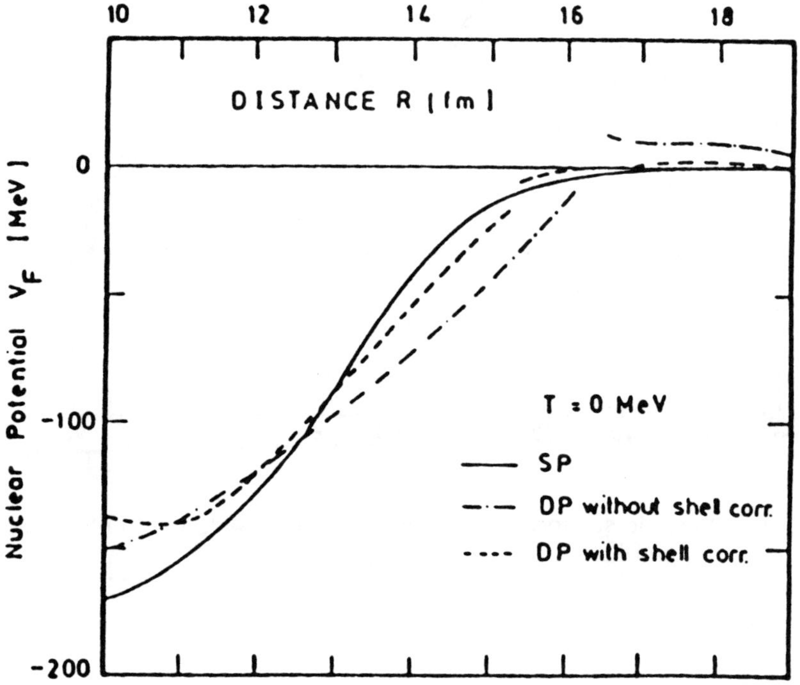

Fig. 3: Model predictions for the ^{208}Pb + ^{208}Pb scattering potential [3].

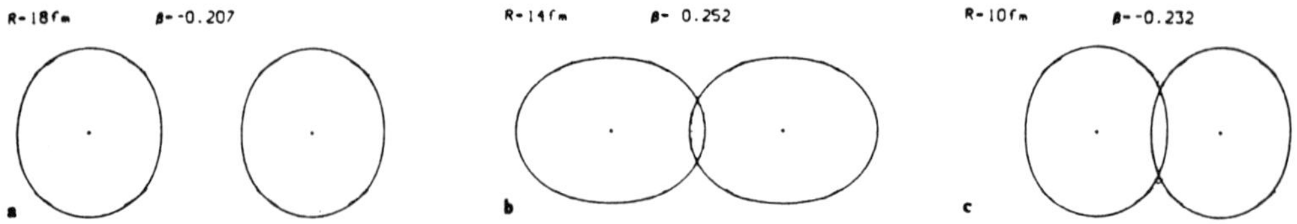

Fig. 4: Half density radii of the deformed distributions for ^{208}Pb + ^{208}Pb which give the lowest energies [3].

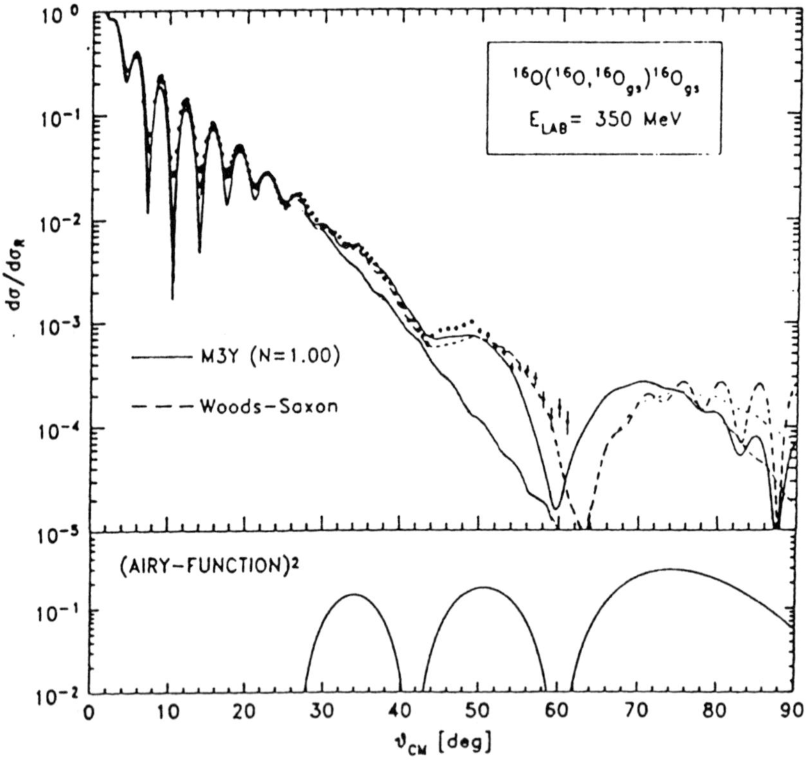

Fig. 5: ^{16}O + ^{16}O elastic scattering at 350 MeV and extending out to the nuclear rainbow region [4].

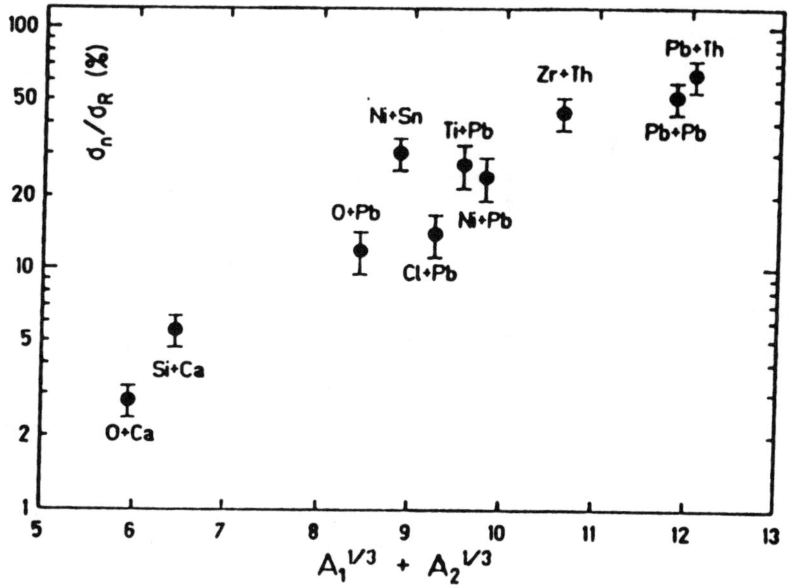

Fig. 6: Quasi-elastic neutron transfer strength at $E/V_B \approx 1.1 - 1.2$ in fractions of the total reaction cross section versus system size.

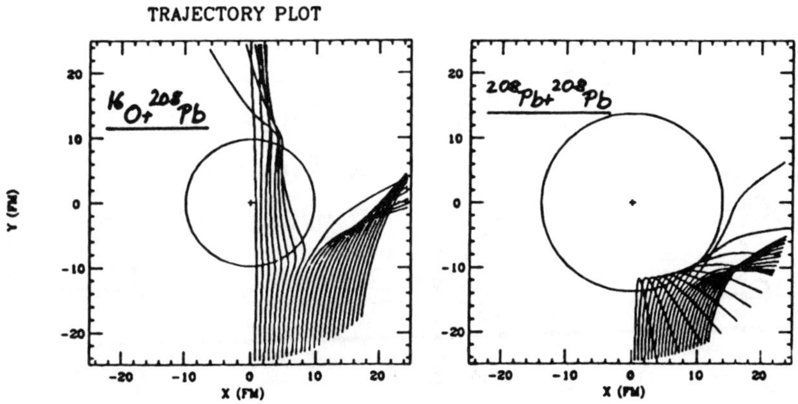

Fig. 7: Classical trajectory calculations for a "light" (^{16}O + ^{208}Pb) and a heavy (^{208}Pb + ^{208}Pb) scattering system at $E/V_B \approx 1.2$.

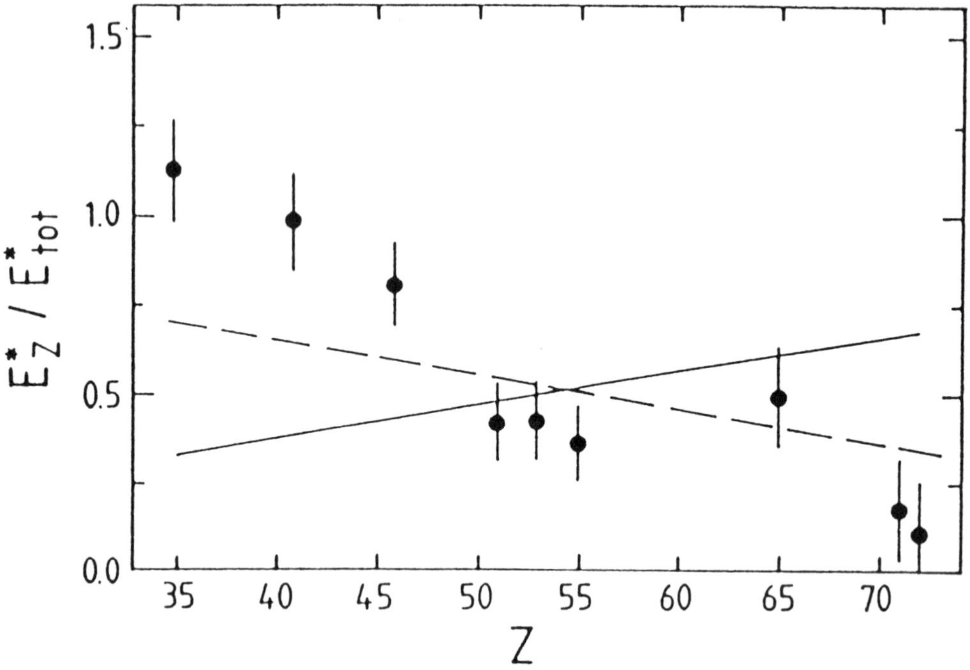

Fig. 8: Division of the total excitation energy between exit-channel fragments for the reaction ^{54}Cr + ^{208}Pb at 4.9 MeV/u [9]. The solid lines are various model calculations [8].

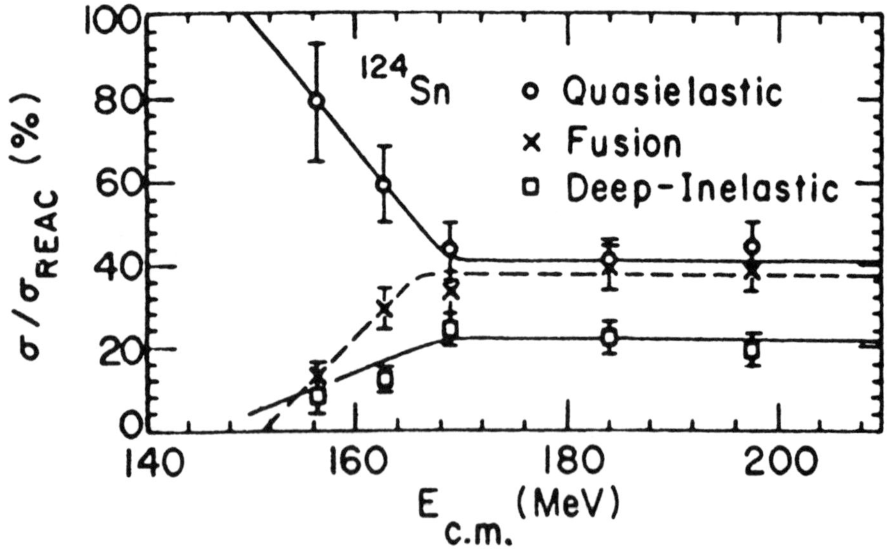

Fig. 9: Strength of fusion, deep-inelastic and quasielastic scattering as a fraction of the total reaction for ^{58}Ni + ^{124}Sn at various energies [10].

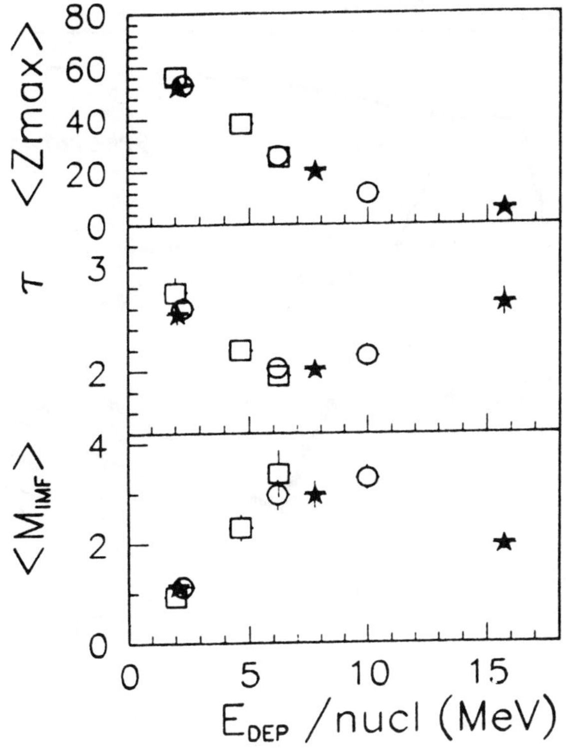

Fig. 10: Average maximum charge, slope parameter τ of the charge distribution, and mean intermediate-mass fragment multiplicity M_{IMF} for a 600 MeV/u ^{197}Au beam incident on targets of C (squares), Al (circles), and Cu (stars) [13].

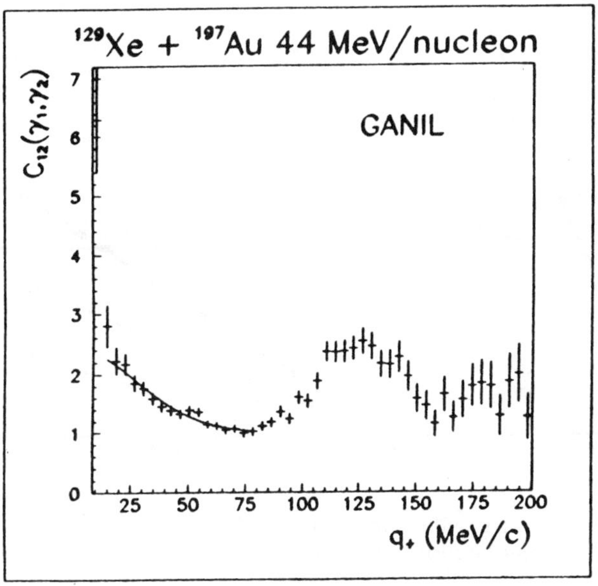

Fig. 11: Two-photon correlation measured for the ^{129}Xe + ^{197}Au system at 44 MeV/u [14].

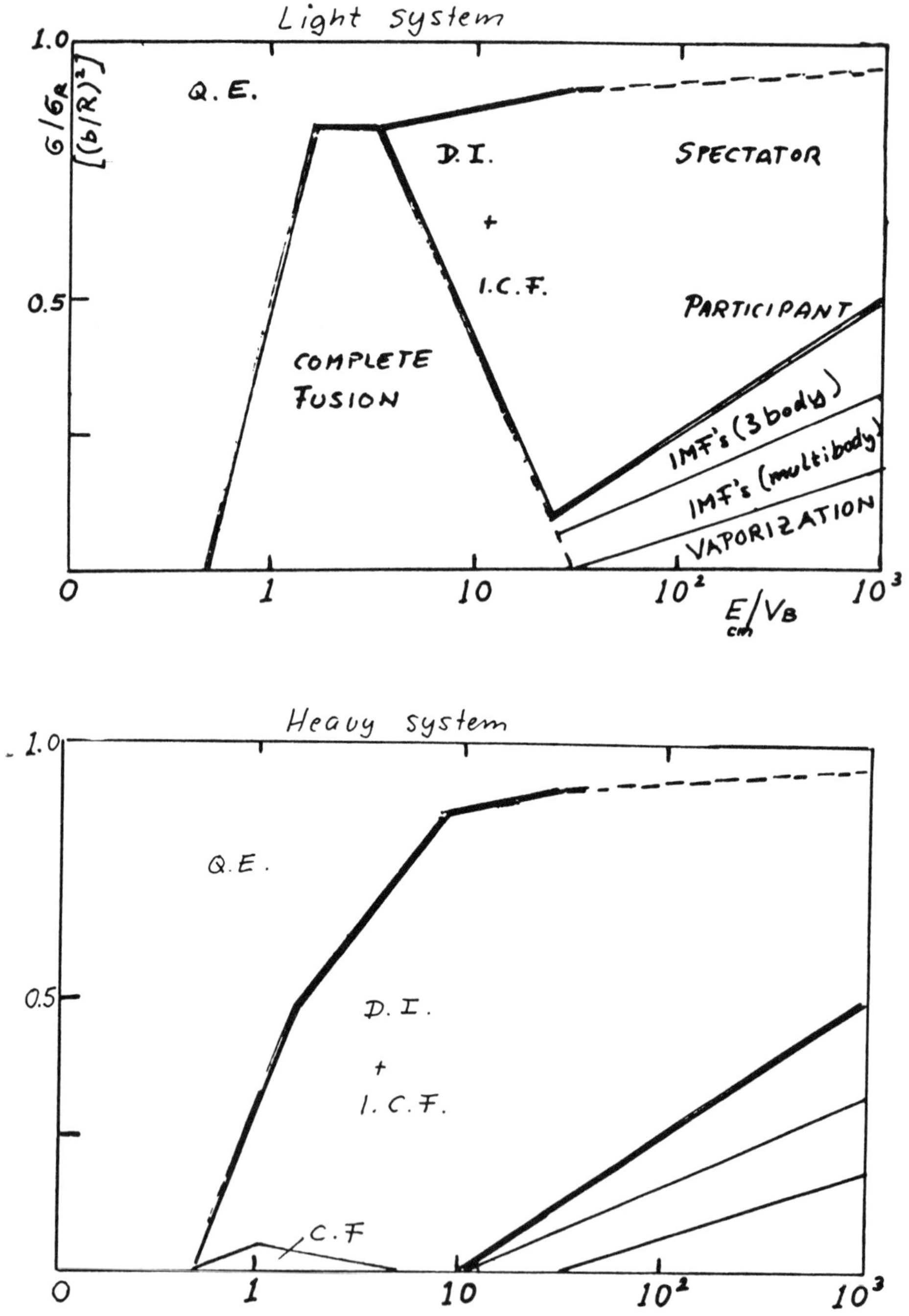

Fig. 12: Distribution of reaction strengths for a light and heavy system as a fraction of the total reaction cross section (see text).

UNIFIED THEORY OF GROUND STATE PROPERTIES OF NUCLEI AND NUCLEAR DYNAMICS

V.A.Khodel
Kurchatov Atomic Energy Institute, Moscow 123182, USSR

V.R.Shaginyan
Leningrad Nuclear Physics Institute
Gatchina, Leningrad 188350, USSR

ABSTRACT

A new microscopic approach to the theory of dynamic correlation, ground state and collective excitations of many-body systems is suggested. These dynamic and ground state properties are known as being interconnected and should be considered on the same base. A useful tool for this goal is a response function formalism. An exact functional equation for effective interaction and linear response function is obtained. The reduction method of the functional equation to a closed set of equations is given. In the simplest case the functional equation may be transformed into nonlinear integrodifferential equation. Few examples of many-body systems are considered on the base of this equation. The developed method is applied to gas–liquid phase transition, for evaluation of transport coefficients and collective excitation width.

1 Introduction

Interest in unified theory of nuclear dynamics has constantly intensified. However up to now, the developed theory for the description of properties of nuclear dynamics and nuclear matter does not exist. This is exceedingly complicated problem because its solution involves simultaneous consideration of different aspects of many-body problem. In other words, one should study the equation of state, excitation spectrum and phase transition on the ground of the same theory. We shall use a simple example to illustrate it. Let us consider an infinite system of particles interacting through a short-range binary potential, so that the system has equilibrium density and energy. Nuclear matter or ^3He may be taken as examples of such systems. At low enough density, we deal with dilute gas. Under this condition gas is stable and its compressibility K is positive

$$K = \rho dP/d\rho.$$

Here P is the pressure and ρ is the density of the system. As the density is increased the compressibility K tends to zero. At some density ρ_1, K is equal to zero and gas becomes unstable. At this stage, the system undergoes a liquid–gas phase transition. Liquid will remain unstable until K becomes positive at some density ρ_2. The system is stable if density $\rho \geq \rho_2$ and it has equilibrium energy E_o at the density ρ_o. It should be mentioned that in the vicinity of the phase transition the system becomes soft and its excitation spectrum appreciably changes. Also the phase transition point is characterized by strong correlations and it has a large influence on the ground state properties of the system. Bearing in mind all this distinguishing features, we have to conclude that the desirable theory should a) be nonperturbative, b) deal with excitation spectrum explicitly, c) have real mechanism for checking phase transitions. The importance of these topics has long been stressed [1–3]. Here we present a new approach to self-consistent description of the dynamic and ground state properties. An appropriate formalism which satisfies the above requirements might be the linear response function one. This formalism is a very flexible and power tool because it allows us to calculate the equation of state, spectra of excitations and points of phase transitions.

The outline of the paper is as follows In sec. 2 we present our formalism and formulate equations for the linear response function and for the effective interaction in terms of bare interaction. Sec.III is devoted to the demonstration of the new version Fermi system approach serviceability. We consider the problem of the evolution of the transport coefficients and resonance width in sec.IV. Paper is concluded by a brief summary.

2 General Formalism

2.1 Functional equation

In this section we derive equations for the linear response function and for the effective interaction. Let us consider a system which is acted by a small external field $\delta v(q,\omega)$. This field induces its own density fluctuations in the system with wave vector $\vec{q}$ and frequency ω:

$$\delta\rho(q,\omega) = \chi(q,\omega)\delta v(q,\omega). \tag{1}$$

Eq.(1) defines linear response function $\chi(q,\omega)$.

In field with imaginary frequency $\omega = i\omega$) the energy transfer is absent, and the system ground state remains the ground one. The system internal energy change

$$\delta^2 E = (0|H|0) - (\tilde{0}|H|\tilde{0})$$

can be expressed in terms of response function [4]. Here H is Hamiltonian of the system, and 0 is the ground state wave function.

$$\delta^2 E = -\frac{1}{2}\sum \delta v(k,\omega)\chi(k,\omega)\delta v(-k,-\omega). \tag{2}$$

By using eq. (1), one obtains

$$\delta^2 E = -\frac{1}{2}\sum \delta\rho(k,\omega)\chi^{-1}(k,\omega)\delta\rho(-k,-\omega).$$

Then one straightforwardly gets

$$\chi^{-1}(k,\omega) = -\frac{\delta^2 E}{\delta\rho(k,\omega)\delta\rho(-k,-\omega)}. \tag{3}$$

In the case of noninteracting particles we have

$$\chi_0^{-1}(k,\omega) = \frac{\delta^2 T_0}{\delta\rho(k,\omega)\delta\rho(-k,-\omega)}, \tag{4}$$

where χ_o is the ideal fermion polarizability, and T_o stands for the energy of non-interacting particles.

Ground state energy of the system is known to be given by the formula [5]

$$E = T_o - \frac{1}{2}\int_o^g \frac{dg'}{g'}\int \frac{d\vec{k}d\omega}{(2\pi)^4 i}\left[\chi(k,\omega) + 2\pi\rho\,\delta(\omega)\right] v(k), \tag{5}$$

where g is coupling constant, $v(k)$ is the bare particle-particle interaction.

An equation determining the linear response function χ is obtained as a result of varying eq. (5) with respect to the density ρ and taking into account eqs. (3), (4):

$$-\chi^{-1}(q,\omega) = -\chi_o^{-1}(q,\omega) - \frac{1}{2}\frac{\delta^2}{\delta\rho(q,\omega)\delta\rho(-q,-\omega)}\int\frac{dg'}{g'}\int\frac{d\vec{k}d\omega'}{(2\pi)^4 i}$$
$$\left[\chi(k,\omega') + 2\pi\rho\,\delta(\omega')\right] v(k). \tag{6}$$

The integration in eq.(5) and (6) over ω' goes along imaginary axis of the complex ω'-plane.

Eq.(6) permits us to determine the linear response function proceeding from the bare interaction v. It is convenient to introduce the function $R(q,\omega)$, which we shall call the effective interaction. In uniform medium it is introduced by the formula

$$R(q,\omega) = \chi_o^{-1}(q,\omega) - \chi^{-1}(q,\omega). \tag{7}$$

Then, upon a simple transformation, we obtain the functional eq. for R [4]:

$$R(q,\omega) = \varphi(q,\omega) - \frac{1}{2}\frac{\delta^2}{\delta\rho(q,\omega)\delta\rho(-q,-\omega)}\int\frac{dg'}{g'}\int\frac{d\vec{k}d\omega'}{(2\pi)^4 i}$$
$$\left[\chi(k,\omega) - \chi_o(k,\omega')\right] v(k), \tag{8}$$

where

$$\varphi(q,\omega) = -\frac{1}{2}\frac{g^2}{\delta\rho(q,\omega)\delta\rho(-q,-\omega)}\int\frac{d\vec{k}d\omega'}{(2\pi)^4 i}\chi_o(k,\omega')v(k) + v(q). \tag{9}$$

Here we have picked out Hartree term $v(q)$ explicitly.

It is not difficult to generalize eq. (8) for the effective interaction to finite temperature. The functional eq. for effective interaction at finite temperature may be constructed in a fashion analogous to our procedure at zero temperature. There are, however, some important differences, which we comment on briefly. First, we deal with the free energy F instead of the ground state energy E. Also there is the other relation between the dynamic form factor and the linear response function [5]

$$S(k,\omega) = -\frac{1}{2\pi} Im\,\chi(k,\omega)\,coth\frac{\omega}{2T}.$$

We may therefore write

$$F(T,\rho) = F_o(T,\rho) - \frac{1}{2}\int_o^g \frac{d\vec{k}d\omega}{(2\pi)^4}[\chi(k,\omega) - \chi_o(k,\omega)]\,coth\left[\frac{\omega}{2T}\right] v(k), \qquad (10)$$

where F_o is the free energy to be calculated in Hartree-Fock approximation, χ_o is the ideal fermion polarizability. And the functional eq. for effective interaction R takes the form:

$$R(q,\omega,T) = \varphi(q,\omega,T) - \frac{1}{2}\frac{\delta^2}{\delta\rho(q,\omega)\delta(-q,-\omega)}$$

$$\int \frac{dg'}{g'}\int \frac{dkd\omega'}{(2\pi)^4}[\chi(k,\omega') - \chi_o(k,\omega')]coth\left(\frac{\omega'}{2T}\right) v(k)$$

$$\varphi(q,\omega,T) = -\frac{1}{2}\frac{\delta^2}{\delta\rho(q,\omega)\delta\rho(-q,-\omega)}\int \frac{dkd\omega'}{(2\pi)^4}\chi_o(k,\omega')coth\left(\frac{\omega'}{2T}\right) v(k) + v(q). \qquad (11)$$

The response function χ is given by eq. (7), or in more usual form

$$\chi(q,\omega) = \chi_o(q,\omega) + \chi_o(q,\omega)R(q,\omega)\chi(q,\omega). \qquad (12)$$

2.2 Local approximation

We have obtained functional equation (8),(11) for the interaction R. Now the basic problem is to find an efficient procedure for solving the above exact equation for R. Strictly speaking, we need scheme for calculation of functional derivatives. The difficulty may be bypassed by using so-called local approximation. Here we consider briefly this approximation in the coordinate space (it was described in detail in ref. [4]). Let $F(\rho,\vec{z})$ be a density functional, which depends on a variable $\vec{z}$, then one has in local approximation

$$\frac{\delta F(\rho,\vec{z})}{\delta\rho(\vec{x})} = \frac{dF(\rho,\vec{z})\delta(\vec{x}-\vec{z})}{d\rho}. \qquad (13)$$

Eq. (13) is exact in an integral sense, because

$$\int \frac{\delta F(\rho,\vec{z})d\vec{x}}{\delta\rho(\vec{x}} = \frac{dF(\rho,\vec{z})}{d\rho}.$$

Upon simple transformation of eq.(8) we obtain the nonlinear integro - differential equation for the effective interaction at zero temperature [4]. Eq. (11) will be discussed in Sec.4.

$$R(q) = \varphi(q) - \frac{1}{2}\frac{\partial^2}{\partial\rho^2}\int \frac{d\vec{k}d\omega dg'}{g'(2\pi)^4}\frac{\chi_o^2(k,i\omega)R(k)}{1-R(k)\chi_o(k,i\omega)}v(\vec{q}-\vec{k}). \tag{14}$$

$$\varphi(q) = -\frac{1}{2}\frac{\partial^2}{\partial\rho^2}\int \frac{d\vec{k}d\omega}{(2\pi)^4}\chi_o(k,i\omega)v(\vec{k}-\vec{q}).$$

The analytical expression for the linear response function $\chi_o(k,i\omega)$ may be found in ref. [4]. We conclude this section with a brief review of essential features of eqs. (8),(11),(14). Solving these equations for the effective interaction, we have found the linear response function χ. Using the function χ one can calculate spectrum of a system and its ground state properties. Then, by making use of eq. (3), one finds [5]

$$\chi^{-1}(q,0)|_{q\to 0} \sim -\frac{\partial^2 E}{\partial^2\rho^2} \sim -\frac{K}{\rho^2}. \tag{15}$$

Hence, if function χ has the pole, then the compressibility $K=0$, and we deal with liquid–gas phase transition. By eq. (15), on passing $q\to 0$, we find

$$\chi^{-1}(T,\rho) = 1 - R(T,\rho)\chi_o(T,\rho) = 0$$

as the equation determining the phase transition point ρ, T.

3 Application to the Model Systems

Let us consider the potential $v(r)$ to contain repulsive components as well as attractive ones. So the uniform system of particles interacting via this potential has an equilibrium energy E_o and the density ρ_o [6].

$$v(r) = [2V_1\exp(-\mu_1 r) + V_2\exp(-\mu_2 r)],$$

where $V_1 = 1458.2\ MeV\cdot fm$; $V_2 = -578.2\ MeV\cdot fm$; $\mu_1 = 3.11\ fm^{-1}$, $\mu_2 = 1.55\ fm^{-1}$.

The results of calculations prove that the system has the equilibrium energy $E_o = 90\ MeV$ at the density $\rho_o \approx 1\ fm^{-3}$. At low enough density, $\rho_2 \approx 0.75\ fm^{-3}$, the compressibility approaches zero. At this point

$$\chi^{-1}(q,0)|_{q\to 0} \approx K,$$

and linear response function χ possesses the pole. On passing from ρ_2 to ρ_1, compressibility K is negative. In accordance with it the function χ (or the kernel of eq. (14)) gets a singularity, and eq. (14) has no solution in this domain. It is at the point $\rho = \rho_1$ that K is equal to zero again. One can prove that there is no instabilities in the region $\rho < \rho_1$. At low enough density, eq. (14) for the effective interaction R may be reduced to the linear one, in the case the function R makes a sense of a vacuum scattering amplitude [4], and we have the dilute gas system. Within this approach uniform systems of particles interacting through the long-range Coulomb potential and short-range Bethe potential have been considered too. All results are in a good agreement with well-known Monte Carlo calculations. The details may be found in Refs. [4,7,8].

4 Frequency-Dependent Effective Interaction

In Sec.3 we have dealt with the effective interaction which does not depend on frequency ω. We need $\omega-$dependent effective interaction to consider collective modes width and transport coefficients. On making use of the functional eqs. (8),(11) and the local approximation one can construct closed equation for $\omega-$dependent effective interaction. For this purpose second functional derivative of the linear response function χ with respect to density is to be evaluated. The problem may be resolved in this way. We perform the functional differentiation of χ_o with respect to the density exactly, local approximation for calculation functional derivatives of effective interaction is used [4,9]. As a result, upon simple transformations, the following closed equation for function $R(q,\omega)$ is obtained.

$$R(q,\omega) = \varphi(q,\omega) - \frac{1}{2}\int \frac{dk d\omega' dg'}{g'(2\pi)^4}\chi''(q,\omega;k,\omega')coth\left(\frac{\omega'}{2T}\right)v(k),$$

$$\varphi(q,\omega) = -\frac{1}{2}\int \chi_o''(q,\omega;k,\omega')coth\left(\frac{\omega'}{2T}\right)v(k)\frac{dk d\omega'}{(2\pi)^4} + v(q), \qquad (16)$$

where

$$\chi''(q,\omega;k,\omega') = 4\frac{\partial R(k,\omega')}{\partial\rho}\frac{\chi^3(k,\omega')}{\chi_o^2(k,\omega')}\chi_o'(q,\omega;k,\omega') +$$

$$+\chi^2(\vec{q}+\vec{k},\omega')\left[\frac{\partial^2}{\partial\rho^2}R(\vec{q}+\vec{k},\omega') + 2\left(\frac{\partial}{\partial\rho}R(\vec{q}+\vec{k},\omega')\right)^2\chi(\vec{q}+\vec{k},\omega')\right]$$

$$+\chi_0''(q,\omega;k,\omega')\left[\frac{\chi^2(k,\omega')}{\chi_o^2(k,\omega')} - 1\right] + \frac{2[\chi_o'(q,\omega;k\omega')]^2\chi^3(k,\omega')R(k,\omega')}{\chi_o^3(k,\omega')}.$$

Here

$$\chi_o'(q,\omega;k\omega') = \frac{\delta\chi_o(k,\omega')}{\delta\rho(q,\omega)} = \chi_o^{-1}(q,\omega)\chi_o^{(2)}(q,\omega;k,\omega')$$

is the functional derivative of the ideal fermion polarizability with $\chi_o^{(2)}$ being the second-order response function [4,9] of independent particles. Analogously, function χ_o'' is defined by equation

$$\chi_o''(q,\omega;k\omega') = \frac{\delta^2\chi_o(k,\omega')}{\delta\rho(q,\omega)\delta\rho(-q,-\omega)} =$$

$$= \frac{1}{\chi_o^2(q,\omega)}\left[\chi_o^{(3)}(q,\omega;k,\omega') - \frac{\chi_o^{(2)}(q,\omega;k\omega')\chi_o^{(2)}(q,\omega;q\omega)}{\chi_o(q,\omega)}\right],$$

where $\chi_o^{(3)}$ is the third-order response function of independent particles [4,9].

Having determined the effective interaction $R(q,\omega)$, which is governed by eq.(16), we can calculate resonance widths and transport coefficients. Of course, ground state properties and spectrum of excitations may be calculated also. The linear response function is of the form (eq.(12))

$$\chi(k,\omega) = \frac{\chi_o(k,\omega)}{1 - R(k,\omega)\chi_o(k,\omega)}. \tag{17}$$

On the passing to the limit $k \to 0$ (so that $kv_F/\omega \ll 1$) one finds

$$\chi(k,\omega) = \frac{\alpha k^2}{\omega^2 - c^2k^2 + ik^2\omega\Gamma}, \tag{18}$$

where c is the zero sound velocity ($T = 0$), v_F is a particle velocity on the Fermi surface, Γ is the resonance width, and α is a dimension factor.

At the hydrodynamical limit ($T \neq 0$), when collisions play essential role, c is the first sound velocity. One should be careful for the passing to the limit $(k,\omega,T) \to 0$ because the response function has an essential singularity at the origin.

By comparing eq. (17) with the result (18), one obtains

$$\Gamma = \lim_{\omega\to 0}\lim_{k\to 0}\frac{1}{k^2\omega}\chi_o(k,\omega) Im\, R(k,\omega).$$

Analogously, the coefficient of kinematical viscosity ν obeying the equation [10]

$$\nu = \lim_{\omega\to 0}\lim_{k\to 0} m^2\omega^3\, Im\,\chi(k,\omega)$$

can be calculated.

5 Summary

In this paper, having used response function formalism and exact relations between ground state energy (or free energy) and linear response function, we have derived a rigorous functional equations for the linear response function and for the effective interaction. The present functional equations combined with the use of the local approximation has been applied to the construction of the closed integro-differential equations for effective interaction. Thus, a new microscopic approach to the description of the main properties of systems is suggested. These properties are parameters of the ground state, spectrum of excitations, resonance widths and phase transition points.

In this paper we are principally concerned with general formulation of the theory to focus our attention on the main features of the approach. We have limited ourselves to the consideration of the density-density response function. Nevertheless, most of the results which have been obtained are easily transported to other types of response functions [11].

References

[1] A.D.Jackson, A.Lande and R.A.Smith, Phys.Rep.,86,55 (1982).

[2] E.Krotscheck, Phys.Rev. B31, 4264 (1985).

[3] M.F.Jiang, J.Heyer, S.D.Yang and T.T.Kuo, Phys.Rev.Lett. 61, 38 (1988).

[4] V.A.Khodel and V.R.Shaginyan, Nucl.Phys.A500, 242 (1989).

[5] D.Pines and F.Nozieres, The Theory of Quantum Liquids, vol.1 (Benjamin, New York, 1966).

[6] E.Krotscheck, Nucl.Phys. A465, 461 (1986).

[7] V.R.Shaginyan, Solid St.Comm. 55, 9 (1985).

[8] S.A.Artamonov, S.G.Ogloblin and V.R.Shaginyan, Yad.Fiz. 46, 1651 (1987).

[9] V.A.Khodel and V.R.Shaginyan, Yad.Fiz.46, 1638 (1987).

[10] D.Forster, Hydrodynamic, Fluctuations, Broken Symmetry, and Correlation Functions (Benjamin, New York, 1966).

[11] V.A.Khodel and V.R.Shaginyan, Phys.of Elementary Part. and Atomic Nuclei 22, 436 (1991).

Mean-field, dissipation and fluctuations in heavy-ion reactions

B. Remaud, B. Benhassine, M. Farine, D. Idier, G. Royer and F. Sébille

Laboratoire de Physique Nucléaire de Nantes
UA IN2P3/CNRS et Université de Nantes N° 57
2, rue de la Houssinière 44072 Nantes, Cedex France

Abstract :

Transport theories, based on the quantum-corrected Boltzmann equation, are widely used in the theory of heavy-ion reactions. These theories are very successfull in the analysis of the average dynamics of heavy-ion reactions ; they allow a microscopic treatment of the balance between mean field and two-body collisions. Two limiting situations are studied : the scission of a nucleus and the characteristics of the nuclear sidewards flow. A study is then presented on the possibility to go beyond this average dynamics and to treat the dynamics of fluctuations.

1. INTRODUCTION

Heavy-ion reactions provide physicists with a unique way to study droplets of Fermi liquid in interaction. When the nuclei are sufficiently heavy ($A > 100$, 150), the finite size effects are lowered and then nuclei can be considered as laboratories to test the theories of strongly interacting quantum N-bodies.

A few tens years ago, theories of heavy-ion reactions were dominantly theories of collective motions dissipatively coupled to the intrinsic degrees of freedom ; they have lead to impressive progress in the knowledge of the nuclear viscosity and have been very successfull in describing a large variety of experimental observations (such as fusion, and deep inelastic reactions)[1] through the dissipative dynamics of collective variables [2,3] (linear and angular momenta, mass, charge and relative motion).

Below 10 MeV per nucleon initial kinetic energy, time scales are such that the decomposition of the N-body hamiltonian in a collective hamiltonian and an intrinsic one was both meaningful and efficient to cope with the complexity of the full N-body problem :

$$H = H_{coll}(Q_j) + H_{int}(q_i) + H_{coupling}(Q_j, q_i). \tag{1.1}$$

Such a decomposition remains phenomenological since there is not yet any theory to find unambiguously the pertinent collective variables of a general N-body hamiltonian. However, theoretical inspection and experimental observations have singled out the role of the relative motion, the linear and angular momenta (and their conjugate variables) in the dynamics of reaction. The long time characteristics ($> 10^{-22}, 10^{-21}$ sec) of these collective variables relatively to the intrinsic time characteristics (10^{-23}sec) made meaningfull the use of equation (1.1), with a coupling term sufficiently small such that one can use the well established linear response theory [4)].

Dissipative reactions around 10 MeV/u can then be viewed as the brownian motion of a few slow collective variables in the swarm of numerous, swift and uncorrelated intrinsic variables. The Fokker-Planck equation is the basic theoretical tool for such a study ; the time evolution of each collective variable Q being schematically described as :

$$\frac{dQ(t)}{dt} = \dot{Q}_{reversible} + \dot{Q}_{dissipative} + \dot{Q}_{fluctuation} \tag{1.2}$$

The reversible contribution comes from the projection of the average nuclear mean-field onto the Q subspace ; the dissipative part is the average result of the coupling with the intrinsic degrees of freedom. The fluctuating contribution (which averages to zero) often treated with a markovian approximation, is strongly connected with the dissipative part via the fluctuation-dissipation theorem[4]. The whole picture is based on a weak coupling hypothesis, which implies that fluctuations do not drive the collective variables away from an "average" stable trajectory determined by the reversible and dissipative contributions. This hypothesis obviously breaks down in cases such as the fission close to the barrier, where the fluctuations may clearly let the system jump on divergent trajectories : (those which lead to fission and those which not). In that case, specific theories have to be worked out[5].

In a parallel way, fully microscopic and (partially) quantum theories of nuclear dynamics have been developed on the basis of the Time-Dependent Hartree-Fock (TDHF[6)])theory. These theories rely on the projection of the N-body density onto a set of Slater determinants, the trajectories of which are followed in time under the action of the nuclear mean field. These independent-particle models however fail at properly describing the nuclear viscosity (spurious low-l window of nuclear transparency) and the dispersion of collective observables around their average values. The residual interaction could theoretically be introduced by a coupling of the TDHF theory with a master equation which could treat the transition of the Slater determinants from a trajectory to another one [7]. However, such a coupling leads to numerical intricacies, difficult to overcome in realistic studies.

The coming into operation of several heavy-ion facilities with beam energies ranging from a few tens to a few thousands MeV per nucleon, has put into evidence the need of new theories which have to be microscopic and have to properly incorporate the two-body collisions in a non-perturbative way. Indeed, microscopic descriptions are mandatory since time scales become so short that a distinction between collective and intrinsic variables is meaningless ; the phase space widening opens room to two-body collisions which are less and less hindered by the Pauli blocking. These strong demands on theories are somewhat smoothed out by the relaxation of the quantum constraints ; at high excitation energies, the quantum correlations such as pairing or shell effects are washed out and semi-classical [8] (or even classical ?) descriptions of nuclei become eligible.

These features have boosted the role of theories based on the (quantum corrected) Boltzmann equation [9–12]. Since the earlier papers, a blooming of published works have demonstrated the ability of the (good old) Boltzmann equation to give insights on the highly dissipative heavy ion reactions [13], even if one has to underline the weakenesses of its formal derivation from the quantum N-body formalism.

The Boltzmann-based approaches (Landau-Vlasov, V.U.U. and B.U.U.) are recognized as the standard theories of heavy-ion reactions ; they provide a natural and smooth link with the above -discussed theories of heavy reactions at low energies. They have a strong predictive power of nuclear properties even at the limits of nucleus stability[13]. However, there are strong interrogations on their reliability when these limits of stability are passed and when nuclear disassembly is expected to appear (and seems to be observed through the growing production of small nuclear fragments). Recently, a few works [14–16] have been published which attempt to overcome the limitations of the average Boltzmann equation. Numerical applications[17] of these recent theories are under way and predictions on experimentally measurable quantities are still expected.

The objective of this paper is to present an overview of the present state-of-art in the transport equations for heavy-ion physics and of their extensions which provide challenging prospects to nuclear theory as well as exciting cross-disciplinary links with the statistical theory of classical and quantum liquids.

The work is organized as follows : in the next section, we present the Boltmann-based transport theories with an emphasis on their solution methods since, apart a few sample cases, the transport equations are more simulated than solved in actual calculations. In sect. 3, we shall analyze from two examples within one of this approach (Landau-Vlasov), the delicate balance between the mean-field and dissipative contributions to the dynamics of observables ; we shall concentrate on the scission of nuclei and on the sidewards flow of nuclear matter in semi-central collisions. Section 4 will present a general discussion of extensions of formalisms in order to treat the fluctuations and possibly the nuclear disassembly through the exponential growth of density fluctuations[18].

2. QUANTUM BOLTZMANN EQUATION

2.1 Overview

The quantum Boltzmann equation (Q.B.E.) gives the time evolution of the one-body density in phase space as a sum of two contributions :

$$\frac{df(\vec{r},\vec{p};t)}{dt} = \dot{f}_{\text{mean field}} + \dot{f}_{\text{collisions}} \tag{2.1}$$

The mean field contribution can be deduced from the Wigner transform of the Hartree-Fock mean field and then includes genuine quantum features as the exchange potential coming from the antisymmetrisation. At the first order of an expansion in powers of $\hbar$, one gets [8)]

$$\dot{f}_{\text{mean field}} = -\{f, \mathcal{H}\} + O(\hbar^2) \tag{2.2}$$

which is the one-body Liouville equation. The well-known properties of the Liouville operator, which keeps constant the densities in phase space, ensures that any initial one-body distributions, which fulfils quantum constraints on phase space occupations, will conserve these properties when propagated by equation (2.2).

The 2-body collisions are taken into account by a Boltzmann collision term with in-medium corrections :

$$\dot{f}_{collisions} = g \int \frac{d\vec{p}_2 d\vec{p}_3 d\vec{p}_4}{h^9} \delta(\epsilon_1 + \epsilon_2 - \epsilon_3 - \epsilon_4)$$

$$\delta(\vec{p}_1 + \vec{p}_2 - \vec{p}_1 - \vec{p}_2)|\vec{v}_1 - \vec{v}_2| \frac{d\sigma}{d\omega} [\bar{f}_1 \bar{f}_2 f_3 f_4 - f_1 f_2 \bar{f}_3 \bar{f}_4] \tag{2.3}$$

The quantum constraints appears in the Pauli-blocking factors $\bar{f}_i = (1 - f_i)$, heuristically introduced by Nordheim [19] and Uehling and Uhlenbeck [20]. The cross section σ is that of the nucleon-nucleon scattering in the nuclear medium, the precise evaluation of which is still an open theoretical question.

Q.B.E. is a powerful tool to investigate the two contributions of (2.1) to heavy ion dynamics. Figures 1 and 2 illustrate the general evolution of one-body densities during the collision of two heavy-ions. Figure 1 shows the spatial densities projected on the reaction plane, the beam direction being horizontal ; the first row shows the two ions at their maximum overlap. Figure 2 displays the velocity distributions at the final stage of the reactions.

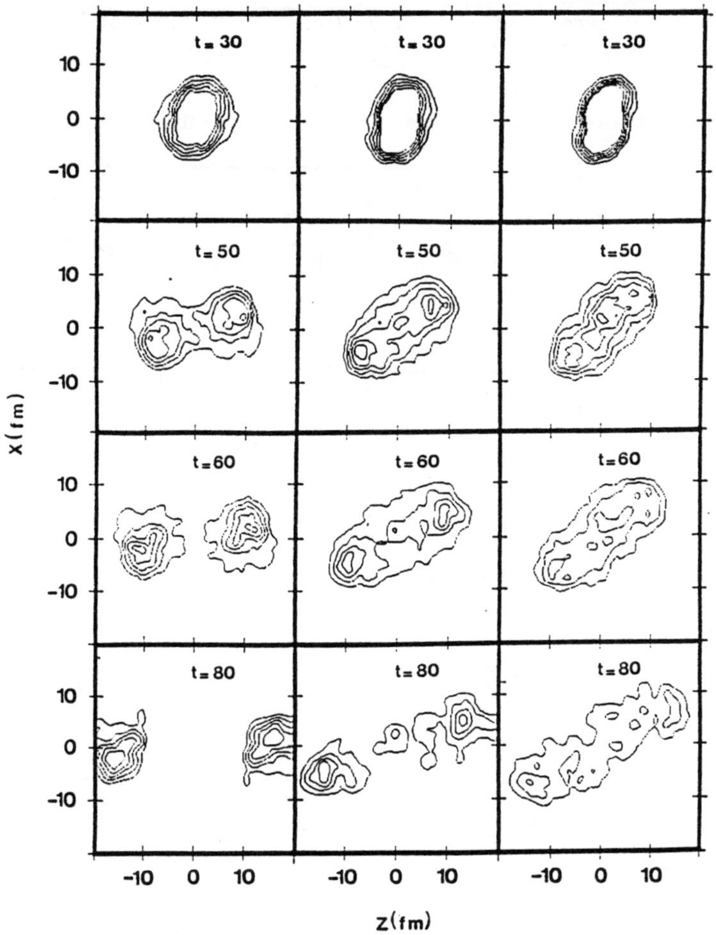

Fig. 1 Time evolution of the density in r-space for a heavy-ion collision. The contour lines show on a linear scale the density projected on the reaction plane. The collision here considered is Nb + Nb (150 MeV/u) at impact parameter b=3fm. The different columns show results for various values of the nucleon-nucleon cross sections : left column, $\sigma_{nn} = 0$; middle column, σ_{nn} takes the standard value of reference 12 ; right column, σ_{nn} is artificially doubled.

The collision here considered is that of two Nobium nuclei at 150 MeV/u beam energy, with an impact parameter of 3 fermis. The figures display the aspects of the one-body densities when are varied the respective weights of the mean field and of the 2-body collisions ; such a variation is achieved by an arbitrary variation of the n-n cross section.

Within the pure mean field approximation ($\sigma = 0$), in the r-space (fig.1, first column) one observes the well-known nuclear transparency ; the two nuclei flow through each other with little interaction ; there is a small dissipation, mainly through surface modes. The initial momentum distributions remain basically unperturbed : two Fermi spheres separated by the initial shift corresponding to the relative velocities of the projectile and of the target (fig.2, first row). Dynamics is dominated in this regime by the pure streaming term of the Hamiltonian.

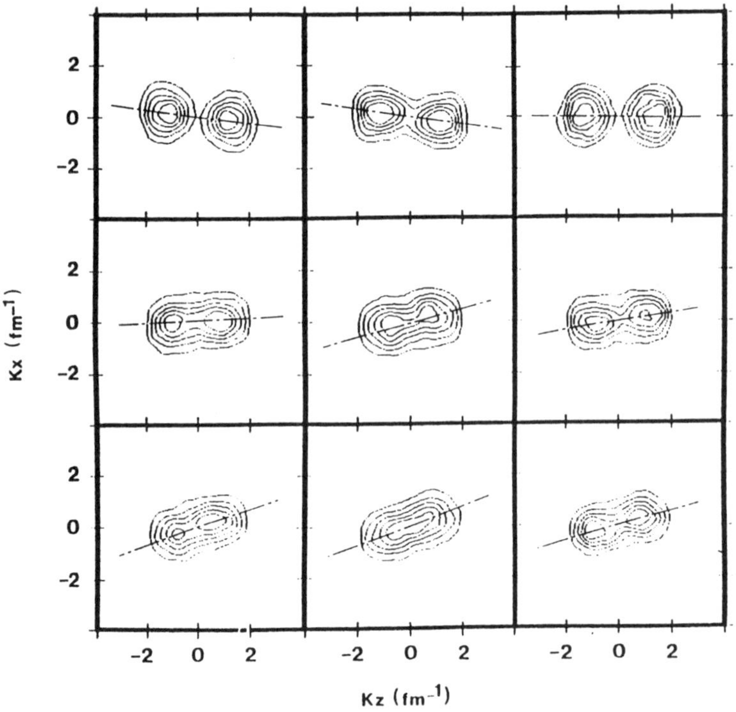

Fig.2 : Time evolution of the density in p-space for the same collision as fig.1. The asymptotic distributions are projected on the reaction plane. The different rows show the results from top to bottom, for increasing values of σ_{nn} (see fig.1) ; the columns are computed with different expressions of the mean field, respectively from left to right, "soft" local interaction, "stiff" local interaction and Gogny momentum-dependent interaction. This figure qualitatively illustrates the respective roles of the mean-field and the collision term in the time-evolution of the momentum distribution.

When one switches on the two-body collisions $\sigma \neq 0$, (2nd and 3rd rows of fig.1), the densities in r-space are strongly affected, due to the increased stopping power of each nucleus relatively to the other. The p-space distributions in the bulk show a smooth and relatively slow evolution towards the equilibrium

$$f(\vec{p}) = \{1 + exp[(|\vec{p}|^2 - p_F^2)/2mT]\}^{-1} \tag{2.4}$$

of the collision term.

The collision term is connected with the repulsive short range part of the nucleon-nucleon interaction ; it induces instantaneous changes in the p-distribution while being local in r-space. The mean field term deals with the long-range part and in the bulk of nuclei has little effect on the momentum distribution. This asymmetry is fully exploited in the actual methods of solutions of the Q.B.E.

2.2. Solutions of the Quantum Boltzmann Equation

The Q.B.E. is a many-fold, non-linear equation which is numerically solvable only for systems with many symmetries ; this is the case of the interstreaming of two infinite, isotropic systems of nuclear matter [21]. These sample cases are very helpful to check the asymptotic properties of the Q.B.E. and to test the accuracy of other solution methods.

To solve Q.B.E. for the asymmetric collisions of two nuclei, one has to deal with the full 6-dimension phase space, which forbides straightforward eulerian numerical methods : methods based on the projection of the solution $f(\vec{r}, \vec{p}; t)$ on a fixed grid of the phase space.

Most successfull methods are lagrangian since they use projections on moving bases :

$$f(\vec{r}, \vec{p}; t) = \int d\tau' w(\vec{r}', \vec{p}')g(\vec{r} - \vec{r}')g(\vec{p} - \vec{p}') \; ; \; d\tau = \frac{d\vec{r}d\vec{p}}{h^3} \tag{2.5}$$

where the time-dependence of f arises from the evolution of the $(\vec{r}', \vec{p}')$ coordinates.

If the g-functions form a (over) complete basis, (2.5) provides a general class of solutions for the mean field part of Q.B.E., provided that the $\vec{r}'$ and $\vec{p}'$ parameters are solutions of Ehrenfest-like equations [12] :

$$\dot{\vec{r}} = < \nabla_{\vec{p}} \mathcal{H}(\vec{r}, \vec{p}) >$$

$$\dot{\vec{p}} = - < \nabla_{\vec{r}} \mathcal{H}(\vec{r}, \vec{p}) > \tag{2.6}$$

The brackets mean that the derivatives of the field have to be averaged on the g-basis. The equations (2.6) are consistent with the conservation of the total energy.

Two sets of g-functions are generally used ; respectively, the coherent states :

$$g_\Delta(\vec{x}) = (2\pi\Delta)^{-3/2} exp(-|\vec{x}|^2/2\Delta) \tag{2.7}$$

and the δ-functions. The coherent states are the basis of the so-called Landau-Vlasov model [12], they have been adopted later by other groups. The models based on test-particles use δ-functions which do not provide a complete basis and then need to be complemented by extra smoothing methods. Although the full complete basis of coherent states has been used (see reference 22 for example), in most calculations, the folding product of (2.5) is discretized and the integral is computed by a Monte-Carlo sampling method :

$$f(\vec{r},\vec{p};t) = \sum_{i=1}^{rA} w_i g_\chi(\vec{r} - \vec{r}_i(t)) g_\phi(\vec{p} - \vec{p}_i(t)) \tag{2.8}$$

Typically r = 50 coherent states per nucleon are found to be necessary for a good reproduction of ground state properties of nuclei ; this emphasizes the fact that the g-functions are not nucleonic wave packets but pieces of one-body phase space, which behave like "classical pseudo-particles" in the one-body nuclear mean field, modified according to (2.6).

Table I

MAPPING RULES BETWEEN THE NUCLEAR PHASE SPACE
AND THE PHASE SPACE OF THE EQUIVALENT GAS OF PSEUDO-PARTICLES

	Nuclear phase space	Pseudo-particle phase space
One body distribution	Wigner transform f of the density matrix	$f(\vec{r},\vec{p};t) = \sum_{i=1}^{rA} w_i\, g(\vec{r}-\vec{r}_i)\, g(\vec{p}-\vec{p}_i)$
Mean - field hamiltonian	Hartree Fock Hamiltonian $\mathcal{H}(\vec{r},\vec{p})$	$<\mathcal{H}(\vec{r},\vec{p})>_g$ H-F hamiltonian folded on form factors g
Equations of motion	$\frac{df}{dt} = -[f, \mathcal{H}]$	Ehrenfest equations
One body observables	$<\mathcal{A}> = \int \frac{d\vec{r}\,d\vec{p}}{h^3} \mathcal{A}(\vec{r},\vec{p})\, f(\vec{r},\vec{p})$	$<\mathcal{A}> = \sum_{i=1}^{rA} w_i <\mathcal{A}>_{g_i}$
two-body collisions	in-medium corrected n-n cross-section σ_{nn}	$\frac{\sigma_{nn}}{r}$

Then Q.B.E. is seldom solved but more generally simulated via a proper mapping of the one-body nuclear phase space onto the phase space of a classical pseudo-particles gas. The rules of this mapping are summarized in table 1.

Such a method based on the use of pseudo-particles to solve transport equations is not unique (see reference 23 for examples in cosmological N-body simulations).

Its validity for the calculation of averaged collective variables has been checked with a good level of accuracy (see reference 24 for the mean field properties and reference 25 for the two-body collisions). However, its relevance to fluctuation studies has been put into question [17], since the fluctuations in a medium are strongly connected with its particle density, which is a free parameter in the pseudo-particle methods.

3. COMPETITION BETWEEN MEAN-FIELD AND TWO-BODY COLLISIONS

As already underlined in the preceding section, the balance between the mean field and the two body collisions may have a deep influence on the nuclear dynamics (see references 13 and 26 for recent reviews). We shall discuss here only two cases : the scission of a nucleus and the nuclear sidewards flow, which illustrate the wide range of applicability of the Q.B.E.

3.1. Nuclear scission

A standing problem of nuclear physics is the role of dissipation in the fission process. The spontaneous fission is out of scope of the Q.B.E., due to the lack of shell and tunnelling effects. We shall study the dynamics of a nucleus initially taken in its ground state, the scission of which is boosted by a proper action of a velocity field[27].

The ground state is prepared as a self consistent form of (2.5), where the weight w is :

$$w = \frac{1}{r}\theta(\epsilon_F - < \mathcal{H} >_g), \tag{3.1}$$

θ is the usual step function, so that the space is uniformly filled up to the Fermi level ϵ_F.

The splitting of the nucleus is generated by a modification of the particle velocity distribution :

$$p_z \rightarrow p_z - (sign\ \ of\ \ z)\Delta p_z. \tag{3.2}$$

Δp_z is added to the z-component of the momentum if the particle has a positive z-coordinate, and is substracted otherwise. For $\Delta p_z = 0.40 \text{ fm}^{-1}$, the density profiles are shown in fig.3, in the pure mean field limit (first column) and with increasing two-body collisions (second and third columns).

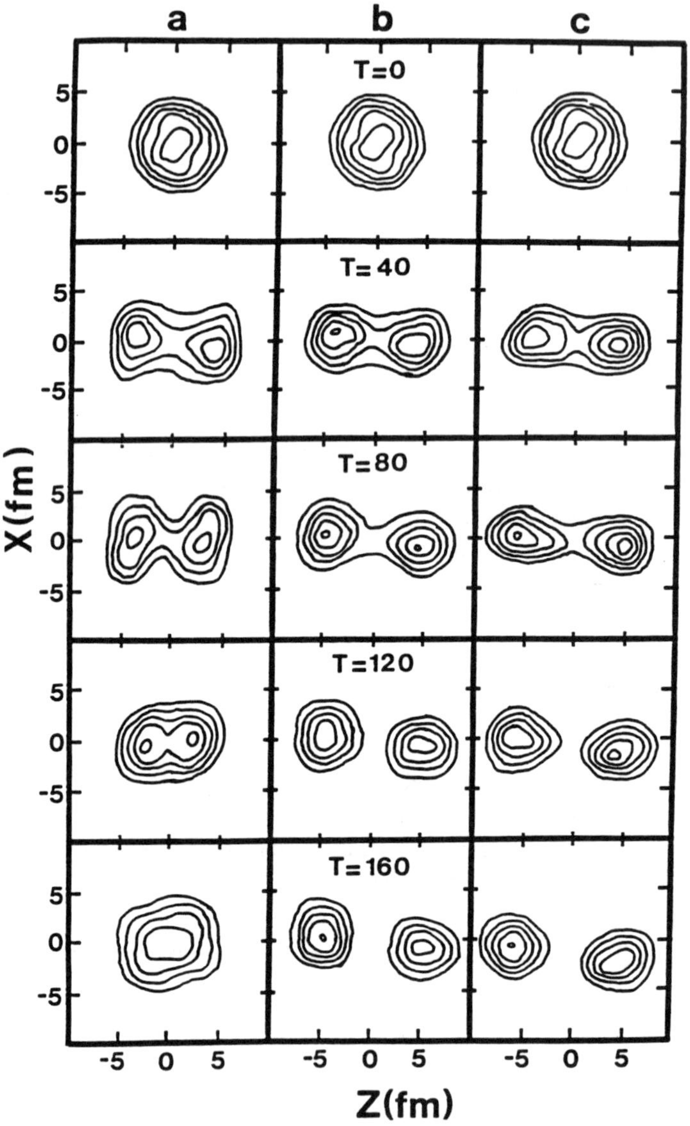

Fig. 3 : Density contours for the scission of a ^{40}Ca with various prescriptions for the collision term (see text).

In the pure mean field limit, the system develops a neck with creviced shapes ; it cannot overcome the scission barrier and swings back towards compact shapes. With collisions, the system explores different shape profiles and undergoes fission ; one should remark that shape vibrations are strongly damped after the scission. To track back the origin of such effects, one can analyze the collective velocity field :

$$v_{coll}(\vec{r}) = \frac{1}{m\rho(\vec{r})} \int \frac{d\vec{p}}{h^3} p_z f(\vec{r}, \vec{p}), \tag{3.3}$$

associated with the scission mode. The corresponding kinetic energy is :

$$E_{coll} = 1/2m \int \rho(\vec{r}) v_{coll}^2(\vec{r}) d\vec{r}, \tag{3.4}$$

$\rho(\vec{r})$ being the spatial density. The variation with time of this collective kinetic energy represents the energy needed to overcome the scission barrier. The dynamic scission barrier can then be defined as :

$$E_{bar} = max_t(E_{coll}(t=0) - E_{coll}(t)) \tag{3.5}$$

In fig.4, the time evolution of the collective energy is shown, with a reversed scale to show the potential barrier ; the initial boost is Δp=0.43 fm^{-2}, slightly higher than this of fig.3.

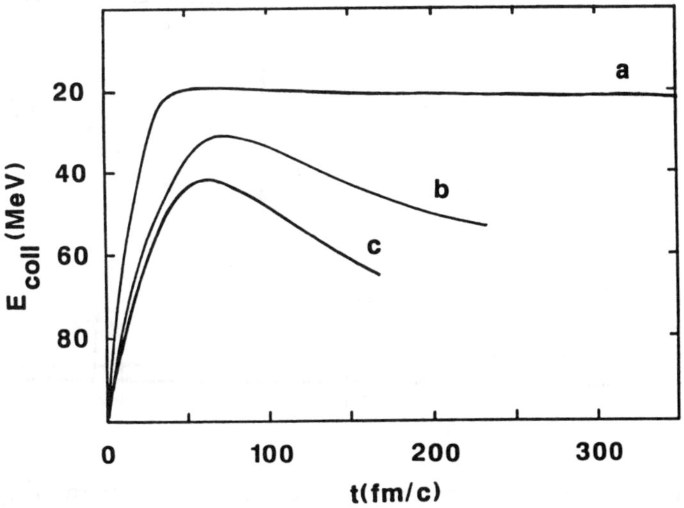

Fig. 4 : Time evolution of the collective energy of a ^{40}Ca nucleus along its scission path. The lines respectively labelled a, b and c, correspond to results with pure mean field, normal two-body collisions and enhanced two-body collisions.

The value of Δp is larger than the critical value then scission is reached even at the mean field limit. However, the two-body collisions are very effective to lower the barrier ; which remains however higher than the macroscopic fission barrier [28] (35 MeV), since the splitting of a nucleus by an initial momentum boost is a rough approximation to the fission process.

The Q.B.E. appears as a tool in view of study the fission of nuclei in regimes were shell effects are negligible. This is the case of hot rotating nuclei [29)] where the competition between fission, evaporation and other decay channels is one of the open problems of heavy ion reactions.

3.2. Inversion of the nuclear sidewards flow

Figure 2 has shown the asymptotic configuration of the momentum space for simulations using various prescriptions of the mean-field and of the nucleon-nucleon cross section. For the same system, one observes very different orientations of distribution symmetry axis relatively to the beam direction. This global deflection of the momentum distribution characterizes the nuclear sidewards flow [30] : i.e. the deviation of participant nucleon trajectories under the influence of attractive forces (mean field) and of repulsive forces (compression, two-body collisions...).

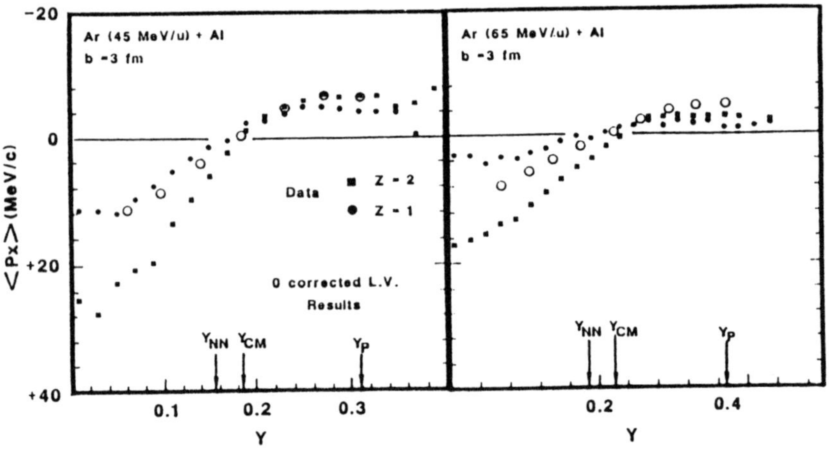

Fig. 5 : Mean transverse momentum as a function of the rapidity. The results are shown for two incident energies of the $^{40}Ar + ^{27}Al$ collision : 45 MeV/u (left) and 65 MeV/u (right). The experimental measurements are those of reference 32. The theoretical calculations are corrected for the detector efficiencies and reaction plane undetermination.

The flow can be defined through the mean transverse momentum :

$$< p_x >_{p_z} = \mathcal{N} \int d\vec{r} \int dp_x dp_y p_x f(\vec{r}, \vec{p})_{p_z} \qquad (3.6)$$

N is the normalization and the z index indicates that (3.6) is computed for a fixed component of the particle velocities relatively to the beam direction.

In fig.5, the mean transverse momentum is shown as a function of the rapidity y ; the theoretical results [31] are compared with the experimental measurements [32] after corrections for detector efficiencies. The flow is negative (notice the reverse scale of ordinate) : nucleons emitted on the opposite side of the projectile are deflected negatively ; this is a signature of the dominance of the attractive mean field. When the beam energy increases the deflection is less pronounced indicating the increasing role of the repulsive forces (2-body collisions/compression).

A simple (and crude) characterization of the $< p_x >$ distribution is the so-called flow parameter :

$$F = \left. \frac{d < p_x >}{dy} \right|_{<p_x>=0} \qquad (3.7)$$

F is then the slope of $< p_x >$ when it crosses the y axis.

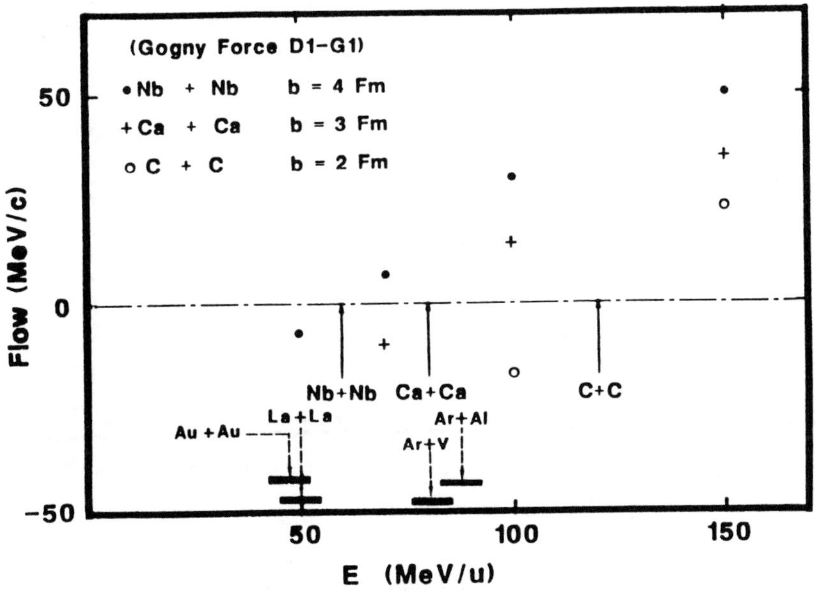

Fig. 6 : Evolution with the beam energy of the flow (see definition in text) for collisions of various symmetric systems. The bars in the lower part of the figures indicate experimental estimates of the energy when the flow vanishes.

Values of F are shown in fig.6 for different systems, at different energies. The arrows indicate energies where flow values change sign. This inversion of the flow is a signature of a change in the collision regime : the progressive opening of the phase space with increasing energies allows two-body collisions to develop and to progressively overcome the global attractive character of the mean field. The strong dependence of the inversion point on the mass of the system is another indication of the role of 2-body collisions ; at first order, the mean field does not depend much on the system size, but the number of collisions scales as the average length of nucleon paths in nuclear matter ; this later quantity varies roughly as $A^{1/3}$, then the 2-body collisions are expected to be enhanced for heavier systems, which is clearly shown by both theoretical and experimental results (see fig.6). The strong sensitivity of the flow to 2-body collisions can be asserted by theoretical calculations with different n-n cross sections. Fig.7 shows how small changes in σ_{nn}, yield shifts in the inversion point which are much larger than the experimental uncertainties.

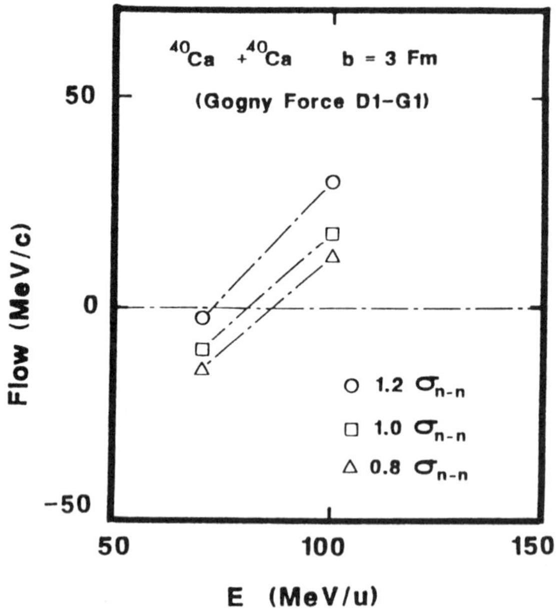

Fig. 7 : Evolution of the flow for different values of the nucleon-nucleon cross-section.

The flow measurements under 150 MeV/u beam energies provide a very powerful access to the properties of the mean field and 2-body collisions in nuclear physics.

4. TWO-BODY COLLISIONS AS A TRANSPORT PROCESS IN MOMENTUM SPACE

The Q.B.E. describes the average trajectories of systems in the one-body phase space. This is a severe limitation to its range of application. The experiments are more and more sophisticated, yielding more and more complete descriptions of events and then they give access to the fluctuations around the average values of observables. Furthermore, the concept of average itself becomes unrelevant when the system is close to critical states (such as spinodal decomposition).

Recently, works have been done to overcome such limitations by extension of the Q.B.E. formalism. Ayik[33] and Ayik and Grégoire [14] have extended the Q.B.E. in the field of nuclear physics, by introduction of a Langevin force to account for the fluctuating contribution of the 2-body collisions to the dynamics. Randrup and Remaud [15] have studied the fluctuations on a basis of a transport equation for phase space occupancies. The backgrounds of the two theories are basically the same. In this review, we shall follow the approach of reference 16. The connection of fluctuations dynamics with transport formalisms in phase space has been developed in classical statistical physics a few tens years ago (see Bixon and Zwanzig[34] or Logan and Kac [35]) ; it provides a natural and well founded extension of the Boltzmann formalism.

We consider a system of N-particles that we suppose to be homogenous in r-space ; the effects of the mean field on particle dynamics can be neglected and the whole dynamics can be studied in the momentum space. The basic idea, apparently first used by Nordsieck, Lamb and Uhlenbeck [36], is to treat the evolution of the system as a random process in momentum space. This process with the Boltzmann's stosszahlansatz can be considered as markovian. Logan and Kac[35] have shown how a microscopic treatment of this markovian process leads to a coherent theory of the Boltzmann equation and of its fluctuations. The momentum space can be divided in K cells which correspond to momentum intervals $d\vec{p}$. The system can be described by the instantaneous occupancies of each cell.

$$\vec{f} = [f_1(t), f_2(t), \cdots f_K(t)] \tag{4.1}$$

The 2-body collisions transfer particles between cells ; if they are scarce enough, each collisions can be treated independently from the others. The time evolution of the coarse-grained space is then described by the transition probabilities.

$$Prob(\vec{f}(t)|\vec{g}(t+dt)) = Prob(\vec{f}|\vec{g}; dt) \tag{4.2}$$

On a short time interval, these probabilities define the transition matrix :

$$Prob(\vec{f}|\vec{g}; dt) = Q(\vec{f}|\vec{g})dt + O(dt^2) \tag{4.3}$$

and the master equation for the process follows :

$$\frac{d\Phi(\vec{f};t)}{dt} = \sum_{\vec{g}} \Phi(\vec{g})Q(\vec{f}|\vec{g}) \tag{4.4}$$

which gives the probability Φ of finding the system in state $\vec{f}$. Logan and Kac [35] have shown how the standard Boltzmann equation can be found from the master equation (4.4) by a proper averaging procedure on the macrostates $\vec{f}$; the deviations from this average then follow a Fokker-Planck equation whose coefficients are connected with those of the Boltzmann equation.

According to the method of reference 16, we derive the general Fokker-Planck associated with the master equation (4.4) :

$$\frac{d}{dt}\phi(\vec{f}) = -\sum_k \frac{d}{df_k}\{V_k(\vec{f})\phi(\vec{f})\}$$

$$+\sum_{k,l} \frac{d^2}{df_k df_l}\{D_{kl}(\vec{f})\phi(\vec{f})\} \tag{4.5}$$

The indexes k and l run on all the momentum cells, the occupation numbers f_k, f_l of which are the components of the "vector" $\vec{f}$. This reference to phase space cells is for convenience, since we shall take at the end the continuous limit, changing discrete sums in functional integrals. The V_k and D_{kl} coefficients are respectively the drift and diffusion coefficients . The average of an observable can be calculated as :

$$< \mathcal{A} >= \int d\vec{f} \mathcal{A}(\vec{f})\phi(\vec{f}) \tag{4.6}$$

In particular, the time derivatives of average occupation probability of a cell and of the correlations between occupation of different cells, follows directly from (4.5) and (4.6.) :

$$\frac{d}{dt} < f_i >=< V_i(\vec{f}) >,$$

$$\frac{d}{dt}\sigma_{ij} =< 2D_{ij}(\vec{f}) > +\sum_k < \frac{\partial V_j}{\partial f_k}\sigma_{ik} + \frac{\partial V_i}{\partial f_k}\sigma_{jk} > \tag{4.7}$$

This set of equations is not closed, since the drift and diffusion coefficients are calculated as averages on the whole distribution of macro-states $\vec{f}$.

For systems far from critical behaviours there is a well defined average state $< \vec{f} >$, around which $\phi(\vec{f})$ is distributed, and one gets the closed set of equations for the first moments :

$$\frac{d}{dt} < f_i >= V_i(< \vec{f} >);$$

$$\frac{d}{dt}\sigma_{ij} = 2D_{ij}(< \vec{f} >) + \sum_k \left(\frac{\partial V_i}{\partial f_k}\sigma_{jk} + \frac{\partial V_j}{\partial f_k}\sigma_{ik} \right) \tag{4.8}$$

The above assumption is consistent with our initial hypothesis of a system homogeneous in space.

4.2. Drift and diffusion coefficients.

The transition matrix Q of (4.3) is connected with the rate of collisions between particles initially situated near the phase space points $\vec{p}_1$ and $\vec{p}_2$:

$$w(12 \iff 34) \equiv w(\vec{p}_1\vec{p}_2 \iff \vec{p}_3\vec{p}_4) = \frac{h^9}{m^2}\delta(\vec{p}_1 + \vec{p}_2 - \vec{p}_3 - \vec{p}_4)$$

$$\delta(\epsilon_1 + \epsilon_2 - \epsilon_3 - \epsilon_4)\frac{d\sigma(12 \iff 34)}{d\omega_{12}} \tag{4.9}$$

This transition rate is valid at the approximation of on-shell particles undergoing only elastic diffusions. The transition matrix has then four distinct components : respectively :

$$W_{12}^{--} = \frac{1}{2}\int \frac{d\vec{p}_3 d\vec{p}_4}{\hbar^6} \bar{f}_3 \bar{f}_4 w(12 \iff 34)$$

$$W_{12}^{++} = \frac{1}{2}\int \frac{d\vec{p}_3 d\vec{p}_4}{\hbar^6} f_3 f_4 w(12 \iff 34)$$

$$W_{12}^{+-} = \int \frac{d\vec{p}_3 d\vec{p}_4}{\hbar^6} f_3 \bar{f}_4 w(23 \iff 14)$$

$$W_{12}^{-+} = \int \frac{d\vec{p}_3 d\vec{p}_4}{\hbar^6} \bar{f}_3 f_4 w(13 \iff 24) \tag{4.10}$$

We have used for simplicity the obvious notations $f_i = f(\vec{p}_i)$. The 1/2 factor in the two first coefficients accounts for the double counting of the nucleon pair. The $\bar{f}_i$ quantities are "blocking" factors in the exit channel of the collision :

$$\bar{f}_i = 1 + \theta f(\vec{p}_i) \tag{4.11}$$

where θ depends on the particle spin assignation : $\theta = 0$ for classical particles, 1 for bosons and -1 for fermions. The drift coefficient on the average trajectory then follows from the difference between the flux of particles leaving a phase space location and the flux of those arriving on it :

$$V_1 \equiv V_1(< f >) = \int \frac{d\vec{p}_2}{h^3}(\bar{f}_1\bar{f}_2W_{12}^{++} - f_1f_2W_{12}^{--}) \equiv \bar{f}_1W_1^{+} - f_1W_1^{-} \tag{4.12}$$

The diagonal part of the diffusion matrix is, as expected in a transport formalism, the sum of the two above-defined fluxes :

$$2D_{11} = \bar{f}_1W_1^{+} + f_1W_1^{-} \tag{4.14}$$

whereas the off-diagonal coefficients are :

$$2D_{12} = f_1f_2W^{--} + \bar{f}_1\bar{f}_2W^{++} - f_1\bar{f}_2W_{12}^{-+} - \bar{f}_1f_2W_{12}^{+-} \tag{4.15}$$

This expression pins out the source of correlation between phase space occupations : the 2-body collisions correlate positively the initial and final phase space locations of the colliding pairs and negatively the final space locations with initial ones. In (4.5), there is a restoring force which arises from the variation of the drift coefficient itself :

$$\frac{\partial V_1}{\partial f_2} = (\theta W_1^{+} - W_2^{-})\delta(\vec{p}_1 - \vec{p}_2)$$

$$-f_1(\theta W_{12}^{-+} + W_{12}^{--}) + \bar{f}_1(W_{12}^{+-} + \theta W_{12}^{++}) \tag{4.16}$$

4.3. Dynamics of fluctuations in phase space.

Since the solutions of the transport equations for the whole phase space is still a challenging open numerical problem [17)], we concentrate on the fluctuations on a single phase space point. The equations for the average phase occupancies and their fluctuations can be easily found from (4.9), (4.13), (4.14) and (4.15) :

$$\frac{df}{dt} = \bar{f}W^{+} - fW^{-}$$

$$\frac{d\sigma_f^2}{dt} = \bar{f}W^{+} - fW^{-} + 2(\theta W^{+} - W^{-})\sigma_f^2, \tag{4.17}$$

where the indexes have been dropped out for simplicity, being understood that these equations have to be solved at each phase space point.

The equation for the average occupancy is the Boltzmann equation as expected, the equilibrium of which is reached when the gain and loss terms are equal : i.e. when

$$f_{eq} = [exp((\epsilon - \tau)/T) - \theta]^{-1} \tag{4.18}$$

In (4.18), one recognizes the well known Maxwell, Bose-Einstein or Fermi-Dirac equilibrium distributions, depending on the value of θ (see (4.12)). The transport coefficients are connected by the detailed balance relation :

$$f_{eq} W^{-} = \bar{f}_{eq} W^{+}. \tag{4.19}$$

and assuming equilibrium for the averages, the equilibrium of fluctuations follows directly :

$$f_{eq}^2 = f\bar{f} \tag{4.20}$$

which is the standard result of classical or quantum statistical physics.

The set of equations (4.17) then provides the dynamics of fluctuations around the average trajectories of the Boltzmann equation, with the correct asymptotic equilibrium solutions. Numerical calculations [16] have been performed to check at which extent the instantaneous solutions for σ_f^2 differ from the equilibrium $f\bar{f}$ of (4.20). An elementary calculation shows that this deviation :

$$\Delta = \sigma_f^2 - f\bar{f} \tag{4.21}$$

is solution of :

$$\dot{\Delta} = 2(\theta W^{+} - W^{-})\Delta. \tag{4.22}$$

For fermions, the two terms of the right-hand side contributes negatively ($\theta = -1$) and the relaxation time can be estimated [16] ; it is of the order of a few 10^{-23}s which is about an order of magnitude shorter than the typical duration of nuclear collisions. Then one can expect that the fluctuations track closely their adiabatic values (4.20).

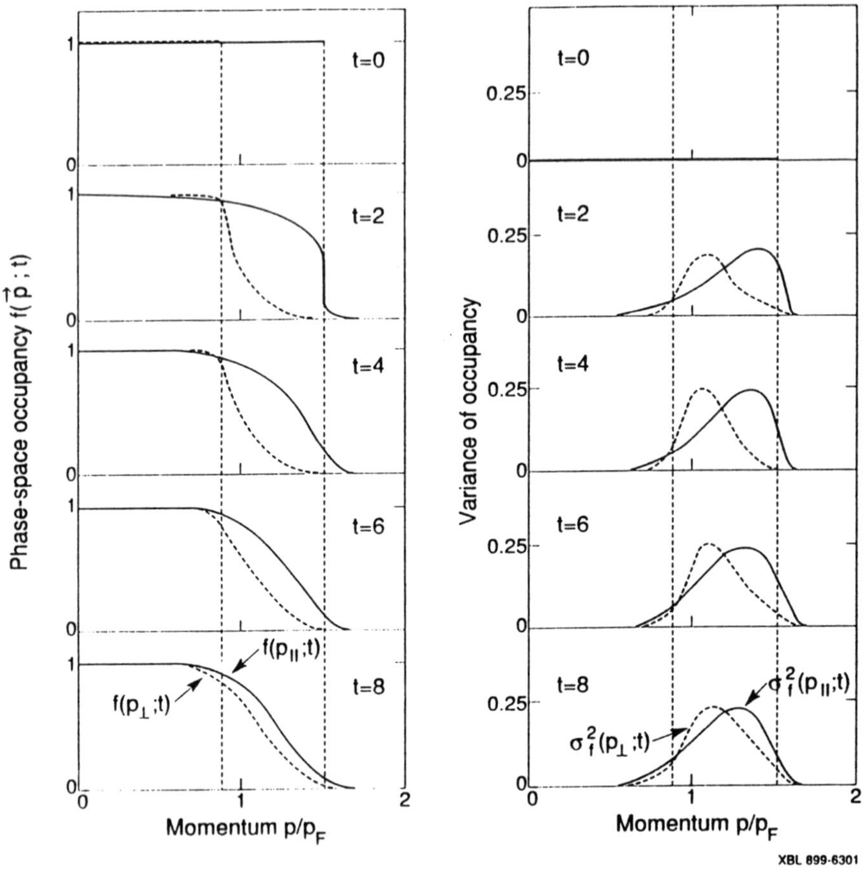

Fig. 8 : For a center separation, corresponding to a bombarding energy of E_{beam} = 37 MeV, the left-hand side part shows the phase occupancy along the parallel and perpendicular to the beam ; the right- hand side part displays the variance of the phase space occupancy.

The behaviour of the fluctuations is illustrated in figures 8 and 9, for an infinite system of fermions, the momentum distributions of which is initially corresponding to two dislocated Fermi spheres. In fig. 8, are shown the time evolutions of respectively the average occupancy and its fluctuations, in the parallel and perpendicular directions. Since initially, the phase space is either fully occupied or empty, at t = 0 there is no fluctuations. The time evolutions illustrate the progressive equilibration of the system and its irreversible motion towards its asymptotic limit given by the Fermi-Dirac momentum distribution.

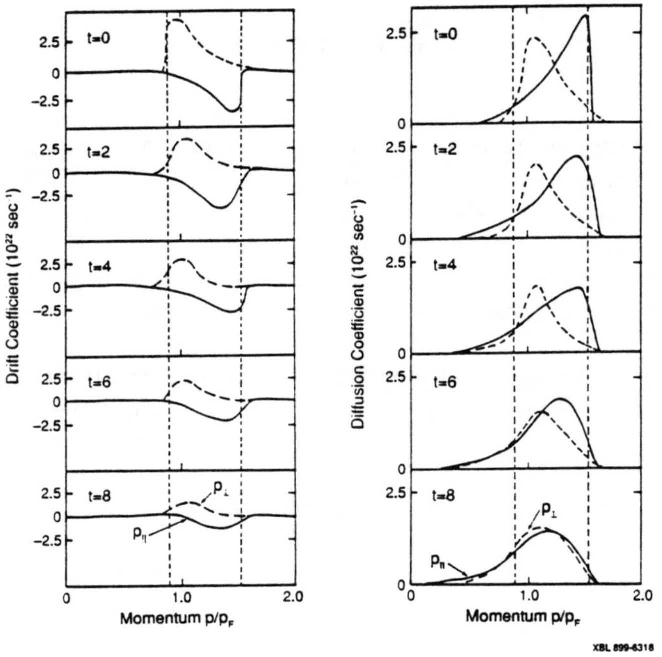

Fig. 9 : For the same case as in fig.8, the figure shows the evolution of the drift (right) and of the diffusion (left) coefficient for the phase space occupancies in the parallel and perpendicular directions.

The drift and diffusion coefficients (see (4.12) and (4.13)) are illustrated in fig.9 ; they reflect the rate of the collisions around the Fermi energy which are the source of the fluctuations.

The whole covariance matrix $\sigma(\vec{r},\vec{p};\vec{r}'\vec{p}')$ gives the correlation between the fluctuations in the single-particle occupancies at two different points in phase space. The detailed equation of its evolution can be found in references 16 and 17 ; let us mention that the diagonal and off-diagonal parts are strongly linked since the fluctuations of the invariants of the collision term should vanish.

At least, we notice that we have discussed the dynamics of the fluctuations in homogeneous systems where one can neglect the mean-field effects. In this case, the transport coefficients are such that no instability can develop and lead to the system disassembly.

The above study is then restricted to the case of systems far from critical behaviours, but it can give access to the dispersion of experimental observables around their average. The regimes where fluctuations grow exponentially need the coupling of such equations with the mean field [18].

5. CONCLUSIONS

In this review, we have sketched the main lines of the transport approaches to the dynamics of heavy-ion collisions. Such approaches are connected with the Boltzmann equation, with corrections to account for the quantum features of the nuclei ; although their theoretical bases have several weaknesses, they have been very successful in the description of the heavy-ion collisions and in the connection of the observed quantities with the basic properties of nuclear matter.

The Boltzmann-like equations for finite nuclei are rather simulated than exactly solved ; we have discussed the general features of the solutions, based on pseudo-particles methods ; they replace the exact solution of the quantum nuclear problem by the study of the transport properties of an equivalent gas of classical pseudo-particles ; we have summarized in Table 1, the rules of this mapping.

We have reported briefly on two limiting situations which illustrate both the flexibility of the theory and the analysis of the competition between mean field and two-body collisions.

The last part of this paper presents currently developed approaches to the extension of the theory to the dynamics of fluctuations. They are based on the similarities of the evolution of phase space occupancies under the influence of two-body collisions, with a random walk. This idea has been fully exploited in the study of classical fluids ; when properly modified to introduce the quantum constraints on phase space occupancies, it can provide a powerful mean of access to the variances of experimentally measured quantities.

Acknowledgments

The authors warmly thank F. Garcias, V. De La Mota and P. Schuck for their contributions to the studies of the rotating nuclei and of the nuclear flow. One of us (B. R.) thanks J. Randrup with whom the project on the fluctuation dynamics has been initiated. Fruitful discussions are acknowledged with A. Hoffmann, S. Hernandez, P. Saïd, C. Grégoire and S. Ayik. E. Gerbaud is thanked for her careful typing of the manuscript.

REFERENCES

1) M. Lefort and C. Ngo, Ann. Phys. (Paris) 3 (1978) 5.
2) J. Randrup, Nucl. Phys. A 383 (1982) 468.
3) C. Grégoire, C. Ngo and B. Remaud, Nucl. Phys. A 383 (1982) 392.
4) H. Hofmann and P.J. Siemens, Nucl. Phys. A 257 (1976) 165 and A 275 (1977) 464.
5) Y. Abe, contribution to this conference.
6)A.K. Kerman and S.E. Koonin, Ann. of Phys. 100 (1976) 332.
7) C. Wolschin, Phys. Rev. Lett. 48 (1982) 1004.
8) For a recent review, P. Schuck, R.W. Hasse, J. Jaenicke, C. Grégoire, B. Remaud, F. Sébille and E. Suraud. Prog. Part. Nucl. Phys. 22 (1989) 181.
9) J. Aichelin and G.F. Bertsch, Phys. Rev. C 31 (1985) 1730.
10) H. Kruse, B. Jacak and H. Stöcker, Phys. Rev. Lett. 54 (1985) 289.
11) B. Remaud, F. Sébille, C. Grégoire, L. Vinet and Y. Raffray Nucl. Phys. A 447 (1985) 555.
12) C. Grégoire, B. Remaud, F. Sébille, L. Vinet and Y. Raffray, Nucl. Phys. A 465 (1987) 317.
13) G.F. Bertsch and S. Das Gupta, Phys. Reports 160 (1988) 190.
14) S. Ayik and C. Grégoire, Phys. Lett. B 212 (1988) 269 and Nucl. Phys. A 513 (1990) 187.
15) J. Randrup and B. Remaud, Nucl. Phys. A 514 (1990) 339.
16) P. Chomaz, G.F. Burgio and J. Randrup, Phys. Lett. B 254 (1991) 340.
17) G.F. Burgio, Ph. Chomaz and J. Randrup, Nucl. Phys. A 529 (1991) 157.
18) C.J. Pethick and D.G. Ravenhall, Nucl. Phys. A 471 (1987) 19c.
19) L.W. Nordheim, Proc. Roy. Soc. (London) A 119 (1928) 689.
20) E.A. Uehling and G.E. Uhlenbeck, Phys. Rev. 43 (1933) 552.
21) C. Toepffer and C.Y. Wong, Phys. Rev. C 25 (1982) 1018.
22) L. Vinet, C. Grégoire, P. Schuck, B. Remaud and F. Sébille, Nucl. Phys. A 468 (1987) 321.
23) E. Bertschinger and J.M. Gelb, Computers in Physics 5 (1991) 164.
24) F. Sébille, Ph. D. Thesis, University of Nantes (1986), unpublished.
25) C. Welke, R. Malfiet, C. Grégoire, M. Prakash and E. Suraud, Phys. Rev. C 40 (1990) 2611.
26) E. Suraud, C. Grégoire and B. Tamain, Prog. Part. Nucl. Phys. 23 (1989) 357.
27) G. Royer, B. Remaud, F. Sébille and V. De La Mota, Phys. Rev. C (1991) in press.
28)G. Royer and B. Remaud, J. Phys. G. Nucl. Phys. 10 (1984) 1057.
29) F. Garcias, V. De La Mota, B. Remaud, G. Royer and F. Sébille, Phys. Lett. B 255 (1991) 311.
30)See for a review, H. Stocker, Int. Conf. on Nuclear Matter and Heavy-Ion Collisions, Les Houches (1989).
31) V. De La Mota, F. Sébille, M. Farine, B. Remaud and P. Schuck, Proceedings of the XXIX Int. Winter Meeting, Bormio (1991).

32) J. Péter, Contribution to this conference.
33) S. Ayik, Z. Phys. A 298 (1982) 468.
34) M. Bixon and R. Zwanzig, Phys. Rev. 187 (1969) 267.
35) J. Logan and M. Kac, Phys. Rev. A 13 (1976) 458.
36) A. Nordsieck, W.E. Lands and G.E. Uhlenbeck, Physica 7 (1940) 344.

AIP Conference Proceedings

		L.C. Number	ISBN
No. 121	High Energy e^+e^- Interactions (Vanderbilt, 1984)	84-72632	0-88318-320-X
No. 122	The Physics of VLSI (Xerox, Palo Alto, CA, 1984)	84-72729	0-88318-321-8
No. 123	Intersections Between Particle and Nuclear Physics (Steamboat Springs, CO, 1984)	84-72790	0-88318-322-6
No. 124	Neutron-Nucleus Collisions: A Probe of Nuclear Structure (Burr Oak State Park, 1984)	84-73216	0-88318-323-4
No. 125	Capture Gamma-Ray Spectroscopy and Related Topics – 1984 (Int'l Symposium, Knoxville, TN)	84-73303	0-88318-324-2
No. 126	Solar Neutrinos and Neutrino Astronomy (Homestake, 1984)	84-63143	0-88318-325-0
No. 127	Physics of High Energy Particle Accelerators (BNL/SUNY Summer School, 1983)	85-70057	0-88318-326-9
No. 128	Nuclear Physics with Stored, Cooled Beams (McCormick's Creek State Park, IN, 1984)	85-71167	0-88318-327-7
No. 129	Radiofrequency Plasma Heating (Sixth Topical Conference) (Callaway Gardens, GA, 1985)	85-48027	0-88318-328-5
No. 130	Laser Acceleration of Particles (Malibu, CA, 1985)	85-48028	0-88318-329-3
No. 131	Workshop on Polarized ^{3}He Beams and Targets (Princeton, NJ, 1984)	85-48026	0-88318-330-7
No. 132	Hadron Spectroscopy–1985 (International Conference, Univ. of Maryland)	85-72537	0-88318-331-5
No. 133	Hadronic Probes and Nuclear Interactions (Arizona State University, 1985)	85-72638	0-88318-332-3
No. 134	The State of High Energy Physics (BNL/SUNY Summer School, 1983)	85-73170	0-88318-333-1
No. 135	Energy Sources: Conservation and Renewables (APS, Washington, DC, 1985)	85-73019	0-88318-334-X
No. 136	Atomic Theory Workshop on Relativistic and QED Effects in Heavy Atoms (Gaithersburg, MD, 1985)	85-73790	0-88318-335-8
No. 137	Polymer-Flow Interaction (La Jolla Institute, 1985)	85-73915	0-88318-336-6
No. 138	Frontiers in Electronic Materials and Processing (Houston, TX, 1985)	86-70108	0-88318-337-4
No. 139	High-Current, High-Brightness, and High-Duty Factor Ion Injectors (La Jolla Institute, 1985)	86-70245	0-88318-338-2
No. 140	Boron-Rich Solids (Albuquerque, NM, 1985)	86-70246	0-88318-339-0
No. 141	Gamma-Ray Bursts (Stanford, CA, 1984)	86-70761	0-88318-340-4
No. 142	Nuclear Structure at High Spin, Excitation, and Momentum Transfer (Indiana University, 1985)	86-70837	0-88318-341-2

No. 143	Mexican School of Particles and Fields (Oaxtepec, México, 1984)	86-81187	0-88318-342-0
No. 144	Magnetospheric Phenomena in Astrophysics (Los Alamos, NM, 1984)	86-71149	0-88318-343-9
No. 145	Polarized Beams at SSC & Polarized Antiprotons (Ann Arbor, MI & Bodega Bay, CA, 1985)	86-71343	0-88318-344-7
No. 146	Advances in Laser Science–I (Dallas, TX, 1985)	86-71536	0-88318-345-5
No. 147	Short Wavelength Coherent Radiation: Generation and Applications (Monterey, CA, 1986)	86-71674	0-88318-346-3
No. 148	Space Colonization: Technology and The Liberal Arts (Geneva, NY, 1985)	86-71675	0-88318-347-1
No. 149	Physics and Chemistry of Protective Coatings (Universal City, CA, 1985)	86-72019	0-88318-348-X
No. 150	Intersections Between Particle and Nuclear Physics (Lake Louise, Canada, 1986)	86-72018	0-88318-349-8
No. 151	Neural Networks for Computing (Snowbird, UT, 1986)	86-72481	0-88318-351-X
No. 152	Heavy Ion Inertial Fusion (Washington, DC, 1986)	86-73185	0-88318-352-8
No. 153	Physics of Particle Accelerators (SLAC Summer School, 1985) (Fermilab Summer School, 1984)	87-70103	0-88318-353-6
No. 154	Physics and Chemistry of Porous Media—II (Ridge Field, CT, 1986)	83-73640	0-88318-354-4
No. 155	The Galactic Center: Proceedings of the Symposium Honoring C. H. Townes (Berkeley, CA, 1986)	86-73186	0-88318-355-2
No. 156	Advanced Accelerator Concepts (Madison, WI, 1986)	87-70635	0-88318-358-0
No. 157	Stability of Amorphous Silicon Alloy Materials and Devices (Palo Alto, CA, 1987)	87-70990	0-88318-359-9
No. 158	Production and Neutralization of Negative Ions and Beams (Brookhaven, NY, 1986)	87-71695	0-88318-358-7
No. 159	Applications of Radio-Frequency Power to Plasma: Seventh Topical Conference (Kissimmee, FL, 1987)	87-71812	0-88318-359-5
No. 160	Advances in Laser Science–II (Seattle, WA, 1986)	87-71962	0-88318-360-9
No. 161	Electron Scattering in Nuclear and Particle Science: In Commemoration of the 35th Anniversary of the Lyman-Hanson-Scott Experiment (Urbana, IL, 1986)	87-72403	0-88318-361-7
No. 162	Few-Body Systems and Multiparticle Dynamics (Crystal City, VA, 1987)	87-72594	0-88318-362-5
No. 163	Pion–Nucleus Physics: Future Directions and New Facilities at LAMPF (Los Alamos, NM, 1987)	87-72961	0-88318-363-3

No. 164	Nuclei Far from Stability: Fifth International Conference (Rosseau Lake, ON, 1987)	87-73214	0-88318-364-1
No. 165	Thin Film Processing and Characterization of High-Temperature Superconductors (Anaheim, CA, 1987)	87-73420	0-88318-365-X
No. 166	Photovoltaic Safety (Denver, CO, 1988)	88-42854	0-88318-366-8
No. 167	Deposition and Growth: Limits for Microelectronics (Anaheim, CA, 1987)	88-71432	0-88318-367-6
No. 168	Atomic Processes in Plasmas (Santa Fe, NM, 1987)	88-71273	0-88318-368-4
No. 169	Modern Physics in America: A Michelson-Morley Centennial Symposium (Cleveland, OH, 1987)	88-71348	0-88318-369-2
No. 170	Nuclear Spectroscopy of Astrophysical Sources (Washington, DC, 1987)	88-71625	0-88318-370-6
No. 171	Vacuum Design of Advanced and Compact Synchrotron Light Sources (Upton, NY, 1988)	88-71824	0-88318-371-4
No. 172	Advances in Laser Science–III: Proceedings of the International Laser Science Conference (Atlantic City, NJ, 1987)	88-71879	0-88318-372-2
No. 173	Cooperative Networks in Physics Education (Oaxtepec, Mexico, 1987)	88-72091	0-88318-373-0
No. 174	Radio Wave Scattering in the Interstellar Medium (San Diego, CA, 1988)	88-72092	0-88318-374-9
No. 175	Non-neutral Plasma Physics (Washington, DC, 1988)	88-72275	0-88318-375-7
No. 176	Intersections Between Particle and Nuclear Physics (Third International Conference) (Rockport, ME, 1988)	88-62535	0-88318-376-5
No. 177	Linear Accelerator and Beam Optics Codes (La Jolla, CA, 1988)	88-46074	0-88318-377-3
No. 178	Nuclear Arms Technologies in the 1990s (Washington, DC, 1988)	88-83262	0-88318-378-1
No. 179	The Michelson Era in American Science: 1870–1930 (Cleveland, OH, 1987)	88-83369	0-88318-379-X
No. 180	Frontiers in Science: International Symposium (Urbana, IL, 1987)	88-83526	0-88318-380-3
No. 181	Muon-Catalyzed Fusion (Sanibel Island, FL, 1988)	88-83636	0-88318-381-1
No. 182	High T_c Superconducting Thin Films, Devices, and Application (Atlanta, GA, 1988)	88-03947	0-88318-382-X
No. 183	Cosmic Abundances of Matter (Minneapolis, MN, 1988)	89-80147	0-88318-383-8
No. 184	Physics of Particle Accelerators (Ithaca, NY, 1988)	89-83575	0-88318-384-6
No. 185	Glueballs, Hybrids, and Exotic Hadrons (Upton, NY, 1988)	89-83513	0-88318-385-4

No. 186	High-Energy Radiation Background in Space (Sanibel Island, FL, 1987)	89-83833	0-88318-386-2
No. 187	High-Energy Spin Physics (Minneapolis, MN, 1988)	89-83948	0-88318-387-0
No. 188	International Symposium on Electron Beam Ion Sources and their Applications (Upton, NY, 1988)	89-84343	0-88318-388-9
No. 189	Relativistic, Quantum Electrodynamic, and Weak Interaction Effects in Atoms (Santa Barbara, CA, 1988)	89-84431	0-88318-389-7
No. 190	Radio-frequency Power in Plasmas (Irvine, CA, 1989)	89-45805	0-88318-397-8
No. 191	Advances in Laser Science–IV (Atlanta, GA, 1988)	89-85595	0-88318-391-9
No. 192	Vacuum Mechatronics (First International Workshop) (Santa Barbara, CA, 1989)	89-45905	0-88318-394-3
No. 193	Advanced Accelerator Concepts (Lake Arrowhead, CA, 1989)	89-45914	0-88318-393-5
No. 194	Quantum Fluids and Solids—1989 (Gainesville, FL, 1989)	89-81079	0-88318-395-1
No. 195	Dense *Z*-Pinches (Laguna Beach, CA, 1989)	89-46212	0-88318-396-X
No. 196	Heavy Quark Physics (Ithaca, NY, 1989)	89-81583	0-88318-644-6
No. 197	Drops and Bubbles (Monterey, CA, 1988)	89-46360	0-88318-392-7
No. 198	Astrophysics in Antarctica (Newark, DE, 1989)	89-46421	0-88318-398-6
No. 199	Surface Conditioning of Vacuum Systems (Los Angeles, CA, 1989)	89-82542	0-88318-756-6
No. 200	High T_c Superconducting Thin Films: Processing, Characterization, and Applications (Boston, MA, 1989)	90-80006	0-88318-759-0
No. 201	QED Stucture Functions (Ann Arbor, MI, 1989)	90-80229	0-88318-671-3
No. 202	NASA Workshop on Physics From a Lunar Base (Stanford, CA, 1989)	90-55073	0-88318-646-2
No. 203	Particle Astrophysics: The NASA Cosmic Ray Program for the 1990s and Beyond (Greenbelt, MD, 1989)	90-55077	0-88318-763-9
No. 204	Aspects of Electron–Molecule Scattering and Photoionization (New Haven, CT, 1989)	90-55175	0-88318-764-7
No. 205	The Physics of Electronic and Atomic Collisions (XVI International Conference) (New York, NY, 1989)	90-53183	0-88318-390-0
No. 206	Atomic Processes in Plasmas (Gaithersburg, MD, 1989)	90-55265	0-88318-769-8
No. 207	Astrophysics from the Moon (Annapolis, MD, 1990)	90-55582	0-88318-770-1
No. 208	Current Topics in Shock Waves (Bethlehem, PA, 1989)	90-55617	0-88318-776-0

No. 209	Computing for High Luminosity and High Intensity Facilities (Santa Fe, NM, 1990)	90-55634	0-88318-786-8
No. 210	Production and Neutralization of Negative Ions and Beams (Brookhaven, NY, 1990)	90-55316	0-88318-786-8
No. 211	High-Energy Astrophysics in the 21st Century (Taos, NM, 1989)	90-55644	0-88318-803-1
No. 212	Accelerator Instrumentation (Brookhaven, NY, 1989)	90-55838	0-88318-645-4
No. 213	Frontiers in Condensed Matter Theory (New York, NY, 1989)	90-6421	0-88318-771-X 0-88318-772-8 (pbk.)
No. 214	Beam Dynamics Issues of High-Luminosity Asymmetric Collider Rings (Berkeley, CA, 1990)	90-55857	0-88318-767-1
No. 215	X-Ray and Inner-Shell Processes (Knoxville, TN, 1990)	90-84700	0-88318-790-6
No. 216	Spectral Line Shapes, Vol. 6 (Austin, TX, 1990)	90-06278	0-88318-791-4
No. 217	Space Nuclear Power Systems (Albuquerque, NM, 1991)	90-56220	0-88318-838-4
No. 218	Positron Beams for Solids and Surfaces (London, Canada, 1990)	90-56407	0-88318-842-2
No. 219	Superconductivity and Its Applications (Buffalo, NY, 1990)	91-55020	0-88318-835-X
No. 220	High Energy Gamma-Ray Astronomy (Ann Arbor, MI, 1990)	91-70876	0-88318-812-0
No. 221	Particle Production Near Threshold (Nashville, IN, 1990)	91-55134	0-88318-829-5
No. 222	After the First Three Minutes (College Park, MD, 1990)	91-55214	0-88318-828-7
No. 223	Polarized Collider Workshop (University Park, PA, 1990)	91-71303	0-88318-826-0
No. 224	LAMPF Workshop on (π, K) Physics (Los Alamos, NM, 1990)	91-71304	0-88318-825-2
No. 225	Half Collision Resonance Phenomena in Molecules (Caracus, Venezuela, 1990)	91-55210	0-88318-840-6
No. 226	The Living Cell in Four Dimensions (Gif sur Yvette, France, 1990)	91-55209	0-88318-794-9
No. 227	Advanced Processing and Characterization Technologies (Clearwater, FL, 1991)	91-55194	0-88318-910-0
No. 228	Anomalous Nuclear Effects in Deuterium/Solid Systems (Provo, UT, 1990)	91-55245	0-88318-833-3
No. 229	Accelerator Instrumentation (Batavia, IL, 1990)	91-55347	0-88318-832-1
No. 230	Nonlinear Dynamics and Particle Acceleration (Tsukuba, Japan, 1990)	91-55348	0-88318-824-4

No. 231	Boron-Rich Solids (Albuquerque, NM, 1990)	91-53024	0-88318-793-4
No. 232	Gamma-Ray Line Astrophysics (Paris–Saclay, France, 1990)	91-55492	0-88318-875-9
No. 233	Atomic Physics 12 (Ann Arbor, MI, 1990)	91-55595	088318-811-2
No. 234	Amorphous Silicon Materials and Solar Cells (Denver, CO, 1991)	91-55575	088318-831-7
No. 235	Physics and Chemistry of MCT and Novel IR Detector Materials (San Francisco, CA, 1990)	91-55493	0-88318-931-3
No. 236	Vacuum Design of Synchrotron Light Sources (Argonne, IL, 1990)	91-55527	0-88318-873-2
No. 237	Kent M. Terwilliger Memorial Symposium (Ann Arbor, MI, 1989)	91-55576	0-88318-788-4
No. 238	Capture Gamma-Ray Spectroscopy (Pacific Grove, CA, 1990)	91-57923	0-88318-830-9
No. 239	Advances in Biomolecular Simulations (Obernai, France, 1991)	91-58106	0-88318-940-2
No. 240	Joint Soviet-American Workshop on the Physics of Semiconductor Lasers (Leningrad, USSR, 1991)	91-58537	0-88318-936-4
No. 241	Scanned Probe Microscopy (Santa Barbara, CA, 1991)	91-76758	0-88318-816-3
No. 242	Strong, Weak, and Electromagnetic Interactions in Nuclei, Atoms, and Astrophysics: A Workshop in Honor of Stewart D. Bloom's Retirement (Livermore, CA, 1991)	91-76876	0-88318-943-7
No. 243	Intersections Between Particle and Nuclear Physics (Tucson, AZ 1991)	91-77580	0-88318-950-X
No. 244	Radio Frequency Power in Plasmas (Charleston, SC 1991)	91-77853	0-88318-937-2
No. 245	Basic Space Science (Bangalore, India 1991)	91-78379	0-88318-951-8
No. 246	Proceedings of the Ninth Symposium on Space Nuclear Power Systems	91-58793	Casebound Pt. 1: 1-56396-021-4 Pt. 2: 1-56396-023-2 Pt. 3: 1-56396-025-7 Set: 1-56396-027-3 Paperback Pt. 1: 1-56396-020-6 Pt. 2: 1-56396-022-2 Pt. 3: 1-56396-024-9 Set: 1-56396-026-5
No. 247	Global Warming: Physics and Facts (Washington, DC 1991)	91-78423	0-88318-932-1
No. 248	Computer-Aided Statistical Physics (Taipei, Taiwan 1991)	91-78378	0-88318-942-9